INSTRUCTIONAL STRATEGIES FOR STUDENTS WITH SPECIAL NEEDS

INSTRUCTIONAL STRATEGIES FOR STUDENTS WITH SPECIAL NEEDS

Dan G. Bachor
FACULTY OF EDUCATION
UNIVERSITY OF VICTORIA

Carol Crealock
FACULTY OF EDUCATION
UNIVERSITY OF WESTERN ONTARIO

Prentice-Hall Canada Inc., Scarborough, Ontario

Canadian Cataloguing in Publication Data

Bachor, Dan G., 1945-
Instructional strategies for students with special needs

Bibliography: p. 485
Includes index.
ISBN 0-13-468000-6

1. Handicapped children—Education (Elementary)
2. Handicapped children—Education. I. Crealock
Carol M., 1941— II. Title

LC4015.B33 1986 371.9'043 C85-099643-0

Prentice-Hall Inc., Englewood Cliffs, New Jersey
Prentice-Hall International Inc., London
Prentice-Hall of Australia, Pty., Sydney
Prentice-Hall of India Pvt., Ltd., New Delhi
Prentice-Hall of Japan, Inc., Tokyo
Prentice-Hall of Southeast Asia (Pte.) Ltd., Singapore
Editora Prentice-Hall do Brasil Ltda., Rio de Janeiro
Prentice-Hall Hispanoamericana, S.A., Mexico
Whitehall Books Limited, Wellington, New Zealand

ISBN 0-13-468000-6

Production Editors: *Deborah Burrett/David Jolliffe*
Design: *Holly Fisher & Associates*
Manufacturing Buyer: *Sheldon Fischer*
Typesetting by *Skymark Graphics*
Cover photo by *Camerique*

2 3 4 5 BG 96 95 94 93

Printed and bound in Canada by Best Gagné Book Manufacturers

The contributions of both authors to this volume are equal. The listing of their names is alphabetical and no suggestion of seniority is intended.

CONTENTS

10. THINKING SKILLS 305

11. ORGANIZATIONAL SKILLS 336

12. SECONDARY SCHOOL ISSUES 361

13. A CONSULTING ROLE FOR THE RESOURCE TEACHER 392

PREFACE

This book is written for Canadian educators who are trying to be many things to Canadian students with special needs. It is our experience that education often is driven locally and that instructional strategies that work have to take into account not only who is being taught but also where, why and for what purpose. We have tried throughout the book to take into account exemplary work from across the country and to describe it in sufficient detail so that our colleagues may use and adapt it to suit their needs.

There are several thrusts to *Instructional Strategies for Students with Special Needs* that make it unique. The first involves the decision-making model. Because assessment and remediation are inseparable components of the teaching and learning experience, we feel that a model which views the two as a continual series of hypothesis development and hypothesis testing can best serve our readers. A second important element of the book is a commitment to the integration and mainstreaming of special needs students. This commitment does not mean that we believe that one must mainstream all students but that we are concerned with how a student can be mainstreamed effectively to achieve specified goals. In this regard, we have deliberately chosen to present instructional strategies that may be used with all students, rather than those that were developed specifically for handicapped students. This approach ties in with our third thrust, which is a non-categorical view of exceptional students. While we appreciate that some exceptional students have unique needs, we are even more aware that they have a great deal in common with both average students and other exceptional students. We suggest that exceptional students are better taught by addressing their individual strengths and weaknesses than by addressing their diagnostic category.

A final distinction of this book is its attempt to present strategies that cover a broad range of school activities. In this regard we prescribe material that is both affective and cognitive; that is core remedial and problem-solving in focus; that speaks to the elementary and secondary school educator; and that reflects the state of the art in technological support for learning. In Chapter 1, we set the stage for our approach and examine several issues that are peculiar to the Canadian population. In

Chapter 2, the decision-making model that provides a framework for the subsequent discussions is described and illustrated, while Chapter 3 presents an overview of the issues involved in assessment. In Chapters 4, 5, and 13, we investigate affective aspects of education in terms of effective behavioural interventions and supportive interpersonal interactions. Chapters 6 through 9 discuss instructional strategies for the core subjects, while strategies that support problem-solving and organizational skills are found in Chapters 10 and 11. In Chapter 12, those techniques that are specific to the secondary school environment are presented, and Chapter 14 explores the contributions of computer technology.

• • •

There are many people who have helped in the development of this book. A very important group has been our students. Their individual contributions are cited throughout the text, but we would also like to acknowledge them generally since they were responsible for the initial development of many of our ideas and the subsequent criticism and revision of others. We also would like to thank Tina Czyzewski, Judee Langley, and Barbara Dillon for the secretarial support they provided and Ida Oya for her help at the final stages, as well as S. A. Pupo, B. Quarrington, and A. Worenklein for reviewing the text at the manuscript stage. Finally, we would like to acknowledge the personal support of our families and friends.

Dan G. Bachor Carol Nugent Crealock

INSTRUCTIONAL STRATEGIES FOR STUDENTS WITH SPECIAL NEEDS

1 CURRENT ISSUES IN EDUCATION FOR SPECIAL NEEDS STUDENTS IN CANADA

This book was conceived and written with two main purposes in mind. The first was to share the authors' commitment to the concept that both students and teachers can benefit from an educational environment that includes as wide a range of students as is feasible. The second was to share the Canadian, American and European experience of educating a wide range of students in the regular stream of education with educators and other concerned individuals across the nation.

There are several major philosophical tenets upon which this book is based. They include 1) a non-categorical approach to understanding and teaching individual students; 2) a belief that most students are best taught all or parts of their school program in the regular class; and 3) a commitment to the idea that educators, families and communities all share in the responsibility for appropriate education of children and youth.

THE NON-CATEGORICAL APPROACH

The non-categorical approach involves seeing students as individuals with individual strengths and weaknesses that must be identified and addressed in the educational environment. Its strength comes from research in classrooms showing that teaching to a categorical description of a type of child rarely does justice to the individual needs of any one child.

There are several reasons for adopting a non-categorical approach. The first reason is the basic problem in the categorical approach — defining the categories themselves. In several areas (learning disabled, behaviourally and socially maladjusted, emotionally disturbed) the definitions are not clear, nor are they agreed upon by researchers. The second reason for adopting a non-categorical approach is the problem involved in relying on categorical definitions to dictate specific delivery of appropriate remedial programming. Rather than relying on such definitions, our orientation in this text is to demonstrate to educators the value of systematically defining and assessing the individual child's learning difficulties in objective, understandable terms as part of the decision-making process that leads to appropriate educational programming and evaluation.

The fields of both regular and special education are beginning to free themselves from the limitations of the causation- and deviency-based medical models and to embrace the behaviouristic and cognitive models that focus on present functioning and individual intervention (Deno 1972; Cartwright and Ysseldyke 1973; Sitko, Fink and Gillespie 1977). Greater attention is now placed on the development of *in situ* task-analytic classroom procedures that measure skills and behavioural patterns that may be related to academic and social learning. There is increasing interest in psychoeducational approaches which view both educational assessment of exceptional students and educational programming as a systematic problem-solving or decision-making process. This process takes into account the interaction of teacher, student, instructional strategy, curriculum and the classroom when making decisions concerning individuals. Diagnosis and instructional programming are increasingly seen as situational and as elements in a continual and ongoing process in which the effectiveness of programs designed specifically for individuals or groups (small or large) of children may be continually evaluated and modified. This approach views the child not in terms of the presence or absence of a handicap or advantage but in terms of his or her specific educational needs (Goguen 1980).

MAINSTREAMING

The second thrust of this book involves a commitment to the integration of students with special needs into the mainstream of the educational experience. For the mildly handicapped child, this will usually mean integration into the regular class; for the more severely handicapped child this will mean integration into a regular school with varying degrees of social integration in the regular class. The emphasis of this book is on regular class integration of the mildly handicapped since this group constitutes the majority of special needs children in the school.

There is considerable evidence that these students are served best in the regular stream (Rosenthal 1980; Madden and Slavin 1983); however, placement in the regular stream must not take the form of "dumping" students. Accordingly, several authors have been concerned about prerequisites for successful mainstreaming (Hundort 1981; Cruickshank 1981; Brown 1983; Wener and O'Shaughnessy 1981; Birch and Wang 1984). These prerequisites include prior planning; professional development of involved personnel; appropriate goal development at the system, school, classroom and individual level; shared responsibility; coordination and resource support; physical plant modifications; systematic evaluation and reinforcement; selective placement of students; reduction in the pupil-to-teacher ratio; and general commitment from the community, school and administration. Where most of these prerequisites have *not* been met, it is extremely difficult for mainstreaming to be successful.

An excellent example of a thoughtful and well-planned approach to mainstreaming can be found in a large urban school board in greater Montreal (Wener and O'Shaughnessy 1981). This board availed itself of government support documents (COPEX

1976), input from local institutions that serviced exceptional students, and their own resource personnel to establish a working committee to plan how mainstreaming would be implemented. A major component of the plan was professional development for everyone. This involved additional training of all educational personnel who would be involved in the process. Trustees, administration, principals and teachers received workshops in identification and intervention procedures using modules that had been carefully developed by their own staff. When the committee felt that a certain segment of the population would resist implementation of a component of the program, they stopped, discussed the nature of the change and its rationale, trained the group on the specific skill or attitude being developed (for example, a positive reinforcement procedure for appropriate behaviour change), got their support and proceeded. A second component of the plan involved goal development. Long-term goals were developed at the system, school and pupil level and then short-term goals were identified and sequenced appropriately. An example of this was the introduction of a social skills program to be given over the four-year high school program. How and at what point individual students or classes could be channelled into the program had to be carefully planned for maximum benefit. Evaluation was built into the plan at several points. All instructional materials went through regular rounds of consultation which provided the users with many opportunities to modify and improve them to suit both general and specific needs. Students also were reviewed regularly with all those who shared responsibility for their program. A final component of this plan was built-in reinforcement. Trustees and superintendents were expected to visit local schools, to know what changes were occurring in them, to support applications for conferences, to support and attend workshops when possible and to express their interest and appreciation to teachers and students. Implementation of the plan has not been without flaw, but after five years the participants feel it is still an exciting ongoing process:

> five years later we can say with confidence that more children with special needs are being better serviced. And because these changes have been planned and effected slowly and thoughtfully, there is good reason to be confident that they will continue to evolve. (Wener and O'Shaughnessy 1981, p. 25)

Having accepted the mainstreaming model and having ensured that the prerequisites for successful mainstreaming are in place, there still are several other issues to be considered. These issues include the questions of 1) who should be mainstreamed and for what purposes; 2) whether mainstreaming works; 3) what its advantages and disadvantages are; and 4) what types of intervention strategies are most efficacious.

Who should be mainstreamed? There is general support for the policy of mainstreaming mildly handicapped students wherever possible for most of their academic work. However, many of these students may need some additional resource support and/or individualized work within the regular class (Meisels 1977; Dardig 1981). Regular consultative support for teachers and a lower pupil-to-teacher ratio (PTR) make success more likely (Wang 1981, 1984).

In discussing who should be mainstreamed, it is essential to point out that the key to effective mainstreaming lies in the degree of severity of the problem, *not* in the category of exceptionality (Clark 1978). Some more severely handicapped children can be mainstreamed successfully if ample support services are provided and there is a strong commitment to integration on the part of the regular class teacher (Bramwell 1981; *Guidelines for Special Education*, Alberta 1981; Snobelen 1981). For these students academic achievement is not a priority goal; rather, social and behavioural skill development are more pertinent.

Does mainstreaming work? Yes, if well planned for by committed educators. However, it is not a panacea for the problems of all exceptional children, nor is it a cheap substitute for the traditional segregated approach (Cruickshank 1981).

Wang recommends that for mainstreaming to work it should be a full-time, rather than a part-time, experience for the student and, therefore, modifications should be made to the regular class to adapt it to the needs of the exceptional student (1981; 1984). Handicapped students enrolled in the Adaptive Learning Environments Program independently completed assigned tasks as well as did average and high-achieving students. The handicapped students used their peers effectively as a source of instructional assistance and successfully availed themselves of instruction in self-management skills to improve their learning rate. However, they functioned at this level only when the PTR was reduced to 15:1 within a comprehensive organizational and resource support system (Wang 1981, 1984).

What are the advantages and disadvantages of mainstreaming? There are advantages in mainstreaming for the exceptional student, the average student and for the teacher. Advantages to the exceptional student include the development of greater independence (Wang 1984), more acceptance from others (Snobelen et al. 1981), and better self-concept (Ballance and Foster 1980). Having interviewed mentally handicapped young adults, Ballance and Foster reported these students felt that integration should be encouraged all through elementary and secondary school and that special classes had been humiliating and had not provided adequate training in the basic skill areas.

Many authors report that there are advantages in mainstreaming for regular students. These advantages are seen primarily in the expansion of their awareness of individual differences, the development of positive attitudes toward handicapped students and the growth of an appreciation that all students can learn something from each other (Bramwell 1981; Wang 1981; Snobelen 1981; Simpson 1980; Gottlieb 1980; Rose 1979).

One advantage to teachers is the development of more positive attitudes toward exceptional learners (Crealock and Nediger 1983; Bramwell 1981). A second advantage is the development of a broader skill base which includes a larger repertoire of strategies required to teach basic skills, to assist children in assuming responsibility and to help children develop realistic attitudes towards self and others (Sowers 1978). This is seen as a development that has benefited not only the teacher but also all the pupils.

These advantages are real but they can only be attained under certain conditions. Such conditions include adequate resource support, careful planning and preparation, a commitment to mainstreaming by both teachers and administrators, and ongoing interaction among all relevant persons.

While the advantages of mainstreaming have proven significant, disadvantages have also been reported. Indeed, many of them are the converse of the advantages. One disadvantage to both exceptional and average students is the exacerbation of negative attitudes towards the handicapped. This can happen when integration has not been carefully planned (MacMillan et al 1976; Simpson 1980). Another disadvantage is the decrease in feelings of self-esteem experienced by the handicapped students when the academic model is a competitive one in which they cannot ever win. Loss of self-esteem also occurs when they are rejected by the average students and/or when they suffer from lack of privacy, feelings of inadequacy and lack of support for their program (Cheong 1980).

Other disadvantages of mainstreaming include increased workload (Hundert 1981; Alson 1977) which is seen as more paper work, a greater need to individualize instruction, more preparation time, more time spent communicating with other teachers and parents, and more effort in managing the classroom environment (Morsink 1981). Posno (1981) refutes these concerns, asserting that development of individual plans is a process that makes teachers more sensitive to student needs and provides an ongoing guideline to curriculum planning.

Inadequate support services and preparation for mainstreaming are also seen as disadvantages (Brown 1983). In many situations, students are "dumped" into classrooms without either physical or educational support being provided. This lack of support and preparation can be seen when there is no reduction in the PTR, no structured individual program plan accompanying the child, few opportunities to consult with system-level experts, little prior or ongoing inservice to help teachers individualize their program, and no programs to help the regular class student become familiar with the strengths and weaknesses of the mainstreamed student(s) (Feitler and Dubasik 1978; Markell 1976; McMurray 1979; Smith 1978; Sokolyk 1981).

A final disadvantage is the need to communicate and coordinate goals, curriculum modifications and ongoing evaluation among regular and special education teachers, principals, consultants, parents and students. This is time consuming and also requires a skill that not all teachers have.

None of these disadvantages is insurmountable. Indeed, with the preplanning and support mentioned earlier, many could be turned into positive features. However, they do reflect the very real experiences that many teachers and students are having as teachers try to come to grips with the reality of dealing with a much more heterogenous group of students than they were expected to teach in a segregated model.

What intervention strategies are most effective? One of the most promising ways to improve academic skills and to foster good social relationships among both average and exceptional students is to provide the opportunity for cooperative learning experiences (Johnson and Johnson 1980, 1983; Sapon-Shevin 1978; Hoben

1980). Competitive or individualistic learning experiences tend to promote rejection and reinforce stereotypes of handicapped persons. Acceptance, on the other hand, is promoted by providing cooperative learning experiences in which handicapped and regular students work together. It is equally necessary that special education and regular class teachers work together as a collaborative team.

A similar approach to fostering a higher level of acceptance of exceptional students is teaming for social or academic purposes (Rose 1979; Turcic and Railton 1983). Rose paired severely mentally handicapped children with non-handicapped peers in a tutoring program in English. Results indicated a significant increase in associative and cooperative play among handicapped and non-handicapped students accompanied by a significant decrease in the level of negative attention-seeking and other inappropriate behaviours on the part of the handicapped. Turcic and Railton developed an integrated community classroom program which was implemented in a regular grade four classroom with 28 students whose achievement levels ranged from grade two to grade nine competency. The class included one student with a severe chronic behaviour problem, two mildly socially maladjusted students, several English-as-a-Second-Language students, several gifted students and one visually impaired student. All the children remained in the regular classroom and received all their instruction there from the home teacher, resource teacher and peers. Students were paired according to academic skills and observed social skills. Examples of the pairing included matching an extroverted student with a shy student, a good on-task behaviour model with a student who was often off-task, a good organizer with a poorer one, and a good speller with a poor one. Through the use of discussion to reach consensus about appropriate behaviours, and through the use of modelling, shaping, and grouping by clusters to expand the social model, and specific remedial support as needed, all students improved both academically and socially. In addition, there were advantages to the teacher which included more available time for individual and small group instruction, more time for socializing with students, for preparation of classroom work, for marking and for giving students feedback more quickly. Generally the teachers felt the changes led to more constructive use of their time and expertise in the classroom.

There are several other successful approaches to intervention. One such approach is to have special educators work with small groups of students in the regular classroom rather than in the resource room on a withdrawal basis (Hoben 1980; Turcic and Railton 1983). Another is to adopt a prevention model (Wilson 1984). This involves catching students with minor difficulties early and correcting their problems before they need more formal remedial help. Prevention could also involve adopting a model (like that of mastery teaching) that ensures that all students learn the current work before going on. Other successful approaches include fostering respect for individual differences and discussing handicaps and their characteristics with regular class students in preparation for mainstreaming.

In summary, integration into regular class is seen as the preferred alternative for most mildly handicapped students. However, this cannot be successfully implemented without considerable planning, regular and reliable support, program modification and continual evaluation. The next section of this chapter will look at the support issue in terms of shared responsibility for the education of all students. The rest of this book will address the issues of planning and program modification in various subject or skill areas.

SHARED RESPONSIBILITY

Traditionally, Canadian teachers have viewed the classroom and the students assigned to them as "theirs." This meant they would shut their door, shape the learning experiences of the group and take full responsibility for whatever curriculum and instruction was required for all their students. If a student had serious problems, the teacher would refer that child to specialists who would confirm or reject the teacher's hypothesis that the child was not able to cope at the regular class level. The child either would be removed from the class into a segregated special learning situation or returned back to the teacher to be handled as he or she saw fit. The system was committed to segregation and most educators were convinced that the regular class environment was not the best place to teach special needs children (Desharnais 1984).

Today we have become committed "integrators" and that has caused a major restructuring of our basic philosophy towards educating special needs children. The change came slowly at the decision-making level but very quickly at the program-delivery level. For well over a decade and a half, educators have been aware that segregation was not meeting the academic and social needs of many handicapped children (Semmel et al. 1979). The growing ethical, philosophical and educational concern led governments in the United States and Canada to pass legislation that either directly mandated the implementation of mainstreaming or strongly suggested that mainstreaming was the best way to go. One of the major changes that has resulted from this approach is that regular class and special education teachers can no longer operate independently and assume full responsibility for the increasingly heterogenous group of students they must teach. Whether an individual teacher wants it or not, many others are involved in the education of the special needs child in his or her classroom. And with incidence levels for identification of special needs children ranging between 10 and 20 percent, most regular classrooms will have 3 to 5 students on individual program plans of a greater or lesser extent. This poses a dilemma. While many regular class teachers complain that they don't get enough consultative support, others are very reluctant to share their authority.

Who are the significant others that share in the responsibility for the education of special needs students? The list is large. It includes, in approximately increasing order of responsibility, the ministries of education in each province (and, to a limited extent, the federal government), the teachers' federations, the community in which the student lives, the school system, the principal and school the student attends, the student himself or herself, the peer group, the teachers, and the parents. Each of these individuals or groups shares responsibility in different ways but the impact of each can be seen by carefully examining the total educational experience of the student.

The provincial and federal governments have the responsibility of providing the laws, regulations and general guidelines that will determine the overall structure of education; they are also responsible for deciding whether or not to provide a large amount of the funding that will make program development feasible across the country.

Teachers' federations are responsible for providing support and inservice to their members where possible; for helping members understand and implement their role within changing legislation; for helping teachers appreciate when that role is appropriate and when it is not; and for helping teachers lobby for the kind of educational support they need to do their jobs effectively.

The community has two major roles. One involves the diagnosis and remediation provided by the many professional groups that are involved with special needs children (psychologists, medical personnel, occupational therapists, speech, hearing and vision specialists, social workers, etc.). The second involves the broader support base offered in the community through service groups like the Big Brothers and Big Sisters, the YW/YMCA, which offer recreational and social programs for a wide variety of children, or the Salvation Army, which offers residential services as well as "boys club" activities for emotionally needy children; through neighbourhoods that support the integration of special needs and regular children on an ongoing basis; through sports and activity programs at local community centres or libraries; and through municipal governments that provide the extra physical supports needed by many handicapped individuals such as paratransit systems and home library services.

The school system is responsible for translating government policy into workable plans at the local level. It is also responsible for providing appropriate inservice so that everyone concerned (from trustees and superintendents through to students in the classroom) has sufficient understanding of exceptional children to interact appropriately with them. The school system is also responsible for consciously providing the personal reinforcement that front-line teachers need as they struggle with changing demands on their time and energy.

The total school environment should provide both leadership and support. Principals must provide motivation, direction and coordination in their schools (Leithwood and Montgomery 1982). They should help teachers see how integration can work to the benefit of all and they should provide ongoing support for the teachers' efforts. Principals must know what is happening in the classroom and be willing and able to give the teacher feedback as appropriate. Cawalte (1981) states that the principal must be a strong positive influence on staff, students and community; must regularly seek feedback in order to improve his or her leadership skills; must constantly try to find ways to make the school a more productive and satisfying place to both work and play; and must be able to present yearly evidence that the school is a better place than it was the year before. Principals also should be willing to work with the teachers to draft the goals and specific strategies for achieving them that would lead to successful programming for the special needs student. The final role of the principal is to coordinate or provide for the coordination of the many activities, efforts and goals of the people that share responsibility for the exceptional student.

Still considering the total school environment, the school should be physically able to support the mobility of all its students. And, at the very least, all the staff and students should be prepared to give integration a chance. One approach to a school environment that provides both leadership and support is described by Askey (1982). In a southern Ontario school, it was decided that the entire staff could and should become involved in a flexible exchange of students in order to provide better programs for all. A systematic plan evolved that included staff development, physical reorganization of the school, an additional materials budget for every teacher, preparation of regular class pupils, and a team approach to discussion and planning for every student in the school. Staff development was seen as the most important thrust of the plan and, accordingly, the principal set the following goals, fulfilling his responsibility to integrate all individual efforts in his school:

> 1. provide opportunities for staff to reflect upon self evaluation checklists which pertain to lesson methodology and curriculum development;
>
> 2. utilize the expertise of the classroom teacher in the design, preparation and presentation of school reports to the Identification, Placement and Review Committee;
>
> 3. encourage individual staff members to develop an on-going plan of proactive and positive public relations;
>
> 4. provide opportunities for each staff member to demonstrate and practice leadership skills through acceptance of chairing major school projects;
>
> 5. provide individual staff members with positive feedback on a continuous basis; and
>
> 6. monitor the extent and involvement of each staff member in his/her professional growth and career development. (Interview with Author)

Exceptional students themselves have many responsibilities for their own programs. These responsibilities include being on time, being prepared, having homework done, behaving appropriately, and being willing to take some risks in recognizing, so far as they are able, the reality and implications of their problems. They should not be allowed to see themselves as passive recipients of others' plans but rather should be taught to see themselves as active participants in their own experience.

Peers have a responsibility to accept and try to understand the exceptional student. They have a responsibility to share their skills and friendship with those who are less fortunate, not only because the exceptional students need to feel they are part of the total school community, but also because the regular students' sense of social maturity and their academic growth can be enhanced by an experience that allows them to share in this manner (Hoben 1980; Turcic and Railton 1983; Johnson and Johnson 1983).

Parents share in the responsibility for their child's education in many ways. The most obvious ones are being aware of and accepting their child's strengths and weaknesses, supporting him or her emotionally, sharing information with others who can use it, participating in school decisions that affect their child, and directly providing additional opportunities for social and academic experiences as needed.

There is no argument over the first three responsibilities, but there has been considerable controversy over the parent's responsibility or even right to become directly involved in their child's academic program. Parents frequently have been counselled against tutoring their children with statements such as the following:

> Many parents try to help their underachievers in reading...such help usually results in upsetting both the child and the parent. I am sure that if the child did not already have a block toward reading, one would develop after a few evening sessions of tears and screaming. (Kasdin 1975, p. 29)

Kasdin, as well as others who take a "hands off" view, does suggest that in the infrequent situation where both the child and parents are comfortable with tutoring it could be attempted. Nevertheless, the primary message is to stay out of the academic area.

Over the past ten years this attitude has changed somewhat, but the message to parents, at best, has remained ambivalent. However, as more teachers realize that they can give the low achieving student neither the amount of one-to-one support nor the additional review of basic material that is needed, they are more willing to accept parents as partners in the educational experience. The fostering of that partnership is aided by programs such as *Teaching Parents to Help at Home*, developed by Alma Harrington (1975). This course contains information on reading skills and suggests home activities that will strengthen their development. The author stresses that teachers do the teaching but parents provide the necessary repetition and reinforcement that the child with learning problems needs in order to master the skill. The materials for each reading skill topic in her program focus on the following four objectives:

> 1. to provide the parents with specific background information for that topic;
> 2. to provide helpful tips for working with skills pertaining to that topic;
> 3. to suggest activities...which can be used and implemented within the home...and
> 4. to refer, only casually, to published materials which parents could buy.... (p. 52)

Programs of this sort help to bridge the gap between the child's academic needs and the parents lack of knowledge in curriculum areas. Ideally, parent and teacher can work together to keep each other informed about the skills being taught at school and reinforced at home and about the continual progress of the child in both milieus.

Many parents also want to have more input into the decision making that affects their children. Lusthaus, Lusthaus and Gibbs (1981) conducted a study in Quebec to determine how parents' actual involvement in decision making compared to their desired involvement. They found that all parents in their study wished to be more involved than they currently were and this was most significant in areas that pertained specifically to their own children as opposed to those that were primarily school issues. This was especially true for parents whose children were in special education classes rather than "free flow" classes (i.e., resource classes, access to which is based on student and/or teacher demand) or regular classes. The specific areas in which they wanted more involvement were transferring schools, keeping information on the child, providing medical services and disciplining the child.

While there is general acceptance of the supportive role of parents, it is not always easy for the parents of a special needs child to accept that role and carry out all their responsibilities. Many parents experience conflicts in dealing with their children and these are aggravated when children have social and academic problems. Being able to read, do mathematics and write are very central to feelings of industry, competence and value in children. When they cannot do these things well, they need a great deal of help finding things they *can* do well. They also need help in understanding why they failed in academic tasks. However, there is a fine line between overlooking some negative behaviours in order to support a child's self-concept and allowing a child to become a tyrant who makes everyone's life miserable. When parents can find the middle road, where they are not nagging their child and focusing on his or her problems, yet are giving the consistent guidance needed to develop socially, the child has a much better chance at a healthy self-image.

Furthermore, it is not easy for parents to know when they have succeeded with their child. Doreen Kronick, the mother of two children with learning disabilities, founder of the Learning Disability Association of Ontario and author of many articles and books on the topic (1981, 1983) feels that a family has succeeded when a child's self-esteem is healthy. Self-esteem can be encouraged through developments of many different kinds; for example, a good sense of humour, a sense of social fairness, good manners, attention to grooming, artistic ability, or swimming skills. Also, the growth of self-esteem often depends in large measure on family expectations and on their definition of success. If parents of children with learning problems can go beyond their traditional values and reward achievement consistent with their child's abilities, then the child has a chance to like and believe in herself or himself.

Many special needs students will require extra support all through their educational experience. Many need to be specifically taught how to organize, categorize, generate solutions and evaluate situations that occur daily if they are to compensate for the information they miss due to faulty perception, poor academic skills or an inability to generalize. This is a large responsibility for those who must help a child to develop these skills. Parents are the first teachers of their children and need to find ways to work with school-based teachers in order to help their children overcome their handicaps and develop into healthy, positive adults.

The last group of educators who have responsibility for the special needs child is that of the teachers themselves. With the increased degree of mainstreaming in the schools, this group involves both regular and special education teachers. Their responsibilities include identifying students with special needs, providing initial remedial opportunities, recognizing the point at which they need to call on specialists to provide a more complete understanding of the student, providing appropriate goals and programs to meet the individual needs of the student, evaluating the program and providing the support and encouragement necessary so that students are able to work to their potential. The degree to which regular class and special education teachers share this responsibility must be sorted out among the personnel involved. Wang and Birch (1980, 1984) suggest that the special education teacher write up the remedial goals, plan the lessons and sequence them appropriately for the child. The regular class teacher should ensure that the student has the opportunity to do the work in class, that he or she is supported and given feedback and reinforcement, and that the social climate is one that encourages cooperative effort among all class members. Both teachers are responsible for maintaining appropriate records and sharing information between themselves and with others who are involved as necessary.

An effective way to create shared responsibility among school staff is to support the concept of in-house teams (Crealock and Nediger 1983; Wilson 1984). An example of such a team has been described by Crealock and Nediger. It includes three staff representatives (one special education teacher, one primary teacher, one junior teacher), the principal, special education and psychology support staff, the nurse, the vice-principal, specific regular class teachers who are presenting a child about whom they are concerned and, occasionally, parents. This team has been meeting once a week from 8:15 a.m. to 8:55 a.m. for the past five years. The teacher presenting the case writes up his or her concerns and action minutes (a record of the decisions taken by the group) are kept. As a result, much of the preliminary work will already have been done at the school level in the event that a child must be considered by the system for special placement. Because the staff representatives change every half year, most of the staff has been through the experience by now. This approach has been found to be valuable both as a staff development experience and as a means of keeping track of all students in the school who have special needs.

This concept of shared responsibility is a new one to many educators but its acceptance is necessary if the full potential of all students in the system is to be actualized. It is hoped that this discussion will make the reader aware of how many others are and must be involved in the full educational experience of the exceptional child.

There have been several excellent outreach programs that demonstrate the value of expanding the network of those who share responsibility for a child's education. One example is a volunteer program, the Step-by-Step Program, that pairs adults in the community with emotionally maladjusted students in the school. It has been functioning in Ottawa for 19 years (Priddle 1981). It was initiated to provide support on a regular basis for children who were experiencing serious problems in school because of emotional and social difficulties. The purposes of the program are threefold.

1. To help an emotionally maladjusted child gain confidence in his or her learning environment through a supportive relationship with an adult.
2. To prevent more serious emotional difficulties by giving the student attention before his or her problems become acute.
3. To assist the principal and teacher in their work with the disturbed student under the supervision of the school psychologist.

In this program volunteers are instructed to conform to the procedural rules of the school in which they serve, to obey rules of confidentiality, responsibility and dependability and to function within established board policy. Volunteers are first screened through a procedure which includes completion of an information sheet, submission of three references and a personal interview conducted by the Co-ordinator of Volunteers. They are then given an orientation which includes one general meeting and two discussion meetings. Only then is each volunteer matched with a pupil and given very explicit instructions concerning his or her responsibility to the child. These instructions deal with such things as the frequency of contact with the child (at least twice a week), suggestions from the school team about approaches to interaction, the need to report any sudden change in the child's behaviour immediately, the need to use tact, and attendance at regular monthly inservice and reporting sessions. There are no academic expectations, so the volunteer is free to select activities that interest the child and which allow a positive experience. The interest and contributions of the volunteers are maintained through successful interaction with the children, monthly inservice and reporting sessions, appreciation through board newsletters and a series of awards. Evaluation of the program has consistently been positive. It continues to attract sincere, conscientious people who see their role within the school system as a meaningful contribution to their community. It has provided sufficient support to the pupils to enable them to remain in the regular program. It serves approximately 100 students a year.

Outreach programs for the gifted have long recognized the value of using the community to enrich opportunities for these exceptional students (Wees 1970). Millar and Rood (1983) describe a mentorship program set up in a rural school area of Alberta that matched 21 gifted grade three students with 24 mentors from both the school and the community (three grade five peer mentors, six parents, six local teachers and nine members of the community). After students were interviewed and their interests identified, mentors and students were matched. Areas and topics to be explored included art, computers, calligraphy, gardening, mathematics, horses, microscopes, music, photography, puppets, political cartoons, sports, travel and oil rigs. Classes were held after school and the group size ranged from one-to-one pairings to 18 children who wanted to participate in a series of sessions featuring local artists. The objectives of the project were to provide contact between gifted students and community experts, to facilitate communication and education of parents and to extend peer contacts for gifted children. The program was evaluated on the basis of both questionnaires to parents and anecdotal responses from all involved. The results were overwhelmingly positive, as the authors' summary comment suggests:

> Spirits soared when this small enrichment class went into the local community to make friends and broaden their interests and abilities. The mentorship program added an important dimension to the pupil's education. (p. 14)

A final example of the concept of shared responsibility being put into action is the program for the mentally handicapped developed in Newfoundland and Labrador (Steer 1981). This program is based on a model which examines four main areas of involvement — education, work, leisure time and residential alternatives — and develops taxonomies from "least restrictive" to "most restrictive." A "most restrictive" profile would describe a student with no school program and no real work, watching TV either at home or living in a large dormitory far from home. A "least restrictive" profile would describe a student enrolled in a school program in a regular class with normal individual attention, having part- or full-time competitive employment, participating in a full range of leisure activities and living in an apartment or home with one to three others. This program tries to change the attitudes of parents, students and teachers so that they believe that the mentally handicapped can function in a more normal environment. It has received inter-ministerial support through the development of group homes, work programs and more individualized school programs. Steer reports many changes in the past five years that have resulted in many more handicapped people reaching the "least restrictive" level of independent functioning.

Having considered the importance both of mainstreaming and the acceptance of shared responsibility for the education of exceptional students, we will now examine the support offered through legislation across Canada.

LEGISLATION

In Canada, education is administered and funded, to varying degrees, by federal, provincial and municipal governments. While there is no ministry at the federal level that can initiate legislation and funding to ensure equal access to education for all students needing special programming, several federal ministries do support programs and contribute to research into special education needs in various ways. For example, the Ministry of Justice and the Solicitor-General have recently supported research to promote a better understanding of the learning-disabled young offender (Crealock 1984; Ross and Fabiano 1983); the Ministry of Indian and Northern Affairs has given grants to support the development of vocational and special education programs in the Northwest Territories and throughout the provinces (Scaldwell 1984); the Ministry of Employment and Immigration has funded special programs; the Ministry of Health and Welfare has participated in cost-sharing programs for seriously handicapped students; the Secretary of State has supported research and the development of teacher training opportunities in special education (Sitko and Harper 1984) as well has having provided continuing research support through the Social Science and Humanities Research Council of Canada (SSHRCC) and the Natural Science Research Council (NSRC).

In other words, involvement at the federal level primarily has been in the form of funding for special research and development projects. Involvement in the form of direct legislation is the prerogative of the provincial governments.

The challenge to legislators to provide more effective education for the exceptional student has come from many sources over the past decade. The passing in the United States of Public Law 94-146 (1975), which mandated appropriate programming for the handicapped, provided a precedent for North American educators and politicians. The concern of communities and families which was often expressed through parent-support associations like the Learning Disability Association of Canada, the Canadian Association for Community Living, the Association for Bright Children, etc., brought the issue home to provincial and municipal leaders. The growing body of research on the inappropriateness of traditional programs heightened the concern of many Canadian educators (Goguen 1980). Also, the grassroots concerns of teachers and students wanting a more comprehensive response to the special needs child added to the demand for a major review of current practice.

The two major areas of review included the availability of equal access to education for all students and problems of appropriate placement and programming for special needs students. To date, six provinces (Newfoundland, Nova Scotia, Quebec, Ontario, Manitoba and Saskatchewan) have passed mandatory legislation (i.e., legislation stating that school boards must provide education for all students, regardless of handicaps), while the remaining four provinces, the Yukon and the Northwest Territories have permissive legislation. This represents a growing appreciation across the country of the need to respond more appropriately to special needs students. In 1980, Goguen reported that neither Prince Edward Island nor the Northwest Territories expressed a concern for mandatory legislation. In 1984, although not yet passed, the Education Ordinance of the Northwest Territories stated:

> the Executive Member shall establish programs of special education designed to provide assistance to students who are unable or unlikely to be served to advantage by the regular education program. (p. 36-7)

The quality of the mandatory aspects of special education legislation varies greatly from province to province; but in each case the legislation includes the equal access or "zero-reject" component, as in the following example:

> 33. Each school municipality in Quebec...shall admit...every child domiciled in the municipality, from the beginning of the school year following the day on which he attains the age of five years until the end of the school year in which he attains the age of sixteen years. (*Gazette Officielle du Québec* 1980, p. 749)

Statements of the necessity to provide the best possible experience for special needs students are typified by the following two passages:

> Societies must provide an enriched form of general education aimed at enhancing the life of those suffering from diverse handicaps.... For lack of interventions of this kind, many defectives may remain socially maladjusted and handicapped to a certain degree and never achieve full development of their abilities. (UNESCO 1973, p. 18)

> ...every person between the ages of six and 21 years shall have the right to attend school in the division in which he or his parents or guardian are residents, and to receive instruction *appropriate* to his age and level of educational achievement.... (Saskatchewan Education 1984, p. 7)

The approach to special education which attempts to meet the needs of individual students in the most appropriate manner possible provides for not just any program, but one of quality. In Ontario there was considerable debate about the inclusion of the word "appropriate" in the Education Amendment Act (1980), but the combined weight of the full legislative committee and very active parents groups ensured its inclusion. Provinces which do include this clause are committed to very specific planning for all exceptional children under their jurisdiction. However, interpretation and implementation of the provincial legislation is the responsibility of municipal governments.

The specific issues that must be addressed at the local level include a) the range of services that either will be provided directly or will be bought from other local boards or provinces, b) the model of program implementation that will be used, c) the allocation of resources and d) the involvement of parents. Leslie and Goguen (1980) suggest that the three main administrative decisions that school boards must make concern the rights of exceptional children to programs based upon their special educational needs, the provision of a range of needed educational alternatives and the appropriation of funds. The trend at the local level is to include all the people influenced by these decisions (parents, regular class teachers, special education resource personnel, students, administrators and school trustees) in the development and ongoing evaluation of programs (Goguen 1980; Wener and O'Shaunessey 1981).

In summary, the development and implementation of individual programs for special needs students in both mainstream and segregated educational situations is the responsibility of all levels of government to varying degrees. The success of such programs, however, still rests with the teachers and students who meet daily in the classroom.

Thus far we have examined the non-categorical approach to education and the concept of mainstreaming — including the issues of responsibility and legislative support for programs for special needs students. We will now address several other general issues related to special needs education.

CATEGORIES OF EXCEPTIONAL STUDENTS

One of the most important issues in the field of special education is that of categorical versus non-categorical approaches to describing special needs students. Earlier, we made our own preference for a non-categorical model clear; however, the categorical approach is still being used across the country for both administrative and program purposes. In Ontario, for example, the five main exceptionality groupings used are intellectual, physical, behavioural, communication, and multiple (Ministry of Education 1984). Other provinces list visually impaired, hearing impaired, trainable mentally retarded, learning disabled, orthopedically handicapped, chronically health impaired, behaviourally handicapped, and multiply handicapped (Saskatchewan Education 1981; Cable 1976).

In the following sections we will define the main categories of exceptional students, indicate the incidence of these exceptionalities in the general population and point out the major characteristics that teachers might look for in trying to understand those students in their classrooms that seem to have special needs (Kneedler

1984; Cruickshank 1980; Hammil, Bartel and Bunch 1984; Saskatchewan Education 1981; Ontario Ministry of Education 1984; Csapo 1981; Halifax Board of Education 1976; Quebec Ministry of Education 1983).

Learning Disabled Students

Students with learning disabilities make up a group whose numbers have increased over the years. The Learning Disability Association of Canada (LDAC) has adopted the following definition:

> Learning Disabilities is a generic term that refers to a heterogenous group of disorders due to identifiable or inferred central nervous system dysfunction. Such disorders may be manifested by delays in early development and/or difficulties in any of the following areas: attention, memory, reasoning, coordination, communicating, reading, writing, spelling, calculation, social competence, and emotional maturation.
>
> Learning disabilities are intrinsic to the individual and may affect learning and behaviour in any individual, including those with potentially average, average, or above average intelligence.
>
> Learning disabilities are not due primarily to visual, hearing or motor handicaps; to mental retardation, emotional disturbance or environmental disadvantage; although they may occur concurrently with any of these.
>
> Learning disabilities may arise from genetic variations, biochemical factors, events in the pre- to postnatal period, or from any other subsequent events resulting in neurological impairment. (CACLD 1982)

This definition contains three essential aspects which are common to all formal definitions of learning disability (LD) used across the country. These are a) a discrepancy clause which states that there is a significant disparity between a student's observed ability or potential ability and his or her demonstrated achievement primarily in language or mathematics; b) an exclusion clause which states that the discrepancy is not primarily due to intellectual, emotional, physical or cultural problems; and c) an etiology clause that either strongly or mildly makes the connection between LD and neurological or biochemical factors. This third clause is somewhat controversial since it is very difficult to prove. The working definition of LD most frequently used across the country is one that points to a marked discrepancy between ability and achievement and, to some extent, excludes other problems as primary causes of the discrepancy (Crealock 1984).

The incidence of learning disability varies greatly across Canada. Definitions that include all three clauses tend to identify approximately two percent of students as learning disabled (Douglas 1972); while definitions relying primarily on the discrepancy and exclusion clauses may identify as many as 20 percent of students as learning disabled. In an Ottawa study, Trites (1979) showed that 14.3 percent of students were considered LD according to teacher reports on hyperactivity. An important factor here is the type of service offered. When the more generous definition is used, most of those identified are mainstreamed since their disabilities usually fall into the mild range which often can be accommodated in the regular classroom if some additional specific support is provided. The more stringent definitions often identify only the more seriously handicapped who require considerable support, often in segregated settings.

There are many symptoms that will alert a teacher to LD in his or her student. The most prevalent is the discrepancy between ability and achievement. This is often first noted when an expectancy of normal or better achievement based on good verbal and comprehension skills is not met when students have to encode through oral reading or writing tasks. A second, often noted symptom is an inappropriate attention span. This is most frequently observed as an inability to attend to oral or written work in areas of difficulty. In other words, while the child may be able to attend to a program on television, he or she may only be able to pay attention to short inputs (sometimes less than a minute) in mathematics, reading or general instructions. A third, frequently noted symptom is found through a careful examination of the student's writing. This is especially relevant for students in the upper elementary grades and in secondary school. Writing problems will be discussed extensively in Chapters 8 and 9, but in general, the writing and spelling of an LD student are often quite different in terms of both quantity and quality from those of a typical student whose low achievement in writing comes primarily from developmental problems. Other characteristics that have helped educators determine whether a student has a learning disability include hyperactivity, poor peer relationships, quick emotional changes, organizational problems, memory difficulties, general coordination problems, impulsiveness, speech and hearing disorders and signs of neurological irregularities.

Intellectually Handicapped Students

These students are normally divided into two groups — the mildly mentally handicapped and the moderately to severely mentally handicapped. Mildly handicapped students are sometimes called general learning disabled (GLD) to distinguish them from the group described in the previous section, sometimes referred to as specific learning disabled (SLD). This division is based on the observation that most GLD students show academic retardation in all areas while SLD students show it only in some specific areas.

The mildly mentally handicapped are generally thought to be capable of academic learning, independent social adjustment and economic self-support. They can function educationally within a regular class if there is aid provided in the form of curriculum modification and support services. Where this is not available, however, they may need a more structured program in a segregated setting.

The incidence for mild mental handicap is approximately two to five percent of the population, depending on where one draws the IQ cutoff. Operationally, students with IQs in the 55 to 80 range are usually felt to be eligible for special support, but this assessment varies somewhat depending on the total pattern of strengths and weaknesses shown by the individual student. In general, however, they have difficulties in all aspects of intellectual functioning — concept learning, language, memory and perception. They learn concepts slowly, need much repetition, show attention problems and develop speech and language competency more slowly and at a less complex level than do children of average or better intelligence. The two most frequent personality characteristics associated with this group are poor motivation to succeed and an external locus of control. The latter becomes evident as an unwillingness to try very hard and as a belief on the part of the students that others or chance are more likely to control their lives than they are themselves. Many also suffer from low self-esteem and have difficulty handling complex social interactions and interpersonal relationships. While there may be strengths and weaknesses within their ability and achievement profile, it is basically characterized by below-average functioning.

The moderately to severely mentally handicapped students comprise a much smaller number of those requiring additional help. Incidence figures for this group are usually less than 0.3 percent. These students are generally felt to have limited potential for academic learning, independent social adjustment and economic self-support. The vast majority require a special education program in a segregated setting that provides them with the time and opportunity to develop as much as possible in motor, language and academic areas. Increasingly, however, they are being integrated into the social aspects of the regular program through shared assemblies, lunch rooms and play grounds. They are also being integrated, to a lesser extent, in subject areas with a recreational rather than conceptual focus, i.e., in music, art, and physical education. This occurs most frequently in the kindergarten to grade six school programs.

The major diagnostic criteria for this group is an individual IQ score between 25 and 55 which is supported by parental and teacher observations of consistently low achievement and poor general adaptive behaviour. Most of these students are diagnosed long before they come to school, but occasionally the extent of a student's disability becomes evident only upon school entry. They are characterized by a serious reduction in their ability to learn concepts, attend appropriately and communicate effectively. The goal of school programs is to enable them to achieve as much independence as possible.

Gifted Students

The gifted are at the other intellectual extreme. In some school areas, programming for them is limited or non-existent since although they are considered exceptional there are often serious debates about whether the schools should spend precious special education dollars on a group that is already blessed with so much. However, the growing consensus does favour some special programming, some of which comes in the form of regular class enrichment and some of which occurs through special resource programming. Theoretically, this group should consist of 2.4 to 4 percent of the population, but, characteristically, many fewer (less than one percent) are identified (Spina and Crealock 1985). They are usually defined as a group of students with an unusually advanced degree of intellectual, artistic or academic ability that can benefit from a differentiated learning experience that goes beyond the breadth and depth normally provided in regular school programs.

The most common diagnostic criterion is an IQ score of 130 ± 3. Other criteria often used include high academic achievement, teacher and/or parent nomination, high scores on tests of creativity, demonstrated artistic ability or other demonstrated talents. Gifted students have been found on average to be socially, emotionally and physically successful as well as academically and intellectually. Nevertheless, there is a wide continuum of social abilities from well-adjusted to poorly-adjusted into which they fall. Gifted students can be culturally deprived, physically handicapped, poorly motivated, language impaired, or learning disabled. Identification of them is much more difficult if a handicap accompanies their giftedness. In a survey that was responded to by over 90 percent of Ontario school boards, it was found that very few boards formally identified the gifted underachiever and when they did, they programmed for them primarily as gifted students and paid little attention either to the reason for the disability or to planning for it (Spina 1983). There is, however, a growing trend to attend to the "high risk" gifted student, that is, the one who is not able to develop his or her skills in the regular program and needs a small, more individualized program.

Behaviourally or Emotionally Handicapped Students

This category includes a broad range of special needs children. The Child Development and Support Services of Manitoba has divided emotionally disturbed students into three categories: psychologically disturbed (0.6 percent), severely to moderately disturbed (2.5 percent) and mildly disturbed (9 percent) (Csapo 1981). This categorization corresponds to the UNESCO *Report of the Advisory Meeting on the Maladjusted* (1973) which listed three similar groups: a small group of deeply disturbed children needing massive help, a small group who need some expert aid for themselves and their parents, and a much larger group for whom adequate programming can be handled at the school level.

In the 1970 report by the Commission on Emotional and Learning Disorders in Children (CELDIC), the incidence figures for all types of emotional handicap ranged from 10 to 15 percent based on the 1966 census. There do not seem to be major changes in incidence since then. Again, however, very few of these students are identified for segregated programming. Most of them are integrated into regular classes with or without some additional support. This seems to be the result of too few resources in many areas. It is also the result of a growing trend in both schools and parent organizations toward the belief that students benefit most from being taught in a setting where normal behaviour and achievement is expected and is modelled by most of the students.

The Ministry of Education in Ontario has defined this group as having a disorder characterized by behaviour problems accruing over such a time period and being of such severity and nature as to adversely affect educational achievement (*Special Education Handbook*, 1981). Frequently noted characteristics of these students include an inability to build or maintain good interpersonal relationships, excessive anxieties, depression, compulsive behaviour, withdrawn behaviour, verbal and/or physically aggressive behaviours, and academic retardation that cannot be explained by other factors.

Visually Impaired Students

This group constitutes about 0.2 percent of special needs students. Those considered partially sighted have visual acuity between 20/70 and 20/200; that is, they can see at 20 feet what the person with normal vision can see at 70 to 200 feet in their better eye with correction. Legally blind students have visual acuity of less than 20/200 and/or a field of vision no greater than 20 degrees in diameter (tunnel vision). While many of these students have traditionally been educated in segregated schools across Canada, the current practice is towards mainstreaming in the local school. This has occurred due to social and familial support considerations and to the growing acceptance of the responsibility to teach such students in regular classrooms. There are still, however, a small group of students who require the more segregated experience for both social and academic reasons.

The most obvious characteristic in identifying this type of child is the degree of blindness. General intelligence does not seem affected by lack of sight, but the students do have serious difficulty learning both abstract and spatial concepts and this often results in slower academic development. They don't tend to be emotionally maladjusted, but some do have difficulties adjusting to society's reaction to their blindness.

Physically Impaired Students

These are students who have a variety of physical disabilities, sufficiently severe to interfere with school attendance or learning. The reported incidence is approximately 0.2 percent. This group includes students with neurological impairments which involve damage to the brain and/or nervous systems such as occur as a result of cerebral palsy, epilepsy or spina bifida. This group also includes students with musculoskeletal conditions such as muscular dystrophy and arthritis, as well as students with a variety of other problems stemming from congenital deficiencies (e.g., thalidamide-caused defects) or from childhood accidents or illnesses (e.g., polio or paralysis resulting from car accidents).

Physically handicapped children range in intelligence from gifted to profoundly retarded and in emotional/social development from healthy to maladjusted. Their physical condition is a source of constant stress but their reaction to this is largely dependent on the reaction of significant others to their condition (Needler 1984).

Hearing Impaired Students

These students constitute approximately 0.6 percent of the population (Suren and Rizzo 1983). They are defined as profoundly deaf if they have a hearing loss of 90 decibels or more (i.e., if they cannot hear a sound of 90 decibels). This means they cannot hear a loud shout at very close range. They are defined as hard of hearing if their loss is in the 60 to 90 decibel range, which is a normal conversation range. Because a hearing impairment has a profound effect on the development of speech and language, serious problems result in the intellectual and achievement areas that depend on verbal skills. Deaf children show no retardation in performance IQ, however. While they may appear anywhere on the continuum of emotional health, many have been reported to be socially immature and lacking in interpersonal skills. They are currently taught in both segregated schools and integrated settings although, again, the trend is towards mainstreaming them as much as possible.

Learning Rates of Exceptional Children

While each of these six categories of exceptional children should be addressed somewhat differently, there is considerable overlap in the needs of the students. For many educators the best approach is to consider the degree of severity of the problem. One way of measuring severity is through an analysis of the child's learning rate. Indeed, for many, the important educational issue in planning for the exceptional child is not primarily the specific characteristics the individual child demonstrates but the rate of learning of which the child is capable.

Most special needs students can learn at least up to about grade four to six levels of competency if they are given enough time to do so. Figure 1-1 shows a learning rate curve that has been used in Saskatchewan to determine the degree of special education support certain classifications of students typically require (Saskatchewan

Education 1981). A possible learning rate line for the gifted has also been included to show how the regular class could fail to meet all their needs unless additional resources were provided. School boards wishing to use such a graph, could plot their own students actual progress by grade (or age) and determine how closely they approach the slopes illustrated in Figure 1-1. This gives a visual representation of the expected academic growth rate that is useful in planning for exceptional students. Students of at least two types can fall along any curve: type one includes students showing reduced or enhanced academic achievement that is consistent with their ability level (e.g., gifted children, retarded students, hard-of-hearing students); while type two would include those who displayed a discrepancy between measured aptitude and achievement (e.g. behaviourally disturbed children, learning disabled children). As can be seen, students whose slope (m) was 1.25 (set A) would be progressing at approximately one and a quarter grade levels per year and therefore could be expected to be functioning at approximately a grade 9 level by age 12, while students in set C (m = 0.75) could be expected to be functioning at approximately a grade 5 level by age 12.

Figure 1-1 **Relationship between Achievement and Chronological Age**

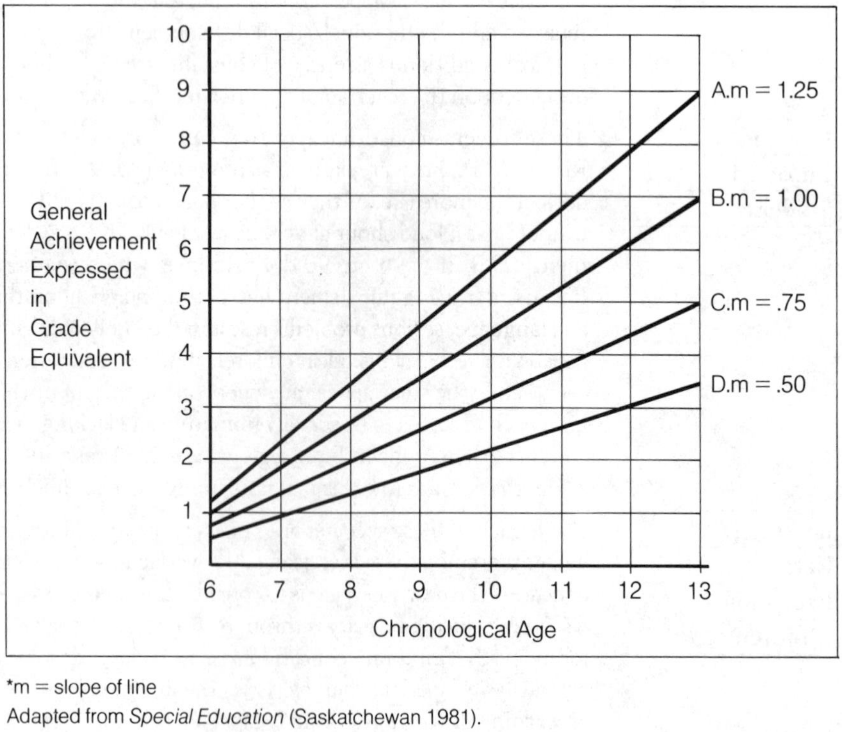

*m = slope of line
Adapted from *Special Education* (Saskatchewan 1981).

This discussion of the categories of exceptional students relates to a major area of concern in the field — whether to treat special needs students as representatives of a category or to see them in terms of their specific strengths and weaknesses. There are other issues to be dealt with by Canadian educators who are attempting to provide individual programs for special needs students. These include language issues, problems in educating native Canadians, geographical issues, professional language issues and approaches to computer technology.

LANGUAGE ISSUES

Language is a concern in two ways. One problem involves the use of one official language in an area where most students speak the other, i.e., the use of English in largely French-speaking areas or the use of French in predominantly English-speaking areas. While all provinces are now much more sensitive to the needs of minority-language students, still only one province officially supports bilingualism: New Brunswick. In Ontario, Manitoba and Nova Scotia, the governments have put large amounts of money into programming in both French and English where warranted by the populations. As well, they have come out strongly on the side of encouraging school programs in both languages. Nevertheless, they have not officially passed legislation to support such programming throughout the province. In all other areas of Canada, second language students are taught in their mother tongue only through good luck and individual demand at local levels.

The second way in which language is of concern has to do with those students whose families immigrate to Canada but who do not speak the official language(s) of the area in which they settle. Many immigrants move to large urban centres where English-as-a-Second-Language (ESL) programs may be available for the first few years of schooling. However, the smaller percentage who live in rural areas are usually offered extra support through special education resources rather than through specific ESL programs. Furthermore, even where there *are* specially funded programs (such as those offered in British Columbia and Ontario), their duration is not usually long enough to ensure sufficient mastery of the language to enable students to fully profit from regular programming.

Cummins (1984), in a longitudinal analysis of the achievement of immigrant children in a large Canadian city, found that while most students had superficial language competence after three years of ESL training, it was not until they had been functioning in English for at least five years that they demonstrated comprehension of concepts at the level that is normally required for more advanced academic work. These findings are cause for concern in two ways. The first is that expectations of normal functioning result from the teacher's observation of apparently good language facility (in other words, from seeing the student converse easily with teachers and peers) when, in fact, the student does not have a good enough grasp of the language to deal with complex verbal syntax and semantic relationships. Cummins' second concern is that this disparity between superficial and substantial competency leads psychologists and teachers to administer English-language tests of intelligence and achievement and to interpret the findings as they would for children whose first language is English. This would appear to be an inappropriate use of such testing, causing the students true ability to be underestimated. This, in turn, results in programming that does not concentrate on the individual's need for more language-related instruction but rather provides instruction at a lower general level than the student might be capable of mastering. These results are corroborated by studies of French immersion programs which show increased competence in comprehension only when tests are administered after several years of French-only language instruction.

NATIVE CANADIANS

There are several important considerations in the education of Native Canadians. They include language, culture, biases in evaluation procedures, health problems, and access to education.

Language is a problem for many Native Canadians. There are several different Native languages and for many Native students, therefore, English is a second language. For example, the Inuit speak *inuktitut*. Accordingly, English should be taught using ESL materials rather than thrust upon these students as the language in which they receive their formal schooling. The problem is compounded when the teacher, as is often the case, speaks very little of the Native language and therefore cannot help the students with language adjustment. For other Native groups, such as the Dene, standard English is a second dialect and so care needs to be taken to ensure they are not penalized for speaking an alternative form of the language. Furthermore, instructional time with Native students should be spent on recovery and strengthening of their Native language.

Cultural differences are a very important consideration as they affect all aspects of the Native students' education. For one thing, oral tradition historically has been the primary method of teaching among many Native groups. This has resulted in less attention being paid to written forms of communication. Among some Native groups, their language is only now being coded in a manner that will standardize the written form and make it available to both Native and non-Native Canadians. Another difference is that Native cultures train and reward visual-motor skills while the dominant culture trains and rewards verbal skills. Furthermore, Native attitudes toward schooling, time, group versus individual goal-seeking and competition all differ from those of mainstream groups and need to be understood and accommodated to improve the Native students' educational experience.

Health issues are also a specific concern in the education of Native children. A major source of poor health is the relative poverty in which many Native Canadians live. A report from the Department of Indian and Northern Affairs (1979) stated that only 50 percent of homes on the reserves had drinkable water, only 45 percent had indoor plumbing and sewage disposal, and that 32 percent of the homes were overcrowded. A survey in British Columbia (Stansbury 1975) found that 67.5 percent of Indians living off the reserves were living below the poverty line. While Indians living closer to large urban areas may fare somewhat better, on average there is a higher incidence of general health problems among Indian students. These problems may or may not be diagnosed and remediated but they certainly do interfere with the learning experience.

A specific health problem that seems more prevalent among Indian students is hearing impairment. As many as 33 percent of students in some schools suffer from otitis media, acute chronic inflammation of the middle ear. Often teachers are not aware of this because the hearing loss that results from this condition is both intermittent and varying in degree. As a result of such daily changes, the condition is difficult to detect but nonetheless devastating to the student trying to learn in a predominantly aural learning environment (Reichman and Healey 1983).

Evaluation and assessment of Native students pose particular problems (Mulcahy et al. 1984). A major difficulty in this regard is distance. Many Native communities are far from urban areas and far from each other. It is costly to fly specialists in or the students out. Occasionally, medical personnel are available to do assessments, but most evaluations of an educational or psychological nature are done by special education personnel. These people do a good job within the limits of their training but are handicapped by the lack of psychologists, psychometrists, social workers or speech consultants to provide additional help in diagnosis and programming. Furthermore, even when extensive evaluation is conducted on an individual student, there are serious problems in interpreting the data. None of the tests commonly used across the country have been normed for Native children. As a result, one is not sure whether a given poor result represents normal development for a Native student or whether that student is functioning at a seriously retarded rate. Such determinations can be made by comparing the student to his/her immediate peer group but it cannot be made in terms of the larger Native groups, nor is it generally fair or useful to compare these test results with those obtained by non-Native students for whom the tests were usually designed (Jackson 1984; Kennedy 1984, Samuda, 1978). One of the very unfortunate results of such comparisons is the finding that proportionally three times as many Native students as non-Native students are labelled as learning disabled (Darou 1982) or as delinquent (Csapo 1981).

Access to education is another problem for Native students when they live in small communities that may or may not have a school. This problem is further aggravated by the fact that students at the secondary level must leave the reserves to attend school. Many are bussed while others are enrolled in residential schools that cut them off from family and friends. In both cases, truancy is high and early school drop-out not unusual. Access to remedial education is even more difficult. There are usually only one or two students in a settlement that require specialized programs and so mainstreaming is the only viable option. Regular class teachers are usually willing to provide this, but they need continuing support from resource personnel that is often not available. Effective mainstreaming also requires home and community support for maximum effectiveness and this too may not be available.

GEOGRAPHY

One of the realities of a country the size of Canada is the vast distances between centres large enough to provide for a heterogenous student population. Outside the heavily populated areas of southern Canada, it is extremely difficult to provide the full range of services for special needs students. Medical, psychological and educational services are generally very difficult to obtain in rural communities and this problem is much worse in the large northern part of the country where communities may be separated by hundreds of kilometres. This results in a complete absence of outside services for most of the students whose exceptionalities are mild and spotty, and infrequent service for those with more severe needs. There have been attempts by several universities (e.g., University of British Columbia, University of Alberta, University of Western Ontario) to provide training to educators through long-distance satellite communication systems. However, these efforts, good though they generally are, reach only a few people for a limited length of time. They do not provide ongoing consultation and support to the regular class teacher who has exceptional students in his/her classroom. Appropriate programming for these students will occur largely through the efforts of the regular class teacher.

PROFESSIONAL TERMS

This is an issue primarily for the readers of this book. Across the country there are no consistent terms used to describe educational personnel or administrative structures. Some of the more pertinent terms are discussed here to familiarize readers with the range of terms used by others across Canada. The first group of terms to be considered are those used to describe and refer to the students about whom this book has been written. Those students are generally referred to as "exceptional," meaning any student who deviates from the average in any way. Thus, this term includes, for example, the gifted, the retarded, the young offender, and the learning disabled. Another term used to refer to the same group is "special needs students." Terms such as "the handicapped" and "low achieving students" refer only to those students who deviate in a negative direction. These terms are used by different school areas and in the literature generally, and thus are used throughout the book generally to refer to the same types of student.

Another frequently used term is "special education personnel." In some areas these specialists are referred to as "learning assistance teachers," as "resource teachers," as "special educators," "learning resource teachers," "withdrawal teachers," "segregated class teachers," or as "special education consultants." While these terms sometimes refer to very precise roles, in most areas they are interchangeable. Therefore, in this book, we will usually refer to such personnel as "resource teachers" or "resource personnel."

Local school jurisdictions are referred to as either school boards or school districts across Canada. Both terms will be used in the text.

COMPUTER TECHNOLOGY

One of the most exciting new thrusts in teaching is the use of computers both in the classroom and as a planning tool. There is a great deal of promise for individualizing instruction for the special needs student. Dr. Stuart Smith, in a recent address to the Conference on Computers in Education (1984), spoke about the need for Canadians to put money and energy into better computer program development to aid in skill learning. He stated that in the skill areas — mathematics, basic literary, spelling, sciences and music — computers could be extremely useful, not because they could save money but because they can individualize instruction.

> The promise is that a student can learn in an individualized way at his or her own pace.... The computer has infinite patience and doesn't embarrass anybody.... The great hope of computerized education... has to do with individualizing the instruction. (p. 10E)

This statement suggests there is great promise for making the special needs student's program in the regular classroom better tailored to his or her individual needs. There are, however, several problems: accessibility of computers, teacher competency in their use, development of appropriate software, and determination of which skills are best taught using the computer rather than traditional methodology (McGrath 1984). These issues will be addressed in detail in Chapter 14.

CONCLUSION

In this chapter we have tried to outline our philosophy concerning the education of special needs students in regular school environments. We have also presented information on legislation concerning exceptional children in Canada, on incidence of categories of exceptional students, and on some of the unique issues that face Canadian educators as they seek to provide appropriate education to all students.

In the remaining chapters we will develop our model for decision making (Chapter 2), and a general approach to assessment (Chapter 3). Chapters 4 and 5 will deal with management of behaviour and development of affective skills in students. Chapters 6 through 9 will discuss the assessment and remediation of the three "Rs," while Chapters 10 and 11 will look at the cognitive development that is required for thinking and organization. Chapter 12 considers these topics specifically as they relate to the secondary school student. The final two chapters are service chapters in the sense that they deal with consulting skills for teachers and provide a thorough discussion of the use of computers in the classroom.

In Chapters 4 through 12, we will follow a common format which includes general considerations, evaluation and remediation. Case histories are included to show how the decision-making model developed in Chapter 2 has been used to plan for effective instruction in the topic area. Please enjoy reading our book and using it to add to your own teaching repertoire.

2 USING A DECISION-MAKING MODEL

In this chapter we will examine a decision-making model for use in planning and evaluating programs for children and adolescents with special needs—i.e., students who for any reason require some supplementary assistance in order to learn, remember and apply what they have been taught. This decision-making model was developed for use with special needs students who typically are placed in a regular classroom. It may, however, be used anytime there is a need to plan a program for an individual student.

Implementing this model involves extensive planning for each student. Individual planning, however, does not mean that students need to be *taught* individually. What it does mean is that part of the assessment phase may need to be completed with individuals rather than small groups.

PRINCIPLES OF THE DECISION-MAKING MODEL

The guiding principle of the decision-making model is straightforward: systematic planning is necessary in working with individuals with special needs. This planning is carried out by moving back and forth between data collection and decision-making.

The decision-making model departs from most other models in two ways—in its approach to assessment and to evaluation. Looking at the role of assessment in other models, we see that traditionally in planning instruction for students with special needs the emphasis has been on determining what strengths and weaknesses they have. While assessing students' abilities is still considered important in the decision-making model, an additional component is added to the process. Assessment is extended to include the planned intervention. In other words, it is argued that "trial lessons" are necessary in order to determine the likelihood of an individual educational program being implemented successfully.

The second departure from other models is found in the evaluation component. In the decision-making model, the process of evaluation begins with looking at the person implementing the program. Next, the instructional materials and methods are considered. Finally, the progress of the child or adolescent is considered. Before any comment is made about the individual receiving assistance, the other two stages of evaluation must have been undertaken.

THE DECISION-MAKING MODEL

Deciding that a student is having academic problems is relatively straightforward for the experienced teacher. Pinpointing a student's specific strengths and weaknesses is more difficult. After determining what a student can and cannot do, a clearly formed instructional plan must be developed. The decision-making model helps to clarify what steps to follow in determining how to proceed with any special needs student.

Figure 2-1 **The Decision-making Model**

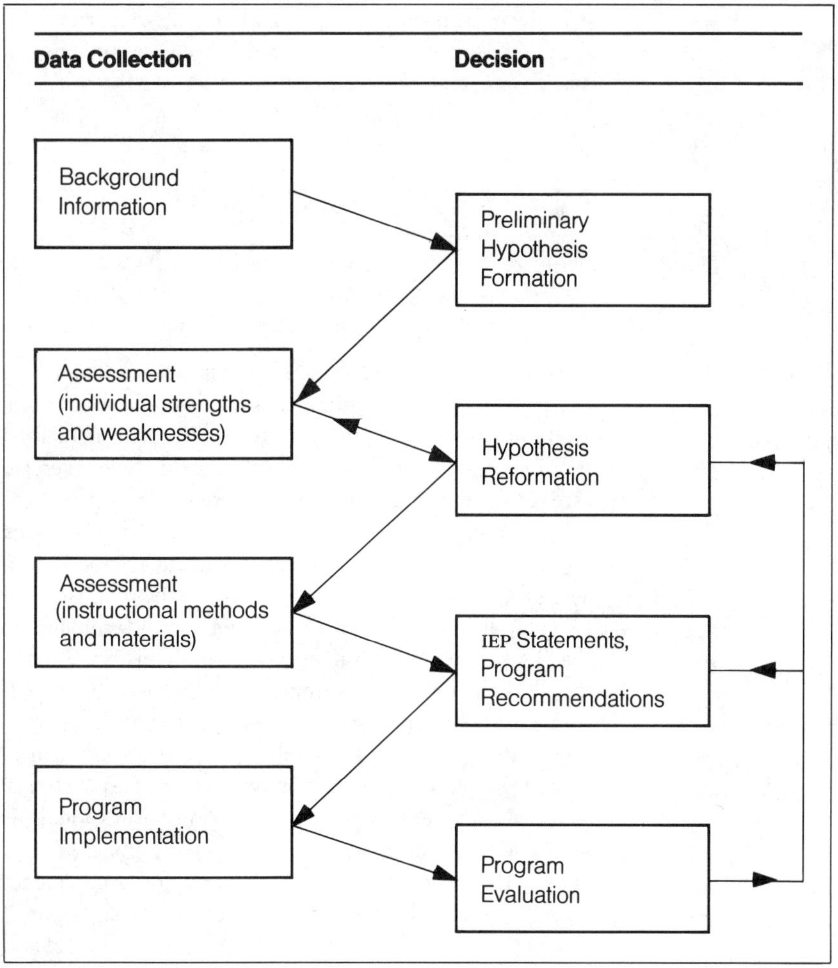

In Figure 2-1 the major steps of the decision-making model are outlined. This model is designed so that the teacher (or any other professional associated with a school-based team) starts with all available background information. Second, some best guesses (preliminary hypotheses) as to why the student is having problems are formulated based on this background data. Third, assessment plans are made, ranging from the more global (e.g., evaluating learning style or intelligence) to the more instructionally specific (e.g., classroom observation and teacher-prepared tests). Fourth, the more tentative hypotheses stated earlier are confirmed, revised or rejected. Fifth, assessment is continued; however, the emphasis is now on determining whether the selected instructional methods and materials are likely to be successful. Sixth, an instructional program is prepared (typically called an Individual Educational Program); and, seventh, the program is implemented. Finally, the instructional program and its implementation are evaluated. At this point, reports are prepared, both to keep other team members informed and to bring parents up to date on their child's or adolescent's progress.

In the remainder of the chapter the details of this model will be filled in. Following the description of the model is a detailed case study in which the model is applied. The chapter ends with a brief explanation of how the decision-making model is used in this text.

USE OF BACKGROUND INFORMATION

In this section, we will examine the types of background information used in making decisions about instructional planning. We will examine the distinctions between these categories and discuss when and how to use each type of data.

Types of Background Information

In general terms, background information is the record of a student's school career and usually is contained in a "cumulative record file." Three types of information are typically classified as background information. First, there is educational/psychological information. This category includes everything from previous report cards and achievement test results to reports by psychologists, child care workers or social workers. Second, background information includes medical/physical information, such as whether or not a student has had the measles. Information on allergies, medications to be taken during the school day, or similar school-relevant data may also be included. Third, there may be ecological information in a student's file. This includes such material as family histories and reports on behaviour and may take the form, for example, of sociograms, graphs of classroom behaviour or summaries of interviews with the student's family.

All three types of background information are found in a variety of formats. These range from physical descriptions and summaries of standardized tests to anecdotal comments and reports prepared by various professionals. For example, background information might include a record of the number of schools attended and a summary of the grades attained. It might also, for example, include any special services' reports that have been completed over time; thus, if the student was assessed by a speech therapist, his or her report would be considered background information.

When looking at background information it is necessary to keep a further distinction in mind. Some information found in case files is factual and some is opinion. Factual information is restricted to descriptions of behaviour or to test scores. Opinions are judgments, such as grades or comments about behaviour. Both types of information are useful. It is important, however, when reading any student file, to separate statements of fact from those of opinion.

In summary, background information is any data or data summary that was completed prior to the presently proposed intervention. It may be about any aspect of a student's past. Finally, it may be either fact or opinion.

Historical and current information The distinction between historical and current background information requires subjective judgment. The person evaluating the background data must decide how much of this information is too old to be considered current. Subjective judgment is involved because some kinds of information will remain current longer than other kinds, depending on the particular case. As a general rule, the more specific the information, the shorter the time it may be considered current. For example, a sample of a student's work is likely not representative of current performance after a few weeks, although some faulty strategies observed in work samples may persist over a longer period of time. The results of a well-standardized, reliable and valid achievement test, on the other hand, may be current for approximately one year.

The differences between data collected from work samples and from standardized achievement tests can be explained by looking at the nature and objectives of those estimates of performance. First, more skills and areas of content knowledge are being measured on an achievement test. As a result of this variety of test items, any test results are more stable over time but less representative of what the student is doing now. Second, achievement test results are considered in comparison to other students or against a fixed set of objectives, rather than against recently completed classroom work. A norm-referenced achievement test, then, will be less sensitive to instruction as students are compared to each other. When progress is measured by comparing an individual's scores to those of a norm-referenced group, student gains must be larger in order to affect relative position within that group. On a standardized criterion-referenced test (CRT) students only show change in relation to a fixed set of objectives, so a CRT probably will not be current as long as the results of a norm-referenced test. CRT tests will be current longer than work sample results, however, as a CRT is not designed to reflect the variations in performance that an analysis of daily work samples may reveal.

Some information will be current for even longer than one year. For example, if it was determined two years ago that a child has a 15 percent hearing loss, that information will probably still be current, although the condition could, of course either worsen or improve. It should not be assumed, however, that all medical information will remain current for a longer duration. Some of this type of information will also quickly become historical. Mild ear infection, for example, will be a temporary condition in most cases and will not result in permanent hearing loss. Thus, reports of such infection usually become historical within one year.

We have been considering educational and medical background to this point. The following guidelines may be used, keeping in mind the qualifications discussed in the last few paragraphs, as general rules of thumb. Educational background usually is historical after one year. Medical information usually is current for one to two years; however, some may be current for a longer period. The other category of information, ecological (for example, specific reports of behaviour), usually will be valid for less than one year. With all three types of information, however, accuracy checks should be made periodically. A review of background information could coincide with the overview of the student's program conducted by the school-based team (or other relevant area or community teams). It could occur as frequently as every 12 to 18 months.

In summary, the first step in using the decision-making model involves reviewing all background information and classifying it as either historical or current, fact or opinion. Historical information is reviewed only to establish the length of time any strength or weakness has existed. A habit that has existed for a long time in all likelihood will be more resistant to change than one that is relatively new. Current information, however, is the main kind of data used in the next step in the decision-making model—preliminary hypothesis formation. In this step both current facts and opinions are used, although the former may be weighted more than the latter.

PRELIMINARY HYPOTHESIS FORMATION

Based on collected background information, preliminary hypotheses are stated. The purpose of stating such "best guesses" is to establish a frame of reference for any further necessary assessment of the student's strengths and weaknesses. Sometimes a student's background information will be current and detailed. In such cases, further assessment may be unnecessary. When no new assessment is necessary an instructional plan may be prepared from the available information. Such cases, however, are rare. More typically, some assessment is necessary. The important point is that background information is to be evaluated and used. Additional assessments should be carried out only when such information is lacking.

When there is no background information—or so little as to say there is none—teacher referral may be the only starting point. In such cases it may not be possible to state helpful hypotheses at this stage. When no hypotheses can be stated, the next step is to begin assessing the student. Such assessments will be global in nature (see below).

More typically, however, an examination of background information will reveal some facts and some opinions. Of this information, some will be current and some will be historical. This information, then, is used in formulating preliminary hypotheses.

The general question to be kept in mind when stating a preliminary hypothesis is "What appears to be influencing the child's or adolescent's ability to proceed?" This question may be used as a frame of reference in deciding which types of assessment procedures, if any, are necessary before possible instructional methods and materials can be examined. The purpose of assessment at this stage is to attempt to specify the student's academic and behavioural strengths and weaknesses.

ASSESSMENT OF AN INDIVIDUAL'S STRENGTHS AND WEAKNESSES

In this step the stated preliminary hypotheses are tested. They are accepted, rejected, revised or expanded upon as new information is obtained. Two types of information are collected in this step: a) global and b) specific.

Global Assessment

The purpose of a global assessment is to discover more about what might be influencing the way in which a student is learning. For example, data may be collected on a student's learning style or data may be gathered on whether the student gives up when faced with what he or she perceives to be a difficult task. The largest percentage of this information is collected by professionals other than the regular teacher. For example, intelligence tests are administered and interpreted by psychologists, family interviews may be conducted by the school nurse or by a social worker or child care worker. In general, the majority of the global medical/physical and psychological information will be obtained by someone other than a teacher. Global educational information, however, is usually obtained by a special education consultant, by a resource room teacher, or by the regular classroom teacher. Teachers may also be expected to provide some ecological data, although most of this type of information usually is obtained by others.

Most of the newly obtained global information, then, is collected by other professionals. This information will be reported to the teacher at a team meeting. There are, of course, cases where no such meeting takes place because no additional global data is needed. In those cases, the decision-making model may be used by teachers to assist them in organizing and collecting educational data without necessarily involving other professionals. If, however, the teacher does find that there is insufficient data, other team members, as suggested above, may be asked to participate in data collection and joint decision making. For example, a student who is having problems learning fractions would not likely be seen by other professionals, whereas a student who requires medication in order to participate in regular stream classes likely will be seen by other professionals.

In summary, students with minor learning problems are not typically seen by anyone but a regular class teacher and a resource room teacher. Students who are suspected of having more complex problems in most cases will be seen by additional professionals both within the school (e.g., school counsellors) and outside it (e.g., social workers).

Specific Assessments

Specific assessments may be completed along with or after global data collection. These assessments are best conducted in four steps: a) curriculum analysis, b) classroom observation, c) work sample analysis, and d) diagnostic testing.

The purpose of curriculum analysis is to determine what the student is expected to know. This information may be obtained from at least two different sources. First, the ministry of education in each province publishes curriculum guidelines. Two examples of this are the British Columbia Ministry of Education's *The Guide to the Core Curriculum* (1977) and *Special Education Core Curriculum Supplement: A Guide for Teachers of the Mentally Handicapped* (1983). Such guidelines will provide a list of general expectations for a given curriculum area. More specific information may be obtained by examining grade/age-appropriate or skill-appropriate textbooks, teachers' manuals, subject-specific ministry guidelines and the interpretation of those guidelines by the local school board.

The next two steps in a specific assessment are completed to determine, in part, what the student does in the classroom. First, classroom observations are completed to monitor how the student behaves. For example, a young student might complete 4 addition problems in a 40-minute period while his or her peers completed 10 to 15. This slow pace may be due to one of a number of reasons; however, it is an important first step to establish that this student is slower than his or her peers. This same child might print twice as many words as his or her classmates, or might be consistently slow in all subjects. It may turn out that this student was slow because he or she was playing with neighbouring classmates or reading a book instead of doing mathematics. When observing the student's classroom behaviour, therefore, it is important to note the time spent working and not working and also any displays of positive or negative feelings. In this way, information is collected about how a student is likely to behave or how he or she might feel about a subject.

Instead of just watching the student in the classroom, a more structured observation may be carried out by interviewing the student. The purpose in doing this is to learn more about the student's likes and dislikes. It is also possible to learn about the student's grasp of basic concepts. For example, students may be questioned about their understanding of number or how they think spelling rules work. Watching a student do a series of subtraction problems during the course of an interview may reveal that the child is working from left to right instead of from right to left. Asking students about ways they like to learn may uncover the student's attitudes towards learning in general and towards specific subject areas.

Work samples provide another indication of what the student does in class. In this case, instead of looking at what the student does while working, the product of independent effort is evaluated. Carefully examining work samples may reveal something about the ways in which a student is approaching assignments. For example, a student may spell all words correctly on a weekly spelling test but be unable to spell or use any of those same words correctly when writing a story a week later. Or, the student may not know how to paragraph and/or may write in sentence fragments or in long run-on sentences. In completing mathematics problems some students have developed unusual algorithms to reach a solution. They may solve fractions by attempting to convert every denominator—or numerator, for that matter—into ten. This error may result from an over-generalization from rules learned for regrouping whole numbers and decimals. Analyzing work samples, then, may provide some hints about the techniques being used to write, spell, do mathematics or complete assignments in other subjects.

After examining what the student is expected to do and learning something about what he or she actually does, a diagnostic test may be prepared to confirm or refine one's guesses. In designing a diagnostic test, it is important to remember that students may perform poorly because they don't know the content, and/or because they are uncertain about some aspect of the task to be done.

To assess content knowledge, strategies observed in work samples and expected curriculum knowledge are used to derive test items. Approximately five to ten test items are required for each aspect of content to be tested. Preparing such a number of items may result in a lengthy test, requiring approximately a half-hour to an hour to administer. Some students who have special needs will not complete a test of this duration. It is important, then, to determine whether they did not complete the test because of a lack of content knowledge, because of uncertainty about how to proceed, because of not wanting to proceed, or some combination of these reasons.

In order to separate content knowledge from task skills, a few test items may be administered at one time. If students can write two paragraphs, they likely have the skills to write a longer passage. If they can complete five to ten intermixed mathematics problems, they likely have the content knowledge to do more questions. They may not, however, have the associated task skills. For example, they may not be able to follow instructions or they may be afraid to try "large" assignments. They may even be afraid of tests or of some specific kinds of test items such as multiple choice. Other students will have some of the task skills and know the procedures required to complete the test; however, they may take an inordinate amount of time to complete it. In summary, a diagnostic test is the final step in a specific assessment of a student's strengths and weaknesses.

HYPOTHESIS REFORMATION

The preliminary hypotheses that were stated may not have been very precise—they were "first guesses." At this point in implementing the decision-making model, more precise statements about what a student does and how he or she does it are possible. However, although the hypotheses stated at this point are "final," they are still hypotheses and are subject to change. For example, despite the best efforts of the test administrator, some children do poorly in all test situations. Furthermore, test items may not have been designed in such a way as to allow some strengths or weaknesses to be revealed. Thus, once the necessary assessments have been performed, the preliminary hypotheses must be reviewed. In light of the new data that have been collected, three possible courses can be taken. The first is to accept the previously stated hypotheses in their initial form. This would indicate that further assessment was likely unnecessary. Even so, assessment data is sometimes useful in establishing the boundaries of a student's content or task capabilities. The second possible course is to revise some of the initial hypotheses which have proven to be partially correct. Using information collected in the assessment stage, these hypotheses can now be stated with more precision. Further assessment was, in such a case, useful. It may have been hypothesized, for example, that students did not know how to do something when, in fact, they had the content knowledge but were afraid or slow to demonstrate it. The revised hypothesis, then, will need to reflect this fear or lack of speed. The third possibility is that some hypotheses were totally incorrect. Such a finding is not a cause for concern. It is better to have found an error at this point than some weeks or months after starting instruction.

ASSESSMENT OF INSTRUCTIONAL METHODS AND MATERIALS

At this point in the decision-making model, background information has been collected, preliminary hypotheses suggested, assessments completed and final hypotheses stated. The focus of the majority of both data collection and the resultant decisions has been on the individual student. Before preparing an individual educational program (IEP), the next step is to "test" to determine whether a particular method is likely to succeed.

There are two main steps in designing test lessons. First, the final hypotheses are translated into areas of instructional need. Each of these is stated, in turn, as a set of tentative instructional objectives. For example, Joan seems to subtract from left to right. She has the idea of regrouping but proceeds as if she were reading instead of doing arithmetic. The tentative instructional objective might be: Joan will complete ten three-digit addition problems which require regrouping by adding from right to left.

Once each objective has been stated, the second step may be taken—the preparation of a trial lesson. These lessons may be thought of as a series of experiments in which, as much as possible, all variables are controlled. That is, the teacher will want to specify the method, the materials, the sequence and the timing of the trial lesson.

Continuing with the example of Joan's mathematics problems, we can see what is involved in a trial lesson. Going back to the background and assessment information obtained by using the decision-making model, we note that Joan is in grade three. Looking at the curriculum, we can establish that she should have mastered two- and three-digit addition and subtraction by this point. For the first "experiment" we decide to try using base-10 blocks as instructional materials. First, we plan to show Joan that she can't obtain the correct answer by moving from left to right when regrouping is required. Joan will be allowed to add using her algorithm, then she will be asked to redo the same problem using base-10 blocks. After showing her that her algorithm doesn't work all of the time, a correct strategy of moving from right to left will be demonstrated. Our purpose, then, is to determine if using base-10 blocks will help Joan to adopt the correct strategy. If this trial lesson is successful, base-10 blocks may be incorporated into the annual instructional goal for addition. If not, the trial lesson will need to be revised, then retaught. Trial lessons, then, provide the data on the basis of which methods and materials are selected, designed or adapted.

INDIVIDUAL EDUCATIONAL PROGRAM

The individual educational program, or IEP, is a planning document which is based on all the information collected so far by following the decision-making model. In addition to being used as a planning document, it serves as a management tool. Comparing an IEP to the plan for a lengthy journey may help to clarify both the components and purpose of an IEP.

First, in planning a trip there are some general considerations. What kind of environment will we travel through? Is the car in good enough shape to make the trip? Similarly, in writing an IEP there are general considerations. To continue with the example of Joan, what did we learn about her that might influence any instructional goals? Is she on medication? Is she multiply-handicapped? If so, there may be modifications to her classroom that are necessary; for example, it may be necessary to install a wheel chair ramp to allow her easier access to the school. Was there some information obtained from other professionals in their contribution to the global assessment that should be considered? She may be a recent accident victim and may still be having difficulty accepting that she is confined to a wheel chair. Are there other factors that need to be considered in her placement? For example, some children become upset if there is a younger sibling in the same classroom in which they are to be placed.

Second, an IEP serves as a kind of road map for instruction; thus, it must be instructionally specific. The information about the specific individual strengths and weaknesses of the student, the final hypotheses and the assessment of instructional methods and materials are all incorporated. The starting point on the map is the student's area(s) of instructional needs. The destination is an annual instructional goal, or, usually, a series of goals. To help move toward this destination, a few time-framed instructional objectives are set. These are analogous to route checks. They provide opportunities to review progress and, if necessary, to select an alternate route to the annual goal.

IMPLEMENTING THE IEP

In writing the IEP many issues were anticipated. In implementing the instructional program, however, some plans will need to be modified. Sometimes students learn more slowly or more quickly than was anticipated. Sometimes methods that appeared to work in the trial lessons either need modification or must be abandoned for other techniques. Thus, the specific plan for implementing an IEP is to be thought of only as a guide, not as a binding contract. It would be folly, however, to abandon the IEP as soon as one component does not follow the pre-set time line, or as soon as a single lesson does not turn out as successfully as planned.

One suggestion that may help to implement the IEP is to treat each of the time-framed intermediate instructional objectives as a major goal. Each lesson, then, becomes a step toward that intermediate goal. Having a shorter-term target allows both the student and the teacher to realize and share the satisfaction of success. In addition, it allows the teacher to monitor and make any changes in IEP specifics on the basis of instructional progress.

In Figure 2-2, a lesson plan format is outlined. First, the annual goal or the intermediate objective is stated. Second, the specific instructional objectives for the daily lesson are given. Third, the evaluation criterion (or criteria) for each objective is

Figure 2-2 **A Suggested Lesson Format**

Subject ______________________ Date ______________________

Goal (Intent) ______________________ Time ______________________

____

____

____

Behavioural objective (what the student will do) and evaluation criterion or criteria

____

____

____

____

____

____

Materials need:

____

____

____

____

____

Lesson sequence

____

____

____

____

____

Evaluation

Effectiveness of teacher delivery ______________________

____

Effectiveness of methods and materials ______________________

____

Comments regarding student reaction ______________________

____

Progress in IEP ______________________

____

stated. Such evaluation procedures may be very informal; for example, "Did the student answer correctly?" or "Was the answer given more quickly this time than earlier?" It may be helpful to list any materials that are necessary for the lesson and to plan the approximate sequence the lesson will take.

Finally, an informal lesson evaluation might be conducted. Each evaluation might best start by asking, "How well did I deliver the lesson?" Without a positive answer to this question, further investigation will be meaningless. Sometimes lessons are delivered effectively to special needs students and sometimes they are not. Acknowledging both poorly presented and well presented lessons results in an increasing awarenesses of what constitutes effective teaching. Next in the lesson evaluation, methods and materials should be examined. Do the selected methods seem to be working? Are the materials useful in facilitating the work habits desired? Finally, in evaluating the lesson student reaction to the method should be examined. Only if both teacher delivery and methods and materials have proven effective, however, does it make sense to evaluate the student's response.

Less frequently, one additional informal evaluative question may be considered: "What progress is being made in meeting the goals set in the IEP?" It is too easy to lose sight of long-term goals even when they are written down. A weekly informal review is useful in reminding oneself of the importance of longer-term goals and in determining what progress has been made toward them.

By conducting such informal checks, both behavioural and achievement goals may be monitored. More formal evaluations should be completed as well; however, these may be done on a more infrequent basis.

Some pitfalls and one concern When implementing the IEP, there are three common pitfalls: a) overteaching, b) becoming program bound, and c) becoming overdependent. The first occurs when, in a one-to-one or small group instructional situation, too much teacher support is provided. Prompts, hints and other suggestions may be necessary to help the student learn; however, some of these aids depend on the teacher being present. Such devices make the student teacher-dependent instead of self-reliant. The second happens when a program is selected and is followed to the letter. Some students become dependent on the prompts inherent in the program. When taken off the particular program, the student may not be able to transfer what has been learned. The third pitfall concerns the nature of teacher-student interaction. It is, in some senses, a variation on the first pitfall. Sometimes the trust that must be established for students to learn is overextended and becomes an emotional dependency. Such a dependency may make the student unwilling to leave the protective custody of the special education teacher.

There is one other relatively frequent concern about implementing IEPs—the students will need individual instruction. Writing an IEP does *not* mean that students must be taught individually; only that they must be planned for individually. One-to-one *instruction* is rapidly becoming an educational dinosaur; but individual *planning* and, to a lesser extent, individual *assessments* are important components of effective small group instruction for special needs students.

PROGRAM EVALUATION

Program evaluation is a necessary step in the successful implementation of any IEP. A formal program evaluation involves two phases: a) performing the interim and final evaluations, and b) reporting the results of these evaluations to parents and other professionals.

Interim Evaluation

Specific evaluation techniques are incorporated in the IEP to determine if the interim instructional objectives have been met. These checks are the "route checks" described above. They must be designed to ensure that the student has mastered each instructional objective. Thus, such checks are used to determine if the student is ready to continue along her or his "lengthy journey."

Final Evaluation

The IEP also incorporates methods of determining if each final (annual) goal has been achieved. Two very important cautions must be kept in mind when planning evaluation procedures: the student must be capable of passing the set criterion of performance without teacher assistance; and the student must be able to apply the newly-learned content, skills, or both outside the context of remedial instruction. The student, for example, should be able to spell "to, too and two" in the following circumstances: a) when the special education teacher dictates a spelling list; b) when the special education teacher requests three sentences written using each word correctly; c) when the student independently wants to use one of the three words in a written composition requested by the special education teacher; d) when the special education teacher requests the student to read one of the three words in a story (in this case, the student should be able to recognize that each word is one he or she spelled correctly previously); and e) when other teachers ask the student to complete all of the above outside the context of special education.

Parent Reports

The second phase of the evaluation process is report writing. Conveying information to a parent or parents and to other professionals on the school-based team is an essential component of a successful IEP.

Reports to parents may be communicated in a number of ways, some more formal than others. Informal reports might be in the form of phone calls and/or little notes such as "Happy Grams." These help involve as well as inform the parent. Such reports could include statements of little things done well during a lesson mixed, occasionally, with a comment about what remains to be accomplished.

More formal approaches to reporting to parents include anecdotal report cards and parent interviews. There are three types of information that should be conveyed in formally reporting to parents. These reports are sent home with the student's report card, which usually is issued four times per year. First, the report starts with a brief program description. The nature of this description will vary depending on which report is being sent home. In the first report, this description may include a general account of the student's placement and of what is being done in each subject or skill area being taught by the special education teacher. As the term proceeds, changes in the program should be described. The second thing that should be conveyed is information about the student's progress in meeting the IEP objectives. On the initial report, this section likely will be limited to reporting the assessment results and plans for the student. Later, it will reflect what has been accomplished. The third type of information to be included is personal comment. Personal comments usually are worded as positive statements about the individual student; for example, "He has been a most pleasant and cooperative youngster" or "Sally's a fine young lady who is eager to do what is asked of her." Such comments are best left to the end of the report so that they don't detract from the message. As well, they provide a good way of ending the report. An example of a parent report is given in Figure 2-3. It was prepared for the third reporting period.

Figure 2-3 **Sample Report for Parents**

Sam continues to receive extra help in composition three times per week with two other students in the resource room.

Sam began the New Year working with sentence combination exercises. By learning and practicing these exercises, Sam learned the advantage of varying sentence length in writing to someone. He also learned different ways to start and stop sentences in doing these same exercises.

In the last two months, Sam has learned and has been practicing three stages in writing his own stories. First, he completes a prewriting phase in which he thinks of what he wants to write and how he wants to write his ideas. In completing this step, Sam has reduced the time he needed to start writng from more than one-half hour to about five to ten minutes, depending on the topic and length of the first draft. Second, Sam prepares an initial draft of what he wants to write. Next, Sam edits his compositions to "make them better." He now recognizes, for example, run-on sentences he has used in his first round of writing and is able with guidance to rewrite them. Sam still continues to need more time and practice in editing. Finally, he is learning a self-checking system so that he can correct his own compositions without teacher assistance.

Both [*teacher's name*] and myself enjoy working with Sam. He has improved his compositions both in the resource room and the classroom over the past few months. In addition he is more confident about his ability to write. Over the next few months Sam will continue writing and editing compositions and practicing his self-checking system. He will also begin preparing reports for other subject areas.

Keep up the good work Sam!

In writing a parent report it is important to accurately describe the extent and context of any instructional successes. This qualification is required when a student is receiving only support services and is spending the majority of his or her time in a normal age-grade or skill-level classroom placement. If the student is improving but is only showing that improvement to the special education teacher, this lack of transfer should be noted. In addition, programs taught by special education teachers may be of a different difficulty level than those offered in the regular classroom placement. Parents need to be informed of such differences so that they will not be confused when one teacher reports larger gains than another.

Other Reports

Reporting is also necessary to ensure program consistency and communication among school-based team members. Such reports should include a number of elements. For one thing, each report must start with an adequate descriptive summary. One of the most common failings of reports found in cumulative records is that they are not dated (day, month and year). In addition, the student's name, birthdate, school, grade, classroom teacher, placement description, and the name of the person preparing the report should be stated on the first page. Depending on the purpose and the occasion of the report, it may include global and specific educational assessments, other relevant assessments completed by the special educator, a description of suggested methods and materials or, if written later in the year, a description of the IEP goals and objectives and an account of the success in meeting interim and annual goals.

General Program Evaluation

Finally, the teacher may wish to conduct a more general program evaluation. This evaluation might include answers to a series of questions. How many students were referred for special education? Which teachers made the referrals? Were these referrals clearly written? Did such referrals assist in selecting or designing appropriate assessment procedures, or an IEP, or both? If not, what can I do to help that teacher (or teachers) write better referrals? How many students were accepted in my program? How many were successfully integrated back in age-grade or skill-appropriate classrooms? How well did I help teachers in preparing for an exceptional student? Were the assessment devices, instructional materials and methods I selected effective? Are there other, better ways of assessing and teaching available? Such general questions may result in a more effective special education program.

Case Study

In this case study, each of the components of the decision-making model is illustrated using the content area of mathematics. The case that is presented is a composite description of work completed with a grade five student over the course of eight months.

Step 1: Background information

Hong speaks English as a second language. She arrived from Vietnam with the rest of her family. She spent the first two grades in another elementary school in an English-as-a-Second-Language Program before she was transferred to her present placement. During grades three and four, Hong received extra assistance in reading. Some aspects of reading are still troublesome for her; for example, she continues to have difficulty understanding some of the nuances of the language. This problem has some influence on Hong's performance in mathematics because she sometimes fails to understand instructions and sometimes has difficulty with word problems.

The only recent historical background information in her file about mathematics was the final report card from her grade four teacher. Hong was judged to be just passing (C-) in this subject area. The only current information is the judgment of her grade five teacher. He is concerned with Hong's classroom test performance in mathematics as well as Hong's fluctuation in performance when she is completing assignments.

Step 2: Preliminary hypotheses

1. Hong's reading skill may be hindering her mathematics performance.
2. Hong may be experiencing difficulty with the grade five curriculum in part because of background weaknesses in mathematics.
3. Given her failure experience with this subject, Hong may be developing a negative attitude towards mathematics.

Step 3: Assessment of the individual

First, although the possibility was not stated as a hypothesis, Hong was observed in the classroom to see whether she was engaging in a behaviour that might interfere with the probability of learning effectively. Hong was observed on three different occasions in three different subject areas, one of which was mathematics. Hong was found to be behaving appropriately approximately 75 percent of the time. The 25 percent of the time that she was not working occurred when she was making the transition from one activity to the next.

Next, Hong was administered a teacher-prepared test in which she completed questions on numeration, addition, subtraction, multiplication, division, reading directions and problem solving. Items were designed to reflect errors found in daily work samples and the grade five curriculum expectations for mathematics. Using error analysis (see Chapter 6), it was determined that Hong had the following strengths and weaknesses:

Strengths

1. In 15 out of 15 problems requiring addition (five single-digit, five double-digit with regrouping, and five triple-digit with regrouping), Hong answered correctly.
2. In 10 problems Hong was able to round numbers to the nearest thousand and the nearest hundred.
3. In five two-digit problems without regrouping, Hong subtracted correctly.
4. In five problems in which there was a two-digit multiplicand and a one-digit multiplier, Hong performed the operation correctly.

Weaknesses

1. Hong in 23 of 42 problems that required subtraction, multiplication, or division (none correct) arrived at the incorrect answer.
2. In five items, Hong was not able to write the place value of an eight- or nine-digit number.
3. In five items, Hong was unable to read or write tenths or hundredths using decimal notation.
4. Hong does not appear to understand the concept of zero when used in multiplication.
5. Hong appears to be using her fingers to work out the answers to most questions.

6. Hong does not appear to have any strategies to solve word problems.

The next step was to interview Hong. This interview was helpful in learning several things about Hong's abilities in mathematics. First, it was learned that Hong's knowledge of mathematical terms is weak. She stated that she did not know what place value was. She thought that difference had something to do with multiplying, and that rounding was when the number was changed to the next closest number. Second, it was determined, as suggested above, that Hong used her fingers to perform all calculations. Further, it was learned that Hong was confused about zero in subtraction and in division as well as in multiplication. Finally, it was learned that Hong had difficulty reading longer numbers (six to eight digits).

Finally, the *Estes Attitude Scales* were administered (see Chapter 4). Hong's attitude to mathematics was lower than in either reading or science.

Step 4: Hypothesis reformulation

1. It does not appear that Hong is behaving in such a way that her behaviour interferes with her classroom work.
2. Hong's attitude to mathematics may be influencing her willingness to try mathematical problems.
3. Hong has no understanding of decimals.
4. Hong cannot perform multiple-digit subtraction problems in which regrouping is required.
5. Hong cannot read the place value of numbers past 100.
6. Hong does not understand the concept of zero in subtraction, multiplication or division.
7. Hong cannot do long division.
8. Hong has few strategies for problem solving.
9. Hong's knowledge of technical terms and/or mathematics vocabulary is weak and this results in her having trouble following directions.

Step 5: Assessment of instructional methods and materials

In preparing a series of trial lessons, it was learned that Hong almost always started out by saying that she could not do whatever it was to be learned. With a little persuasion and some demonstration, she would try. Within a very short time, Hong was doing correctly whatever was to be learned (this pattern continued over the next few months). Concrete teaching aids (Dienes blocks, for example) and diagrams were very useful in helping Hong to understand.

Step 6: IEP goals and short-term objectives

1. Decrease negative attitude toward mathematics.

   a) Given the successful completion of any predetermined operation, Hong's negative self-statements about her ability to complete these operations will decrease to one statement per lesson. This objective will be evaluated by observation and will be met within four months.

   b) Given the successful completion of any predetermined operation and having independently read the necessary instructions, Hong's negative self-statements will decrease to one per lesson. This objective will be evaluated by observation and will be met within six months.

c) Given the successful completion of any set of teacher-selected word problems involving any operation, Hong will decrease her negative self-statements to one per week. The objective will be evaluated by observation and will be met within eight months.

2. Increase Hong's ability to read and write the place value of numbers to one hundred million.

   a) Given a four-digit number, Hong will be able to read and write it in standard and expanded notation with 95 percent accuracy. This objective will be evaluated by observation and will be met within two weeks.

   b) Given a six-digit number, Hong will be able to read and write it in standard and expanded notation with 95 percent accuracy. This objective will be evaluated by observation and will be met within four weeks.

   c) Given a nine-digit number, Hong will be able to read and write it in standard and expanded notation with 95 percent accuracy. This objective will be evaluated by the use of a teacher-prepared test and will be met within six weeks.

3. Increase Hong's knowledge of decimals so that she can read and write decimals from whole numbers, fractions or mixed numbers.

   a) Given a grid of 10, 100 or 1000 parts, Hong will identify any specified portion and write the fractional and decimal numbers with 95 percent accuracy. This objective will be evaluated by observation and will be met within two weeks.

   b) Given a fractional number, Hong will write it as a decimal with 95 percent accuracy. This objective will be evaluated by observation and will be met within four weeks.

   c) Given a fraction, a whole number or a mixed number, Hong will write it as a decimal number with 95 percent accuracy. This objective will be evaluated by the use of a teacher-prepared test and will be met within six weeks.

4. Increase speed of recall of subtraction facts to 20.

   a) Given 30 subtraction questions where the minuend is less than 10, Hong will complete all problems correctly within one minute. This objective will be evaluated by observation and will be met within six weeks.

   b) Given 40 subtraction questions where the minuend is less than 20, Hong will complete all problems correctly within one minute. This objective will be evaluated by observation and will be met within eight weeks.

   c) Given 50 subtraction questions where the minuend is less than 21 and where 0 is included, Hong will complete all problems correctly within one minute. This objective will be evaluated by observation and will be met within ten weeks.

5. Increase Hong's ability to perform the subtraction algorithm to answer questions involving four minuends and four subtrahends with regrouping required.

   a) Given a two-digit subtraction problem with regrouping required, Hong will perform the correct operations with 95 percent accuracy. This objective will be evaluated by observation and will be met within six weeks.

b) Given a three-digit subtraction problem with regrouping required, Hong will calculate the answer with 95 percent accuracy. This objective will be evaluated by observation and will be met within eight weeks.

c) Given a four-digit subtraction question with regrouping required, Hong will calculate the answer with 95 percent accuracy. This objective will be evaluated by the use of a teacher-prepared test and will be met within ten weeks.

6. Increase the speed of recall of multiplication facts to 12.

a) Given 30 multiplication questions with numbers less than 6, Hong will complete all problems within one minute. This objective will be evaluated by observation and will be met within 11 weeks.

b) Given 40 multiplication questions with numbers less than 10, Hong will complete all problems correctly within one minute. This objective will be evaluated by observation and will be met within 13 weeks.

c) Given 50 multiplication questions with numbers less than 13, Hong will complete all problems correctly within one minute. This objective will be evaluated by observation and will be met within 15 weeks.

7. Increase Hong's ability to perform the multiplication algorithm to a four-digit multiplicand and multiplier.

a) Given a four-digit multiplication question with no digit greater than five, Hong will calculate the answer with 95 percent accuracy. This objective will be evaluated by observation and will be met within 11 weeks.

b) Given a four-digit multiplication question with no digit greater than seven, Hong will calculate the answer with 95 percent accuracy. This objective will be evaluated by observation and will be met within 13 weeks.

c) Given a four-digit multiplication question with no digit greater than 9 and including 0, Hong will calculate the answer with 95 percent accuracy. This objective will be evaluated by the use of a teacher-prepared test and will be met within 15 weeks.

8. Increase the speed of recall of division facts with divisors to nine.

a) Given 30 division questions with a divisor less than 6, Hong will complete all problems within one minute. This objective will be evaluated by observation and will be met within 17 weeks.

b) Given 40 division questions with a divisor less than 7, Hong will complete all problems correctly within one minute. This objective will be evaluated by observation and will be met within 19 weeks.

c) Given 50 division questions with a divisor less than 10, Hong will complete all problems correctly within one minute. This objective will be evaluated by observation and will be met within 21 weeks.

9. Increase the ability to perform the division algorithm to a four-digit dividend and a two-digit divisor with remainders.

a) Given a question containing a four-digit dividend and a two-digit divisor with no digit greater than 5, Hong will calculate the answer with 95 percent accuracy. This objective will be evaluated by observation and will be met within 17 weeks.

b) Given a question containing a four-digit dividend and a two-digit divisor with no digit greater than 7, Hong will calculate the answer with 95 percent accuracy. This objective will be evaluated by observation and will be met within 19 weeks.
c) Given a question containing a four-digit dividend and a two-digit divisor with no digit greater than 9 and including 0, Hong will calculate the answer with 95 percent accuracy. This objective will be evaluated by the use of a teacher-prepared test and will be met within 21 weeks.

10. Increase strategies to solve word problems.
a) Given a word problem in which addition, subtraction, multiplication or division is required, Hong will preview, reread and state the operation or operations required and the numbers needed.
b) Given a word problem in which addition, subtraction, multiplication or division is required, Hong will calculate the answer correctly and write it in sentence form. Both of these objectives will be evaluated by observation over the course of the eight months as operations are introduced and mastered. In addition, a teacher-prepared test will be administered in the eighth month.

Steps 7 and 8: Implementation and evaluation of the IEP

Each of the above goals was implemented based on the time-line stated in the IEP. For most topics, numeration, place value, subtraction, multiplication and division, it was necessary to start instruction using teaching materials that are concrete (e.g., base-10 blocks) or semi-concrete (e.g., number grids). In this particular case, almost all objectives were met. There was only one exception, which is not always the outcome of working with a student who has special needs! Hong's improvements, reported as scores on teacher-prepared tests, in which items increased in difficulty over the term of the IEP, were as follows: numeration from 66 to 100 percent, addition from 80 to 100 percent, subtraction from 20 to 100 percent, multiplication from 31 to 94 percent, division from 0 to 81 percent, and problem solving from 40 to 80 percent. In spite of these significant gains, however, Hong was not completely confident in her abilities to do mathematics. She continued to express one or two negative self-statements right to the end to the time period for the IEP.

SUMMARY

In this chapter, the decision-making model has been detailed. In using this model, the teacher (or other professional) alternates between data collection and decision making. Information is selected and used to make assessment decisions. This data is used, in turn, to make other instructional planning decisions. These plans are then implemented, evaluated, and reported.

Chapter 3, "Collecting Information to Help Make Decisions," expands on one element of the decision-making model. In most of the remaining chapters of this text, the framework of the decision-making model is used to present an approach to programming in specific subject/skill areas. The information given in some chapters is intended to add an additional skill to the repertoire of the resource room teacher, other special education or regular classroom teacher, or other professional who might be part of a multi-disciplinary team. Some of the information on consulting skills presented in Chapter 13 also falls into this category. In other chapters, the information given is intended to allow better use of the entire model. Such is the case in Chapters 6 and 7, which discuss the subject/skill areas of reading and mathematics respectively.

We believe that both the philosophy and the structure of the model allows two goals to be met: first, the use of the model leads to a better understanding of individual students by expanding the evaluative context, second, the use of the model leads to better instructional planning and evaluation for any individual student.

3 COLLECTING INFORMATION TO HELP MAKE DECISIONS

Collecting information starts with the decision that something is amiss with a student. Once it is determined that some action should be taken, teachers must decide whether to involve others in the decision-making process. If the answer is "yes," a referral will be made to the school-based team, if one is functioning, and perhaps later to an area or regional team. Before deciding to refer, however, the teacher may elect to collect information to assist in decision-making or to prepare the referral, or both. The resource room teacher may be enlisted in data collection at this point.

Recognizing that a student has special needs and collecting the data needed both to make the decision and prepare the referral, make up the first stage of the assessment process. In terms of the decision-making model in Chapter 2, the assessment process provides the data upon which to make the decisions that may result in an Individual Educational Program being prepared.

ASSESSMENT: COLLECTING PRELIMINARY INFORMATION

In general terms, assessment may be thought of as collecting information using a number of methods. Assessment, as applied to preparing instructional strategies for special needs students, uses a series of techniques to sample students' academic or behavioural repertoires or any other school-relevant experiences. Such samples may be obtained, for example, by administering a commercially prepared test. Sometimes, it is more appropriate to interview a student. Other times, the better choice will be to construct test lessons that provide information about teaching techniques, instructional materials or other aspects of instructional programs. In every case, the purpose is to collect some information which is used to make a decision about what to do next.

Of course, assessment in itself includes some decision making. These decisions, however, are limited to two main purposes: a) determining how to collect a specified set of information and b) determining if the collected information is both accurate and reasonable. The latter point is especially important because any assessment technique provides no more than an estimate of a student's capability and because each technique is subject to error. Therefore, in all likelihood decisions based on assessment results will need to be revised as more complete information becomes available. Nevertheless, the probability of making an inappropriate decision may be reduced in two ways: a) by selecting or designing the most appropriate assessment technique and b) by using and scoring that technique correctly.

Preliminary Questions

When deciding whether or not an assessment technique is appropriate, the obtained background information and/or the information provided on the referral may be analyzed using the following general questions:

1. What is the main area of concern? Is this area of concern something that should be responded to by the classroom teacher? Should another person, such as the resource room teacher, see the student? Should other professionals become involved and, if so, to what extent?
2. Who is to be assessed? The selected assessment technique may differ if one is interested in learning about a large group of students as opposed to learning about one student.
3. What outcomes are required? The purpose of this question is to try to determine which approach to assessment to try: testing or observation. The technique selected will be a function of the kind of information required. For example, if quantitative information is required for a placement decision, a standardized norm-referenced test usually is selected.

First Decisions

The first decisions faced by a teacher are whether to address the area of concern in the classroom context and whether to involve others in seeking a solution. If others are to be involved, the teacher must then decide whether the area of concern should be addressed in the classroom, outside the classroom, or both inside and outside the classroom simultaneously.

There are no pre-set criteria for deciding when to, or whether to, involve others. If others are to be involved, a referral is written. Who will receive the referral is dependent on the area of concern and the severity of the problem. Interventions for students with less severe academic difficulties, or behaviour problems, or both usually occur in that student's regular classroom placement. The exception occurs when there are complicating factors which require services that can be provided only outside the regular classroom placement. In most cases, therefore, the pattern is to start at the top of the cascade of services given in Figure 3-1. As can be seen in this figure, there is an inverse relationship between the extent of services provided and the number of students served. Thus, the decision to increase the restrictions (in terms of access to regular classrooms) placed on the learner is a function of the nature of the services provided. That is, as you move from bottom to top in the cascade of services, the student interacts with students who have similar special needs. In the more restricted settings, the cost of services usually increases as does the number and types of professionals involved. For example, medical personnel may consult regularly, or even be part of the staff of the special school, whereas they rarely are involved with students in the regular classroom. In addition, the decision to refer outside the classroom will depend on the extent to which the teacher is prepared to work with learners with special needs.

Figure 3-1 **Service Alternatives for Students with Special Needs.**

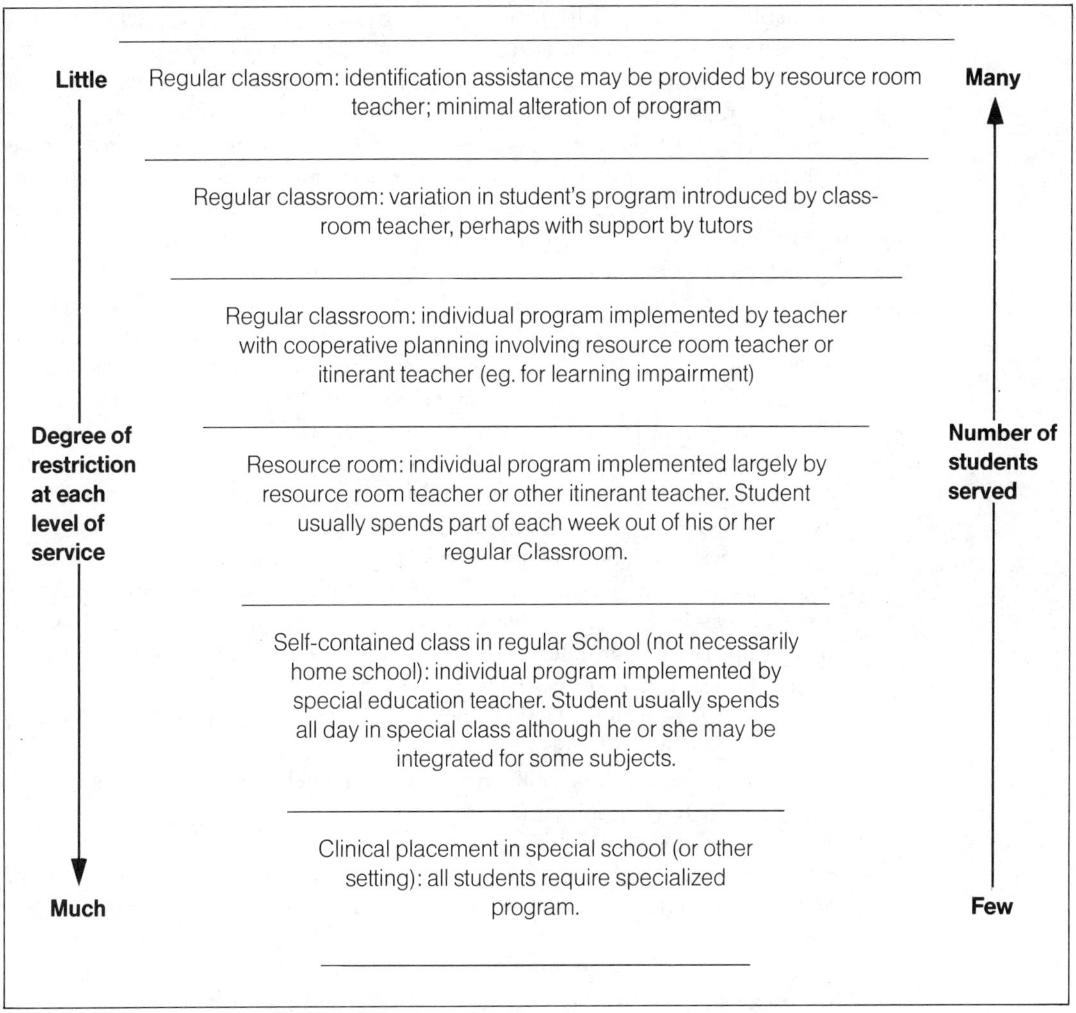

Adapted from C. Crealock and M. Sitko, "Special education in the secondary school: Selected issues" in R. Clark, R. Gidney and G. Milburn (Eds.), *Issues in Secondary Schooling* (London, Ontario: Faculty of Education, University of Western Ontario, 1983), 41.

In summary, subjective judgment plays an essential role in deciding where to provide service to any student with special needs. In most cases, initial data collection and intervention will occur in the regular classroom. The decision to refer will be based on a number of factors including the nature, severity and frequency of the area of concern.

Reasons for Referring

The major reasons for teacher concern are unacceptable behaviour (or, to a lesser extent, social incompetence) and academic difficulties. Both concerns may be school- and teacher-specific. For example, what is acceptable behaviour in a class for students with behaviour problems may not be acceptable behaviour in a classroom for normal achievers. Similarly, there may be varying academic standards set within and between schools. A second factor that affects referral probability is teacher tolerance. Some teachers accept a wider range of behaviour than do others. Similarly, with academic concerns, some teachers are more comfortable with students who are experiencing relatively serious academic difficulties than are others. Within these rather broad boundaries, two categories of behaviour (conflict and sudden change) and three areas of academic concern (reading, mathematics and written composition) traditionally have been the most common reasons for referral.

Students are referred, most often, when they are in conflict with other class members or with the teacher. This conflict may range from the threat of violence (swearing or taunting) or actual violence against others (fighting) to refusal to participate in any classroom activity (which may itself range from not following instructions to not talking at all). Such conflict may be the result of frustration over academic difficulty, or it may be related to personal or family crises.

A second common reason for referral is a sudden change in the way in which a student behaves in the classroom or with others. Sudden changes in behaviour may also be attributed to either academic problems or personal crises. In the majority of cases, however, the timing of sudden change varies from that of conflict. Sudden changes tend to occur at the onset of, or shortly after the onset of, a family or personal crisis such as divorce, death, physical or sexual abuse, or after failure at a highly valued subject or skill. The result is usually a sharp decrease in school performance. This pattern is not typical of students in conflict.

Academic reasons for referring students are related to grade level. With young children in grades one to three, the most typical reasons for referral are associated with problems in decoding skills that emerge when children are learning to read. Starting in about grade four, referrals for reading comprehension become common. At about the same time, students are referred for problems in mathematics. These referrals are the typical ones until junior high school. Starting about grade seven, students' problems associated with written composition become very common reasons for referral.

Summary In the early grades, learning to read is the most common area of concern. Later, in the middle grades, referrals tend to be divided between reading comprehension and mathematics. Still later, written composition is added as a main concern, if not *the* main area of concern from about grade seven forward. Referrals for students' behaviour continue throughout the school years. The reasons for these referrals range from "acting out" to vandalism and theft (e.g., Jones 1980) on the one hand to shyness and difficulty coping with death (e.g., Knowles and Reeves 1981) on the other.

Preparing a Referral

If the decision is made to refer outside the classroom, the teacher should try to provide as much salient information as is possible about the student. Referral forms vary from location to location in the type and specificity of information required. In general, teachers may want to include the following data in any referral:

1. As complete a description of the student as is possible (including, for example, the student's name, age, telephone number, the teachers' and parents' name, and any relevant medical history).
2. A summary of all current relevant test information.
3. A list of typical error patterns or work samples in which such error patterns are demonstrated.
4. An indication of the frequency and duration of a behaviour, an anecdotal description of the behaviour, or both.
5. The day, month and year in which the information was collected. In addition, as much data as possible should be provided as to where and when the area of concern is being displayed.

ASSESSMENT: GETTING STARTED

The next issue faced in conducting an assessment is how to select an assessment technique. Different techniques will be chosen for different purposes. For example, one may want to learn about the general achievement of a class rather than how an algorithm is being used in mathematics by a particular student. In the first case, if the intention is to compare the achievement of a group of students relative to the standards provided in a test manual, the *Canadian Test of Basic Skills* (Hieronymous, Lindquist, Hoover and King 1978, 1979, 1980, 1981;Scannel, Haugh, Schild, Ulmer and King 1978, 1981) or the *Canadian Achievement Test* (CTC, McGraw-Hill Ryerson 1977, 1978, 1981) are appropriate choices. If, on the other hand, the purpose is to learn about the instructional progress of the class as a whole, a carefully prepared teacher-made test would be an appropriate choice. If the intention is to learn about a particular problem experienced by an individual student, a more appropriate choice would be a diagnostic interview (e.g., Liedtke 1979) or an error or work sample analysis (e.g., Bachor 1979, 1983).

A second and related factor in choosing an assessment technique is the nature of the required outcomes. If norm-referenced scores are required, for example, to assist in a placement decision, a different technique would be selected than if the goal were to prepare a specific instructional program.

In selecting a technique the first consideration is what has already been done. Both historical and current background information are reviewed in order to reduce the possibility of assessment redundancy.

Second, there are two basic options once the decision to test a student has been made. One may choose either to administer a teacher-designed test or to administer a commercially prepared test.

We are using the distinction teacher-prepared and commercially prepared as opposed to informal and formal. Both teacher- and commercially prepared tests may be formal in the sense that standardized conditions are imposed on all students in formal tests. Similarly, both commercially and teacher-prepared tests may be used as informal tests; for example, as the basis of a structured interview. The assumption in such use is that a standardized interpretation will not be followed.

Finally, one may choose to use an observational strategy, which in this case refers to interviewing, recording behaviour, and/or analyzing errors. These choices are diagrammed in Figure 3-2.

ASSESSMENT: TEST SELECTION GUIDELINES

Once a choice has been made—and this choice may include all the techniques in Figure 3-2, or only one, or some combination of them—the next step is to increase the probability that the collected information will be accurate and reasonable. No matter which technique is chosen, steps must be taken to ensure test adequacy. The specific criteria applied will vary slightly depending on the assessment technique being used. Relevant criteria for different types of assessment will be discussed in the following sections.

In this section general guidelines are provided for test selection. Following the discussion of these guidelines, is a summary of two Canadian achievement tests. Other tests being used in Canada are reviewed in subsequent chapters in connection with particular content areas. For example, the Canadian edition of the *Key Math Diagnostic Arithmetic Test* (Connolly, Nachtman and Pritchett 1977) is reviewed in the chapter on mathematics and the *Gates-MacGinitie Reading Tests* (MacGinitie 1978, 1980) are reviewed in the chapter on reading. Finally, some guidelines that may be helpful in designing teacher-prepared tests will also be presented in this chapter.

Four general criteria may be used to evaluate the appropriateness of any selected test: a) layout, b) content, c) technical adequacy and d) administrative considerations. Some of these criteria may be applied to both commercially prepared and to teacher-prepared tests. These test selection guidelines are elaborated below. They are derived from four sources: a) the standards proposed by the American Psychological Association, The American Educational Research Association and the National Council on Measurement (1974); b) the guidelines for criterion-referenced measurement suggested by Hambledon and Eignor (1978); c) the criteria proposed by McGuigan (1982); and d) the reactions of special needs students as they wrote a variety of tests.

Layout

The criterion of layout may be applied to both the answer booklet (for both commercially and teacher-prepared tests) and to the test manual.

The following questions will assist in evaluating the test booklet:

1. Are sufficient practice items provided?
2. Do the practice items require the student to answer in the same way as on the test proper?
3. Are the test instructions clear for the student to follow or likely to be confused easily (perhaps because they are illegible)?
4. Are the test items well spaced or crowded together?
5. Do students answer directly in the test booklet or on an answer sheet?
6. Is the test booklet or answer sheet well laid out or is it likely to be confusing for students taking the exam?

Figure 3-2 **Data Collection Options**

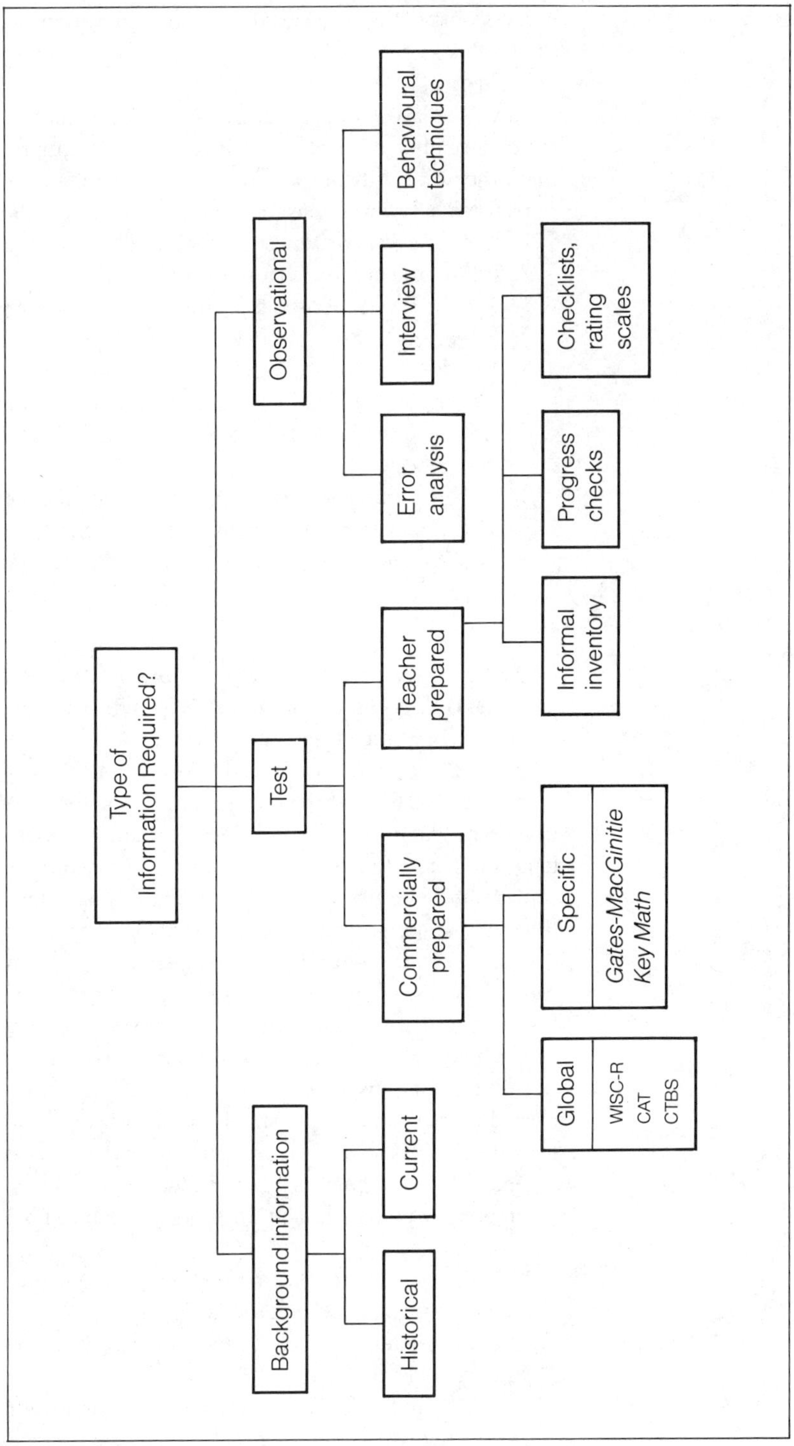

For students with special needs, the above questions usually are considered first. If the test booklet is not carefully prepared, the probability of the results being valid is reduced greatly.

Next, the layout of the manual should be evaluated. If the manual is confusing, the chance of error in administering or interpreting the test increases. Once again, then, the probability of the results being accurate is reduced. Answering the questions listed below will assist in evaluating the test manual:

1. Is there a well-written statement of rationale provided?
2. Is (are) the purpose(s) for test use listed?
3. Are the applications and the limitations of those purposes listed?
4. Is there a clear description of the development of the test?
5. Are the directions to be followed in administering the test clear?
6. Are any special qualifications necessary to administer the test listed?
7. Are the directions likely to be understood by the special needs student(s) taking the test?
8. Is there sufficient information to evaluate the technical adequacy of the test; or, is there a separate, readily available technical manual?
9. Are the criteria to score and interpret the test clearly specified and easy to read and follow?

Content

Once the criterion of layout has been satisfied, the next consideration is the appropriateness of the content of the test. Again, this criterion may be applied to teacher-made and commercially prepared tests.

Normally content is evaluated when technical adequacy is considered. It is our opinion, however, that the content of any test for students with special needs is so important that it should be considered separately. For example, if a student has been taught to read by a language experience approach then, given a test which requires considerable phonics knowledge, the probability is low that student will do well. In this case, a poor performance should not be seen as a student weakness. If the student is to continue in a language experience program in which phonics is downplayed, the test choice or test design was inappropriate. If the student is expected, however, to learn phonics, the test has revealed program weaknesses. In the latter case, phonics will be included as an annual goal in any ensuing Individual Educational Program. It is important to keep the distinction between student and program clear. Program weaknesses should not be attributed to students. Nor should poor test choice or test design be interpreted as a student's weakness.

Questions as to content, then, should be asked of all tests, whether teacher- or commercially prepared. The three questions listed below may be used to evaluate test content:

1. When was the test last revised (i.e., is the content still representative of current curriculum expectations)?
2. Is the content presented in such a way as to allow content knowledge to be separated from task skills?
3. Are there sufficient items and do these items sample the curriculum or skill area under consideration well?

Technical Adequacy

As we noted above, technical adequacy usually is considered first. For students with special needs, however, it may be best to delay evaluating technical adequacy until determining if the layout and content of a test are appropriate. An example may illustrate this point. Suppose a test has been standardized in the United States but is being used in Canada. If the content of that test parallels Canadian or, more appropriately, provincial or territorial content, it may be acceptable to use it in the following circumstances: a) the test is criterion-referenced; b) the test is norm-referenced but local, regional or national norms have been developed (typically, separate norms are necessary for English and French speakers); or both of the above are true in the case where the test is norm-referenced and criterion-referenced. Thus, if the layout and content criteria are met, the test may be evaluated further. If they are not met, the test is best abandoned at this point.

The criteria to be applied to criterion-referenced tests (CRTs) are somewhat different than those to be applied to norm-referenced tests (NRTs).

Questions for CRTs

1. Do the test items match the listed objectives?
2. Does each test item measure a stated objective by using an equivalent format in each case?
3. Are all test items for each objective written at the same level of difficulty?
4. Is there a sufficient number of test items per objective?
5. Over the short duration, does a student pass the same items when retested on them?

Questions for NRTs

1. Is the test standardized in Canada?
2. Are Canadian norms included?
3. Do the norms presented refer to a clearly defined and well described population?
4. Are the limitations of the normative data listed?
5. Are the norms based on a well-planned sample of students as opposed to a readily available sample?
6. Are the sampling procedures clearly described?
7. What is the average item difficulty?
8. What are the reliabilities for each sub-test and for the test as a whole?
9. Are there any indicators of concurrent or predictive validity reported?
10. What is the standard error of measurement?

Administrative Considerations

There are two categories of decisions to be made that might be termed administrative. The first is the broader category. It contains decisions to be made about the assessment program itself. The second set of decisions concerns test administration. For example, when all other considerations are equal, factors such as the amount of time taken to administer the test battery become important.

Questions regarding the assessment program

1. What kinds of assessment instruments are being employed by members of the school-based team or by other teams?

2. Do tests meet the criteria specified for layout, content and technical adequacy?
3. What is the placement availability for all programs for special needs students?
4. Are the settings being used for assessment suitable?
5. Is a case manager—the person responsible for any particular student while he or she is being seen by any team—being assigned for each student when he or she is referred to the school-based team?
6. Is the case manager assignment appropriate given the needs of the student and the parent(s)?
7. How much out-of-class time does the assessment program take per student and per school-based team member?

Questions regarding assessment selection

1. How much time does it take to complete a specific test?
2. Is (are) the test(s) in question available in the school or in the district or board.
3. What is the purchase price of the test and accompanying manuals?
4. Are there comparable tests that are cheaper and/or easier to use?
5. What are the costs associated with the administration of a test or tests (personnel costs, etc.)?

ASSESSMENT: TEST REVIEW

There are, at present, only two achievement tests that are standardized for use across English-speaking Canada. These are the *Canadian Test of Basic Skills* (CTBS) (Hieronymus, Lindquist, Hoover, Schild, Ulmer and King 1978, 1980, 1981;Scannel, Haugh, Schild, Ulmer and King 1978, 1981) and the *Canadian Achievement Test* (CAT) (CTC, McGraw-Hill Ryerson 1977, 1978, 1981). Both tests will be described and reviewed. In addition to these two tests, there are other regionally-specific achievement batteries which have been developed in Canada. One example is the B.C. QUIET (Wormeli 1983). While it may be appropriate to use such tests, they should be subjected to the same scrutiny as other tests and accepted or rejected for use accordingly.

The Canadian Test of Basic Skills

The CTBS is based on the *Iowa Test of Basic Skills* (ITBS) (Hieronymus, Lindquist and Hoover 1978) and has been revised three times since being introduced to the Canadian market. It consists of three separate batteries designed to measure skills in, or related to, reading, language, work study and mathematics. Not all skills are included in each battery.

The primary battery is intended for use from kindergarten to mid grade three and consists of four levels. The second battery is organized into six levels, each of which is intended for one grade, starting with grade three and ending with grade eight. The high school battery is a recent addition to the CTBS. It is intended for use with grade nine through twelve and includes four levels, one for each grade. Of the three batteries, parallel test forms have only been included for the second one.

Layout Depending on the battery, students are required to respond in the test booklet or on a separate answer sheet.

Children completing the primary battery answer directly in the test booklet. In the first two levels of this battery, practice items are included; however, these items are not found in any other levels of the test. The practice items that are included parallel test items found in Levels 5 and 6 (i.e., in the primary battery, which is made up of Levels 5, 6, 7 and 8).

Test items in each of the remaining two batteries are presented in one large booklet. Students respond on separate answer sheets. Some students may find these last two batteries difficult to complete as item answer spaces are crowded together.

There are two types of manuals for the CTBS: the *Teacher's Guide* (one for Levels 5 and 6, one for Levels 7 and 8, one for Levels 9-14, and one for Levels 15-18) and a technical manual (Hieronymus et al. 1984). Each of the teacher's guides contains a summary of relevant technical data (test development, reliability, etc.) for the specific levels covered in that manual. Information is also provided on test administration, scoring and interpretation. In addition, a few instructional suggestions are provided. The technical manual contains the details of test development, normative data, standardization information, and other information about technical adequacy.

Both NRT and CRT scoring procedures are possible. Both hand scoring and the publisher's scoring service can be used to achieve a norm-referenced interpretation. Four scores may be calculated or are available from the publisher's scoring service: raw scores, grade equivalents, standard scores and percentile ranks. Criterion-referenced scoring must be completed by hand. The CRT interpretation is not a true one in that items are not matched to objectives. Instead, individuals' performance on test items may be matched to the skill that item references by reading a table. This procedure is time consuming.

Test development The CTBS test items are not, strictly speaking, derived from Canadian curricula. The authors make extensive use of the test development completed for the ITBS. Based on the final version of ITBS items, Canadian educators were surveyed regarding item appropriateness, the representativeness of skills covered, and the sequence of skill development. On the basis of this opinion survey, inappropriate test items were revised.

Test content

Listening The listening subtest appears only in Levels 5-8. Students are presented with three pictures, which increase in detail across levels. They listen to a series of instructions and then choose the picture that best reflects those instructions.

Vocabulary This subtest is found in Levels 5-14, varying slightly in format in some cases. In Levels 5 and 6, students are shown three pictures. They are required to select the most appropriate picture after listening to some words. In Levels 7 and 8 the task changes. In the first part, children are presented with 17 pictures, each having 4 descriptive words beside it. The task is to select the best word-picture match. In the second part, students must select one of three or four words that best complete a phrase. For Levels 9-14, students must read a phrase and select one of four words that complete it.

Word analysis This subtest is found only in Levels 5-8. In the first two levels, students are asked to identify letters; to identify letter-sound correspondence; and to identify pictures that match beginning, ending and rhyming sounds. Students completing Levels 7 and 8 are required to identify vowel sounds, silent letters, medial sounds, affixes, inflections and compound words.

Reading (reading skills, reading comprehension) Reading tests, of varying formats, are included in all levels, except for the first one. Starting with Level 6, students match words to pictures, select one of three words to complete a sentence, and match sentences to pictures. The format for Level 7 is similar to Level 6 with two additions: "yes" or "no" answers are given to written questions, and other questions are answered based on a series of stories.

In Levels 9-18, students are required to read a variety of passages, which are mainly prose but on occasion are poetry, and then they are to answer from 4 to 15 questions on each passage. Students may be required to recall or locate specific information in the passage; they may be required to relate information found in various parts of a passage; or they may be required to draw an inference by doing such things as stating a definition or creating a title.

Language skills (written expression) In the first 14 levels, this subtest is named "Language Skills." The name is changed to "Written Language" for the high school battery. It should be noted, however, that in no case is composition required.

In Levels 5 and 6, there is one subtest, "Language." Students are required to listen to instructions and then select the appropriate picture from among three.

Starting with Level 7 and continuing to Level 14, four subtests are included: spelling, capitalization, punctuation and usage. In the spelling subtest, four words are presented, one of which may be misspelled. In the capitalization subtest, students determine if there are capitalization errors in one of two sentences. The same procedure is used in the next section in which students identify punctuation errors. Finally, students must select from among three sentences the one that contains grammatical errors. In Levels 7 through 14, a fifth choice is added to each question—a sentence with no mistakes.

For Levels 15 to 18, students are presented with definitions, sentences, paragraphs, etc. They are required to identify the most appropriate organization, grammar and usage. They are also asked to identify errors in spelling, capitalization and punctuation.

Work study skills (using sources of information) Starting with Level 7 and continuing to Level 18, students are required to complete subtests in which they locate, interpret and use information. They are required to answer questions after reading a variety of visual data, such as maps, charts, graphs and tables. In addition, students are presented with questions in which they may be required to use spelling, alphabetization and classification skills in order to read a table of contents, use a dictionary, read an index and use an encyclopedia.

Mathematics The number of mathematics subtests varies across levels. There is one subtest in Levels 5 and 6. Items measure such things as the concepts of number, money and time. In Levels 7-14 there are three subtests: concepts, problem-solving and computation. Students may be required to answer questions about the number system or mathematical terms, to complete operations, or to listen to or read word problems and then complete any necessary computations.

Technical data

Reliability Internal consistency reliability coefficients, Kuder-Richardson Formula 20 (KR-20), and standard errors of measurement are reported in the technical manual. Internal consistencies range from 0.61 to 0.91 for individual subtests, with the majority being over 0.77. Composite internal consistencies across the CTBS are satisfactory, ranging from 0.84 to 0.98. Test-retest reliability is not reported.

Validity Considerable space is taken in the technical manual to address content validity. It is pointed out, however, that validity is ultimately determined by the test user (see the questions given earlier for this purpose).

Criterion-related validity is discussed but not established.

Standardization procedures and norms The CTBS was standardized on Canadian school children in the fall of 1980. According to information provided in the technical manual, a stratified random sample of schools was selected. This sampling was quite restricted, consisting of English-speaking students in grade three by province.

The standardization procedures are not well described. Information about the sample, such as socio-economic status, ethnic background, demographic characteristics, or representation of special needs students, is lacking.

The CTBS was normed in 102 schools on a total of 32 137 students from grades 1 to 12. Schools were included from every province and from both territories, although the latter were grouped together.

Administrative considerations

Simultaneous testing The time limits and directions for administration are the same within Levels 7 and 8, Levels 9-14, and Levels 15-18. Such commonalities allow the administration of different levels to different students in the same class. Time limits differ in the primary battery. There are also separate instructions for Level 5 and Level 6, and for Levels 7 and 8.

Time The CTBS takes approximately four hours to administer in total.

The Canadian Achievement Test

The CAT is a derivative of the *California Achievement Test* (CTB, McGraw-Hill 1977, 1978) and is new to the Canadian market. While not specifically organized as such, the CAT may be seen as consisting of two separate batteries designed to measure skills in, or related to, reading, language, mathematics and using reference materials. Construing the CAT as two batteries may be helpful because slightly different skills are assessed in each and the time required to complete them varies.

Two "locator" tests are included as part of the CAT. Each of these tests consists of 20 vocabulary and 20 mathematics items and was designed to assist in matching students to appropriate test levels. Locator Test 1 is intended for use with students in grades three to six, while Locator Test 2 is intended for use in grades six to twelve. The locator test option is not available for students in grades one and two.

The first battery is divided into two levels that are intended for use in grades 1.6-3.9. Each level consists of nine subtests and an optional practice test. The second battery contains six levels that are intended for use in grades 3.6-12.9. Subtests on reading, spelling, language mechanics and expression, and mathematics are common to both batteries. The batteries differ in that the first has a subtest on word attack while the second has one on reference skills.

Layout Each test level is presented in a separate booklet. Separate answer sheets are provided for Levels 14-19. However, students respond in the test booklet for the first two levels and could do so for all levels. Allowing students to respond in the test booklet after the first two levels would require that the tests be hand scored. This extra effort, however, may be necessary for some special needs students.

Practice tests are provided as options for the first five levels of the CAT. These practice items are formulated in the same way as the test items proper.

Seven different manuals are available: three examiner's manuals, directions for locator tests, a class management guide, a technical bulletin and a coordinator's handbook. Separate from these manuals are eight different norm books, which are necessary if it is desirable to do hand scoring.

Scoring may be done by hand, or the answer sheets or test booklets may be submitted to the publisher's scoring service. Both norm-referenced and criterion-referenced scoring is possible.

If the publisher's scoring service is employed, three types of scores are available that may be of interest to teachers: raw scores, norm-referenced scores (percentile ranks, stanines, grade equivalents and scale scores) and criterion-referenced scores (objective performance scores). National percentiles and stanines, plus local percentiles and stanines are available.

Hand scoring is quite straightforward, although it is time-consuming. With the exception of the norms, the same scoring procedures are possible as above.

Test development The initial item pool for the CAT was derived from the *California Achievement Test* (CTB, McGraw-Hill 1977, 1978). All items were reviewed and, if necessary, revised to reflect current Canadian educational objectives. The author states that over 1000 items were rewritten, tried out and reviewed. The details of such item rewrites are not provided in any of the three examiner's manuals or in the technical manual (CTC, McGraw-Hill 1983).

Test items were tried out for each level with two grades. For example, Level 12 was administered to 381 grade two students and 166 grade three students. A total of 4198 students were tested during item tryouts.

Test content

Reading (phonic analysis and structural analysis) In Levels 12 and 13 there are two subtests on word analysis. Phonic analysis is a measure of beginning, medial and ending sounds in which students must identify words containing various consonant or vowel combinations, or both. In the subtest called "Structural Analysis," students are required to identify the number of syllables in a word; to select two words which form a compound word, to identify words making up a contraction, and to identify root words and affixes.

Reading (vocabulary, comprehension) There are three components to this subtest. First, students must select one word from a set of four that parallels the meaning of an underlined word in a phrase. Next, following the same format, students must select the word that has the opposite meaning. Third, students are provided with one definition of a word that has multiple meanings. They must select one of three sentences that best matches the given definition.

Students are required to read a variety of types of prose, which range from sentences to lengthy passages, and an occasional poem to provide an estimate of reading comprehension. Students must answer one to ten questions based on each reading. These include multiple choice cloze exercises, questions testing text-explicit recall, questions requiring the students to draw inferences, and others requiring the students to relate information found in the reading passages.

Spelling In this test, students must read a number of sentences in which one or more words is underlined. The underlined words may be misspelled. Students must decide if errors are present. There are 20 items at each level and the number of underlined words per item varies from one to three.

Language (mechanics and expression) There are two separate tests for language skills: mechanics and expression. In the first part of the mechanics test, students are required to examine sentences which have been marked in phrases and decide if capitalization is necessary. Next, students must determine if additional punctuation is required in a second set of sentences.

The language expression subtest varies as a function of level. At Levels 12 and 13, the student chooses the word or words that best complete(s) a sentence. Starting at Level 13, students are required to identify the subject or verb of a sentence as well. From Level 14-19, items are added that require students to select the most appropriate sentence wording, to organize sentences into a paragraph, to select from some sets of sentences the ones that best amplify a given topic sentence, and to identify the most appropriate concluding sentence for a short paragraph.

Mathematics (computation, concepts and applications) Two mathematics subtests are found in the CAT. In the first, students are required to complete basic fact items. Level 12 has 26 items which were designed to measure addition, subtraction and multiplication. There are 40 items in each of the other levels. Items increase in complexity in each higher level to include such topics as fractions, decimals and algebraic expressions.

The second subtest, which deals with concepts and applications, contains a wide mixture of types of items. Some of these items duplicate items in the first subtest. For example, at Level 18 there are similar questions containing algebraic expressions in both subtests. There are, however, marked differences between the two tests. In the second, there are items that are intended to assess problem solving, graph reading and estimation.

Reference skills This 40-item subtest is found in Levels 14-19. It was designed to measure students' ability to read for information. Included are items that require using dictionaries, card catalogues, maps, diagrams, indexes, tables of contents and common forms. Items are adjusted, included or excluded depending on the level of the test.

Technical data

Reliability Internal consistency coefficients (KR20) and standard errors of measurement are reported in the technical manual. With the exception of the spelling subtest at all levels (0.51-0.79), the structural analysis section in Levels 12 and 13 (0.58-0.70), and the language sections in Levels 12 and 13 (0.69-0.81), the reliability coefficients are satisfactory (0.80-0.97). Total coefficients for each level range from 0.93 to 0.97.

Test-retest reliability is not reported.

KR20 coefficients are also reported for the category objectives. Reliability coefficients for some of these categories are very low: 0.09, 0.18, 0.24. One factor contributing to these low coefficients is the restricted number of items per category. Some categories are, however, consistently low at all levels while others fluctuate depending on level and still others are quite acceptable at all levels. Performance in each category is estimated and reported as a performance band. Performance bands are derived from a number of factors (CTC, McGraw-Hill Ryerson 1983, p. 40). To simplify this concept, however, it can be understood as a prediction of future performance on a specific criterion-referenced test item. Further technical details regarding calculation and interpretation of these estimates of performance are provided in the *Class Management Guide, Canadian Achievement Tests, Form A* (CTC, McGraw-Hill Ryerson 1983).

Validity Considerable space is given in the technical manual to the procedures used to validate the category objectives, item tryout and item selection. Content validity is left to the judgment of the test user (see questions given above).

Standardization procedures and norms The CAT was normed during late spring on 76000 Canadian children from grades 1 to 12.

Careful and complete standardization procedures based on a ten-cell, stratified sampling of school districts by geographic region and degree of urbanization are described at length in the *Technical Bulletin* (CTC, McGraw-Hill 1983). Breakdowns by geographic region, socio-economic status, minority group membership, special education needs, grade level, school and community characteristics, etc., are presented.

Administrative considerations

Simultaneous testing Multi-level testing is possible for Levels 14-19 but not for

Levels 12 and 13 as the total testing times for these two levels are different both from each other and from the remaining levels.

Time and potential extra costs The CAT takes about three hours to administer.

Having separate norm tables, a technical bulletin and coordinator's handbook makes the hand-scoring option more expensive and time-consuming than might otherwise be necessary.

ASSESSMENT: TEACHER-PREPARED TESTS

In a three-year training program designed to help teachers develop effective informal assessment techniques for mainstreamed students, Plata and Cavin (1984) found that the teachers involved liked the program. Teachers responded positively because they believed the techniques allowed them to pinpoint student problems accurately, to assist students when needed, to record continuous progress, and to substantiate other materials evaluation with performance progress charts. Students liked the non-traditional approach to assessment because their grades improved, they could chart their own progress, and they could ask for help without fear of reprisals.

Frame of Reference

Teacher-prepared tests and observation strategies are derived from a combination of historical and current background information and from the global assessments that other professionals have already completed. Global assessment information may include for example, results of an intelligence test, an indication of learning style, a measure of anxiety, or all three. In effect, all the data collected to this point becomes background information for the resource room teacher (and the other members of the school-based team). This information is used to state hypotheses which are used in turn to plan specific assessments. For example, if it was determined that the student was highly anxious about tests, he or she should be observed completing two types of assignments: a) those the student does well in or enjoys, or both; and b) those he or she does poorly in, or dislikes, or both. The student may behave differently in the first case than in the second (see below for guidelines on observation). As another example, the current background information in combination with an individual intelligence test may be interpreted to indicate a specific subject area weakness. These data may lead to further work-sample analysis and to the preparation of a teacher-made test.

In the discussion of the decision-making model in Chapter 2, it was suggested that there are two main sources of data for test items: a) work samples, which in this case include everything from previously completed teacher-prepared and commercially prepared tests to previously completed assignments from two days to two months ago; and b) curriculum guidelines, which range from specific instructor guide books to the more general ministry guidelines. Second, it was suggested that it might be necessary to adjust item formats to keep content knowledge and task skills separate. It is also useful in preparing test items to observe students as they are completing assignments.

The order in which these steps should be completed for any particular student may vary depending on the area of concern and on whether other professionals are involved. The order of proceeding, then, is subjective. In the following section, observation will be discussed first.

Observation, Work Sample Analysis, and Test Item Construction

Observation guidelines Once potential areas of concern have been isolated, i.e., once preliminary hypotheses have been stated, the next step is to observe the student completing the two types of assignments suggested in the last section (i.e., assignments that the student does well in, or enjoys, or both, and those that he or she does poorly in, or does not enjoy, or both. Specific techniques for observing students are provided in Chapter 4. The following questions, however, may help to guide classroom observations:

1. Does the student seem happier (smiling or requesting extra work) when working in a subject area he or she is successful at or enjoys than when he or she is working in the area of concern?
2. Does the student work slower or faster when completing assignments in the subject or skill of concern?
3. Does the student spend less time completing assignments in the area of concern?
4. Does the student appear to be compliant but think of reasons to avoid work (e.g., having stomach aches or headaches, needing to use the toilet, needing to go to the library to find a reference, forgetting the assignment at home, etc.)?

If the answer to these and other similar questions is "yes," inadequate task skills may be masking content knowledge. For example, some young children take up to three to four times the amount of time to complete their assignments as do their peers. One young boy was observed at the Learning Assistance Center completing single-digit addition problems at the rate of 1 per 15 minutes (Bachor 1983). Precision teaching (see Chapter 7) was used to improve his speed of performance.

Observation may reveal that a student has either test or task anxiety. In the former case, the student performs well except when being tested. In the latter case, the student seems to have the skills and content knowledge when the teacher is present but feels incapable of working independently. Such students may be so certain that they do not have any skills that they tell themselves that they are "dumb" and "can't do math," "can't spell" or "can't compose," etc. In such cases, rather than focusing on strategies to teach content knowledge, trial lessons should focus on building a positive self-image or on establishing appropriate work habits while the student is completing work in the subject in question.

It is useful to incorporate interviews, which may be considered a variation of observation, quite early in the specific assessment procedure. Two general purposes should be kept in mind when interviewing a student. First, you may want to learn about the student's academic and personal interests. Such interests can form the subject matter for trial lessons or, later, could provide the content around which an annual goal is designed. Interests like music, for example, might be used as the theme for some reading material or might form the basis of mathematics word problems. A second goal in interviewing is to learn about the procedures that a student uses to complete test items and also to learn what the student understands when he or she "talks through" a test item. For this type of interview the student is first asked to complete several test items. Performance during such diagnostic interviews may be best understood by comparing it to performance shown in work samples. The reason such a comparison is useful is that an interview is not a sample of independent performance. Work completed during an interview may not typify work completed without the presence of an adult. Samples of what the student does when working alone may provide different data than those provided by an interview. Both are useful either in preparing further test items or in stating reformulated hypotheses to be used in planning trial lessons.

Test item preparation The first specific source of information that the test constructor may wish to use is the data collected by observing and interviewing. The second source is previously completed work samples. For example, if a child shows a particular error pattern, the child's performance on that type of item may be examined in order to determine what he or she is doing both correctly and incorrectly. Examples of test items based on work-sample analysis are repetitions of previously incorrectly completed mathematics test items, or paragraph dictations containing words that were misspelled in science or social studies assignments.

In constructing test items, the guidelines for selecting commercially prepared tests are to be followed for layout, content and, where applicable, technical adequacy. Specific rules for test construction are listed in such sources as Gronlund (1981) or Stanley and Hopkins (1972). Items prepared at this stage should reflect current levels of performance. If they are written following instruction in order to evaluate the program or instructional methods, the items should also test for transfer (see section on program evaluation, Chapter 2).

SUMMARY

The first decision made by the classroom teacher is whether or not to involve others in working with any student with special needs. This decision may be made with the assistance of the resource room teacher. Once a positive decision is reached, a referral is prepared and decisions as to the most appropriate assessment procedures are made either by the resource room teacher, the school-based team, or another broader based team. The decision-making model given in Chapter 2 provides the frame of reference within which to make those decisions.

4 INTERFERING BEHAVIOURS: ASSESSMENT AND INTERVENTION

In this chapter the decision-making model is applied to the issue of behaviour, and the importance of taking an ecological approach in applying the model to behaviour is stressed. Before discussing each of the steps of the model, however, some issues related to concerns about behaviour are explored. In the final section of the chapter, a case study is presented.

PERSPECTIVES ON BEHAVIOUR

Regular classroom teachers and special education teachers alike allocate some proportion of their time to classroom management. As long as this proportion is small, such activity is considered part of one's normal obligations as a teacher. In some cases, however, behaviour management becomes a preoccupation. For example, Deutsch (1960) estimated that as much as 80 percent of a teacher's time in some urban schools in "deprived areas" was spent in trying to keep classroom order. More typically, however, teachers appear to spend about 10 percent of their time engaged in management activities (Crealock 1983).

Teachers, then, devote different proportions of the instructional day to management activities. The proportion of management time will be adjusted according to the teacher's perception of what constitutes acceptable behaviour in the classroom. As well, within any one classroom, the amount of time within which individual students are doing what they are supposed to be doing, or are "on-task," will vary. Systematic observation is one method of demonstrating the degree to which on-task time varies. For example, university students being trained at the Learning Assistance Center (LAC) have observed children with learning problems in language arts, or in mathematics, or in both language arts and mathematics. Over the past four years, reported time on-task over the three completed 30-minute observation periods has ranged from 10 percent to 99 percent (Bachor 1983). In another situation, Crealock (1983) reports that special needs students are off-task 21.5 percent of the time when in a regular classroom and 17.5 percent of the time when in a special education classroom. If time on-task as reported to the Learning Assistance Center and as reported by Crealock is typical of students with learning problems, the amount of on-task time varies greatly among children and adolescents with special needs. Time on-task may be a concern but it will need to be monitored on an individual basis.

An example will help to illustrate the effects of inefficient time management. If one assumes that a child is to complete 30 problems of some kind—the content area is irrelevant for the moment—and further assumes that each problem should take one minute to finish, the impact of the amount of time spent off-task becomes more meaningful. The student who is on-task all the time would complete 30 problems in 30 minutes. The student who is on-task 80 percent of the time would complete the 30 problems in approximately 38 minutes. The student who is working only 6 minutes out of 10, however, would take approximately 50 minutes to complete the same amount of work. Finally, the student who spends only 10 percent of his or her time working would complete only 3 problems in 30 minutes. To finish all 30 problems would take 300 minutes. Thus, in the more typical examples, there is a variation of 20 minutes between the most efficient worker and the least efficient one. If we add the extreme case, the difference in completion time would be 270 minutes, which is a phenomenal difference.

The number of students who fall into the extreme case, that is, those students who complete their various assignments less than 20 percent of the time, appears to be small. For example, only 5 out of a total of 230 children and adolescents referred to the LAC over the past four years fell into this category. Although this number is small, students such as these five can become a major preoccupation if an appropriate IEP is not planned for them. Many others seen in the LAC spent somewhere between 40 and 60 percent of their time in the regular classroom on-task. Looking back to the last paragraph, it is easy to see that 40 to 60 percent still is a low rate of on-task time. It is important to realize, however, that there was a sizable group of students with learning problems who were seen at the LAC (about one-half) who spent as much time doing what they were supposed to be doing as all other students in their various classes.

Children or adolescents who display various learning problems, then, will vary in the percentage of time they spend on-task. A second factor needs to be considered when one is evaluating the relative degree of on-task or off-task behaviour displayed by children or adolescents with learning problems: the total number of children or adolescents displaying off-task behaviour may be quite large. Swift et al. (1972) report that 25 to 35 percent of the children they sampled, who were enrolled in elementary schools in France and the United States, were displaying behaviours that impeded efficient classroom learning. Not all these students had learning problems. Thus, it is important to determine whether the behaviour of a student with a learning problem is typical of that of others in the classroom. If he or she is off-task but no more so than other students in the classroom, the problem may rest somewhere other than with the student. Unfortunately, the student with learning problems likely cannot afford to be inefficient in his or her time management.

We recognize that almost all students will spend some small part of the school day off-task. No one would expect 100 percent time on-task. On the other hand, it is important that students' classroom "engaged time" (time spent working) should be high over the course of the school day.

Time spent working is only one concern, however. A second concern is the individual student who is not learning effectively. For example, a student may solve addition problems by counting; however, the amount of time taken will be much greater than would be the case if the student were using a better (i.e., faster) algorithm.

Another sort of problem occurs when a student elects not to engage in academic activities at all, or at least to minimize them.

Two separate management issues have been raised to this point: classroom discipline and individual behaviour. Classroom discipline refers to the management procedures teachers employ that facilitate instruction and learning. In the case of the behaviour of the individual, the emphasis is on how well a student manages time and how well he or she learns when working.

Classroom discipline is an obvious and on-going concern of teachers. Some of the techniques discussed in this chapter may be helpful to the teacher or other professional who is concerned with large-group management strategies. We, however, are more concerned about the second case, the student who avoids an academic task by engaging in off-task behaviour, or who accomplishes little although he or she is working hard. Like Hobbs (1967, 1982), we suggest that in our society a person's worth is determined to a large degree by his or her ability to perform. Thus, there appears to be a relation between not learning to read, write or do math well and the development of negative feelings about self. These negative feelings, in turn, may be manifested in maladaptive behaviour.

In cases where a child or an adolescent engages in some pattern of behaviour that appears to prevent learning, that behaviour may need to be addressed specifically as part of any IEP. As we suggested in Chapter 3, the interfering behaviours displayed by students varies to some degree with age and grade level. Typically, teachers are concerned with the class of behaviours that might be termed "defiant." Examples include talking out, hitting others, or failing to attend classes. Equally interfering, however, are another class of behaviours, which might be termed "passive resistance." Examples include shyness or being afraid to answer. These behaviours are equally disrupting in the sense that that they decrease the likelihood of learning occurring.

APPLYING THE DECISION-MAKING MODEL

In evaluating a student's behaviour, the decision-making model presented in Chapter 2 may be applied. First, background information should be reviewed and tentative hypotheses suggested. Once this has been done the next step is to complete any necessary assessment. Typical assessment strategies include observing, interviewing and completing checklists.

As a general rule of thumb, we suggest that if a student is engaged in behaviour that significantly decreases the probability of task completion, that behaviour is an area of concern. We propose that the criterion of 80 percent time on-task over the whole school day be used to determine acceptable classroom performance. Individual students, however, may need to have different expectations set for them. One note of caution: some students have developed effective techniques of appearing to use their time well. When, however, one checks the amount of work accomplished, it is found to be insignificant in relations to the amount of time spent "working."

Once it has been determined that the student is engaging in behaviour that is interfering with his or her potential to learn, the next step is to prepare trial lessons. It may be that some individuals will be off-task only in certain subject areas. In such cases the proportion of time on-task may increase dramatically with changes in teaching technique or instructional materials, or both. In other cases, the student may have given up completely. He or she may have become a "master of failure" (Bachor 1984). In such cases trial lessons may not be helpful in further planning as the student may be too distrustful of any adult. (For suggestions in working with such behaviour, see the sections beginning with "Intervention Techniques for Interfering Behaviours," below.) Even in such cases, however, the student's behaviour in school may be a function of his or her perceived success in other activities, both school and non-school related.

Finally, once trial lessons have been undertaken and assessed, an IEP is prepared, implemented and monitored.

FORMING HYPOTHESES ABOUT PROBLEM BEHAVIOUR

To reiterate, the first question that should be asked about children or adolescents with learning problems is: "Is behaviour interfering with the probability of learning occurring?" Not all students with learning problems display a significant proportion of off-task behaviour. However, when a student does display such behaviour, the first step is to examine available background information.

Five Clusters of Behavioural Patterns

The categories of behaviour suggested by Stott (1971, 1972, 1978) may be helpful when reviewing the background information of students referred for inappropriate behaviour. He suggests five clusters of behavioural patterns may be hypothesized: unforthcoming behaviour, inconsequential behaviour, hostility, withdrawal and depression. Stott argues that these patterns represent the

> "originating causes," in the sense that they are antecedent to and calculated to produce faulty learning attitudes. If any of them is operating strongly enough to seriously hinder the child's learning, a secondary, reactive type of cause emerges. (1978, pp. 29-30)

Each of these patterns will be described in more detail below. The first two are seen by Stott as the most common patterns.

Unforthcoming behaviour The primary characteristic of this pattern of behaviour is being afraid to commit oneself to an answer. In addition, new or strange situations are feared. "Children of this type often use the strategy of giving a tentative half-answer and awaiting the adult's confirmation if correct, or silence if wrong, thus getting the adult to provide the answer for them" (Stott 1978, p. 29). Such children may answer only when they are absolutely positive of being correct. Even in such cases, it may take considerable persuasion to obtain any response from them.

Inconsequential behaviour This type of behaviour is typified by individuals who act without taking time to think through what they are going to do. Kagan (1965) refers to this same pattern of behaviour as "impulsive." Children who exhibit this behaviour may frequently respond by guessing and giving multiple answers to the same question. According to Stott (1971), guesses are so commonly made they appear to be the preferred response.

One result of this pattern of behaviour is failure. Stott adds that the student seeks ways to compensate for this failure by being mischievously clever. He or she may take short cuts which defeat the purpose of the activity, or may attempt to mask consistently incorrect answers with his or her "cute remarks."

Hostility Closely related to inconsequential behaviour is the third pattern, hostility. Indeed, hostility seems to follow from inconsequential behaviour. It is suggested that hostility results when parents and teachers react impatiently and antagonistically (Stott 1971). Stott (1972) further suggests that hostility alternates with anxiety in many children. He suggests that a good way to express hostility at home is to do poorly at school.

There are two main types of hostility (Stott 1971). The first is that the child becomes sullen, moody and refuses to participate. The second is that the child displays "angry refusal" and threatens or actually leaves the room. This second group of children have outbreaks of unpredictable behaviour. They seem to seek out excitement and may display violent outbursts or run away even when there is no obvious provocation or source of frustration.

Withdrawal and depression The last two patterns of behaviour are not as well described by Stott as the previous three. Withdrawal refers to the tendency to be alone and to react passively to instructional or other materials. Depression is described in terms of its causes. According to Stott (1978), depression "may be the result of lack of sleep, exposure to a highly disorganized home environment, or chronic ill-health" (p. 29).

Other Indicators of Problem Behaviour

Suggestions are given below which may be useful, along with the patterns described by Stott, in reviewing the available background information for any student.

Checking the number of days absent from school is often informative. Frequent absence may be due to chronic ill-health. In such cases, it is important to review what was supposed to be taught during this period of absence. It may be that the academic concern and the absence match, which would suggest instruction never occurred.

In some cases a fall-off in performance may occur following illness. If the reason for absence was meningitis, hearing impairment may be a problem (Kirk and Gallagher 1979). Conductive hearing loss, especially in younger children, may occur following high fevers and chronic respiratory infections. In other cases, the reason for the performance fall-off may be less traumatic. Children or adolescents returning to school may still be somewhat run-down and simply not performing as well as they might. Nonetheless, it is important to monitor the performance of students who behave differently after illness.

A high absence rate may be indicative of other things besides illness. A student who believes school serves no useful purpose may attend seldom. Absence from specific classes may suggest a subject difficulty or a conflict with a teacher. Such patterns usually are observed in older students. Similar patterns, however, may be identified in elementary grade children as well. For example, a child may be "sick" to avoid attending school. Such "sickness" may be indicative of some unhappiness at school.

Ancedotal records and reports of interviews with parents and teachers may contain useful information. Records should be checked to see if there is a pattern in some form of complaint (headaches, stomach upset, etc.). These complaints may be reported just prior to a subject or task that a child wants to avoid; for example, mathematics. Searching for repeated names may suggest a conflict with a classmate, a teacher, a sibling or a parent. Repeated incidents in specific or related subject areas may indicate, for example, frustration with composition or difficulty with reading comprehension.

Another source of information is the work completed by students. Evaluating students' compositions or poetry may reveal personal distress. In some cases, adolescents have hinted strongly that they were going to commit suicide in their compositions before they attempted suicide or were successful in taking their lives (Johnson and Ray 1983). However, in many cases violence is a common, normal theme, one which seems related to the exploration of the world of fantasy. For example, young male adolescents may want to write about some kind of war-related activity. They suggest topics ranging from the design and use of military clothing and vehicles to elaborate explanations of the nature of past, present and future weaponry. Other times, they may suggest what appears to be a neutral topic like sports and end up wanting to write, for example, about the role of fighting in hockey. Such themes are safe ways to explore feelings about violence without risk of physical injury to anyone.

Perhaps the best indicator in determining whether the topic of any theme or poem should cause alarm will be the openness with which the topic is discussed. The child or adolescent who is considering suicide is likely to be reticent about his or her flirtation with death, whereas the child or adolescent who is fantasizing about violence will talk to anyone who will listen for as long as he or she will listen. Indeed, a student who is fantasizing may even wear clothing that makes the exploration seem to have a basis in reality. However, these rules of thumb are not completely reliable. Given that there may be a basis for serious concern, all themes or poems that hint of personal distress should be treated as important indicators until there is evidence to the contrary.

ASSESSING CURRENT BEHAVIOUR

Some Cautions About Assessing Behaviour

After reviewing the background information and formulating some tentative hypotheses, the next step is to select a method of assessing current behaviour. In formulating any hypotheses about the behaviour of a student, the word "tentative" will need to be kept clearly in mind. Children and adolescents behave differently under different circumstances, or even under the same circumstances on different days. This variability of behaviour over the short term should not be surprising. Each one of us varies in the way we behave with others to some degree, depending on circumstances. In the case of students who seem to be having behaviour problems, it has been suggested that little is known about the long-term course of such problems in children or adolescents.

> A high proportion of children who have been on clinic waiting lists without receiving any treatment and public school students identified by classroom screening methods are, on later follow-up, not judged to have serious problems of personal adjustment. If emotional disturbance were a relatively stable characteristic of a child, it seems unlikely that "spontaneous remission" would occur in two thirds to three fourths of the cases studied. (Hobbs 1982, p. 185)

In short, any behaviour problem may be of a transient nature (Csapo 1981). What is a "problem" now may not be a week later or even later in the day when the child or adolescent is in a different environment.

In the case of the student engaging in interfering behaviours, however, there may be some stability to his or her pattern. If a child or adolescent is afraid of mathematics, for example, his or her reactions are unlikely to vary as long as the task is perceived to be difficult. On the other hand, if the task is judged easy, there may be no difference between the way this child or adolescent completes this particular assignment than any other student who does not find mathematics frightening. One factor influencing the student's behaviour will be the strategy he or she uses to attempt to solve the problem. Another factor is the complexity of the assignment. Yet, there is no one particular variable that may be isolated as being the determiner of how a child will react to any subject area.

Assessment Techniques: Two General Questions

In assessing the behaviour of any student, two general questions should be asked: a) What important persons and places are involved? and b) What kinds of behaviour are important? Each of these questions will be answered in different ways. The first one will be answered by data collection, whereas the second requires that a series of decisions be made.

What important persons and places are involved? Answering this question requires completing both global and specific assessments. The question itself is derived from the ecological approach taken by Hobbs (1966, 1967, 1982) and Rhodes (1967, 1970). One basic assertion advanced by both these men is that children and adolescents are part of a total ecology. Therefore, more than the situation-specific behaviour must be taken into account when trying to effect change. A child's or adolescent's behaviour may be affected by who is in the environment, how these people interact with one another, and by what limitations the physical environment imposes. Such information may be gathered, in part, by identifying the important people and places in the student's life. This can be done by a) reading the current background information; b) interviewing both the individual in question and his or her parents, siblings and friends; and c) administering assessment devices such as the Q-sort (Kroth 1973). After discovering the people, events and places that are functionally important in the student's life, you may then organize this information in a diagram.

When diagramming the individual's functional environment, Hobbs (1982) argues that the ecosystem must be drawn with the student as the defining member. To do this, the student is placed in the centre of the diagram. Then all other important places or individuals are clustered around the student, or defining member. In completing the ecosystem diagram, we suggest that the other components be clustered according to their relative importance to the defining member. This step may be accomplished by drawing the circles for the more salient people or places larger, or by drawing those circles closer, or by both drawing them larger and closer. The diagram presented in Figure 4-1 illustrates this type of schematic representation of a child's ecosystem.

In reading any diagram of a child's or adolescent's ecosystem, it must be remembered that time is an important component of the system. During any assessment procedures or any intervention, there are changes occurring in the individual's ecology, whether or not he or she is present. Since that student must continue to function in his or her ecology, ecosystem diagrams include a large arrow indicating the passage of time. This arrow is a reminder that salient people and places change.

What kinds of behaviour are important? When deciding if someone is behaving in an intolerable way, it is important to evaluate our own expectations. Hewett and Taylor (1980) suggest that four considerations might be used to evaluate problem behaviour: a) frequency of occurrence, b) degree of severity, c) duration, and d) clustering. The first factor that affects tolerance of a behaviour is the frequency with which that behaviour occurs. An appropriate question for evaluating this would be, simply, the following: Is a particular student doing something on a regular basis? The second question that might be asked requires that a judgment be made about the severity of the problem: Is the behaviour in question likely to cause harm to others or to the individual in question? The third consideration is the duration of the behaviour. To determine duration one might ask: How long does a student behave in a particular way? Finally, a judgment must be made about clustering. Is the child or adolescent engaged in several behaviours that, taken together, exceed our (i.e., the teacher's or other involved professional's) range of tolerance? Asking these four questions may help to clarify how much importance should be attached to the way in which the student is behaving.

We pointed out earlier that we were not so much interested in behaviour that might be judged unacceptable in a classroom as we were concerned about behaviour

Figure 4-1 **Schematic Representation of a Child's Ecosystem**

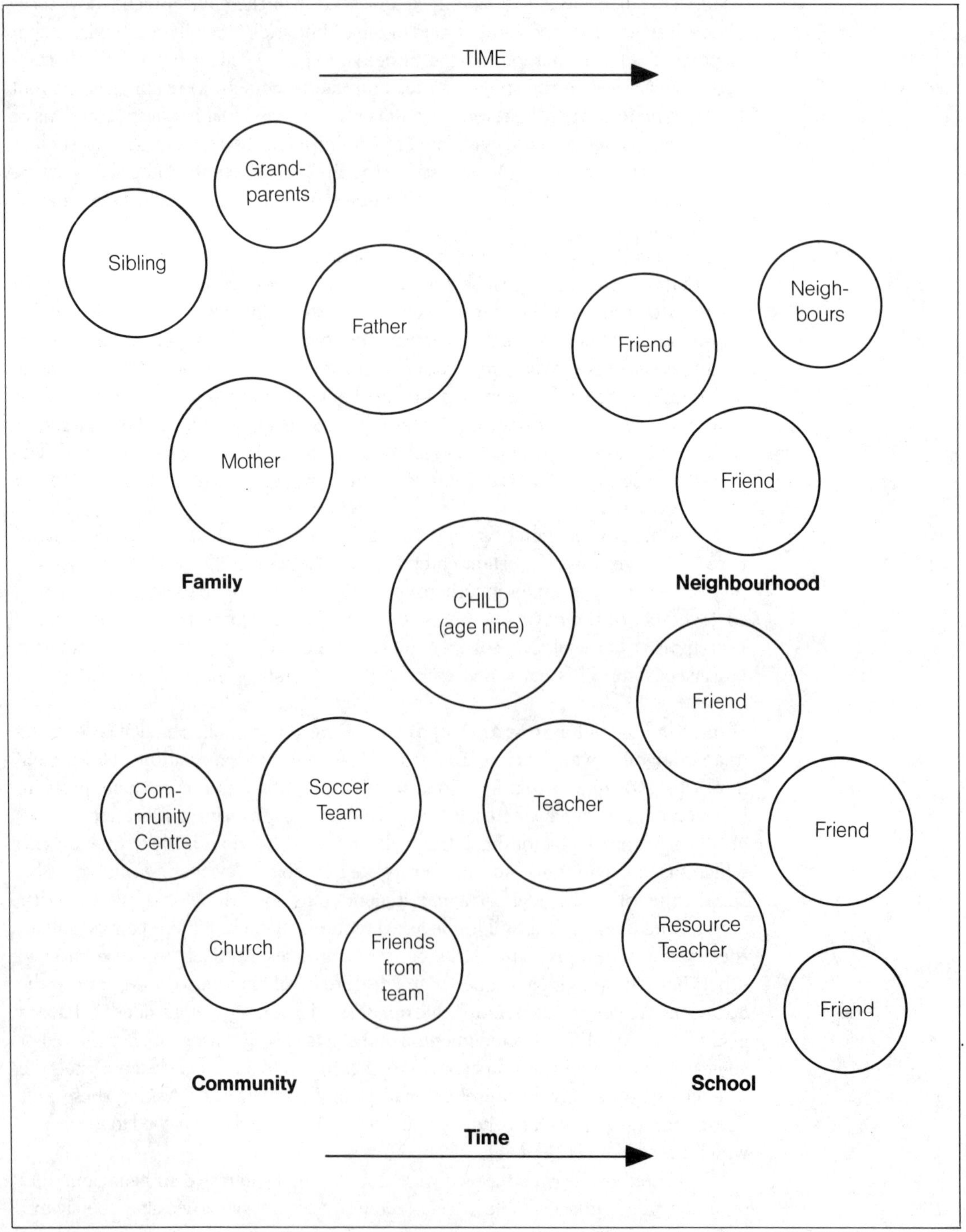

that might be seen as interfering. The above four questions are useful in both cases. In the classroom, the teacher may want to ask questions about his or her range of tolerance when making the decision to refer a child for counselling. These same questions may be asked when trying to decide if a student requires extra assistance with classroom work.

Consider, for example, a teacher who decides that a child is not reading well based on the results of a test obtained early in the year. There is some chance that a low test score is due to factors other than poor reading skills. The teacher might ask how frequently this individual has performed in this manner. In asking about severity, the focus will be on how weak the students' skills are. For duration, the question will be how long the student has had difficulty with reading (one year, two years, or longer). In considering clustering, the focus is on the influence of one set of skills on other academic expectations. Can the student who has been identified as not reading well be expected to maintain grade level performance in other subject areas such as science or mathematics? In this way all four considerations may be taken into account for both general behaviour and behaviour specific to academic performance.

ASSESSMENT TECHNIQUES: METHODS OF DATA COLLECTION

There are four potential methods of obtaining data that may be used when making decisions about students' behaviour. One simple, straightforward method is to ask students about themselves. This may be done either directly, through interviews, or indirectly, by asking the student to complete a paper and pencil measure which is then scored in some manner. The second method is to observe students in various settings. In using this second technique, behaviour may be either observed directly, recorded and then graphed; or it may be observed indirectly through checklists or rating scales, which requires that the teacher make an inference based on what he or she has observed in the classroom. The third method is to ask peers to give their opinions about a student's behaviour. This technique usually requires that students answer a series of questions about one of their fellow students. The fourth method is to ask others who are outside the classroom context but who are part of the student's ecology to provide information. This approach also makes use of interviews, checklists and/or rating scales. There may be some overlap between this source of information and previous ones; for example, questions worded in a similar manner may be asked of the referred individual, his or her peer group, the teacher and the parent(s).

The Student as a Source of Data

It was suggested above that one method of collecting data was to interview the student. Interviews may be general in nature or very specific. General interviews may be used, for example, to collect information about a student's family background, or about personal preferences and interests. Some sample questions that could be used as part of a general interview are given in Figure 4-2. In general, the tone of any interview should be carefully balanced so that selected items do not affect responses. The majority of items should be worded so that their tone is positive—for example, "I am happiest when...."—or neutral—"If I were my teacher I...." Some may be worded negatively, but these should be in the minority.

Figure 4-2 **Sample Items for a General Interview**

1. Attitude toward the family:

a. Compared to other families, mine is...........
b. One nice thing about my family is.............
c. How do you think you will feel about your family when you are older?
d. What are some things you enjoy doing with your family?

2. Personal interests and preferences:

a. I learn best when........
b. If someone makes fun of me, I.............
c. I am happiest when..........
d. What are some of your favourite things?
e. What do you like to do when you come home from school?

Specific, rather than general, interviews may serve as a source of information about how a student thinks about a particular subject such as mathematics (Liedtke 1979, 1982), or how that student thinks reading test questions, for example, should be answered (MacGinitie 1983). In a specific interview, the student completes a small number of items, talking through each one as he or she does it. The purpose in completing such specific interviews is to estimate both what students do as they work out assignments and how they do it.

Interviews are a form of self-report; that is, the interview provides a forum that permits the student to report on himself or herself. In some cases, students will initiate the interview process themselves. Students may feel safer talking about specific difficulties in subject areas. They will, however, also report to a trusted person many other kinds of feelings, some of which are very personal. In such cases, the students should be given the opportunity to share their feelings. After listening, there may be some concerns for which the teacher will want to involve other professionals. Examples include sexual abuse, physical abuse, or problems associated with chemical addiction.

Paper and pencil measures Another way of collecting data is to have students complete a checklist or rating scale. When completing checklists, students indicate the presence or absence of some characteristic. They do this by, for example, answering "yes" or "no" or marking those responses which apply. In the second case, students are required to rate the degree to which they hold a particular feeling.

Checklists and rating scales may be either teacher-prepared or commercially prepared. In addition, items may be of a general nature, as, for example, when students are asked to indicate their interests; or items may be of a specific nature, as when students are asked whether they like to listen to the teacher read.

Teacher-prepared checklists and rating scales In Figure 4-3 examples of items are presented that may be used to construct a teacher-prepared checklist or rating scale. Items may be global or specific, in the sense that they may require general information or may ask for detailed reactions. The same or similar items may be used to construct either a checklist or a rating scale. The major difference between the two techniques is in the type of answer wanted. If the data required is dichotomous, a checklist is the most appropriate instrument. If, on the other hand, an indication of degree of preference is required, a rating scale will yield the needed data.

Figure 4-3 **Sample Checklist and Rating Scale Items**

Sample checklist items

1. School is my favorite place. **yes** **no**
2. Holidays are fun.
3. Dreaming is a productive use of time.

Sample rating scale items

1. Mathematics is my favourite subject.

Not true for me	Sometimes true for me	Always true for me
1	2	3

2. Reading is an enjoyable way to spend time.

Never ______________________________ Always

As may be seen in Figure 4-3, two variations on checklist answer forms may be constructed. Students may be required to indicate either "yes" or "no" to each item, or they may be asked to check only those items that they feel are applicable to them. It is easy to see that one is a variation of the other. For both methods of obtaining responses, the student is indicating an absolute preference—in the first case by responding with some variation of yes or no to each item and in the second case by the presence or absence of an answer. Unfortunately, we can not offer a general rule of thumb to assist in deciding when to select between the two answer-sheet formats.

The decision as to whether to use a checklist or a rating scale is, in our opinion, easier. Checklists are more appropriate when trying to determine the presence or absence of a skill. Rating scales are more appropriate when looking for an indication of feelings. For example, a boy may say that he does not always feel school is his favourite place: sometimes it is, sometimes it isn't. It would be easier, then, for this boy to indicate his feelings on a rating scale.

When designing rating scales, one may choose between two main variations of answer form. These variations are also illustrated in Figure 4-3. In the most common type of answer form, there are two further variations in format. First, the words or phrases selected as descriptors may be varied. In Figure 4-3, for example, "Not true for me" and two others are used. Another example of appropriate terms would be the following: "Most Like Me," "Like Me," "Uncertain," "Unlike Me," and "Most Unlike Me." In selecting a set of terms, it is important that they parallel one another, as in the above examples. In addition, the alternatives should be selected to match the question or statement given. For example, it would not have made sense to use "Most like me," etc., in the sample statement listed for mathematics in Figure 4-3. Second, the number of gradations of ratings may be varied. For instance, in the example in Figure 4-3, students were given three alternatives. This scale could be expanded to four alternatives by adding "Usually true for me." Anywhere from three to seven choices are typical in rating scales.

The second type of answer form is also given in Figure 4-3. This type of form requires respondents to mark on a line to indicate the degree to which they believe something applies to them. Thus, if someone believed that dreaming was a productive activity half of the time, he or she would make a mark through the middle of the line. The advantage of this second type of scale is that it does not force the respondent to choose a specific response. Best results appear to be obtained when the line is premeasured to a specific length, say ten centimetres. In this instance, any mark within the first centimetre would be scored one, in the second, two, and so on until the final centimetre which would be scored ten.

For younger children the examples given above likely will not yield valid data. The idea of the rating scale is still useful, however, especially if younger students could respond to pictures instead of words. Using this idea, Ollila (1981) has designed a pictorial rating scale to assess attitude to reading. Sample items from this scale are given in Figure 4-4. Similar pictorial rating scales could be developed to assess attitudes towards other subject areas.

As can be seen, children are presented with pictures of moose posed in various ways: one happy, one neutral, and some unhappy. Perhaps the best feature of this instrument is that it provides a standardized set of pictorial responses. There are, however, no predetermined questions or statements. Thus, this pictorial rating scale may be used to enable children to rate their attitude towards any content area or other area of concern.

Commercially prepared checklists and rating scales Instead of one of the above options, commercially prepared checklists or rating scales may be selected. The advantage of selecting a commercially prepared checklist or rating scale is that the teacher does not have to construct the items.

Caution, however, must be taken in selecting one of these measures. Only a few will meet the standards for tests discussed in Chapter 3. The major weakness in selecting such techniques is that they call for the interpretation of a person's behaviour (Salvia and Ysseldyke 1981), even if, in this case, the person making the inference is the student himself or herself. In other words, the possibility of error exists because the students must judge which of the given alternatives on the checklist or rating scale is most appropriate. If none of the alternatives accurately describes their behaviour as they see it, they may mis-rate their behaviour in their effort to arrive at the best choice.

Figure 4-4 **Sample Items from a Pictorial Rating Scale**

How do you feel about going to school in the morning?

How do you feel when it is time to do your worksheet or workbook in school?

How do you think your friends feel when you read aloud in your class?

From L. Ollila, A *Primary Reading Attitude Inventory* unpublished, University of Victoria, (1981). Reprinted with permission of the author.

Fairly comprehensive reviews of self-evaluation checklists and ratings scales may be found in the *Eighth Mental Measurements Yearbook* (Buros 1978) in the section on personality tests. It is recommended that readers consult Buros when they want to select a commercially prepared self-evaluation or other behaviour scale. This source, however, does not give samples of how items are worded, nor does it illustrate the item format.

A fairly complete listing of self-evaluation scales and other instruments that may be used to evaluate various aspects of behaviour is provided by Csapo (1981). She presents illustrations of items and comments briefly on the characteristics of each scale.

Two examples in which the focus is on self-perception of school-related behaviour will be reviewed briefly in this chapter: a) the *Student's Perception of Ability Scale* (Boersma and Chapman 1977) and b) the *Estes Attitude Scale: Measures of Attitude Toward School Subjects* (Estes, Johnstone-Estes, Richards and Roettger 1981). The first example is a checklist of the yes-no type and the second is a rating scale.

The *Student's Perception of Ability Scale* (SPAS) consists of 70 items in which the student is to answer "yes" or "no." Items are presented that relate to reading, spelling, language arts, arithmetic, penmanship, and to school in general. Sample items are given in Figure 4-5.

Two types of reliability are reported for the SPAS. Cronbach Alpha, a measure of internal consistency, is reported to be 0.92 for the full scale with a sample of 642 children. Test-retest reliability, over a period of four to six weeks, is given as 0.83 for the full scale.

Figure 4-5 **Sample Items from the Student's Perception of Ability Scale for Children (SPAS)**

I always understand everything I read	**yes**	**no**
I usually finish my school work	**yes**	**no**
I am poor at subtraction	**yes**	**no**
I always do neat work	**yes**	**no**
I have trouble telling others what I mean	**yes**	**no**

Reprinted with permission of the publisher, Psi-Can Consulting, Edmonton (1977).

The *Estes Attitude Scale* consists of two forms, one for use in grades three to six (elementary form) and the other for use in grades seven through twelve (secondary form). The elementary form contains 14 items in each of three separate areas for a total of 42 items. The topics to be rated are mathematics, reading and science. Students select their ratings from a three-point rating scale. The secondary form is longer (75 items) and covers more subject areas (15 items on each subtest). Students are asked to rate their attitudes toward English, mathematics, reading, science and social studies using a five-point rating scale. Sample items from both the elementary and secondary scales are given in Figure 4-6.

Internal consistency measures are reported as ranging from 0.76 to 0.88 for the elementary form and from 0.76 to 0.93 for the secondary form (Estes et al. 1981).

Figure 4-6 **Items from the Estes Attitude Scale**

Elementary Scale
1. People who like math are often odd.
2. Most teachers really know how to teach math.
3. Reading is rewarding to me.
4. Reading is something I can do without.
5. Science classes are usually fun.
6. Elementary students should not have to study science.

Secondary Scale
1. Work in English class helps students to do better work in other classes.
2. Working math problems is fun, like solving a puzzle.
3. Reading is for learning but not for enjoyment.
4. A deeper love of nature comes from the study of science.
5. Much of what is taught in social studies is not important.

The Resource Teacher as a Source of Data

While having students describe or evaluate their own behaviour is useful, resource teachers or other professionals may want also to observe what those students do in the classroom and other school settings. Observation provides an opportunity to see how well a student works or which peers he or she interacts with. It is also useful in gaining an indication of the classroom teacher's expectations.

Like any other kind of assessment, observation should be subjected to those standards for tests provided in Chapter 3 that are applicable. In other words, the resource teacher should think of observation as a way of assessing. Accordingly, steps must be taken to make sure the observational technique chosen is the most appropriate and that the resultant data is valid and reliable.

To review, it was recommended that in sampling classroom behaviour, resource teachers should obtain estimates from at least two sources: a) a subject, preferably an academic subject, in which the student is proceeding reasonably successfully, and b) the subject in which performance is below expectation and has been identified as an area of concern. The more diversified the sample of behaviour taken, the higher the probability that a reasonably accurate estimate of the student's behaviour will be obtained.

A second precaution that will increase the chance of the observations being valid is consultation with the classroom teacher both before and after observing. Before observing, some indication of classroom expectations should be obtained. What kind of behaviour is considered acceptable, what is unacceptable? The purpose of determining the classroom teacher's expectations is to increase the probability of the data being valid. The observer's expectations may differ from those of the classroom teacher. After completing an observation, the observer should ask the classroom teacher if the student's behaviour was typical. Students are very much aware that they are being observed despite one's best efforts to be unobtrusive. Just the presence of another person in a classroom affects the ecosystem of that class and may affect any individual's behaviour. That is why it is so important to determine if, in the teacher's opinion, classroom behaviour was typical.

Another useful technique is to occasionally check the behaviour of others in the classroom. If Alfred isn't doing his reading assignment, is his behaviour different from that of anyone else in his reading group? If Susan is taking five minutes per mathematics problem, is she behaving differently than the others? Thus, collecting peer group data provides another measure of both classroom expectations and the difficulty that students in general may be having with an assignment.

Types of observation techniques Once having decided to observe a student or students in some school-related activity, one must next decide how to collect the data. Depending on how frequently any behaviour occurs, one of two types of data collection may be selected. If a child or adolescent acts in a certain manner on a frequent basis, some counting technique may be appropriate. If, however, a particular behaviour is displayed infrequently, describing what occurred may be the most appropriate method. Three techniques for counting and two ways of describing behaviour are given below. In addition, some suggestions as to when to use each technique are provided in each section.

Figure 4-7 **A Sample Chronologue**

10:55-11:00	Looks at notebook for a few seconds, sharpens his pencil, returns to his desk and starts to drum.
11:01-11:05	Questioned by fellow student, "Don't you ever have to do spelling?" Michael ignored question and lept over his desk. Sitting down, he retorts, "Who me?" and then laughs.
11:06-11:18	Takes glue bottle out of his desk and squeezes the air out of it.
11:19-11:20	Locates math assignment and opens his book.
11:21	Looks at math assignment as teacher is approaching.

From D.G. Bachor, "From strategy deficiency to instruction," paper presented at the Learning Assistance Teachers' Association Annual Conference, Vancouver (1984).

A chronologue: A useful starting point When first deciding to observe, often it is difficult to determine which technique to select. If you are not familiar with the student, one useful technique is to start with a general classroom visit in which a chronologue is completed. A chronologue is a time-referenced description of the activities taking place in the classroom from the perspective of the referred student.

In completing a chronologue, the observer writes down all behaviours that are occurring. In addition, the time the child or adolescent starts and stops a sequence is noted. An example of part of a chronologue is found in Figure 4-7. In that example, Michael, the student being described, is a thirteen year old adolescent in grade seven who is not working very effectively in school. He prefers instead to attempt to joke his way through activities and, in general, to avoid anything that may involve work.

Anecdotal records: A second descriptive technique Anecdotal recording is useful when a student displays a particular action or set of actions quite infrequently or without a particular discernible pattern. In writing an anecdotal report, it is useful to attempt to describe three things: a) what happened just prior to the incident, b) the incident itself, and c) what happened after the incident occurred. In addition to a description of the incident, the following data should be included: a) the student's name, b) the date of the incident, c) the time the incident occurred, and d) a description of the setting in which the incident occurred. An example of an anecdotal record is given in Figure 4-8.

Figure 4-8 **An Example of an Anecdotal Record**

Name: Billy Watson Date: March 5, 1983

Observer: Doug Jones Time: 10:25 a.m.

Setting: The end of mathematics period, students were working individually on assignments.

Just prior to the start of recess, Billy complained of a stomach ache with the result he was to stay in at recess.

Event recording: A behaviour count strategy When the behaviour to be observed occurs very frequently and is of short duration, an event-recording strategy is an appropriate choice. In event recording the observer records each time a particular behaviour occurs. The behaviour to be observed (referred to as the "target behaviour") will need to be carefully defined so that the probability of accurate recording is increased. Using this technique it is possible to observe several target behaviours at the same time by keeping separate tallies on a piece of paper.

This technique may not be very useful from the viewpoint of the resource teacher. Usually the resource teacher needs more data than this technique can provide. In some cases, however, event recording is useful. For example, the resource teacher may want to know how many comprehension questions are answered during reading period. He or she may want to further analyze this data and estimate the number of questions done per minute. This ratio may be calculated by dividing the number of questions completed by the minutes available to do the assignment. It may be informative to calculate this ratio for several other students in the class to estimate what the typical classroom performance is and compare the ratio of the student being observed to this average.

Duration recording: A second count strategy Earlier it was suggested that a criterion of 80 percent on-task was a reasonable expectation for students. Duration recording is the behaviour-recording technique that allows an estimation of time on-task. The observer records the number of minutes that the student actually spends working over a given period. The most convenient way to estimate the duration of time working is to use a stop watch that allows cumulative time to be recorded. The observer notes the time the student starts working and the time he or she stops, recording the total elapsed working time on the stop watch. A percentage is then calculated by dividing elapsed working time by the total time of the observation period. We have found that it is possible to use a stop watch and to keep notes on the referred student at the same time.

Interval recording: A third count technique Interval recording allows the resource teacher to estimate the degree to which the teacher interacts with the referred student in the classroom as well as the percentage of time the student spends on-task.

As was the case with duration recording, on-task behaviour must be defined before observation begins. Once this behaviour is defined, decisions must be made about the length of the interval and the scoring procedure to be used. Before starting to record, the total amount of planned observation time is divided into intervals of 5 to 30 seconds. This interval size is typically set at 10 seconds. With an interval smaller than 10 seconds, there is some chance that scoring will be inaccurate. With a larger interval size, information loss is high. In other words, the reliability of scores may be affected when the interval is too small and the validity of scores affected when the interval is too large.

Once the interval size and the duration of the observation period have been established, the next decision is how to score each interval. A plus (+) or a minus (–) is to be recorded in each interval depending on whether the individual is on-task or not. Jobes and Hawthorne (1977) recommend that plus should be recorded if the student is on-task any time during the ten-second interval. A minus is recorded if the student is off-task for the entire ten-second period. The problem with this scoring technique is that the percentage of on-task behaviour is over-estimated. A second scoring procedure is to record a plus only when the child or adolescent is on-task for any complete ten-second interval. A minus is recorded if off-task behaviour is displayed any time during the interval. This presents the opposite problem to the one noted in the Jobes and Hawthorne scoring procedure: the percentage of off-task behaviour is over-estimated. In other words, there is some degree of error in both scoring systems. A decision as to which one is appropriate is best made on an individual case basis. In general terms, however, a simple criterion may be used to help make a choice as to which method to employ. Look carefully at each case and decide which error is tolerable. Thus, when trying to motivate a student to continue some task, it may be better to over-estimate on-task behaviour at first. Later, as self-responsibility develops, the recording procedure could be modified, over-estimating off-task behaviour, resulting in a more accurate, if conservative, estimate of on-task behaviour.

One benefit of selecting the interval recording system is that it is possible to record both the individual's behaviour and the nature of the teacher's interaction with that student. Hops, Walker and Hutton (1974) suggest that two teacher behaviours may be noted. If the teacher approves of some student action, a circle (0) is drawn in the interval in which the approval occurred. If the teacher disapproves of something, a slash (/) is drawn in the appropriate interval. In this way it can be determined when and if the teacher approves or disapproves of what a student is doing. An example of a partially completed interval record is given in Figure 4-9.

Figure 4-9 **A Partially Completed Interval Recording Form**

Name James Jones Date December 20, 1986

Definition of On-task Behaviour Sitting in seat completing assigned work.

Number of Minutes

+	+	+	+	+	+	+	+	+	+	+	-	-	-	-	-	-	+	3
+	+	+	+	+	+	+	+	+	+	+	-	-	-	-	-	-	+	6
																		9
																		12
																		15
																		18
																		21
																		24
																		27
																		30

Estimated Percentage On-task

The Student's Peers as a Source of Data

As any child goes through school, peers come to be an important component of his or her individual ecosystem. Starting as early as grade one, children begin to interact with one another differentially (Chapman, Larsen and Parker 1979). It has been pointed out that children and adolescents with learning problems are less popular than students without learning problems (Bryan 1978; Bryan and Bryan 1978). Thus, one factor that may be important in working with children and adolescents with learning problems is peer rejection (Bryan 1976; Gottlieb 1974; Gresham 1981;Richardson and Royce 1968).

The problem is how best to determine peer social behaviour. Greenwood, Walker and Hops (1977) point out that the usual choice, sociometric measures, have yielded reliability scores too low to be useful for identification purposes. Further, they argue that teacher ratings may be more accurate estimates of student interaction than peer ratings are. While it may be true that teacher ratings are more accurate, they are limited to classroom interactions. Brown (1984) reviews evidence that leads to the opposite conclusion; that is, sociometric ratings are the more accurate measure. In a more qualified statement, Csapo (1981) points out that sociometric measures are reasonably good estimates of peer acceptance, which is what they were intended to measure.

In summary, two sources of data may be used to provide estimates of peer social interaction. Teacher ratings appear to be a reliable estimate of classroom interaction. Sociograms seem to be a reasonable choice as an indicator of peer acceptance. Only the second technique will be discussed here. Given the uncertainty as to the reliability of social acceptance measures, however, all sociometric techniques should be used with caution.

The sociogram The sociogram is an assessment technique that allow peers to identify those students that they like and those that they dislike. Moreno (1943), who developed the sociogram, makes the following recommendations about the use of them:

a) The questions should be stated clearly and simply to prevent confusion.
b) The questions should be designed so that they are limited in scope and so that children or adolescents are reacting to a question that will be affected by their decisions.
c) The number of allowed choices should not be limited and children should make those choices in private.

Further, Brown (1984), in reviewing various recommendations about sociograms, adds the following three limitations:

a) A class size of about 20 is necessary before the sociogram is accurate.
b) Classes should be in session for about six weeks before a sociogram is attempted.
c) Answers from one question may not be generalizable to another question; that is, choices about who one might want to play with after school may not relate to who one would want to receive assistance from on an assignment.

Brown and Hammill (1978) recommend that questions on a sociogram be paired so that positive and negative opinions are sought at the same time. Some sample questions might be:

a) Which (of the children in your class) (of your classmates) would you like (dislike) to help you work (during reading) (on an English assignment)?
b) Which (children in your classroom) (of your classmates) would you like (dislike) to help you (draw a picture) (complete an assignment that involves art work)?
c) Which (children in your class) (of your classmates) would you invite to have lunch with you?

The wording in the first and last sets of parentheses indicates wording changes for individuals of different ages. Like or dislike would be used in alternating questions, following the recommendation of Brown and Hammill (1978) that questions be paired.

There is no standardized method of scoring the choices students make in answering questions about the students with whom they would choose to work. Anyone who has ever completed a sociogram knows that diagramming student choices is very time consuming. An alternative is to set up a matrix table for each set of questions. Popeil et al. (1983) use such a technique, developed by N. Long. In setting up the matrix, all students' names are listed in alphabetical order along the top and left-hand margin of the page, as in Figure 4-10. The number of the choice made (first, second, etc.,) is then recorded in the grid. Reading across the grid allows the reader to determine who was chosen by each individual. By reading the columns, the number of times each individual was chosen, the ranking of choices, and the total number of choices may be determined.

Other Individuals as a Source of Data

There are several types of data that may be gathered from other individuals: their perceptions of an event, their estimation of the student's behaviour in a particular situation, and their general expectations for the referred student. The appropriate techniques for collecting such data are similar to those already discussed in this chapter; i.e., these data may be collected by interviewing appropriate individuals or by administering checklists and rating scales to them. If data is collected from several individuals, the results may be compared to determine the degree of similarity and difference in expectations. This may be useful, for example, in determining school-based team members expectations of IEP goals.

There is one additional technique that may be useful in comparing the expectations of parents, teachers and the student: the behavioural Q-sort (Kroth 1973). Brown (1984) suggests that the behavioural Q-sort is ideally suited to the purpose of ecological assessment and thus to assessing the perceptions of other individuals in the ecosystem. In addition, Kroth (1973) suggests that it is possible to compare the individual's view of "real" self and "ideal" self using a set of items written in behavioural terms. The advantage of using this technique is that the individual's progress towards a set of goals, the ideal self, can be measured. Further, an individual's perception of his or her behaviour can be validated by having a second person collect data to determine the frequency of the behaviour-in-question. For the latter two reasons, we describe the behavioural Q-sort in the next chapter on affective education and self-concept. The behavioural Q-sort is mentioned in this section only to ensure that its use as an ecological measure will not be overlooked.

Figure 4-10 **A Matrix of Choices for "Liked the Most"**

	Bill A.	Sheila A.	Lorena B.	Susan C.	Robert C.	Howard D.	Barbara D.	Mary D.	Anne H.	Ian H.	Gary K.	Patricia M.	Marilyn P.	Deborah P.	Lynda P.	Carl P.	Kevi R.	Andrew S.	James S.	Terry W.
Bill A.		3				2														1
Sheila A.	1			2				3												4
Lorena B.									1			2								
Susan C.		2						3									1			
Robert C.	3					2										1				
Howard D.	2					1														
Barbara D.			3										2		1					
Mary D.	4	2		1																3
Anne H.			1									2								
Ian H.					1						3							2		
Gary K.										2								1		
Patricia M.			2				3		1								4			
Marilyn P.							1		4			2			3					
Deborah P.													3		1	2				
Lynda P.			3				2						4	1						
Carl P.					1					2										
Kevi R.				1								2								
Andrew S.												2	1						3	
James S.												2						1		
Terry W.	1									3						2				
Number of times selected	5	3	4	3	2	3	3	3	3	4	2	4	3	1	3	3	2	3	1	4

INTERVENTION TECHNIQUES FOR INTERFERING BEHAVIOURS

Once it has been determined that a student's behaviour is interfering with his or her opportunities to learn effectively, a decision must be made about what the next step should be. There are three possible choices: a) an instructional intervention, b) an instructional intervention in combination with one directed at the student's behaviour, or c) an intervention directed at some aspect of that behaviour.

If the behaviour problem seems to be a consequence of school failure, it might be appropriate to ignore the behaviour while directing attention, instead, only to the student's ineffective learning strategies. For example, revisions could be made in the student's academic program to enable him or her to benefit more from classroom instruction. Our clinical experience leads us to suggest that, in the majority of cases, an instructional intervention will result in a cessation of interfering behaviours with children and adolescents who are experiencing difficulty learning. Thus, as a first strategy, we recommend an instructional intervention.

If instructional intervention is chosen, one may then follow the steps of the decision-making model presented in Chapter 2, beginning with the preparation of trial lessons. The content of individual trial lessons will vary depending on the area of concern. For example, lessons in reading comprehension may result in a reduction in test anxiety for a student when he or she is answering questions in science lessons. Techniques and materials that may be used in trial lessons and later in an IEP are described in Chapter 6 through 12.

In some instances, the interfering behaviour may become so prevalent that an instructional intervention alone no longer would be effective. When this happens, an intervention will need to be directed at the student's behaviour as well as at his or her instructional needs. Some strategies that may be used to address interfering behaviour in the context of instruction include cognitive behaviour modification (Meichenbaum 1976, 1977; Meichenbaum and Genest 1980), self-instruction or self-regulation training (Nelson and Polsgrove 1984) and behavioural contracting (Kirschenbaum and Flanery 1983).

There are at least three cases in which the sole focus of the intervention may be on the child's or adolescent's behaviour. First, Kronick (1981) argues that, until recently, the social skills of children and adolescents with learning problems have been ignored and that these skills should be one focus of attention. Second, behavioural intervention may be necessary in some cases so that the child or adolescent is capable, physically, of learning. Examples include cases of alcohol and drug abuse (Lotsof 1978) and anorexia nervosa (Maloney and Klykylo 1983; Rollins and Piazza 1978). Third, some type of intervention likely is necessary if a student has reached the point of being emotionally incapable of learning. These individuals have been called, among other things, "children who hate" (Redl 1966; Redl and Wineman 1952).

In the sections that follow, strategies for reducing interfering behaviours are reviewed. Before presenting these strategies, however, one general consideration is discussed.

Selecting Behavioural Intervention Techniques

From the viewpoint of the adult, a behavioural intervention technique may be seen only in terms of its tool function. In such circumstances, the most powerful technique available is the one likely to be chosen. In other words, the adult may ask, "What is the fastest, most effective way to stop a particular unwanted behaviour?" The more powerful technique satisfies the decision-maker's need to change, reduce or eliminate specific patterns of behaviour as quickly as possible. However, taking the child's or adolescent's needs into account, the issue becomes one of determining the impact of the selected intervention. In thinking of the child or adolescent, two questions should be answered before a behavioural intervention is recommended: a) Is there any stigma associated with the chosen technique? and b) What effects will the technique have on self-esteem?

For example, Polsgrove and Reith (1983) point out that punishment suppresses problem behaviour; but, it may also reinforce other unwanted behaviours and even teach aggressive behaviour through modelling. Another example of a technique which may have an unwanted effect is the use of response cost. This is a mild form of punishment which involves the removal of rewards that were previously earned or awarded. It is analogous to imposing a fine. The purpose is to eliminate unwanted behaviour. Although, this technique has resulted in gains when carefully employed; negative side effects may also result (Polsgrove and Reith 1983; Walker 1983).

In the case of aversive techniques, there may be ready agreement about the need for caution. There are other techniques, however, which we may think unlikely to have negative side effects but which can, in fact, cause problems. For example, our clinical experience suggests that children and adolescents develop negative attitudes toward a subject because of the method of presentation of material. For example, some students will find the pace too quick for them and will start to dislike a topic if the pace is not moderated. Other children find that they can't understand material unless they can "see it." When the style of presentation is maintained, perhaps because the teacher cannot present the material in such a way as to allow the student to "see" the material, the student may learn to dislike either the topic or the teacher, or both. Children's and adolescents' attitude towards composition (e.g., Shook 1982) is only one example.

In summary, the first consideration in selecting an intervention should always be the potential for negative impact on the student.

Cognitive Behaviour Modification or Self-instructional Strategies

The idea behind cognitive behaviour modification and self-instructional training is that effective learners have a strategy in mind as they are working. Brown (1975, 1978) has suggested that such strategy awareness or self-monitoring be referred to as "meta-cognition." Metacognition, then, is the awareness we have of whether we are performing effectively.

Metacognitive skills seem to be important in successful school performance (e.g., Gerber 1982). A recently completed study by Taylor and Samuels (1983) illustrates this point. They presented elementary children with scrambled and normal texts (normal in this case meant arranged in English sentences and logically developed paragraphs). Good readers—those who were aware of text structure—recalled normal text better than poor readers—those who were not aware of text structure. When the passages were scrambled, the recall of good readers was similar to that of poor readers on both scrambled and normal text. Thus, knowledge of content-specific or task-specific metacognitive skills appears to be important in instructional success.

Cognitive behaviour modification This is one technique that may be used to teach children and adolescents to monitor their own instructional strategies more effectively (Meichenbaum 1977). The steps of this procedure are detailed in Chapter 10 so they are reviewed only briefly here.

First, a specific task is performed by an adult while he or she states the steps of the procedure out loud (cognitive modelling). Next, the child performs the same procedure under the adult's supervision (overt external guidance). Third, the child repeats the procedure, saying the instructions out loud (overt self-guidance). Fourth, the child whispers the instructions while going through the task (faded, overt guidance). Finally, the task is performed by the child using private, inaudible self-talk (covert self-instruction).

Self-questioning procedures Self-questioning is a technique similar to cognitive behaviour modification. Students are taught specific types of questions so that they can ask more effective questions. Using questions to self-monitor has been demonstrated to be useful in learning to read. For example, Wong, in a series of studies (1979, 1980; Wong and Jones 1982) has demonstrated that question-asking improved the recall of main ideas and increased the awareness of textual units.

One simple model of questioning (Sitko, Slemon and Bachor 1983) is presented in Chapter 6. Students usually learn this questioning system quite quickly as there are only four categories in it.

Behavioural Contracting

Sometimes children and adolescents require a more structured set of prompts to remember to complete assignments. Behavioural contracts are one tool that can be used to remind a student of what he or she has set out to do.

We recommend that first "contracts" be very informal. Referring to them as "plans" or "agreements" might be more appropriate. These plans should be designed in such a way that the student has an excellent chance of succeeding. Younger children enjoy being made to feel that such plans are realistic by being reminded that this is what adults do when they make an agreement. To emphasize the importance of following the plan, the agreement might be sealed by a handshake. A similar technique may be followed with adolescent, although in their case the role-playing component should be de-emphasized.

In implementing these informal agreements, it is very important that the adult remember to follow up. Since plans may be made with a number of individuals, it is a good idea to record the main points of the plan or agreement. Such a record may be worded as simply as the following note:

> Oct. 14. 1983
>
> Susan agreed to complete her assignments in math for the next two days.
> Check with her tomorrow to see if she is on track.

Some children do not find informal agreements sufficient to prompt the appropriate behaviour. In such cases, more formal contracts likely will be necessary. If the child or adolescent has not met a previously agreed upon informal contract, reaching a new contract should begin with a discussion of the previous incomplete plan or agreement. The following guidelines may be useful in such negotiations:

1. Find a private or personalized place to begin the discussion. For example, with younger children a rug may be set in the corner and called "the problem-solving rug."
2. Define the incomplete agreement as a problem to be resolved by mutual problem solving.
3. Creative problem-solving methods such as brainstorming for ideas (see Chapter 10) might be introduced in order to reach a workable plan.
4. Draw up a new contract. Depending on the student, some positive consequences for living up to the contract or negative consequences for not meeting the contract may be introduced at this point. If negative consequences are selected, the risks associated with punishment discussed earlier in this chapter should be considered.

An example of a behavioural contract is given in Figure 4-11.

Figure 4-11 **Sample Behavioural Contract**

Statement of Contractual Agreement

Effective Dates: <u>January 7, 1985 to February 28, 1985</u>

Persons Involved: <u>Susan Jones (student) and Barbara Smith (resource teacher)</u>

Susan Jones and Barbara Smith agree to fulfill the following responsibilities during mathematics period.

IF, Susan completes her regular classroom mathematics assignments with 80 percent accuracy for TEN consecutive days,
THEN, Barbara will arrange for ONE-HALF HOUR of computer time.

This contract is to be reviewed in TWENTY days.

____________________ (Signature of student)

____________________ (signature of resource teacher)

From D.S. Kirschenbaum and R.C. Flannery, "Behavioural contracting: outcomes and elements" in M. Hersen, R.M. Eisler and P.M. Miller (Eds.), *Progress in Behaviour Modification, Volume 15* (New York: Academic Press, 1983). Adapted with permission of the authors and Academic Press.

Social Skills Training

Kronick (1981) has pointed out that programs to address social skills have appeared only in the last two decades. Yet such skills are very important and may be difficult to learn. One of the problems. Kronick suggests, is the lack of direct feedback. She points out that most people respond "to social awkwardness with silence, avoidance, and subtle, implicit disapproval" (1981, p. 122).

Walker et al. (1983) have developed a social skills curriculum for handicapped children. Three general goals were set: a) to facilitate social development, b) to assist handicapped children in meeting the behavioural demands of least restrictive settings, and c) to increase the probability of acceptance of handicapped individuals by their non-handicapped peers. ("Least restrictive settings" is used to describe one type of placement made for a student. In this case, it refers to the best possible class placement in which the student likely will learn most effectively. For a parallel discussion, see the section on mainstreaming in Chapter 1.)

The teaching scripts are divided into 28 peer-to-peer social skills and five classroom behaviours. Examples of peer skills include non-specific affective skills (smiling and touching), interactive skills (starting and answering), and coping skills (expressing anger, refusing requests politely). The five classroom behaviours are doing work of acceptable classroom quality, following classroom rules, compliance, making assistance known, and listening to instructions and directions.

For each of the skill areas, the teaching scripts are divided into a series of nine steps. In the first steps, the skill is defined, examples are given, the definition is reviewed, and further examples are provided by the students. This definition phase is followed by a series of activities in which the teacher models the social skill and the students role play it. Finally, the students, in an informal contractual agreement, practice the newly acquired skill in, for example, the school cafeteria.

Programs to Address Behaviour

Several different philosophies have emerged as to the most effective way to help children or adolescents change their behaviour if they are in conflict. Two approaches are reviewed in this chapter: an adaptation of reality therapy (Glasser 1965, 1969; Bassin, Bratter and Rachin 1976), as an example of the personal responsibility model; and Project Re-Ed (Hobbs 1967, 1982), as an example of an ecological model.

A primary goal in using either of these techniques is to assist the student in becoming what de Charms (1967, 1972, 1975) terms an "origin"; that is, someone who is responsible for his or her own behaviour. The goals of the teacher in assisting someone to become an "origin" are as follows:

> help the person, (a) to determine realistic goals for himself; (b) to know his strengths and weaknesses; (c) to determine concrete action that he can take now that will help him to reach his goals; and (d) to consider how he can tell whether he is approaching his goal, that is, whether his action is having the desired effect. (de Charms 1972, p. 97)

Personal responsibility models The crucial concept in reality therapy and other similar child-oriented behavioural techniques (e.g., Dreikers 1964, 1972) is that children and adolescents must be trusted with the responsibility of decision-making. If they are not able to assume this responsibility, they are to be provided with opportunities to learn how.

There are four main steps in implementing such personal responsibility models with children and adults.

1. The child or adolescent needs to believe that the people with whom he or she is interacting will act in a caring and fair manner. As part of this process, warm and personal relationships are to be developed in which the corner stone is trust. Such relationships, however, must stop at trusting ones; that is, more personal entanglements are to be avoided.
2. While feelings are to be respected, currently displayed, specifically defined behaviours are the focus of the intervention. That is, what the student is doing presently, as opposed to what he or she has a history of doing, is to be emphasized. The individual must be aware of what he or she is doing well and not so well so that he or she may learn to make any necessary or desirable changes.
3. An informal or formal contract is devised in which the child's or adolescent's commitment to the agreement is obtained. Commitment is the key to this step. The student must agree to be responsible for completing the terms of the agreement. Contracts may be one of the types described earlier in this chapter.
4. If the contractual agreement is not kept, the consequences should follow in such a way that the individual begins to learn to be responsible for his or her actions.

An ecological model Many of the principles of Project Re-Ed, the selected example of an ecological model, are the same as those of the personal responsibility models. The list of important components described by Hobbs (1982) includes working with present behaviour, building on trust and teaching self-control. In addition, Hobbs argues that intelligence and competence can be taught and that time is an ally in any intervention. Perhaps, however, the key tenet in this ecological model is that behaviour change should be understood as an interaction among individuals.

In an ecological model the child or adolescent is construed as the central figure in a system that is maladaptive. Taking a systems viewpoint, however, does not mean that the individual is not considered responsible for his or her own behaviour. Rather, it means that any changes in behaviour made by the individual-in-question must be seen in terms of their impact on the system in which the individual functions.

In working with individuals, an "enablement plan," which is similar in intention and design to an IEP, is prepared. The purpose in preparing this plan is to return the ecosystem "to a level of functioning that is satisfactory to its principal members" (Hobbs 1972, p. 233). Six types of responsibility are specified by Hobbs as being part of an enablement plan. These responsibilities are best defined by a series of questions:

1. Who will be responsible for getting the task done?
2. Who will actually provide service, if any is required?
3. By what date will the service be completed or the goal be met?
4. What costs will be incurred and what funding is available?
5. What criteria will be used to determine success?
6. What follow-up steps may be necessary?

These six questions, then, are used as criteria in setting up an enablement plan. This plan is used to help individuals to learn to acquire adequate coping behaviours.

Any program should be planned for the shortest possible time. The shortest possible time has passed when the point is reached at which the individual is learning to cope effectively with the demands of his or her ecosystem. The rationale for this suggestion is that the impact on the individual's ecosystem is to be kept to a minimum.

Case Study

The final section of this chapter is a case study of a boy in grade two who is considered to have a behaviour problem. In reviewing this case it will be apparent that more than the student's behaviour is at issue. In this sense Tim is typical. Inappropriate behaviour often is associated with weaknesses in academic areas. In Tim's case, the weakness is in reading.

Step 1: Assessing background information

Tim's background information is helpful in understanding why he might be feeling some degree of frustration with school. First, this year is Tim's second attempt at grade two. He was held back due to problems in reading and poor school performance in general, accompanied by "unacceptable" behaviour. In addition, Tim has been receiving supplementary assistance of some kind since kindergarten.

At the end of the second grade in the previous year, Tim was given an individualized intelligence test as part of the decision-making process for determining where to place Tim the next year. The results of this test were interpreted as indicating that Tim was in the high average ability range. On the basis of this data plus some educational assessment, the decision was made to have Tim repeat the second year.

There is an additional complication. Tim is the youngest of two children. His older sister is diagnosed as having brain damage. She is extremely demanding of the parents' time as she displays what has been termed "bizarre" behaviour at home. School staff suspect that Tim is modelling at school behaviour similar to that being displayed by his sister at home. She receives considerable attention at home and it is believed that Tim is trying to gain a similar degree of attention at school.

Step 2: Preliminary hypotheses

1. Tim is displaying interfering behaviour in a wide variety of school contexts.
2. Tim's behaviour worsens in situations where he is required to read or in other situations where he is unhappy.
3. The degree, amount and frequency of Tim's behaviour may be related to the amount of attention Tim is receiving at home.

Step 3: Assessment of the individual

Observation, interviews and individual testing were used to determine levels of behaviour and the degree of reading difficulty.

Tim was observed in three situations. First, he was observed in the regular classroom. He averages just over one off-task behaviour per minute. These behaviours include singing, whistling, verbalizations ("I can't do this." "Hey, you." "This is hard!") and other miscellaneous noises; and moving out of seat and clowning to get attention (making faces and throwing a book up in the air are two examples). On occasion—approximately 10 minutes out of a possible 30—Tim appears to be attempting to work.

Second, Tim was observed on the playground at recess. Tim approaches another child who runs away from him. Tim calls for the boy to stop. He then chases and catches this second child, throwing him against the fence, asking why the second boy didn't stop. This action is greeted with cries of "Let me go" by the second child and threats of telling the teacher. Tim releases this boy saying that he hates him. Next, Tim pushes his way into a group of boys who are playing together. He elbows one of the boys, swears at them all, and then walks away without looking back. After a verbal exchange, Tim hits one of the boys in this group; follows another boy and pushes him to the ground; and runs away swearing, returning to the fence from where he had started. During this ten-minute period, he had displayed 11 behaviours that were deemed inappropriate.

For the third observation, Tim was watched while he was in a resource room. He is receiving extra assistance in reading by himself due to his frequent outcries. As the session begins, Tim is reporting that he does not want to be in the resource room; rather, he wants to be in the regular classroom where his friends are. Next he states he hates reading, hates being in the resource room. Finally, he starts to cry. After a few minutes, he reveals that the reason he feels upset is that he has extra work to do as a result of coming to the resource room. Next, he directs his anger at the resource teacher, slamming down his pencil, all the while reaffirming his dislikes to the teacher. Finally, when asked if he is ready to read, he responds with "Shit! I guess I have to now" and then begins to attempt to read. This exchange occurred during the first six minutes of the working session.

Next Tim was interviewed. He reported that he enjoys playing sports, especially baseball and soccer. His favourite colour is blue and he loves video games. Reading is fun when comics are being read. Math is just "Okay." Further, Tim states that school is boring, that the work is too hard since grade two is for older students. Lastly, he suggests that if he were placed in grade one his behaviour would be appropriate and he would get his work done. He closed with the following summary: "I don't know how I passed grade one because I always walked around and didn't do any work."

An analysis of his daily work samples over a period of two weeks revealed that Tim appeared to have a good notion of what a sentence and a paragraph were; his ideas in stories followed logically; and his letter formation was appropriate for his age and grade. Tim did omit some words from sentences on occasion, made some spelling errors, and reversed some letters in a few words.

Finally, Tim was administered a teacher-prepared test. He was required to read some shorter passages at different levels and to answer questions. He was also required to spell some words in isolation and in the context of a dictated sentence. Tim had difficulty with reading materials that had been estimated to be about the grade one level of difficulty by a readability analysis. Spelling skills paralleled his reading skills.

Step 4: Reformulate hypotheses

1. Tim is engaging in behaviour unacceptable to the setting in which he is placed.

He perceives this behaviour to be related to his inability to succeed.

2. Tim reads materials at about the grade one level successfully. Reading is accompanied by protestation and increased unacceptable behaviour. Tim's inability to read appears related to his poor word recognition.

Step 5: Assessment of instructional methods and materials

After a few trial lessons, it was discovered that Tim would try hard to behave appropriately if the material was reduced in difficulty and if he was given baseball cards for acceptable behaviour in class.

Step 6: Development of IEP goals

1. To decrease the amount of verbal hostility and hitting during peer interactions.
2. To increase the proportion of time spent completing class assignments.
3. To increase sight vocabulary and word recognition to a beginning grade three level.
4. To increase reading comprehension to a beginning grade three level.

Steps 7 and 8: Implementing and evaluating the IEP

The IEP goals were implemented over a period of eight months. During that time there were periods when Tim's behaviour worsened and periods when his behaviour was not a problem. The relationship between Tim's feelings about himself, his abilities and his family continued to play a role in progress. On three occasions Tim's actions resulted in crisis situations at school. Over the entire program, however, progress was made.

The first goal of decreasing hostile verbalizations and hitting during peer interactions was accomplished partially. As suggested above, Tim's behaviour continued to vary as a function of his feelings and personal experiences over the period that the IEP was in effect.

Two parallel programs were established. Tim's parents received "happy grams" when Tim had kept his contract of no hitting or swearing for the entire day. The accumulation of ten happy grams resulted in a family treat in which Tim was the centre of attention. Sometimes this treat was given without his sister. At school, Tim's love of sports was incorporated into his contract. Baseball cards were used as reinforcers for appropriate behaviour in the classroom and on the playground. It was, however, difficult to monitor playground behaviour. In the classroom, Tim's behaviour improved considerably, according to his classroom teacher. Periodic systematic behavioural observations (event recording) provided further documentation of decreases in the rate of verbal hostility.

Progress in the second goal was highly correlated with gains in the first one. As the amount of verbal and physical aggression towards peers decreased, a parallel program was introduced for "work time." Again, behaviour was recorded on a periodic basis, using duration recording. The classroom teacher monitored Tim's behaviour on a regular basis over the course of the eight month program. Tim received baseball cards and teacher praise for work accomplished. Gradually the frequency of giving cards was decreased. This decrease was matched with an increase in Tim's confidence in his ability to accomplish school tasks.

The third and fourth goals were worked toward at the same time. Word recognition and word attack were introduced through rhyme, and reading comprehension was taught by using the beginning exercises for logical and structural comprehension (see Chapter 6). By the end of the repeated grade two year, Tim was reading material at a mid-grade three level of difficulty.

5 SELF-CONCEPT AND AFFECTIVE EDUCATION

In providing assistance to students who have learning problems the focus recently has been on skill deficiencies within specific content areas. Feelings, when they have been taken into account at all, usually have been considered within the context of behaviour. That is, they are considered in terms of what a child or adolescent *does* as a result of feeling a particular way. In the last chapter we suggested that feelings also may be considered slightly differently. They may have an indirect effect on behaviour; e.g., feelings, such as uncertainty about how to solve a problem or fear of trying a task, may interfere with the probability of learning occurring.

There is, however, another point of view. Professionals are starting to emphasize the importance of the affective domain as a separate area of study. Kronick (1979, 1981) has pointed out that affective skills often are weak in students with learning problems. Bryan and Bryan (1981) provide evidence to suggest that affective aspects of learning problems should be an important independent area of study. Morse et al. (1980) state that planning for a child's or adolescent's emotional well-being should be given as much weight as planning for the attainment of both cognitive and physical goals.

Parents are also concerned that their children master affective and social skills. For example, the top 13 goals that parents have for schooling, as reported by Hardy (1981), are as follows: 1) responsiveness to learning, 2) creativity in learning and living, 3) basic knowledge and skills, 4) fitness and health, 5) artistic expressiveness, 6) a sense of self-worth, 7) understanding of the family in society, 8) practical life skills, 9) a sense of responsibility in society, 10) cultural understanding and tolerance, 11) good work attitudes and skills, 12) respect for the environment, and 13) a developed set of values. A large number of these goal statements either directly or indirectly refer to affective outcomes. For example, numbers one and six centred on feelings about learning or about self. In several of the other items, an affective dimension is implied.

Thus, both professionals and parents alike are beginning to agree that affective goals should be given more emphasis. Finally, looking at learning from the perspective of the child or adolescent, once again the same conclusion is reached. For example, a student who believes she or he cannot make sense of a subject may stop answering questions in class or stop trying to complete any assigned work. In more extreme cases, some students come to believe that they cannot learn anything related to the subject at all. Some students generalize this feeling of "I can't" to such an extent that they believe that there is something wrong with them. That is, they change their attributions from "I can't do (mathematics)" to "I am no good."

The way individuals feel about themselves the majority of the time has been termed "self-concept." When individuals generally are content with themselves as persons, they are referred to as having a positive self-concept. Conversely, when they feel badly about themselves the majority of the time they are referred to as having a negative self-concept.

We are concerned about the latter case, individuals who feel badly about themselves. More particularly, we are concerned with individuals who feel badly about themselves and who have learning problems. Sometimes this complication makes the solution easier because negative self-concept often is tied to learning problems. That is, for some individuals with learning problems, a negative self-concept can be overcome in the context of working on the subject area in which the failure occurs. Often this is the case for young children who may have difficulty differentiating subject specific learning problems from their concept of themselves as persons. Teaching them the content or skills that are troubling them, such as that numbers make sense or that the letter *b* is always the same letter, often will result in improvements in self-concept.

Sometimes, however, the solution is not found so readily. In some cases negative feelings about self have become so pervasive that they must be addressed separately. Such is the case more often with older students who have experienced failure for a long period of time. Problems with self image are not restricted to older students, however. There are cases where younger children, as early as kindergarten in our experience, have learned to believe that they are "no good."

With the above considerations in mind, we turn our attention in this chapter to the issue of the affective aspects of learning. Before proceeding further with this topic, however, two precautions must be stated.

First, there will be situations, such as in the case of personal crisis, in which the involvement of other professionals will be required. For example, there usually is a legal requirement to involve others in cases where neglect or physical or sexual abuse are suspected. In still other cases, for reasons of personal discomfort or lack of the necessary skills, referral to other professionals will be appropriate. In many other cases, judicious emotional support or offered assistance from the regular classroom or resource room teacher will be timely, welcomed and accepted by the individual with special needs.

Second, we suggest that cognitive interventions be attempted as the strategy of first choice if such interventions are necessary. In the context of working on academic problems children and adolescents often learn to trust. In such cases, they usually will begin to share or to display their lack of confidence in themselves. Given this first step, affective goals may be introduced either in the context of providing academic assistance or as a separate program if necessary.

In the remainder of this chapter we will discuss four topics. First, we present some hypotheses about factors that may affect self-concept. Second, we suggest methods of assessing self-concept. Third, we provide some suggestions for stating affective goals in behavioural terms. Finally, we suggest some techniques that may be used in working toward affective goals.

HYPOTHESES ABOUT SELF-CONCEPT

Several hypotheses may be advanced about immediate or longer-term influences on self-concept and how such influences may result in changes in behaviour. First, positive and negative feelings have been demonstrated to influence behaviour. Second, feelings may have an impact on the progress made towards a goal or may be used as an explanation when something does not work out as intended. Finally, cognition and affect (when examined from the point of view of a development theory) are both as important in the development of self.

The Role of Positive Feelings

Upon first consideration it would seem logical that positive feelings, as in when one is happy, should contribute to a positive self-concept. The relationship, however, may not be as straightforward as one first would imagine.

We start by suggesting how a positive self-concept usually develops. When children or adolescents face a variety of tasks, dilemmas and other situations that they perceive themselves able to resolve in a satisfactory manner they begin to believe in themselves and to attain a positive self-concept. The precise reasons children attain a positive self-concept are not clear. At least two factors, however, are important.

First, success must occur across several areas and it must occur more than once. Success in one area may not result in a positive self-concept. Children or adolescents may report that they are "pretty good at science" or "not bad at fixing cars" but they themselves are "not good." Similarly, if success only occurs a limited number of times it may be attributed to other factors, such as "I was lucky" or "It was easy."

Second, children or adolescents must believe that the success they achieve is due to their efforts. In working with students with special needs, we often find that they deny there own successes even when presented with evidence to the contrary. For example, one boy refused to believe that he could do math even when he was shown graphs of his improvements over the course of a few months. Instead, he noted points where his performance had fallen off, claiming these decreases represented his true abilities. Considerable effort was required to convince him that he could do math before he slowly began to believe in himself, that is, in his own ability. This pattern is very typical. The relationship between success and self-confidence for many students with learning problems may be summarized as follows: Confidence follows competence.

Next, we examine some hypotheses about the relationship between temporary positive feelings and the overall development of a positive self-concept. The connection between the two should be obvious, yet when this relationship is examined in the context of learning it is shown to be more complicated than one might think. Bruner, Goodnow and Austin (1956), for example, suggest that when a person feels good he or she avoids cognitive strain. Having positive feelings may lessen one's desire to try and therefore may influence what a child or adolescent does when presented with something to be learned.

Isen et al. (1981) suggest three variations on the cognitive strain theme. They completed a series of studies in which they explored the relationship between positive affect and learning. Their findings may be stated in the form of hypotheses as follows:

1. Positive feelings may play an important role in one's willingness to try. Their argument is that, in most cases, "feeling good" leads one to adopt the simplest possible strategy where more than one is available.
2. When feedback is given to the individual who is expressing positive feelings in the context of a problem-solving solution, the probability that a correct solution will be found is increased.
3. When a person is feeling happy and they are given an "important" task, their performance likely will be more cautious.

These hypotheses may be helpful in thinking about some students with learning problems. For example, selecting the simplest strategy is a commonly observed pattern among students who have learning problems. Sometimes such simple strategies are selected because more complex ones have not been learned. In other cases, however, if these hypotheses are valid, the selection of the simple strategy may have more to do with how individuals feel than what they know. In working with students who have learning problems, we have observed students modifying their selection of strategy because of their desire to avoid failure. It may be that in some cases avoidance occurs because such students want to avoid stress, not because they don't know how to solve the work in question.

In summary, a person who is feeling happy may attempt to avoid either difficult tasks or tasks likely to lead to failure and may therefore choose the simplest strategy. If the task is perceived as important and if feedback is given as the person solves the task, more taxing strategies may be adopted.

The Impact of Negative Feelings

A second and better described influence on self-concept and behaviour is negative affect. The impact of negative feelings, such as fear or anxiety, may be restricted to a specific context or may be generalized to all contexts. In some cases, these negative feelings become internalized and generalized, resulting in a negative self-concept.

Specific negative feelings may have a debilitating effect on school performance. For example, students may not hand in an assignment because they feel they will get a failing grade. In this case the anticipation of a poor evaluation is crucial. In other cases, worry may be associated with doing a task as opposed to anticipating the outcome of the performance. It is also possible that both concerns are operating at the same time. Test anxiety is a commonly reported example that is found among students with learning problems that may involve fear of failure or worry about doing the task, or both.

Leibert and Morris (1967) suggest that test anxiety has two components: worry and emotionality. Worry is that aspect of test anxiety that is cognitive; for example, the prediction of failure. Emotionality, on the other hand, refers to the physiological aspects that may be associated with an anxious event or time period, such as muscle tension. In a recent review, Morris, Davis and Hutchings (1981) suggest that worry is more strongly associated with poor performance than is emotionality. Further, Tryon (1980), in a review of the literature, noted that treatments directed towards worry were more effective in terms of anxiety reduction and increased grades than those directed towards emotionality.

Test anxiety may become generalized to such an extent that it affects more than just performance on tests. For example, Summers (1984) reports that test anxious intermediate aged children in comparison to others who were not test anxious had significantly lower academic self-concepts. In addition, these children attributed their failure to learn to lack of ability, and they tended to persist at assignments for a shorter period of time.

On a positive note, even when negative aspects of self-concept become generalized to a number of contexts, students may perceive that they are better in some areas than in others. For example, Winne, Woodlands and Wong (1982) compared the self-concept of students identified as learning disabled both with students identified as normally achieving and with those identified as gifted. As expected, they report that the self-concept of learning disabled students is lower in academic areas than that of students identified as normally achieving or as gifted. On the other hand, students identified as learning disabled as compared to those classified as gifted reported a higher self-concept on two scales "that might be crudely lumped together as involving contact with others" (Winne, Woodlands and Wong 1982, p. 474).

Further Hypotheses Concerning the Function of Feelings

Another hypothesis is that feelings do not appear to be important in interrupting behaviours or in goal setting; rather, such feelings may influence progress towards any goal (Scheier and Carver 1981). In other words, once a goal has been set, feelings appear to be an important influence in the speed with which individuals advance toward that goal. For example, an adolescent's goal is to learn to drive a car; however, he or she is afraid of taking the driving test. The speed with which he or she attempts to learn to drive may be affected by the fear of taking the driving examination. Similarly, if individuals feel extremely positive about striving towards a particular goal, the speed with which they press toward that goal likely will increase as a successful conclusion is approached.

Another way of thinking about the effects of feelings on behaviour is to see feelings as explanatory or, in other words, as the cause of something that has occurred (Dweck and Repucci 1973). For example, Pearl, Bryan and Donahue (1980) report that students identified as underachieving were more likely to attribute success to task ease than to effort. In another study, Pearl (1982) reported that children in the third and fourth grades identified as having learning disabilities attributed their failure more to luck than to lack of effort; however, these attributions were made only in academic tasks. Furthermore, such explanations appear to be more readily sought when an outcome is unexpected (Weiner 1981). Thus, a student who had been failing, for example, mathematics may attribute an unexpected success to luck.

This search for causal explanation has been referred to in various ways; however, one key distinction is the degree to which cause is attributed to things within the individual's control (internal focus) or to factors outside the individual's control (external focus).

The Developing Sense of Self

Finally, we may think of self-concept as changing as the child learns more about the surrounding world. There have been several different theories advanced as to how changes in self-concept occur over time. We are presenting only one of these explanations. In this theory, cognitive and affective development are considered together.

Kegan (1982) proposes a developmental theory of "evolving self" in which feeling and thought interact in the determination of self-concept. This process of self-determination is a lifelong process of "differentiating and integrating what is taken as self and what is taken as other" (Kegan 1982, p. 76).

An example may help us to understand this interplay of feeling and thinking and its impact on behaviour. Kegan (1982) relates an incident in which a four year old was complaining because he had not received two pieces of pastry as his older brother had. The mother first tried to explain logically that older children should receive more pastry because they are bigger; when the four year old was ten he too would receive more pastry. When this logical approach proved to be of no avail, the mother, in frustration, cut the smaller piece in half. Much to the surprise of both mother and older brother, the four year old was satisfied. He then felt he had the same amount as the older brother and stopped complaining.

Kegan's point is that children of different ages have different realities, different perceptions and different feelings about the same event. He would argue that four year olds react differently than do ten year olds because ten year olds have developed different ways of thinking. These differences in thinking, in turn, influence the perception of events and any subsequent affective reactions. In Kegan's example, the four year old child believed that the action of cutting the pastry changed it into the same amount as his older brother had. He could not differentiate size from amount, that is, he could not differentiate the object from his view of it. His affective reactions changed as a result of his new view of the pastry; he was no longer upset. Thus, in Kegan's terms the child's subjective view of the object (his subject-object perception) had changed although the object remained the same.

In addition to affective changes occurring due to immediate changes in perception, affective reactions vary over time as the child acquires different perceptions. When the four year old becomes ten, he also will see that one piece cut in half does not equal two larger pieces. His affective resolution will be different, he likely will want a fair solution; that is, he likely will want each person to have two equal sized pieces of pastry. The child learns to differentiate and integrate his or her perceptions of external influences with his or her internal representation of them. He or she gradually changes in the way in which the world is differentiated from self. It is through this process of learning about self-world relations that the child attains a sense of organization as a person in a social context.

According to Kegan, this process follows a series of cycles in which the self is defined and redefined. In Figure 5-1, the cycles of self-definition and redefinition are presented. Kegan argues that individuals move back and forth between a need for attachment and integration (what he terms psychologics favouring inclusion) and a need for independence and separation (what he terms psychologics favouring independence). In following these cycles an individual is thought to be striving for a sense of equilibrium between inclusion and independence. Kegan argues, however,

Figure 5-1 **Kegan's Cycles of Evolving Self**

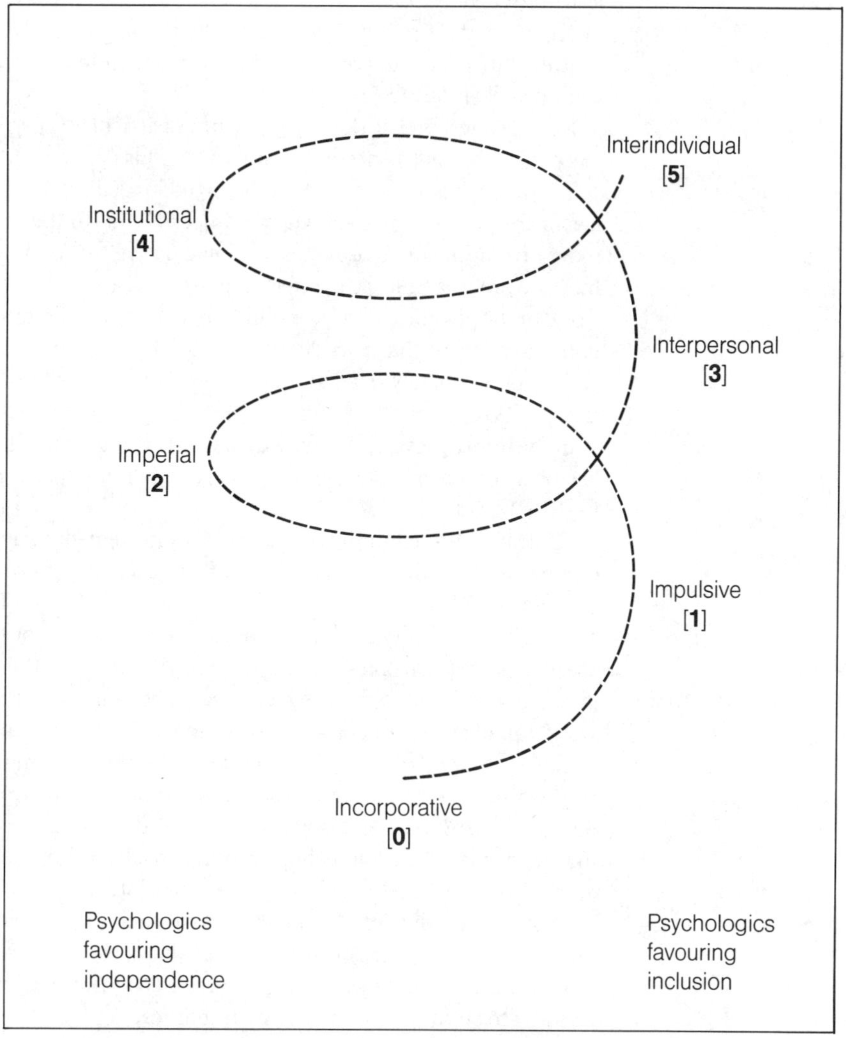

From Robert Kegan, *The evolving self: Problem and process in human development* (Cambridge, Massachusetts: Harvard University Press, 1982), 109. Reprinted with the permission of Harvard University Press. Copyright 1982 by the President and Fellows of Harvard College.

that no person ever completely succeeds in reaching a balance between the need for independence and the need for attachment. There always is a slight imbalance in our ongoing efforts to find this equilibrium. This imbalance is represented in Figure 5-1 as unevenness in the diagram of the spiral.

The cycles or balances that individuals attempt to find are termed "truces." Kegan argues that there are six cycles that occur from childhood to adulthood. These cycles are summarized in Table 5-1.

Kegan uses two other terms in Table 5-1 that may require explanation. Psychological embeddedness refers to the individual's primary sense of self during any period; in other words, it refers to what typically is going on inside the individual's mind. Cultural embeddedness refers to the environment in which the child, adolescent or adult usually is found.

Kegan suggests that the truces occur at particular (approximate) ages; however, learners with specials needs may not follow this pattern.

Given the age groups that are being addressed in this book, the most relevant cycles are the impulsive self and the imperial self. In the impulsive cycle, the child moves from imaginative fantasy to imaginative reality. In other words, he or she moves from imagining the impossible from within a private world to imagining a world of the possible. The older child, about aged six, discovers that a word is a word, no matter where that particular word is found. In the same way, a slightly older child, approximately aged seven, moves from a sense of "I am" to "what I am" (Kegan 1982).

In the imperial cycle, the individual is struggling with self-esteem, self-display and personal competence. Depending on age, different aspects of the imperial cycle are likely to be at the forefront.

Central to the beginning of this cycle is a movement away from placing parents in a primary role. This diminished parental role also means that the child's egocentric behaviour will be tolerated less at school than at home. The child is learning to assume responsibility for himself or herself in a social context. In addition, Kegan speculates that learning to participate in the world of the school provides a good developmental match for the elementary aged child. The child loves and wants a world full of rules, ritual and order, which the elementary school provides.

Toward the end of the imperial cycle, as the child becomes an adolescent, another shift occurs. Early, the emphasis was on moving away from parents and towards independence. Now there is a culture expectation that influences the newly acquired independence. The adolescent is expected to take into account the feelings of others even while pursuing his or her own self-interest.

In summary, during the imperial cycle there is a movement away from being self-centred and towards learning to consider others in a cultural context. In addition, however, there is a continued emphasis on learning to be independent and on learning to be confident in one's own abilities.

Summary

Affect, then, may be considered in several different ways. These range from focusing on current feelings about self to examining the changes in self-concept experienced over the course of development.

In looking ahead to the next issue, it should be pointed out that hypotheses about self-concept are important to two phases of assessment. First, what is hypothesized about affect will influence what is measured. For example, if the focus is on attribution, measures of internal-external control are likely to be selected; that is, we are likely to look for what we have hypothesized to be influencing the individual's behaviour. These guesses, of course, still are subject to rejection, revision or acceptance during the data collection phase. Second, hypotheses held about affect will influence

Table 5-1 **Kegan's Affective Stages**

Evolutionary balance and psychological embeddedness	**Culture of embeddedness**	**Function 1: Confirmation (holding on)**	**Function 2: Contradiction (letting go)**	**Function 3: Continuity (staying put for reintegration)**	**Some common natural transitional "subject-objects" (bridges)**
(0) INCORPORATIVE Embedded in reflexes, sensing and moving.	Mothering one(s) or primary caretaker(s). *Mothering culture.*	Literal holding: close physical presence, comfort and protecting. Eye contact. Recognizing the infant. Dependence upon and merger with oneself.	Recognizes and promotes toddler's emergence from embeddedness. Does not meet child's every need, stops nursing, reduces carrying, acknowledges displays of independence and willful refusal.	Permits self to become part of bigger culture, i.e., the family. High risk: prolonged separation from infant during transition period (6 mos. — 2 yrs.).	Medium of 0-1 transition: *blankie, teddy,* etc. A soft, comforting nurturant representative of undifferentiated subjectivity, at once evoking that state and "objectifying" it.
(1) IMPULSIVE Embedded in: impulse and perception.	Typically, the family triangle. *Parenting culture.*	Acknowledges and cultures exercises of fantasy, intense attachments and rivalries.	Recognizes and promotes child's emergence from egocentric embeddedness in fantasy and impulse. Holds child responsible for his or her feelings, excludes from marriage, from parents' bed, from home during school day, recognizes child's self-sufficiency and asserts own "other sufficiency."	Couple permits itself to become part of bigger culture, including school and peer relations. High risk: dissolution of marriage or family unit during transition period (roughly 5-7 yrs.)	Medium of 1-2 transition: *imaginary friend.* A repository for impulses which before *were* me, and which eventually will be part of me, but here a little of each. E.g., only I can see it, but it is not me.

Evolutionary balance and psychological embeddedness	Culture of embeddedness	Function 1: Confirmation (holding on)	Function 2: Contradiction (letting go)	Function 3: Continuity (staying put for reintegration)	Some common natural transitional "subject-objects" (bridges)
(2) IMPERIAL Embedded in: enduring disposition, needs, interests, wishes.	Role *recognizing culture*. School and family as institutions of authority and role differentiation. Peer gang which requires role taking.	Acknowledges and cultures displays of self-sufficiency, competence and role differentiation.	Recognizes and promotes pre-adolescent's (or adolescent's) emergence from embeddedness in self-sufficiency. Denies the validity of only taking one's own interests into account, demands mutuality, that the person hold up his/her end of relationship. Expects trustworthiness.	Family and school permit themselves to become secondary to relationships of shared internal experiences. High risk: family relocation during transition period (roughly early adolescence, 12-16).	Medium of 2-3 transition: *chum*. Another who is identical to me and real but whose needs and self-system are exactly like needs which before *were* me, eventually a part of me, but now something between.
(3) INTER-PERSONAL Embedded in: mutuality, interpersonal concordance.	Mutually reciprocal one-to-one relationships. *Culture of mutuality.*	Acknowledges and cultures capacity for collaborative self-sacrifice in mutually attuned interpersonal relationships. Orients to internal state, shared subjective experience, "feelings," mood.	Recognizes and promotes late adolescent's or adult's emergence from embeddedness in interpersonalism. Person or context that will not be fused with but still seeks, and is interested in, association. Demands the person assume responsibility for own	Interpersonal patterns permit relationship to be relativized or placed in bigger context of ideology and psychological self-definition. High risk: interpersonal partners leave at very time one is emerging from embeddedness. (No easily supplied age norms.)	Medium for 3-4 transition: *going away to college, a temporary job*.... Opportunities for provisional identity which both leave the interpersonalist context behind and preserve it, intact, for return; a time-limited participation in institutional life (e.g. 4 years of college).

Evolutionary balance and psychological embeddedness	Culture of embeddedness	Function 1: Confirmation (holding on)	Function 2: Contradiction (letting go)	Function 3: Continuity (staying put for reintegration)	Some common natural transitional "subject-objects" (bridges)
			initiatives and preferences. Asserts the other's independence.		
(4) INSTITUTIONAL Embedded in: personal autonomy, self-system identity.	*Culture of identity or self-authority* (in love or work). Typically: group involvement in career, admission to public arena.	Acknowledges and cultures capacity for independence; self-definition; assumption of authority; exercise of personal enhancement, ambition or achievement; "career" rather than "job," "life partner" rather than "helpmate," etc.	Recognizes and promotes adult's emergence from embeddedness in independent self-definition. Will not accept mediated, nonintimate, form-subordinated relationship.	Ideological forms permit themselves to be relativized on behalf of the play between forms. High risk: ideological supports vanish (e.g., job loss) at very time one is separating from this embeddedness. (No easily supplied age norms.)	Medium of 4-5 transition: *ideological self-surrender (religious or political); love affairs protected by unavailability of the partner*. At once a surrender of the identification with the form while preserving the form.
(5) INTER-INDIVIDUAL Embedded in: interpenetration of systems.	*Culture of intimacy* (in domain of love and work). Typically: genuinely adult love relationship.	Acknowledges and cultures capacity for interdependence, for self-surrender and intimacy, for interdependent self-definition.			

how test results are interpreted. For example, if the developmental viewpoint suggested by Kegan is accepted, what is attributed to "inappropriate behaviour" may change depending on the age of the person being examined. Looking further ahead, it should also be noted that hypotheses held about affect will be important when intervention techniques are selected.

ASSESSMENT OF AFFECT

There are two points that should be taken into consideration when deciding to attempt to assess affect. First, feelings about self are not readily translated into measurable outcomes. Second, the correlation between achievement and feelings about self may vary with grade level and age. Each of these points is discussed briefly below.

Regardless of the technique chosen, it is difficult to measure an individual's feelings about self because a considerable amount of inference is required. The nature of the inference is a function of the measure employed. In general, one of two assumptions is made when one is attempting to measure various aspects of a student's self-concept. Either it is assumed that students accurately describe their feelings about self on paper and pencil measures or it is assumed that observers accurately characterize students' feelings about themselves. No matter which type of measure is selected, there will be some degree of measurement error in the obtained scores due to the necessity of inference and the difficulties associated with quantifying measures relating to the self.

Even when a measure of the child's self-concept has been obtained, we still have to take into account that self-concept has a greater or lesser influence on behaviour at various times during development. This is not to say that feelings about self do not have an impact on what individuals do; rather, there is evidence to suggest that the degree of this influence may vary over time. For example, in a recent review of studies of self-concept, Byrne (1984) suggests the correlation between self-concept and achievement may be strongest at about the grade six level. Scheirer and Kraut (1979) set that grade level at seven, one higher than Byrne. Hansford and Hattie (1982), in a more general review related to self-esteem measures, suggest that the correlation between self-esteem and achievement changes with age (rather than grade levels) and that the correlation gradually increases between the ages of 9 and 15. After that age, the correlation decreases to a lower level through to university (Hansford and Hattie 1982).

In the next section, we will describe some of the techniques employed to measure feelings about self. Following that, we will describe the behavioural Q-sort—a special case in that it is designed to allow goals in self-concept change to be specified and progress toward those goals to be measured and validated. This feature is useful in designing an IEP and in measuring affective change.

Strategies to Measure Affective Variables

As in Chapter 4, two variations on methods to assess feelings about self are presented: teacher-prepared techniques and commercially prepared techniques.

Teacher-prepared techniques The teacher has basically the same choices in designing techniques to assess self-concept as he or she did in the last chapter. The only variations are the inferences made and, in some cases, the content of the prepared items. Thus the strategies described in Chapter 4 generally are appropriate techniques for estimating feelings about self.

The change in the wording of items is very important in allowing inferences to be made about the student's self-concept as opposed to the student's behaviour. In interviewing, for example, questions should not be about the behaviour of the individual; rather, attention should be focused on the feelings of that individual. Some of the items suggested in Chapter 4 already were designed to call attention to feelings about self. Questions concerning behaviour and others concerning feelings were mixed together because, in our opinion, it is better to assess behaviour- and feelings-statements at the same time. The only additional requirement is that approximately three to five behavioural items will be needed and another three to five feelings-statement items.

Commercially prepared assessment techniques As is the case with teacher-prepared assessment techniques, there is considerable overlap between instruments available to measure behaviour and those available to measure feelings-statements. Again, the instruments reviewed in Chapter 4 contain some items that may be more appropriate as indicators of feelings-statements than of actual behaviour.

Table 5-2 lists some of the instruments that are available as measures of feelings about self. It will be noted that none of these instruments has been normed in Canada. To what extent children and adolescents in different countries vary in how they feel is unknown. However, these types of measures have been used widely in several English-speaking countries. It is left to the reader to judge whether the measures listed or any other measures of this type are valid in particular testing circumstances.

The Behavioural Q-sort The Behavioural Q-sort (Kroth 1973) is one technique that may be used to compare the expectations of teachers, parents and the students themselves. As will be explained in detail later in this section, it is possible for an examiner to compare the individual's view of his or her "real" self to his or her "ideal" self. This comparison is made by having the child or adolescent rate himself or herself by responding twice to items written in behavioural terms. Kroth argues that the advantage of writing Q-sort items in behavioural terms is that the individual's progress towards a set of goals, the ideal self, can be measured. Further, an individual's perception of his or her behaviour can be validated by having a second person collect data to determine the frequency of the behaviour in question. Thus, in using this tool in conjunction with observation, data may be collected on both feelings about self and behaviour and, to some extent, that data can be verified. In this way the interaction between feelings about self, both real and ideal, and behaviour may be charted.

In completing the Behavioural Q-sort, individuals are presented with items like the ones in Figure 5-2. The items listed in this figure are more appropriate for children in the elementary grades. They would need to be modified for older students and may need to be totally rewritten for others, such as parents.

Individuals are required to decide how well they think an item represents how they might behave. They are required to rate this feeling in two ways using the form board

Table 5-2 **Inventories Used to Assess Affect**

Measure	**Age/Grade**	**Reliability**		**Format**	**Domain**
		Internal Consistency	**Test-retest**		
***Animal Crackers* (Adkins and Balif)	Preschool-grade 1	0.90			General
**Children's Locus of Control* (Bialer-Cromwell)	Grades 1-6	-0.01 to 0.87	0.22 to 84	Agree-disagree	General
**Children's Nowicki-Strickland Internal Control Scale*	Ages 9-18	0.24 to 0.63	0.63 to 0.71	Agree-disagree	General
**Children's Picture Test of Internal-External Control* (Battle)	Elementary school aged			Open-ended	General
***Coopersmith Self-Esteem Inventory*	8-10 years		0.78 to 0.88		General
**Intellectual Achievement Responsibility Questionnaire* (IAR)	Grades 3-12	0.54 to 0.60	0.47 to 0.74	Choice of attribution	School
**Internal-External Locus of Control Scale* (Rotter)	Adolescents & adults	0.65 to 0.79	0.49 to 0.83	Agree-disagree	General
***Piers-Harris Children's Self-Concept Scale*	Grades 3-12	0.79 to 0.93	0.72	Agree-disagree	General

* Reliabilities are those reported in studies reviewed by Stipek and Weisz (1981).

** Reliabilities are those reported in test manuals.

Adapted, with permission, from D.J. Stipek and J.R. Weisz, "Perceived Personal Control and Academic Achievement," *Review of Educational Research* 51 (1981): 104-105. Copyright 1981 by the American Educational Research Association, Washington, D.C.

given in Figure 5-3. On the first occasion, the individual reads the items and rates each one in the context of how he or she behaves at the present time. On the second reading of the items, ratings are assigned in terms of how he or she wishes to behave in the future.

In scoring the Behavioural Q-sort, the individual's responses are entered in the record form given in Figure 5-4. Responses from the first rating, the "real" self, are entered in Column S-1. Responses from the second rating, the "ideal" self, are entered in Column S-2. Differences between the two columns are calculated and that score is squared. The greater the agreement between the two columns the better the match between how the individual behaves now and how he or she wishes to behave.

Kroth (1973) offers two helpful suggestions in interpreting and using the Behavioural Q-sort. First, any squared difference equal to or greater than 16 is a potential area of need. Thus, if a child rates himself or herself as 8 on "As I am now" and 2 on "As I would like to be" when answering about smiling, there would be a squared discrepancy of 36. Kroth would suggest that "smiling" could be selected as an appropriate goal for an Individual Educational Program. Kroth warns the reader that just because there is a discrepancy does not mean that the individual wants to change. Second, Kroth points out that a low score does not mean necessarily that the individual has a poor self-concept. Completing the Behavioural Q-sort procedure at a later time, however, will provide an indication of change in agreement or disagreement between perceived "real" and "ideal" self.

WRITING BEHAVIOURAL AFFECTIVE OBJECTIVES

Once data has been collected and the decision has been made to include feelings about self in an IEP, the next step is to set a series of objectives. At this point feelings about self have been translated into either scores or descriptions. These data are incorporated into the IEP as the present levels of performance. Projecting forward from this data, annual goals are set. If the Behavioural Q-sort is the basis of the data, both present levels of performance and an annual goal are implied. In some cases, however, other more general goals may be desirable, either in addition to or instead of the ones derived from Behavioural Q-sorts. For such circumstances, Krathwohl, Bloom and Masia (1964) have defined five categories of affective educational goals that may be useful in preparing IEPs. Descriptions of these categories of affective goals are given in Table 5-3. Gronlund (1985) composed a list of instructional objectives and behavioural terms that may be helpful in writing IEPs; these objectives and terms are given in Table 5-4.

INTERVENTIONS FOR AFFECTIVE GOALS

We begin our discussion of interventions for improving self-concept by discussing three general issues. Next, we describe a two-stage model of intervention. Third, we present some of the more specialized techniques that are being suggested as affective interventions, such as peer counselling and bibliotherapy.

Table 5-3 **Major Categories in the Affective Domain of the Taxonomy of Educational Objectives**

1. **Receiving.** Receiving refers to the student's willingness to attend to particular phenomena or stimuli (classroom activities, textbook, music, etc.). From a teaching stand-point, it is concerned with getting, holding and directing the student's attention. Learning outcomes in this area range from the simple awareness that a thing exists to selective attention on the part of the learner. Receiving represents the lowest level of learning outcomes in the affective domain.

2. **Responding.** Responding refers to active participation on the part of the student. At this level he not only attends to a particular phenomenon but also reacts to it in some way. Learning outcomes in this area may emphasize acquiescence in responding (reads assigned material), willingness to respond (voluntarily reads beyond assignment), or satisfaction in responding (reads for pleasure or enjoyment). The higher levels of this category include those instructional objectives that are commonly classified under "interest"; that is, those that stress the seeking out and enjoyment of particular activities.

3. **Valuing.** Valuing is concerned with the worth or value a student attaches to a particular object, phenomenon, or behaviour. This ranges in degree from the more simple acceptance of a value (desires to improve group skills) to the more complex level of commitment (assumes responsibility for the effective functioning of the group). Valuing is based on the internalization of a set of specified values, but clues to these values are expressed in the student's overt behaviour. Learning outcomes in this area are concerned with behaviour that is consistent and stable enough to make the value clearly identifiable. Instructional objectives that are commonly classified under "attitudes" and "appreciation" would fall into this category.

4. **Organization.** Organization is concerned with bringing together different values, resolving conflicts between them, and beginning the building of an internally consistent value system. Thus the emphasis is on comparing, relating and synthesizing values. Learning outcomes may be concerned with the conceptualization of a value (recognizes the responsibility of each individual for improving human relations) or with the organization of a value system (develops a vocational plan that satisfies his need for both economic security and social service). Instructional objectives relating to the development of a philosophy of life would fall into this category.

5. **Characterization by a value or value complex.** At this level of the affective domain, the individual has a value system that has controlled his behaviour for a sufficiently long time for him to have developed a characteristic "life style." Thus the behaviour is pervasive, consistent and predictable. Learning outcomes at this level cover a broad range of activities, but the major emphasis is on the fact that the behaviour is typical or characteristic of the student. Instructional objectives that are concerned with the student's general patterns of adjustment (personal, social, emotional) would be appropriate here.

Reprinted, by permission, from David R. Krathwohl, B.S. Bloom and B.B. Masia, *Taxonomy of Educational Objectives. Handbook II: Affective domain* (New York: Longman Inc., 1964)

Table 5-4 **Examples of General Instructional Objectives and Behavioural Terms for the Affective Domain of the Taxonomy of Educational Objectives**

Illustrative general instructional objectives	Illustrative behavioural terms for stating specific learning outcomes
Listens attentively Shows awareness of the importance of learning Shows sensitivity of human needs and social problems Accepts differences of race and culture Attends closely to the classroom activities	Asks, chooses, describes, follows, gives, holds, identifies, locates, names, points to, selects, sits erect, replies, uses
Completes assigned homework Obeys school rules Participates in class discussion Completes laboratory work Volunteers for special tasks Shows interest in subject Enjoys helping others	Answers, assists, complies, conforms, discusses, greets, helps, labels, performs, practices, presents, reads, recites, reports, selects, tells, writes
Demonstrates belief in the democratic process Appreciates good literature (art or music) Appreciates the role of science (or other subjects) in everyday life Shows concern for the welfare of others Demonstrates problem-solving attitude Demonstrates commitment to social improvement	Completes, describes, differentiates, explains, follows, forms, initiates, invites, joins, justifies, proposes, reads, reports, selects, shares, studies, works
Recognizes the need for balance between freedom and responsibility in a democracy Recognizes the role of systematic planning in solving problems Accepts responsibility for his or her own strengths and limitations Formulates a life plan in harmony with his or her abilities, interests and beliefs	Adheres, alters, arranges, combines, compares, completes, defends, explains, generalizes, identifies, integrates, modifies, orders, organizes, prepares, relates, synthesizes
Displays safety consciousness Demonstrates self-reliance in working independently Practices cooperation in group activities Uses objective approach in problem solving Demonstrates industry, punctuality and self-discipline Maintains good health habits	Acts, discriminates, displays, influences, listens, modifies, performs, practices, proposes, qualifies, questions, revises, serves, solves, uses, verifies

Figure 5-2 **Sample Items on the Behavioural Q-Sort (Elementary Level)**

1. Gets work done on time.
2. Pokes or hits classmates.
3. Out of seat without permission.
4. Scores high in spelling.
5. Plays with objects while working.
6. Scores high in reading.
7. Disturbs neighbours by making noises.
8. Is quiet during class time.
9. Tips chair often.
10. Follows directions.
11. Smiles frequently.
12. Often taps foot, fingers or pencil.
13. Pays attention to work.
14. Works slowly.
15. Throws objects in class.
16. Reads well orally.
17. Talks to classmates often.
18. Scores high in English.
19. Talks out without permission.
20. Rocks in chair.
21. Scores high in arithmetic.
22. Asks teacher questions.
23. Uses free time to read or study.
24. Works until the job is finished.
25. Walks around room during study time.

From Roger Kroth, "The Behavioural Q-Sort as a Diagnostic Tool" in *Academic Therapy* 8 (1973): 320. Reprinted with the permission of the publisher.

Figure 5-3 **The Q-Sort Formboard**

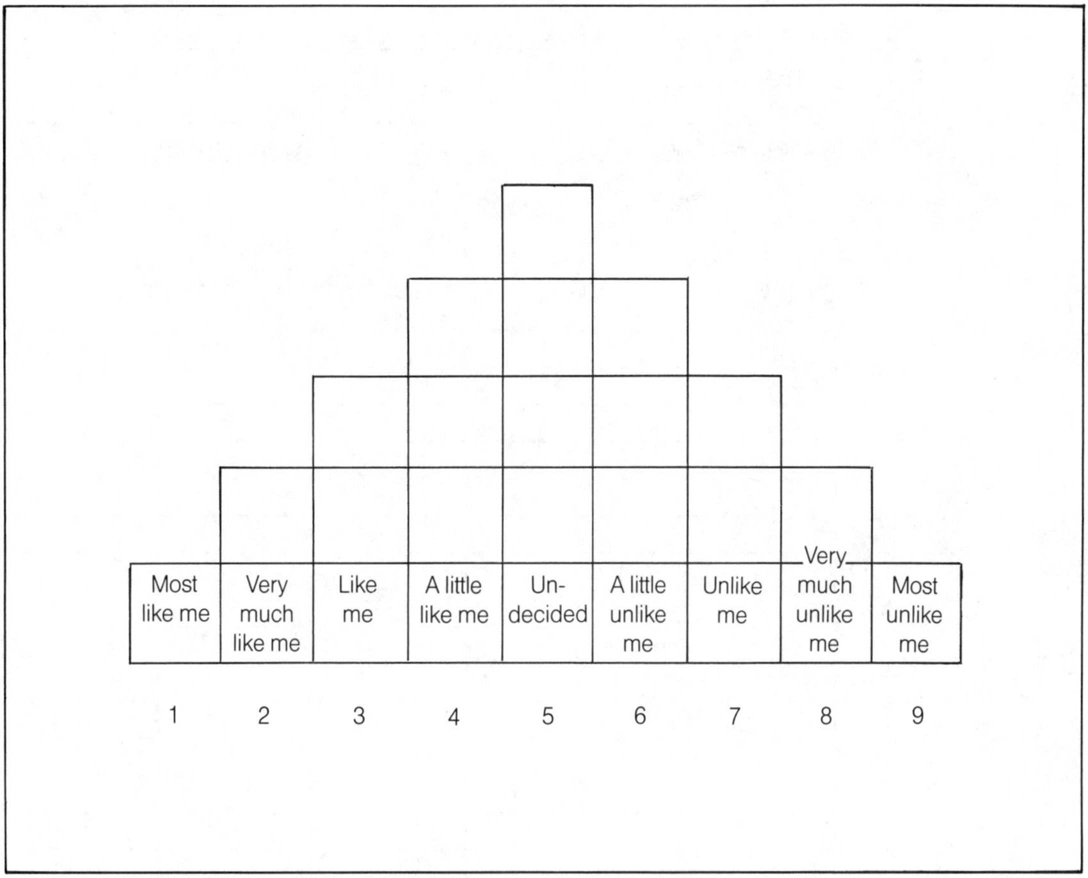

From Roger Kroth, "The Behavioural Q-Sort as a Diagnostic Tool," *Academic Therapy* 8 (1973).

General Issues

There are three areas of concern which should be considered before discussing specific interventions. These are 1) the prerequisites of intervention in the affective domain, 2) the degree of self-disclosure to strive for in such intervention, and 3) the role of value judgments in affective interventions.

It was suggested earlier that trust is one of the prerequisites of affective interventions. In attempting to establish trust, the person providing help must be very careful not to establish a dependence. Rather, the goal is that the person receiving help will feel comfortable and will want to assume responsibility for setting and meeting his or her affective goals.

The second general concern is the degree to which the person giving help should self-disclose. In trusting, both the person helping and the one receiving help become vulnerable to some extent; that is, both individuals share feelings with each other in the process of discussion. In part, such a sharing is necessary in order that trust be

Figure 5-4 **Q-Sort Record Form**

Name of Subject ______________________ Sex ________ Date Tested ______________

Address ______________________ Phone ______________ Date of Birth ______________

School ______________________ Teacher ____________ Grade ________ Age ______________

Name of Examinee ______________________ Relationship to Child ______________

$$r = 1 \frac{\sum D^2}{200}$$

Card No.	Column S-1	Column S-2	D	D^2
1				
2				
3				
4				
5				
6				
7				
8				
9				
10				
11				
12				
13				
14				
15				
16				
17				
18				
19				
20				
21				
22				
23				
24				
25				
			$\sum$ =	

From Roger Kroth, "The Behavioural Q-Sort as a Diagnostic Tool" *Academic Therapy* 8 (1973): 323.

established. A useful distinction in determining how much personal information to reveal in this situation is the difference between self-disclosure and personal-disclosure. The former is desirable in this context, while the latter is not. In self-disclosure, a helper might say something like "I watched the late night movie last night so as a result I'm tired." In personal disclosure, more detail, background or explanation is added. Thus, the above example would be elaborated on in some way, making it more revealing; for example, "I watched the late night movie last night since I couldn't sleep because I had a fight with my friend yesterday. As a result, I'm tired today."

The third general issue that should be considered is the role of judgments in interventions. In thinking about value judgments, it is important to make two separate yet interrelated distinctions between 1) the types of interventions and 2) the person asked to make the value judgment. In the case of behavioural interventions, the child or adolescent is the one, in our opinion, who should make a value judgment about his or her own behaviour. The professional (teacher, child care worker, social worker or other person in the helping professions), however, needs to judge when someone is behaving in a way that is harmful to himself or herself or to others. In the case of cognitive interventions, the professional will need to judge when a student needs assistance. Even in these cases, we suggest that professional judgment of need should be followed by having the student evaluate his or her own learning behaviour or, if possible, skill deficiency. This type of evaluation is inherent in diagnostic interviews. In the case of affective interventions, however, value judgments are to be avoided if possible. Feelings do need to be acknowledged; however, it is important not to confuse recognizing feelings with making value judgments about them. Thus, one might say something like "I see that you are happy," rather than "You have no right to be happy."

In summary, it is necessary to keep one very important difference between affective interventions and cognitive and behavioural interventions in mind—insofar as feelings about self are part of an IEP, value judgments are to be avoided; whereas value judgments are necessary in assisting students to reach behavioural or cognitive goals. In other respects, affective interventions are similar to cognitive and behavioural interventions in that a) trust is desirable, though dependence is to be avoided, and b) some degree of helper self-disclosure, though not personal-disclosure, is necessary.

A Two-cycle Model of Affective Intervention

In keeping with Kegan's model of affective development in which affect and cognition are considered together, we are suggesting a two-cycle model of affective intervention. In the first cycle, teachers or other professionals adopt strategies that are designed to improve the student's self-concept. In the second cycle, the teacher takes steps to integrate instructional goals in which students' feelings are the target with the new goal of differentiating personal feelings from those of others in the context of instruction.

Cycle One: Self-discovery In cycle one activities, the emphasis is placed on helping the child or adolescent develop self-esteem. Methods of improving self-esteem range from inviting participation of shy children to being tolerant of adolescents' fads. In general, guidelines for improving self-esteem vary to some extent with age and grade. The number of suggestions that have been offered as ways of influencing affect are too numerous to detail in this chapter. For example, Canfield and Wells (1976) list over 100 practical suggestions for enhancing self-concept. Included are suggestions in the following categories: 1) building an environment of positive support, 2) encouraging strengths in children and adolescents, 3) helping individuals to accept their bodies, and 4) the language of self.

We suggest that the following three general strategies will help to encourage the development of self-esteem. They are appropriate for both children and adolescents.

1. Create a fair climate in which students are fully informed of the classroom rules and, if feasible, have some say in making those rules. In this context, allow students to assume responsibility and to demonstrate that they can be trustworthy.
2. Attempt to create a climate that is psychologically and emotionally safe. For example, in striving for psychological safety, try to be consistent in the way in which students are treated in the classroom and avoid having "victims" and "favourites."
3. Try to establish a feeling of safety or comfort with regard to physical differences or handicaps. In attempting to establish a safe physical climate, try to allow for size differences in desk selection and seating position in the room. In addition, remember that some children or adolescents with special needs may not be as physically capable as other individuals. Try to avoid circumstances in which a student's awkwardness or physical handicap will be highlighted.

Cycle Two: Self-discovery in Relation to Others In cycle two, the emphasis changes slightly. Now importance is placed on both self-discovery and on integrating and differentiating self from others. Two additional strategies are appropriate in this cycle. First, more use is made of comparison and contrast; that is, students are encouraged to look for similarities and differences between themselves and others. Second, the focus of discussions is expanded to include learning about others within the context of the curriculum. For example, in *Developing Understanding of Self and Others* (Dinkmeyer 1970), children are provided with opportunities to explore the feelings of others in a series of lessons in which puppets are used as thematic characters. In another case, children and adolescents might be asked to think about a social studies theme in terms of how individuals in that circumstance might feel, and they could then be asked how such imagined feelings compare and contrast with their own experiences.

Thus, the difference between the two cycles is one of emphasis. They are not distinct programs. Put in perspective, we are suggesting that affective education for children with special needs may need to progress a little more gradually and in a more structured manner than for normally achieving students.

In both cycle one and cycle two activities, the objective is to improve the self-esteem of students with learning problems. Such training usually is limited to group interventions, although it does not have to be so restricted. In cases where individual interventions are required, other more specialized techniques may be used. Examples of such interventions are discussed in the next section of this chapter.

Other Affective Interventions

Weinstein and Fantini (1970) suggest that there are three broad categories of affective concerns. These are concern about 1) self-image, 2) disconnectedness and 3) control over one's life. The first concern is reflected in comments that state or imply "I don't like myself." The second concern, disconnectedness, is reflected in statements that reveal a loss of foundation or anchor; for example, "My parents can't be trusted." The last concern is often expressed as frustration or as helplessness; for example, "There is no hope for me, so leave me alone." To some degree, each person experiences these feelings. In such circumstances the counsel of a friend may be sought. Children and adolescents may use a similar tactic or they may seek the aid of a trusted adult. In some cases, however, the degree of alarm being raised by the child or adolescent will warrant more intensive assistance than an untrained person is able to provide.

In the paragraphs that follow, we briefly describe some strategies that may be incorporated into a special education or classroom context when students want help in areas related to self-concept. Eight techniques are described: peer counselling, puppetry, role playing, play therapy, music therapy, bibliotherapy, art therapy and movement program or dance therapy.

Peer Counselling Peer counselling is a strategy in which individuals are encouraged to help other individuals of approximately the same age group (Carr and Saunders 1979; Varenhorst 1978). It could be seen as a formalization of the process of friendship carried out by someone in a role similar to that of a paraprofessional.

The peer counsellor is trained in a variety of skills, such as interpersonal relations, decision-making and problem-solving (Carr 1983; Sussman 1983). He or she then interacts with others in both formal and informal settings to assist individuals in dealing with a variety of concerns. For example, this technique has proven effective in helping students with such problems as drug and alcohol abuse, loneliness and teenage pregnancies (Carr 1983). In addition, it has been used with a wide variety of age groups, ranging from elementary students (Gumaer 1973) to the aged (France and McDowell 1982; France and Gallagher 1984).

Use of Puppets and Masks Two techniques that are useful in helping children and adolescents work through issues in a protected environment are puppetry and the use of masks. With younger children especially—although the use of puppets is not restricted to younger children—puppets are very effective. For example, they may be used to allow children to "pretend" their way through a set of feelings, such as in cases of neglect or abuse. Puppets may be used extemporaneously or incorporated into lessons as appropriate, or both. In addition, puppets are very effective in role-playing situations (see below) with some children. Puppets are useful in these pretend and role-playing situations because children feel it is safe to talk when using them.

For children or adolescents who feel puppets are inappropriate for them, masks make an excellent substitute. Masks allow children and adolescents all of the same freedoms as puppets.

Neither puppets nor masks need to be made professionally. Part of the project may be to have children or adolescents create their own puppets or masks. Artistic ability is unimportant in such cases; rather, the emphasis is on the child's or adolescent's acceptance of the puppet or mask in expressing feelings.

Role-playing Some individuals may reject both puppets and masks as age inappropriate. In such cases, role-playng techniques may be used. In role-playing (Chesler and Fox 1966; Creekmore and Madin 1981), students make use of some form of play acting, usually to find alternative solutions to specific problems. Students usually are required to play specific roles, such as a class clown. An analogous situation is the kind of performance required of the character actor. An important difference, however, is that acting *per se* is de-emphasized in role-playing.

Role-playing usually consists of somewhere between four and nine steps (Creekmore and Madan 1981;Dergousoff 1984; Dreikurs, Grunwald and Pepper 1971;McIntyre 1974). The difference between the various models of role-playing is the degree to which the steps are specified. In each case, however, the basic procedure is the same. The following six-step procedure covers the main elements of role-playing:

1. Warm up time

This is a period lasting about two minutes in which students are given a formal "get ready" time.

2. Posing the problem

In this step a problem is posed to the students. It may be drawn from a variety of sources ranging from personal experience to materials found in novels (see bibliotherapy, below).

3. Brainstorming

In this step, potential solutions are suggested but not evaluated.

4. Selecting actors

Two to four actors are selected for the role-play. It is important not to typecast individuals in roles that they are "playing" in real life. That is, shy students should not play shy students.

5. The action (and stopping it)

The length of the role-play will vary depending on what problem is being presented. Action is stopped if one of the "actors" starts to have trouble continuing or if the "play" breaks down.

6. Discussion

In this step the role-play is discussed. We suggest that the discussion focus on two components: 1) the brainstorming session and 2) the role-play. Solutions from both steps may be compared and evaluated, and a series of action plans prepared (see Chapter 10, Creative Problem-Solving, for more details).

Play Therapy In play therapy children are encouraged to express their pent-up feelings through the use of toys (Schaefer 1979). It may take place anywhere that is private and where the therapist has control over the environment.

The toys used in play therapy should be selected carefully. For example, toys should not be chosen that would be difficult to manipulate. Some examples of appropriate toys are 1) sand, water, paints and clay, 2) cannons, soldiers, guns and punching bags and 3) doll or puppet families. The purpose in selecting any toy should be that through its use children can express how they feel.

There are two main types of play therapy: non-directive (Axline 1977) and directive (Burt and Myrick 1980). In non-directive play therapy the child directs the play. The child decides what to play with and how much time to spend, while the therapist clarifies or reflects the child's feelings. In directive therapy the play is highly structured and follows a developmental pattern. In this approach the referred child engages in different kinds of play in both a one to one situation (with an adult) and in group play (with other children).

Music Therapy Charlesworth (1982) points out that music has been referred to as "a therapeutic device in the Bible, Greek and Roman literature and Egyptian records" (p. 191). He further suggests that the effects of music are both physiological and mental.

When used in music therapy (Michel 1971), different types of music are selected and played depending on the mood or effect that is desired. The purpose of playing music is to help the child or adolescent to verbalize his or her feelings. The use of carefully selected music as a background for treatment sessions appears helpful in increasing the amount of verbalization and in facilitating creative story writing during group treatment sessions.

Soothing, classical, background music resulted in increased discussion when played quietly to groups of four to eight people, as compared to selecting quiet stimulating music or playing no music (Stratton and Zalanowski 1984). In addition, the use of music was more facilitating when the discussions were more focused rather than more free-flowing. Finally, there was some indication that background music may be effective only until individuals in the treatment groups become more comfortable with one another.

Froehlich (1984) also conducted a study in which the effects of music therapy were examined. Children selected were patients in a hospital and ranged in age from 5 to 12 years. The amount verbalization that occurred when music was added to play therapy was compared to the amount of verbalization that occurred in medical play therapy. In the music therapy condition, original lyrics were used to explore hospitalization and illness. In the medical play therapy condition, pictures were used to prompt discussion.

Wells and Stevens (1984) used music as a stimulus to encourage young inpatient adolescents to write stories. Specific musical themes were selected, such as Count Basie's "Prime Time" and Jean-Luc Ponty's "Egocentric Molecules," so that the nature of the feelings expressed could be examined. Wells and Stevens claim that the advantage of using music with young adolescents is that they are less likely to be rejected by their peers for writing emotional statements than in other forms of therapy.

Bibliotherapy Bibliotherapy is a technique which encourages individuals to read about specific resolutions to personal problems in literature. In this way it is hoped that readers will learn strategies to solve their own problems or will start to understand problems their friends are experiencing.

Rather than listing books that may be used for bibliotherapy, we are suggesting two source books which will be of great assistance to anyone wishing to use this technique. Both sources provide extensive listings of books, are recent, and will be helpful when selecting reading material for particular situations.

Dreyer (1981) has prepared two volumes in which over 400 psychological, behavioural and developmental topics of concern to children and young adolescents are described. In the first volume, books prior to 1975 are included and in the second one, books published between 1975 and 1978 are listed. Materials are indexed by topic, author and title.

Tway (1981) has prepared a cumulative annotated list of books which is now in its sixth edition. Tway only includes material published since the last edition in each volume. Books are categorized under age groupings and cross-referenced under categories. The introduction includes a statement of purpose and selection criteria for the material included.

Other authors (e.g., Jaskoski 1984; Lippin 1982) are suggesting the use of poetry instead of novels. Poetry usually is selected by the therapist. However, at least one therapist (Bell 1982) writes his own poems for use with clients. The key criterion is that the focus is on the person rather than the poem itself. Poems, like novels, are selected to reflect the personal issue in question.

In addition, the form and length of poetry are such that children and adolescents can be asked to write poems instead of read them. In this way the individual engages in a form of self-expression, much as in art therapy.

Art Therapy In art therapy (Feder and Feder 1981;Kramer 1958, 1971), painting and drawing are used as manipulative vehicles with which to explore themes that may be difficult for some people to verbalize. Often such themes are drawn from intense emotions directed towards others, such as love or hate. At other times these themes are drawn from more inward concerns such as self-image.

Erikson (1979) argues that the main advantage of art in therapy is the medium itself. The neutrality of the materials makes them useful. "Materials don't get upset, angry, or indignant.... They neither encourage nor discourage your efforts; they do not promise one thing, then change their minds" (Erikson 1979, p. 78).

Art therapy has been used in a number of different ways. For example, Crowl (1980) reports a case study in which this technique was used successfully in relieving a nine year old boy's fear of surgery. More generally, Stern (1952) and Ulman (1975) report that undergoing art therapy has a positive effect on self-esteem.

Movement Programs or Dance Therapy The use of movement in a systematic way has been referred to both as movement programs and dance therapy (Docherty 1982; Feder and Feder 1981). There is no one philosophy surrounding the use of movement or dance therapy, and as a result there are many variations in programs as implemented. Programs range from scripted ones to ones in which great individual freedom among students' movements are encouraged.

Although there is no one prescribed program, there is a fairly common pattern in the types of activities. According to Wiener and Lidstone (1969), creative movement or dance therapy is unlike dancing in that there are no steps or patterns to follow. Rather, students are given time to 1) warm-up, 2) explore, expand and repeat an identified movement and 3) cool down (Owens 1981). Boorman (1969, 1971) provides two practical guides for teaching movement in the first six grades. Docherty (1975) is another practical source of lesson plans.

Both children and adolescents may benefit from the opportunity to engage in a movement program or in dance therapy. Kraft (1978) and Schurr (1975) stress that the use of movement programs can make a significant impact on self-concept. Kraft, for example, reports that children exposed to movement programs were more capable of making positive feeling statements about themselves and about their school than were others who were deprived of this type of program.

Summary and a Note of Caution Eight techniques for influencing self-esteem and dealing with other personal problems have been described. None of the eight should be implemented by other than a qualified, well-trained individual.

In addition to the techniques described in this section there are other similar strategies available. For example, a relatively new technique is to make use of photography (Cosden and Reynolds 1982). Also, several of the described therapies have been combined with promising results; for example, art therapy has been combined with movement (Grodner et al. 1982; Moore 1983). In still other cases, these techniques have been used in concert with non-affective interventions; for example, peer counsellors have been used as peer tutors (Sussman 1983).

With each of these techniques, with the possible exception of peer counselling, there has been some debate about how they are to be used. There are two main suggested strategies. The first is based on analytic interpretation; that is, the art or music, etc., seen as a vehicle for symbolic content. The second strategy emphasizes the process of the technique itself. It is assumed that the process of drawing, for example, is helpful in improving self-esteem. The latter, while more conservative, appears to us to require fewer inferences in its use than the former.

Regardless of the technique selected, considerable caution must be used in implementing an affective intervention. To start, affective interventions with children or adolescents with special needs may require extra planning and likely need to be more structured than would be the case for normally achieving children or adolescents. In addition, affective interventions for students with learning problems likely will need to be conducted more slowly than those with normally achieving students. Usually it is necessary to move slowly as these students typically have lost some degree of trust for adults.

Finally, it must be remembered that monitoring the effects of an affective intervention is more difficult than monitoring those of an academic or a behavioural intervention. In the latter cases, there are observable signs of progress, either in skill improvement or further knowledge in the first case or in changed behaviour in the second case. With affective programs, monitoring progress relies on either student self-report or on second-person observation. As we stated earlier, there is a good chance that such measurements will be inaccurate. Even if reasonably accurate, inferences still are necessary to determine if desired outcomes are occurring.

6 READING: ASSESSMENT AND INSTRUCTION

In the earlier grades, the proportion of instructional time devoted to reading exceeds that of any other subject. While the ratio of instructional time decreases in later grades, reading is valued throughout school. Further, we continue to read throughout life. And even though the amount of time spent reading varies from person to person, reading, occupies some small part of the daily lives of almost everyone. For example, the ability to read is necessary to order dinner in a restaurant. Reading provides us with information as we scan the daily newspaper. In addition, the type and level of reading necessary to cope in society is increasing (Harris and Sipay 1980). For example, tax forms get increasingly more complicated to read. Perhaps because of the obvious importance and the frequently required use of this tool, reading has been considered the first "R" in education.

Students who do not read well, then, are a primary concern of parents and professionals alike. It would not be an exaggeration to suggest that children who cannot read well receive attention earlier and faster than children who display any other school-related academic problem. Yet, in spite of the importance attached to it, much confusion surrounds reading.

The process or processes by which we acquire reading skills remains an enigma (Bartel, Boyd and Tobia 1984; Durkin 1978; Vernon 1971). Some authors (e.g., Jenkins and Larson 1978) argue that reading is a bottom-up process; that is, reading is an automatic result of word recognition. Others (e.g., Smith 1978) suggest that reading is a top-down process; that is, reading is dependent on the knowledge brought by the reader to the text. The ability of poorer readers appears to vary according to three factors: a) how they feel about the materials, b) how familiar they are with the content of the material, and c) the difficulty of the text.

The practice of teaching reading is also surrounded by much debate (e.g., Swaby 1984). Currently there are a wide variety of assessment and instructional techniques from which to select. In part, the amount of choice available is a reflection of the variation in theory since the practical tools of reading assessment and instruction have been derived from the various perspectives on reading. It is from these tools that regular classroom teachers, resource teachers and other school-based personnel must choose when working with students with reading problems. It is our belief that following the decision-making model presented in Chapter 2 is useful in working with readers who need extra assistance.

THE DECISION-MAKING MODEL APPLIED TO READING

In applying the decision-making model to reading, the steps outlined in Chapter 2 are followed. First, the background information available on the student is reviewed. Second, preliminary hypotheses are formulated. Third, the individual is assessed. Fourth, tentative hypotheses are revised and reformulated as preliminary instructional goals. Fifth, materials and methods are tried out. Sixth, an IEP is written and implemented. Finally, the program is evaluated on an on-going basis.

As a first step, the student's historical and current reading performance records should be reviewed. However, in reviewing these records it is important to remember that the information they contain varies considerably. As a result, records may not contain helpful clues and may have to be abandoned in favour of, for example, classroom observation and/or curriculum analysis.

The main purpose in reviewing student records is to obtain hints both about the student and about what has been attempted in the past. First, records may be examined to find hints about an individual's strengths and weaknesses. What knowledge does he or she bring to reading? Is the student from another culture so that much of what he or she currently is being presented is unfamiliar? Is the language spoken at home different than the language of instruction? Are there hints contained about the parent's and the child's attitudes toward reading? Is there a record of large periods of absence from school? Second, records may be a source of information about the nature of the instructional program that the student has experienced. Has the student frequently been changed from one reading program to another? Is there something in the nature of the reading program that contributes to potential difficulties? Does the collected test data reflect the instructional program? Are the instructional programs age-appropriate?

The answers to such questions will influence greatly the preliminary hypotheses formulated about the nature of the student's reading difficulties. (Please note, however, that the specific questions asked will vary with individual students, so the examples presented here will not be relevant in all cases.) In some cases, reviewing current and historical information will yield clear and straightforward hypotheses. In other cases, as we suggested above, the results of this effort will be puzzlement. Children and adolescents do not always fall into simple, predictable patterns. In such cases, the best course may be to start fresh.

If records are abandoned as not helpful, a starting question should ask what the student is expected to do in class. This question may be followed by a second one which asks what the student is doing now. Regardless of the data found in records, the answers formulated at this point are only tentative hypotheses. Additional information about previously acquired reading habits may be gathered through interviews with parents, previous teachers (if available) and the students themselves. In this manner more up-to-date background information is gathered.

Next, the process of assessing the student begins. The choice of assessment strategies depends on the data collected in the last step. If there is considerable current background information available, global assessment may not be necessary. There is little point in administering a group reading test when it is already known, for example, that a student is reading three grade levels below expectation. What is needed at this point is more specific information about reading performance. If, however, there is little information about the way the student reads or how he or she feels about reading, global assessment is a logical starting point. Global data likely will provide a basis for hypothesis formation and for the collection of more specific information about reading performance.

Next comes a crucial step in the assessment process, one that relates directly to instructional planning. Once an estimation of individual strengths and weaknesses has been obtained, an equal amount of information is necessary about teaching strategies and instructional materials. For example, if a student is reported to "hate" reading but "loves" the current popular music star, a magazine about this star might be used as the basis of a trial lesson. If the reaction to this material and the selected teaching technique is favourable, such magazines may provide a starting point for instruction. Of course, an IEP cannot be built exclusively around too narrow a theme. The point of the trial lesson is not to plan instruction for the remainder of the available time period, be it six months or a year; rather, the trial lesson should make it possible to decide where to start reading instruction. Information is needed about whether the student likes a topic, will respond well to an instructional technique, or appears to benefit from a particular support system. If the trial lesson yields answers to these questions, it is possible to estimate the potential usefulness of instructional strategies and materials.

Once a reasonably clear assessment picture is completed, the next step is to prepare an IEP. It is quite likely that the IEP will have to be modified in the longer term. There is also some chance that aspects of the IEP will have to be changed in the short term. An IEP is a planning document which helps map out a strategy for reading instruction. As such it may need reworking and it definitely will need monitoring throughout the course of instruction.

Summary Thus far we have presented an overview of the application of the decision-making model to reading. In the balance of this chapter some hypotheses about reading failure are reviewed, assessment strategies are discussed and, finally, instructional techniques are presented.

HYPOTHESES ABOUT READING FAILURE

As we have seen, the first step in using the decision-making model with students who are referred for reading is to review any available background information. At that time tentative hypotheses may be formed about the factors contributing to poor reading performance.

There are several general hypotheses as to why reading failure occurs. Some of these focus on the student, taking into account the following seven factors: a) absence from school during crucial instructional periods; b) early home experience with books and, later, the amount of time spent reading; c) anxiety or fear associated with reading; d) lack of enabling skills, such as word attack, and lack of appropriate background knowledge and vocabulary; e) lack of text-specific processing and interpretation skills; and f) general problems associated with language, thinking and reasoning. In addition to considering the student, there are hypotheses that focus on the teacher and on the instructional program. No one hypothesis, however, will explain adequately the lack of fluency in reading; rather, each problem suggested above may contribute to disfluency. These hypotheses are discussed in more detail below, starting with those which focus on the student.

Is the student's reading problem the result of having missed instruction? This can be tested by checking for lengthy absences, especially during the first four grades. Children who have been ill for a month or longer may not have had the opportunity to learn specific sounds in a phonics program for example, or they may not have learned specific vocabulary introduced during their absence. Usually such absence(s) will have been noted quickly and the missed components already will have been gone over with the student. In some cases program components are missed because the student has moved from one province to another. This may result in lost instructional time. Less obviously, it may cause the student to miss blocks of skills if the order in which they are taught differs from province to province.

Parents may play a major role in what and how much their children read. Certainly the recent assessment of reading conducted in British Columbia (Jeroski 1984) indicates the importance of the home environment, showing that parents have a strong impact on leisure reading. In order to assess a student's reading time at home, interviews with parents or the students themselves may be conducted. Parents of younger children usually are helpful in describing a) the amount of time spent reading or looking at books and b) the nature and variety of material selected both by themselves and the children. Children as young as four or five may be interviewed about reading. Although they may not describe fully their reading repertoire, a reasonable indication of feelings about reading usually will be obtained. Older students often are the best source of information about their own reading habits (or lack of them). A few will be defensive about their perception of books, but most are very frank about their opinion of the value of reading.

While interviewing parents and/or the student about out-of-school reading, it is possible also to check whether the student has anxiety or fears about reading. Having students complete attitude surveys also may be useful in determining if students are afraid or anxious about reading.

That the lack of enabling skills may inhibit reading fluency is one of the most frequently stated positions in reading research; yet it is perhaps the most difficult to demonstrate. For example, Jeroski (1984), in a province-wide assessment of reading abilities in British Columbia, omitted enabling skills, such as word attack methods, in favour of measuring such things as reading comprehension and study skills. More significantly, there is little agreement about the importance of any specific enabling skill in either reading acquisition (Hammill and McNutt 1981;Smith 1978) or skill maintenance (Schumaker et al. 1983). Phonics, for example, may be identified as a skill possessed by some fluent readers; however, it cannot be argued on that basis that phonics is a necessary enabling skill for all readers. In practice, a decision must be made whether or not to assess a particular enabling skill and how that assessment should be carried out. Perhaps the best approach is to examine a) the reading program in place and b) whether the student has mastered the skills appropriate to his or her age and level of instruction.

Inadequate reading performance may be the result of a lack of text-specific processing and interpretation skills. Some of the major problems in this area are in separating main ideas from ones of lesser importance, being able to predict what will come next, and having a sense of what to expect in a story. Wong (1982) points out that poor readers lack effective reading strategies. Taking the opposite perspective, Anderson (1980) suggests that efficient readers know how to a) monitor their reading to determine if they understand and b) use text clues, such as backtracking or scanning ahead, when they have trouble understanding. Owings et al. (1980) tested the ability of fifth grade students to find incongruity in a text. They found that more fluent readers are able to find the mixed up story spontaneously, while less fluent readers cannot, unless told there is something amiss in the story. To determine if readers have or do not have text-specific processing and interpretation skills, the assessor may try several tactics, as long as the emphasis is on recognition of meaning or text-order. Two examples illustrate the suggested tactics. First, short passages of text may be given to a student in which words that have multiple meanings are used, and the student then may be asked to identify the meaning of those words in context. Second, short passages of scrambled text may be administered to students to determine if they recognize that something is amiss and if they are able to correctly reorder the text.

Finally, some students have general problems with thinking, reasoning, language, or all three which affect their reading ability. Problems in thinking and reasoning have been discussed in terms of what have come to be called metacognitive skills (e.g., Baker and Brown 1980). For example, Wong (1982) distinguishes between comprehension and metacomprehension: comprehension is understanding instructions; metacomprehension is knowing that the instructions have been understood. Some students with learning problems manage the first, but then fail at the second. Equally importantly, they fail to believe that they have understood even when it is demonstrated that understanding has occurred. The last concern, language, which was described at the beginning of the paragraph, is of obvious importance. For example, even if students cannot read a word, they may be able to state the meaning of that word or use it correctly in a sentence when asked to do so orally. Other times students will not know the word nor will they be able to use it in an oral context, yet they will be able to decode it. In this case, students will lack a meaningful (language-based) context into which to place the decoded word. It is likely that each time this word is faced it will need to be decoded until they have a meaningful context. In order to assess thinking skills, the suggestions in Chapter 10 may be followed. To assess language proficiency informal techniques such as interviewing may be used or work samples may be analyzed to determine if language is written inappropriately.

The techniques described in Chapter 8 may be helpful in assessing written language.

Apart from problems which students may have, both instructional techniques and selected reading materials may contribute to inadequate fluency in reading. Chall (1967), in an excellent older text, puts the issue clearly in perspective by reviewing changes in reading methodology from 1890 to 1955. During this period, various methods of teaching beginning reading were popular by turn. First, an approach based on synthetic phonics was preferred. This was followed by reading silently for meaning. Finally phonics were reintroduced along with context and picture clues, structural analysis, and dictionary skills. Language experience was introduced during the period in which silent reading was popular, and it was maintained as a component of many programs in the following period. What is of note is that in each period previously used methods were criticized as being inappropriate or unsuccessful. For example, Gray (1922) suggested that phonics had been carried too far to be useful in recognizing words and understanding them and thus contributed to failure to learn to read. Later Gates (1922) argued that the whole-word method of learning to read contributed to problems in word recognition. Orton (1937) similarly suggested that whole-word methods did not provide the student with sufficient sounding-out skills (auditory clues). Thus, it would appear that all methods have some inherent weaknesses which if ignored may contribute to a student's reading failure. Returning to the decision-making model provides the appropriate technique to answer questions about instructional procedures. We suggest that trial lessons be prepared prior to writing an IEP in order to evaluate instruction. In addition, trial lessons can be used any time there is a question concerning the effectiveness of instruction.

READING ASSESSMENT

The next step in the decision-making model is to assess the student's strengths and weaknesses. Two sources of information are useful in deciding what type of assessment to conduct. First, as discussed above, a review of the student's background information will provide data on what has been done. Second, and equally importantly, a review of general curriculum expectations or expectations of other competencies provides an estimate of what remains to be accomplished. General curriculum expectations may be used also as tentative guidelines in setting annual goals for many students.

Once guidelines for assessment have been established, techniques to assess current reading performance must be selected. Assessment measures can be either teacher-prepared or commercially prepared.

In the following section, some general expectations for readers are provided as a frame of reference for setting program expectations. Next, commercially prepared and teacher-prepared assessment techniques are reviewed.

General Expectations for Readers

In order to assess students' reading strengths and weaknesses it is helpful to know what is generally expected of students at a particular age or grade level, both in terms of acquired competencies and curriculum-specific achievements. However, two cautions are necessary: a) acquisition of general competencies by any particular student will vary as a function of the teacher's choice of method; and b) the acquisition of competencies must be reviewed in the context of the specific reading material being used. Reading ability is a function of the person's reading skills, the teacher's teaching skills, and the readability of the text (Bormuth 1974). In Tables 6-1 and 6-2 general expectations of student reading ability are provided.

Table 6-1 **Basic Knowledge and Strategies for Reading**

Grade level	Student competence
Grades 1-3	Has a large speaking vocabulary Uses spoken vocabulary appropriately Recognizes and reads all upper case and lower case letters Recognizes that letters make words Recognizes and uses commonly occuring words Identifies the constancy of any one word or set of words, regardless of location in text Reads from left to right consistently Recognizes that words go together to make sentences Uses a picture dictionary or other dictionary to find the meaning of unfamiliar words Uses structural clues or phonics to make sense out of unfamiliar words Uses context clues to find meaning of unfamiliar words
Grades 4-9	Identifies pronoun references Recognizes and identifies irregular verb forms Recognizes and responds to words implying affective states (e.g., fear, anxiety, happiness) in text Recognizes and interprets antonyms, homonyms, and synonyms Recognizes and interprets other words that have multiple meanings or ambiguous referents Identifies and responds to subject-specific technical vocabulary

It is recognized that not all students with special needs will become fluent readers. There likely will be a considerable range in reading fluency among individuals with special needs. With this in mind, the Ministry of Education, Province of British Columbia has suggested a series of reading competencies for use in special education, ranging from functional literacy to complete fluency. These competencies include developing functional literacy skills, applying functional literacy in the home and community, using word recognition and word analysis skills, finding meanings of words, demonstrating oral reading, comprehending various kinds of materials (oral, visual or written), reading differentially depending on purpose, and using resource materials (Ministry of Education, Province of British Columbia 1983).

When assessing any student's reading ability, however, it is necessary to consider more than just the acquisition of specific competencies. Downing (1982; Downing and Leong 1982) suggests that there are three phases to becoming a skillful reader.

Table 6-2 **General Reading Comprehension, Search and Application Strategies**

Grade level	Student competence
Grades 1-3	Identifies, classifies and describes objects and pictures Identifies meaning of words in story or other text when read orally to him or her Uses pictures and surrounding sentences to make sense out of a word, phrase or sentence Recognizes and understands time sequences (past, present and future) in a textual context Recognizes main ideas and key story details Answers specific questions of different types about text Retells passage in own words Identifies meaning associated with a comma and a period
Grades 4-6	Identifies meaning associated with all forms of punctuation Identifies where in a passage (beginning, middle, end) information is likely to be found Predicts outcomes of passages of varying lengths
Grades 7-9	Uses different reading techniques according to difficulty and type of text Can locate information using reference clues (title page, table of contents, headings and other index clues) Uses other reference materials (card catalogues, encyclopedias) and specialized sources (maps, graphs, atlas)

First comes an initial cognitive phase in which the beginning reader must determine what needs to be done. Downing notes that it is normal for most beginning readers to be confused about the purpose of reading. Second, comes a mastering phase in which readers work to perfect their reading performance. Finally, comes the automaticity phase in which the reader practices beyond the point of mastery until overlearning has occurred. Downing (1982) argues that too much attention is paid to the featural aspects of reading (the written symbols) and not enough attention to the functional concepts (understanding the purposes of reading). Following this line of reasoning leads to a different notion of what needs to be included in any assessment of reading. It may be just as important to determine the child's understanding of the purpose of reading as to assess whether he or she has acquired particular competencies.

Selecting an Appropriate Estimate of Reading Ability

Once the student's background and the curriculum expectations have been determined, the next step in the decision-making process is to select an appropriate assessment technique. Commercially prepared measures may be chosen, or the teacher may decide to design his or her own measures.

In beginning a reading assessment it may seem at first that there are many techniques and tests from which to choose. However, careful analysis soon reveals that the variety of selections is only an illusion of choice. Applying the criteria suggested in Chapter 3 to the many commercially prepared measures shows that a great number of them fall short in one way or another. For example, Salvia and Ysseldyke (1981) reviewed some better known reading tests for technical adequacy. They report only one test in which test-retest reliability is demonstrated. They also report one case where no reliability data of any kind is given.

Commercially Prepared Tests

There are three main types of commercially prepared reading tests on the market. First, there are standardized, group administered measures of reading. Second, there are standardized individual measures of reading. Third, there is a group of informal reading tests. Tests in this third group usually are administered individually but are neither criterion-referenced nor norm-referenced. Some examples of each of the three types of reading tests are summarized in Table 6-3. The table does not include all available tests, however, as we have not attempted to provide a comprehensive review of available reading tests. Instead we have summarized the tests more commonly used in Canada, including those normed in the United States but used frequently in this country. In addition, we have included a few tests that are atypical in some way. The atypical tests are those that include passages to the college level. Most tests end at grade 8, a few go to grade 12, but tests with passages to college level are rare. The other atypical test is the one designed specifically to assess reading comprehension. It is atypical because of the content addressed.

The *Advanced Reading Inventory* (Johns 1981) is a newer informal inventory which is intended for use with older students. It may be administered to a group or to an individual, although the test is better suited to individual administration. There are two forms of each of the three test components: word lists, graded passages and cloze passages. According to information in the inventory, the word lists were selected randomly from a list compiled by Taylor, Frackenpohl and White (1979). The difficulty of each graded passage was determined by applying two readability formulas (Fry and Dale-Chall), by conducting field tests, and by the authors' judgment. The comprehension questions in the graded story passages are multiple choice and are easy to score. Some older students, however, become quite anxious when asked to complete multiple choice. This reaction may affect the validity of the results for these students. The cloze tests are based on the graded story passages. No reliability estimates are provided.

The *Basic Reading Inventory* (Johns 1981) was developed for use with younger students. This inventory is intended for use in grade one (starting at pre-primer level) through grade eight. No information is provided about the selection of the word lists. Three forms of the graded story passages are included. These are approximately parallel when judged by applying three different readability formulas (Fry, Spache and Dale-Chall—although only two formulas were applied to any one paragraph). The inclusion of three parallel forms is quite useful because it allows both retesting and testing for different purposes. With the exception of the first passage for grade one, each story is about 100 words in length and is provided with 10 questions which assess comprehension. At the pre-primer level, the length is about half that of the later passages and there are only five questions. No estimates of reliability are provided.

Table 6-3 **A Summary of Reading Tests Used in Canada**

Test	Category	Grade level	Content	Technical adequacy	Administration
Advanced Reading Inventory (Johns 1981)	Informal	7-College	Parallel forms including graded word lists, graded reading passages with 10 questions per passage, and cloze tests	No information provided as to reliability	Individually administered, group administration possible
Basic Reading Inventory (Johns 1981)	Informal	1-8	Three parallel forms including graded word lists and graded reading passages with 10 questions per passage (excepting the pre-primer level where there are five)	No information provided as to reliability; readability estimates are included	Individually administered
Diagnostic Reading Scales (Spache 1981)	Standardized	1-8	Word recognition paragraphs for oral and silent reading, and phonics skills subtests	Internal consistency adequate, test-retest variable from poor to good; inadequate demonstration of validity	Individually administered
Doren Diagnostic Reading Test of Word Recognition Skills (Doren 1973)	Standardized	Ungraded, approximate use for grades 1-4	Emphasis on enabling skills such as letter recognition, beginning sounds, whole word recognition, words within words, speech consonants, ending sounds, blending, rhyming, vowels, discriminate guessing, spelling, and sight words	Correlational data provided as estimates of reliability and validity	Group administration
Durrell Analysis of Reading (Durrell and Catterson 1980)	Standardized	1-6	Graded word letters, four set of paragraphs and seven sub-tests with emphasis on enabling	Internal consistency and criterion validity are poor to good	Individually administered

Test	Category	Grade level	Content	Technical adequacy	Administration
			skills (listening vocabulary, sounds in isolation, spelling, phonic spelling, visual memory of words, identifying word sounds and estimates of phonic prereading skills)		
Gates-MacGinitie Reading Tests, Canadian Edition (MacGinitie 1978, 1980)	Standardized	1-12	Vocabulary and paragraphs with questions	Internal consistency very good, validity adequate	Group administration
Gates-Mckillop-Horowitz Reading Diagnostic Tests (Gates, Mckillop and Horowitz 1981)	Standardized	1-6	Paragraphs, sentences, word recognition, word attack, visual form of words, auditory blending and discrimination, spelling, and an informal writing sample	Test-retest on paragraphs only, .94. No other reliability data reported. Validity not reported, content validity poor	Individually administered
Gilmore Oral Reading Test (Gilmore and Gilmore 1968)	Standardized	1-8	Two forms of ten paragraphs, each set forms a continuous story	Alternate form reliability inadequate; no evidence of validity	Invidually administered
Gray Oral Reading Test (Gray and Robinson 1967)	Standardized	1-College	Four forms of 13 paragraphs	Alternate form reliability good; no test-retest reported; no validity data	Individually administered
Metropolitan Achievement Tests, Reading Instructional Tests (Farr, Prescott, Balow and Hogan 1978)	Standardized	K-9	Two forms; content is adjusted according to level of test; subtests include visual discrimination, letter recognition, auditory discrimination,	Internal consistency good; criterion validity ranges from poor to good; content validity left to test user	Group administration

Test	Category	Grade level	Content	Technical adequacy	Administration
			sight vocabulary, phoneme-grapheme (consonants, vowels), vocabulary in context, skimming and scanning and reading comprehension		
Stanford Diagnostic Reading Test (Karlsen, Madden and Gardner 1976)	Standardized, criterion-referenced and norm-referenced	1-2	Content adjusted according to level, included are subtests on auditory discrimination and vocabulary phonetic analysis, structural analysis, word reading, sentence reading, paragraph reading and comprehension	Internal consistency good; criterion validity ranges from poor to good; content validity left to test user	Group administration
Test of Reading Comprehension (Brown, Hammill and Wiederholt 1978)	Standardized	1-12	General vocabulary, mathematics vocabulary, social studies vocabulary, science vocabulary, syntactic similarities, paragraph reading, reading the directions of schoolwork and sentence sequencing	Internal consistency and test-retest adequate; validity adequate	Small group or individual administration
Wide Range Achievement Test (Jastak and Jastak 1978)	Standardized	Ages 5 to adult	Reading component consists of letter identification and word recognition; spelling and arithmetic subtests are included as well	Split-half reliability good; validity inadequate	Individual administration
Woodcock Reading Mastery Tests (Woodcock 1973)	Standardized	1-12	Letter identification, word identification, word attack, word comprehension and passage comprehension	Reliability variable; validity of some subtests questionable	Individual administration

The *Diagnostic Reading Test* (Spache 1981) is designed to be individually administered to students in grades one through eight. It consists of word recognition lists, two sets of reading paragraphs (one for silent reading and one for oral reading) and a series of phonics subtests. Paragraphs vary in length from approximately 25 to 185 words. Standardization procedures are not well described. The internal consistency data seem adequate, although the method of calculation may inflate the correlations. The reliability of some subtests is quite low for specific grades; therefore, caution will be necessary when using those tests. Validity is inadequately demonstrated.

Doren (1973) argues that measuring what she terms "reading functions" is insufficient. Her rationale is that it is necessary to know about specific skills, such as rhyming or beginning sounds, in order to diagnose reading deficiency. The *Doren Diagnostic Reading Test of Word Recognition Skills* was prepared on the basis of this position. Items were written to address enabling (assumed to be prerequisite) skills for reading. Standardization procedures were conducted in the United States. The reliability and validity information is rather sketchy. Reliability (internal consistency) ranges from 0.53 to 0.82. Three subtests have low internal consistencies: letter recognition (0.53), words within words (0.62) and beginning sounds (0.66). No test-retest reliability estimates are provided. Validity is estimated by reporting correlations between this test and an unspecified reading achievement test, which may make one want to question the criterion. Correlations, calculated for grades one to four, ranged from 0.77 to 0.92.

The *Durrell Analysis of Reading* (Durrell and Caterson 1980) is an individually administered test for use in grades one to six. There are two graded word lists, four sets of graded paragraphs, and seven other subtests designed to provide estimates of enabling skills. Three of the four sets of paragraphs are to be used as estimates of oral reading, silent reading and listening comprehension, and the fourth is a set of supplementary paragraphs. The paragraphs vary in length from approximately 20 to 140 words. The key criterion used to determine reading level is the speed at which a passage is read. The seven other main subtests include listening vocabulary, sounds in isolation, spelling, phonic spelling, visual memory of words, identifying sounds in words and some measures of phonics. The standardization sample is described poorly—the only indication is that the sample was randomly chosen. Furthermore, no Canadian locations were included in the standardization sample. Correlations of reading times between oral passages were 0.85 and between silent passages they were 0.80. Using reading speed as the criterion, the reliabilities for the reading paragraphs are good. No estimates of reliability are provided for the comprehension questions. Internal consistency reliability (Kuder-Richardson 21) for all other subtests ranged from 0.63 to 0.91. The reliabilities of three subtests (the primary spelling test and both tests of visual memory of words) are suspect, but the remainder are adequate. Validity is established by correlation with the *Metropolitan Achievement Tests* and ranges from 0.22 to 0.86, the higher scores being for the reading paragraphs. The procedure of using reading speed as the key criterion may make the validity of this test somewhat suspect.

The *Gates-MacGinitie Reading Tests* (MacGinitie 1978, 1979, 1980) is one of the few reading tests to be standardized in Canada. There are seven levels of this test.

The first level is used to estimate reading readiness and therefore varies in format from the remaining levels. It was designed for use as a group test for grades one to twelve, but may be administered individually. For levels A to F, two subtests are included in each level, one for vocabulary and one for reading comprehension. There are between 40 and 45 items in each of these subtests. The reading comprehension passages are variable in length and include between one and four questions. Practice items are included at all levels. Internal consistency estimates (Kuder-Richardson 20) are good. These estimates are reported as 0.85 to 0.94 for the vocabulary subtests and as 0.85-0.92 for the comprehension subtests. Using the United States edition, however, Johns (1984) reports that Level E is not consistent between the two forms of the test. Whether this criticism is generalized across all levels or is applicable to the Canadian edition is not known. Caution should be exercised in treating the alternative forms as equivalent. Content validity seems adequate.

The *Gates-McKillop-Horowitz Reading Diagnostic Tests* (Gates, McKillop and Horowitz 1981) are intended for use in grades one to six. The emphasis given by the authors is on enabling skills. Several subtests are included in this oral reading test: paragraph reading, sentence reading, word recognition (both flash and analysis conditions), word attack, recognizing the visual form of words (identifying vowels in nonsense words pronounced by an examiner), auditory tests, spelling and an informal paragraph. The authors recommend that subtests be administered separately. Before any subtest in this test is administered independently, it must meet the technical adequacy criteria for norm-referenced tests given in Chapter 3. Readers will be hard pressed to apply these criteria to any subtest due to a lack of information. In addition, many of the criticisms directed at the earlier edition of this test apply to this edition as well (e.g., Salvia and Ysseldyke 1981). Standardization information is sketchy. Reliability data is reported only for paragraph reading based on errors made in oral reading (test-retest 0.94). The absence of both internal consistency data and other test-retest data may be cause to question the technical adequacy of this measure, especially given the author's recommendation to administer subtests on an individual basis. Interpretation of this test is difficult and requires considerable clinical skills.

The *Gilmore Oral Reading Test* (Gilmore and Gilmore 1968) consists of two sets of ten paragraphs that form a continuous story. The paragraphs range in length from about 25 words to 260 words. It is intended for use in grades one through eight. Reading score is determined by error rate only. Standardization procedures are inadequately described. Reliability data (alternative form) is very limited and no validity data is provided for the current edition.

The *Gray Oral Reading Test* (Gray and Robinson 1967) is intended for use from grade one through to college. It consists of 4 sets of 13 paragraphs, which is useful. Scoring procedures are based on both speed of reading and error rate. Separate norms are provided for each gender. Reliabilities (alternative form) are good, ranging from 0.96 to 0.98. Test-retest reliability is not reported, nor is evidence of validity.

The Metropolitan Achievement Tests: Reading Instructional Tests (Farr et al. 1978) is one of three tests designed to provide an estimate of a particular subject area, the other two in this instructional series are tests of a) language instruction and

b) mathematics instruction. There are two forms of the reading test, which is designed to measure reading from kindergarten to grade twelve. Test content varies considerably across levels. One test may be of interest to readers working with older students in grades five and up. It is called skimming and scanning. In it students are required to find specific information under speeded conditions. The test is well standardized for use in the fall and spring. Internal consistencies (Kuder-Richardson 20) for the fall standardized test range from 0.88 to 0.95. No test-retest reliabilities are reported. There is extensive reporting of the content validity data for the instructional objectives. Canadian users should be careful in using several subtests in which the content is quite specific to the United States.

The *Stanford Diagnostic Reading Test* (Karlsen, Madden and Gardner 1976) is a multi-level test similar to the *Metropolitan* in that the content is adjusted according to level. It may be used from grades one to twelve. Both criterion-referenced and norm-referenced scoring and interpretations are possible. Standardization procedures are well described in the manual. Internal consistency coefficients are 0.90 or better for all but the auditory vocabulary subtest (0.84 to 0.92). Alternate form reliability coefficients range from 0.64 to 0.95. Criterion validity is reported against the *Stanford Achievement Test* and this ranges from 0.39 to 0.94. Content validity is left to the reader.

The *Test of Reading Comprehension* (Brown, Hammill and Wiederholt 1978) consists of eight subtests: general vocabulary, syntactic similarities, paragraph reading, mathematics vocabulary, social studies vocabulary, science vocabulary, reading the directions of schoolwork and sentence sequencing. The authors argue that a reasonable estimate of the factors that relate to reading comprehension is provided for grades 1 to 12 when these tests are taken together. The standardization procedures are not well described. Internal consistency ranges from 0.50 to 0.97, with the majority of coefficients being over 0.80. The internal consistency of science vocabulary at grade 2 (0.53) and general vocabulary at grade 11 (0.50) is suspect. Using a very small sample in one city, test-retest reliabilities were reported between 0.64 to 0.91. Validity is thoroughly discussed in the manual and appears adequate.

The *Wide Range Achievement Test* (Jastak and Jastak 1978) is used frequently as a quick screen. The reading subtests consist basically of a word recognition test. Thus the content validity as a measure of reading is poor; however, it may be valid as a measure of word recognition. The internal consistency (split-half) is good. No test-retest reliabilities are reported. The standardization procedures stated by the authors in the manual are inadequate.

The *Woodcock Reading Mastery Tests* (Woodcock 1973) are intended for use in grades one through twelve; however, at least one reviewer is of the opinion that its use should be restricted to an upper limit of grade six (Haggard 1981). Subtests included are letter identification, word identification, word attack (pronunciation of nonsense words), word comprehension and passage comprehension. Standardization procedures are not well described in the test manual. The split-half reliability data reported is adequate (0.79-0.99) for all but one subtest, letter identification (0.02-0.2). Test-retest reliability is reported for a limited sample for each of the subtests. It is adequate (0.68 to 0.97) for all subtests except letter identification. The use of this particular subtest is very suspect. The author goes to considerable effort to demonstrate validity; however, the content validity of the passage comprehension subtest appears questionable. The passages are very short cloze exercises, which do not allow a complete picture of comprehension to be obtained.

Summary It is possible to identify some limitations in the reading tests we have reviewed. These limitations fall into three general categories. First, the population for which such tests are intended generally is restricted. Second, the range of techniques used to sample reading ability is somewhat narrow. Third, what is actually measured varies from test to test. Each of these general points is discussed in turn below.

First, the population that is sampled by most test constructors is restricted. The main target group is the elementary grades, in particular grades three to six. While this group might be extended to include grades two, seven and eight, the vast majority of the material is directed at the first group.

Second, there are certain restrictions built into the techniques used in these tests. The majority of the passages selected are short, ranging in length from approximately 20 words to 300 words. While the rationale for such short passages is readily apparent, the length does pose at least two possible problems. First, in sampling shorter passages the reliability of the obtained estimate may not be sufficiently high. High reliability is possible, however, when several shorter passages are included. Second is the more serious question of validity in estimating comprehension. It is acknowledged that children and adolescents need to know how to read and extract information from passages of the length typically included in reading tests. The problem is that such estimates do not appear to provide a complete picture of comprehension. Often children who can read and answer questions on passages found on the majority of reading tests may have a great deal of difficulty reading texts that are considerably longer.

Furthermore, the nature of what is measured varies from test to test. For example, on one test there may be items to assess phonics while on another this skill is missing. To make matters more confusing, the name assigned to a subtest and what it actually measures vary from test to test. For example, passages from which reading or reading comprehension is estimated appear on most tests; but in some cases reading is estimated by the number of errors made, while in other cases it is estimated by speed, and in still other cases reading is estimated by a combination of word errors and speed. In addition, the number and type of questions included on any test vary considerably. The point is that it is very important to be aware of exactly what is being estimated when any test is chosen.

Finally, as a cautionary note it seems necessary to point out that one should not be mislead by the name attached to the test itself. Just because a test is labelled diagnostic does not mean that it is any better or any worse than one simply termed a reading test. There may be little difference in the way in which the main components of the tests are constructed or in the number of items that are found on specific subtests. What may vary is the name given to the subtests and the diversity of the subtests included. The technical adequacy of additional subtests found on some diagnostic tests may be poor.

In short, it is important to be very careful in selecting and interpreting commercially prepared reading tests. There are limitations inherent in any of these and the test user must be aware of the resulting limitations of the obtained data. Thus, it is important to keep in mind when using these tests as part of the decision-making process that any hypotheses based on test results are tentative.

Word Recognition Lists

In addition to the commercially prepared inventories, there are a large number of word recognition lists available. Administering any one of these lists, as long as it is representative of the high frequency vocabulary being taught, will provide some useful assessment information. First, an estimate of the individual's sight vocabulary may be obtained. Second, some information is provided about word attack, although this information is limited because some word attack clues are not found when lists are presented in isolation.

As a word of caution, it does not follow from the decision to administer a standard word list in order to estimate sight vocabulary that words missed on such lists are to be taught in isolation. Nor is it recommended that such lists be memorized. Rather, we suggest only that a standard word list may be used to estimate sight vocabulary and, to some extent, word attack skills.

There are several of these word recognition lists available. The best known is the one originally prepared in 1936 by Dolch (1950). This list is still representative of words frequently found in basal readers (Mangieri and Kahn 1977). In addition, there are several other commonly reprinted word lists. Examples are the list prepared by La Pray and Ross (1969) and the one constructed by Hillerich (1974). The latter list has 158 words (out of a total of 240) in common with the Dolch list. Such lists are found so commonly that we have not reprinted any of them in the present volume.

The emphasis in the majority of such high frequency lists is placed on the elementary grades. Not so common is a core vocabulary list that is extended into secondary school. In Figure 6-1, the *Quick Word Test* is given. It is divided into three parts: a) general vocabulary, b) general subject vocabulary and c) seven sections of subject-specific vocabulary. The words on this test were extracted from 37 textbooks assigned to grades 8, 9 and 10. It should be noted that the criterion of pass is adjusted for each of the three parts of the test.

Teacher-prepared Assessment Techniques

Teacher-prepared assessment techniques may also be used to confirm, revise, or discard tentative hypotheses about a student's reading problems. Commercially prepared measures do not always test the necessary areas (Harris and Sipay 1980). Indeed, Bliesmer (1972) has suggested that an estimate of the individual's ability to read in a functional and realistic sense can be gained only through the use of teacher-prepared measures.

In devising techniques to estimate reading abilities, it is important to remember that the rules for test construction given in Chapter 3 apply to teacher-prepared materials as well as to commercially prepared materials.

Assessing interest and attitude A useful starting point in assessing a student's reading ability is to test the student's interest in reading. This is important because estimates of a student's ability may vary considerably according to the degree of

Figure 6-1 **Quick word Test**

Part I – **General Vocabulary (90% Mastery Criterion)**

a	for	it	the
and	in	of	to
are	is	that	you
about	had	not	time
all	has	on	two
also	have	one	up
an	he	only	us
any	him	or	used
as	his	other	very
at	how	see	was
be	I	should	we
been	if	so	were
but	into	some	what
by	its	such	when
called	like	than	which
can	many	their	who
could	may	them	will
do	more	then	with
each	most	there	work
first	must	these	would
from	no	they	your
good		this	

(87 words)

Part II – **General Subject Vocabulary (70% Mastery Criterion)**

account	fabric	meal	river
add	family	measure	sales
air	feet	metal	sentence
American	figure	money	set
amount	food	north	small
area	forest	numbers	speed
book	forum	office	steel
British	fuel	oil	surface
business	games	order	table
Canadian	gear	paid	temperature
care	government	paper	tools
cells	heat	part	trade
child	job	pattern	triangle
children	lake	pay	unit
city	land	people	value
copper	language	piston	vegetables
copy	length	point	water
cylinder	light	population	whole
different	line	power	wire

Part II – continued

earth
energy
engine
English
example
lines
long
made
margin
mass
pressure
rational
record
region
represent
wood
word
world
years

(95 words)

Part III – **Subject Vocabulary (60% Mastery Criterion) COMMERCE**

assignment
balanced
bill
business
buys
calculated
carriage
centered
cheque
clerk
compute
consumer
copies
correction
credit
currency
customer
debt
deduct
dictaction
divider
due
employee
employer
entries
errors
exemptions
factory
gross
insurance
journal
ledger
loan
mail
merchandise
operator
parcel
payment
purchases
receipt
receives
salary
salesman
savings
selling
shorthand
spaced
stops
subtract
transactions
type
typewriter
vowels
wage
worker

(55 words)

Part III – **Subject Vocabulary (60% Mastery Criterion) ENGLISH**

adjectives
antecedent
apostrophe
audience
author
auxiliary
awkward
characters
clauses
climax
clues
coherence
colloquial
comedy
communicate
compositions
conclusion
contemporary
critic
define
describing
diary
drama
figurative
formal
grammars
handwriting
homonyms
inflections
interpretation
interrogative
languages
listening
literature
message
metaphor
modifiers
novel
paragraphs
performed
phrase
players
plays
plural
poetry
predicate
presence
prose
reader
rewrite
rhythm
scene
semicolon
singular
slang
speakers
spelled
tale
teach
television
theatre
tone
verbs
vocabulary
writers

(65 words)

Part III – **Subject Vocabulary (60% Mastery Criterion) HOME ECONOMICS**

abdomen	closet	labeled	sew
apron	collar	meats	shelf
attractive	cook	menu	show
babies	cottons	napkins	sink
bakery	crisp	nursing	skirts
bedtime	disease	nutritious	slash
behavior	dishes	nylon	sleeves
beverage	drawers	oven	socks
birth	dresses	packaged	soiled
blend	eating	physician	strained
blouses	entertaining	pockets	suit
brand	fitted	pork	tender
bread	flavored	powders	textile
budget	freezing	pregnancy	texture
bust	grooming	recipe	utensils
buttonhole	guests	rinse	wardrobe
calories	hems	salad	wash
canned	hose	salted	wearing
cereal	hostess	satisfying	yarn
chocolate	juices	saving	yeast

(80 words)

Part III – **Subject Vocabulary (60% Mastery Criterion) INDUSTRIAL EDUCATION**

alloy	equipped	locking	rocker
artificial	fir	machining	sanding
automotive	forging	mahogany	shaping
bearings	frame	manufacture	shops
blades	gauge	mold	socket
boiler	gears	muffler	stroke
buffer	glued	nails	thickness
cable	grained	outlet	throttle
casting	grooves	pine	timing
chisels	hobby	pipe	valves
connecting	horn	piston	vapor
craftsman	horsepower	plywood	velocity
curve	joints	repair	workshop
drag	layout	resin	

(55 words)

Part III – **Subject Vocabulary (60% Mastery Criterion) MATHEMATICS**

algebra	denominator	intersect	pyramid
binary	divisible	multiplication	representing
congruent	equals	notation	segment

Part III — continued

cubes	fraction	numeral	subset
decimal	geometric	plus	

(19 words)

Part III — **Subject Vocabulary (60% Mastery Criterion) SCIENCE**

absorbed	digestive	melt	scientist
apparatus	earthquake	meteorite	seismic
astronomer	electromagnet	microscope	sensors
balloon	equalize	molecular	solutions
beaker	experimental	neutrons	stimuli
burn	fission	nuclei	substances
calculate	flame	organ	telescope
chemists	flask	photosynthesis	theory
chloride	gram	pigments	thermometer
circles	granite	platinum	tissue
coefficient	gravity	protons	vein
compounds	images	radiation	vibrating
conductor	insects	radioactive	wavelength
constant	intestine	react	weight
crystal	investigation	reflecting	
decay	lenses	response	
deposited	lunar	samples	

(65 words)

Part III — **Subject Vocabulary (60% Mastery Criterion) SOCIAL STUDIES**

allies	elected	legislation	resources
arable	empire	majority	revenue
Asia	exploitation	manufacturers	route
boom	exploration	medieval	sea
boundaries	feature	metropolitan	seasonal
Caribbean	fertile	migration	settlements
climatic	feudal	neighborhood	sits
collective	forage	opposition	society
colonists	freight	overseas	sovereignty
commerce	government	passenger	tariff
commodities	hemisphere	peasant	territory
confederation	highlands	pioneer	transportation
continents	historical	plots	treaty
council	immigrants	prairie	tribes
cultivation	independent	progress	tropical
declaration	institution	rail	urban
depression	invention	rainfall	wealth
domestic	labor	reform	zone
economics	latitude	representation	

(75 words)

From Edward G. Summers, unpublished paper, University of British Columbia. Reprinted with permission of the author.

interest he or she has in a particular reading passage. The content of a passage does not necessarily influence how well or poorly a student will read; however, many students with reading difficulties will not try as hard when they are not interested in the material and as a result their reading skills may be underestimated.

Interest can be assessed in a variety of ways. One technique is to interview the student. The set of questions used in the interview can be adjusted to some extent to suit the situation of the student being interviewed. Some sample questions are given in Figure 6-2.

Another strategy, which may be used along with or as a supplement to an interview, is to have the student complete a questionnaire. Figure 6-3 presents a questionnaire used with a student who had difficulty with reading comprehension. Along with questions that focus specifically on reading comprehension, the questionnaire also lists other items. For example, items on testing are included because often reading comprehension problems are related to poor test-taking skills. Similarly, comprehension problems may be compounded by inadequate enabling skills such as word attack in context.

As suggested in Chapter 5, younger students' attitudes towards specific subjects may be measured using pictures. This type of inventory can be used to assess a student's attitude towards reading. The questions on the inventory suggested by Ollila (1981) may be adapted to particular situations, or it may be used as given in Figure 6-4.

Estimating reading speed Another useful kind of information to obtain is estimates of students' reading speed. In estimating this it is important to note that the reading speed of fluent readers differs depending on what they are reading and the purpose for which they are reading. Individuals with learning problems must learn to adjust their reading rates in the same manner.

There are three general categories of reading purpose and appropriate speed. First, a fast, skimming rate may be used to locate information, review familiar material, or for other similar purposes. Second, a careful, detailed type of reading is necessary to distill a technical text, to read any source in which the syntax is difficult, to read passages containing so many ideas that it is difficult to comprehend the text, or to read any unfamiliar content. Third, a "normal" rate is employed when the purpose of reading falls in between the first and second cases as long as the material is of approximately average difficulty for the specific person doing the reading. Examples include reading for pleasure and reading familiar texts for which detailed reading is unnecessary. Clearly the "normal" rate of reading is very subjective if it is based on the individual's ability. Using it means that material that one person would read using a normal rate may be read at a slower or faster rate by another person.

It would be difficult to provide guidelines for the appropriate speed for skimming or for detailed reading. Even within these categories, speeds will be adjusted in each instance according to the individual's purpose in reading. However, some guidelines are available which may be used to estimate whether the normal reading rate of a special needs student approximates that of other readers. Using median rates reported in different standardized tests, Harris and Sipay (1980) report the range of reading rates for grades 2 to 12. These are given in Table 6-4. In using these guide-

Figure 6-2 **Some Sample Questions in a Reading Interest Interview**

1. Why do you think you need to know how to read?
2. When you are grown up/ an adult (wording adjusted according to age of student being interviewed), how often do you think you will read?
3. If you could choose any two kinds of things to read, what would you pick? (Next questions will vary depending on answers given.)
   a) What novels, non-fiction, comic books, magazines, newspapers do you read?
   b) What do you like about "x"?
   c) Do you have any favorite characters in "x"?
4. Do you have some favorite movies or television shows?
   a) What are they?
   b) What do you especially like about "x" movie or television show?
5. If you could tell your new friend about some interests that are special to you, what would they be?
6. If you had lots of money to buy things to read, what would you buy?
7. If you were to write your own book what would it be called?
   a) What main ideas would you have in it?
   b) Would other people want to read it? Why?
8. When do you like to read (in the morning, evening, before bed time)?
9. Do you like to listen to other people reading?
10. (For older or better-read students.) Are you interested in what other people are reading? Would you read something recommended by someone else?
   a) Would you like to read to other people? Do you recommend books to others to read?
   b) What would you select to read? Do you talk about what you recommended or what was recommended to you afterwards?

Figure 6-3 **Questionnaire for Estimating Problems in Reading Comprehension**

1. Some people are smart enough to do well in school but still have problems with their school work.
   a) Always a problem for me.
   b) Often a problem for me.
   c) Not often a problem for me.
   d) Never a problem for me.
2. When asked to write a social studies report, some people have trouble making it long enough.
   a) Always a problem for me.
   b) Often a problem for me.
   c) Not often a problem for me.
   d) Never a problem for me.

3. Although some books are easier than others, some people read all books at the same speed.
   a) Always a problem for me.
   b) Often a problem for me.
   c) Not often a problem for me.
   d) Never a problem for me.

4. Some people miss questions on tests because they don't check their answers.
   a) Always a problem for me.
   b) Often a problem for me.
   c) Not often a problem for me.
   d) Never a problem for me.

5. It is hard for some people to break up long words into their parts: for example, candidate into can-did-ate, or reporter into re-port-er.
   a) Always a problem for me.
   b) Often a problem for me.
   c) Not often a problem for me.
   d) Never a problem for me.

6. Some people have trouble writing essay tests.
   a) Always a problem for me.
   b) Often a problem for me.
   c) Not often a problem for me.
   d) Never a problem for me.

7. Some people cannot sound out hard words.
   a) Always a problem for me.
   b) Often a problem for me.
   c) Not often a problem for me.
   d) Never a problem for me.

8. Some people do not know the meaning of big words like photosynthesis, radius or irony.
   a) Always a problem for me.
   b) Often a problem for me.
   c) Not often a problem for me.
   d) Never a problem for me.

9. Some people do not recognize "give-away" questions on tests.
   a) Always a problem for me.
   b) Often a problem for me.
   c) Not often a problem for me.
   d) Never a problem for me.

10. Some people do not look over a chapter in a book before reading it more carefully.
    a) Always a problem for me.
    b) Often a problem for me.
    c) Not often a problem for me.
    d) Never a problem for me.

11. Some people do not take good notes in class.
    a) Always a problem for me.
    b) Often a problem for me.
    c) Not often a problem for me.
    d) Never a problem for me.

12. Some people miss questions on tests because they did not read the directions carefully enough.
    a) Always a problem for me.
    b) Often a problem for me.
    c) Not often a problem for me.
    d) Never a problem for me.

13. Some people cannot guess the meaning of a word by reading the rest of the sentence.
    a) Always a problem for me.
    b) Often a problem for me.
    c) Not often a problem for me.
    d) Never a problem for me.

14. Some people do not review their notes and book work before a test.
    a) Always a problem for me.
    b) Often a problem for me.
    c) Not often a problem for me.
    d) Never a problem for me.

15. Some people do not know how to look things up or use library material, such as card indexes, encyclopedias or maps.
    a) Always a problem for me.
    b) Often a problem for me.
    c) Not often a problem for me.
    d) Never a problem for me.

16. Some people do not know how to ask questions about new materials.
    a) Always a problem for me.
    b) Often a problem for me.
    c) Not often a problem for me.
    d) Never a problem for me.

17. It is hard for some people to read a chapter and then tell what it means in a few words.
    a) Always a problem for me.
    b) Often a problem for me.
    c) Not often a problem for me.
    d) Never a problem for me.

18. Some people do not use their study time the way they should.
    a) Always a problem for me.
    b) Often a problem for me.
    c) Not often a problem for me.
    d) Never a problem for me.

19. When some people come to a word they can not read they do not look for clues to help them sound it out.
    a) Always a problem for me.
    b) Often a problem for me.
    c) Not often a problem for me.
    d) Never a problem for me.

20. Before a test some people do not ask what it will cover.
    a) Always a problem for me.
    b) Often a problem for me.
    c) Not often a problem for me.
    d) Never a problem for me.

21. Some people find it hard to read sentences like this next one: The palm nut, from which oil is made, is one of the Congo's leading exports.
    a) Always a problem for me.
    b) Often a problem for me.
    c) Not often a problem for me.
    d) Never a problem for me.

22. What percentage of people who have some of the problems listed on this form have major difficulties with school work _____?

Adapted , by permission, from a questionnaire prepared by Alley, Deshler and Warner, Institute for Research in Learning Disabilities, *Research Monograph 10* (Lawrence, Kansas: University of Kansas, 1981).

lines, one should keep in mind at least two qualifications. First, the majority of the passages found in reading tests are relatively short. Our clinical observations indicate that children and adolescents with reading difficulties make adjustments in their reading rate depending on their perception of the length of the material. If they perceive the material to be too long, they may refuse to read at all. Second, we do not know if children sustain their reading speed in non-test situations. It is likely that they do not read consistently at one speed even given texts of normal difficulty. In short, the guidelines reported in Table 6-4 should be treated as tentative ones.

Informal reading inventory One technique that is described frequently is the teacher-prepared informal reading inventory. Normally, informal reading inventories are constructed from specific reading programs or from basal readers. They are seen to have the advantage of providing data that more closely represents the student's reading performance in the classroom reading program. In order to accomplish this, however, the teacher-prepared inventory must be well constructed (see Chapter 3).

The guidelines which follow differ slightly from those traditionally associated with a teacher-prepared reading inventory. The intention of these guidelines is that the resulting inventory can be used to test specific hypotheses, rather than as a general classroom-based teacher-prepared reading inventory. This modified inventory may be constructed by following the four general guidelines given below:

Figure 6-4 **A Primary Reading Attitude Scale.**

Name________ **Room**_____________ **Teacher**____________

Circle the picture that best describes how you feel.

1. How do you think your good friends feel about reading?

2. How do you feel when you are asked to read aloud in your class?

3. How do you feel when you are asked to read aloud to your teacher in your class?

4. In school how do you feel when you come to a new word in reading?

5. How do you feel about going to school in the morning?

6. When you read your book in school, how do you feel?

7. How do you feel when your teacher reads a story aloud to your class?

8. How do you feel when it is time to do your worksheet or workbook in school?

9. When you read to your friends, how do you think they like it?

10. How do you feel when your parents read to you at home?

11. How do you think you'll feel about reading when you're grown up?

12. How do you feel about talking to your teacher about a book you have just read?

13. How do you feel when someone gives you a book for your birthday present?

14. How do you feel when you are watching TV?

15. How do you feel when you are listening to a radio?

From L. Ollila, *A Primary Reading Attitude Inventory*, unpublished, University of Victoria (1981). Reprinted with permission of the author.

1. a) If the inventory is being based on a graded basal series, paragraphs should be sampled from approximately the middle of each reader across the range of readers likely to be encountered by the referred students. The middle of the text is recommended because it is most representative of the particular level being taught. Usually in basal readers familiar material is reviewed at the beginning and new, preparatory material is introduced near the end. When using source material other than basal readers to prepare an inventory, the criterion of selecting a passage from the middle likely is not applicable. For example, if the inventory is prepared from a series of novels or from readers that have a controlled vocabulary the choice of passages from the middle of the text will be less important as there likely is no graduated increase in difficulty.

   If the material being used is non-graded or taken from some other curriculum (mathematics, science, social studies, etc.), several samples of material should be taken from throughout the text. Readability formulas should be applied in order to estimate difficulty before selecting the most representative passages. Applying such formulas may reveal inconsistencies between sections of the sampled text. Furthermore, applying different formulas may result in variations in difficulty estimates (Harris and Jacobson 1980).

b) Although a deviation from practice, it may not be necessary to type out the selected passages, depending on the estimated frequency of usage and the purpose in assessing. It may be more practical and more realistic to have students read material as it is found in the various sources. The main reason for typing the inventories is so that they may be used again. If one is testing a specific hypothesis with a specific individual, preparing such tests may be wasted effort.

c) Regardless of the source of the material, the next step is to select passages of different lengths. Typically, passages of 25 to 400 words are selected for use. The length of the selections should be adjusted according to two main factors: i) the age and grade of the students completing the inventory, and ii) the purpose in conducting the assessment. If, for example, the purpose in preparing the inventory is to assess the information search strategies of a grade nine student who is having trouble in social studies, the passage selected could be a section of 20 to 30 pages in a reference text.

2. Most commercially prepared inventories include vocabulary lists of about 20 words, selected for each level of reading material included. Similar lists may be compiled for a teacher-prepared inventory. The specific vocabulary should be selected according to the purpose in administering the test. For example, it may be suspected that a student is failing science because he or she lacks the technical vocabulary specific to the subject area or to a unit of study; in this case, science vocabulary would make up the selected words.
3. Finally, five to ten questions usually should be written in order to estimate reading comprehension. Again, the number of questions depends on the purpose in administering the inventory. On some occasions it may be that no questions are necessary; if, for example, the purpose of assessment is to estimate reading speed or to ascertain how well main ideas can be retold. Whatever the number, questions should be written in such a way that specific hypotheses may be tested. Pearson and Johnson (1978) have developed a very useful approach to preparing questions. They suggest using three types of questions: text explicit, text implicit and script implicit. Questions that are text explicit are ones in which the information is stated directly in the material to be read. The second type, text implicit, are questions in which the information is stated but requires the use of inference. Script implicit questions are those that cannot be answered unless the student has some preknowledge of the material to be read. Raphael (1984) has summarized studies done to date using Pearson and Johnson's method and suggests that a) students can learn to discriminate between types of questions and b) training on text explicit and text implicit questions seems to be of benefit to students who are poor achievers.
4. The scoring procedures recommended for classroom based inventories are not appropriate when specific hypotheses are being tested. It makes little sense to

Table 6-4 **Median Rates* of Reading for Different Grade Levels**

Grade	2	3	4	5	6	7	8	9	12
Highest test	118	138	170	195	230	246	267	260	295
Median test	86	116	155	177	206	215	237	252	251
Lowest test	35	75	120	145	171	176	188	199	216

*These rates have been determined by several standardized reading tests. The number of tests included in the various grades are as follows: 7 for grades 2 and 3; 8 for grades 4, 5, 6 and 7; 6 for grades 8 and 9; and 3 for grade 12.

classify readers as reading at the independent level, instructional level or frustration level, as is the case when most commercially prepared inventories are used, when the objective is to determine whether a reader does or does not have a particular skill. Therefore, specific scoring procedures are not recommended; rather, separate criteria will have to be set to determine if hypotheses are confirmed, to be revised, or rejected.

If constructed following the above guidelines, the teacher-prepared reading inventory has two potential advantages when used with students with special needs: a) specific hypotheses about the way the student reads may be stated and tested; and b) this measure can be constructed to match the particular kind of curriculum that students are required to read.

Cloze procedures This technique has been suggested as a measure of reading comprehension (e.g., Page and Pinnell 1979); however, there is some debate as to whether cloze procedures provide estimates of comprehension (Ashby-Davis 1985).

Cloze and modified cloze techniques do seem useful in estimating word sense; that is, the degree to which students are able to read a specific text of a given difficulty and determine which words fit into the text.

In constructing cloze passages for assessment purposes, every *n*th word is omitted, usually the fifth word. Passages of various lengths may be used although most selections are 200 to 300 words long (Ekwall and Shanker 1983).

Bormuth (1967, 1968) suggests that cloze passages may be used to calculate whether reading material is of approximately the right level of difficulty to assign to a particular student. He recommends that the material is too hard if 35 percent or less of the student's answers are correct; it is about right if 45 percent are correct; and it is too easy if 55 percent or better are correct (Bormuth 1975).

One variation on cloze techniques used for assessment purposes is called modified cloze or maze. Instead of providing blanks where students must supply words, three or four possible choices are provided. If words are selected carefully, this modification could provide some indication of sight word knowledge or knowledge of other vocabulary.

Techniques to assess comprehension Comprehension is one of the most puzzled about components of reading. One frequently asked question is whether it is possible to determine if something read has been understood. In order to answer this question, we suggest that reading comprehension must be seen as a set of overlapping activities. Reading is done with different purposes in mind and, accordingly, different comprehension strategies are employed depending on the reason for reading.

In addition to the teacher-prepared reading inventory and the cloze procedures described above, the following suggestions may be useful in determining if comprehension has occurred.

Asking students to paraphrase a reading passage may indicate whether they can state the key ideas. Sequence of occurrence may be added as a requirement if it is important to the meaning of the text. In any case, it is often useful to note how material is organized and recalled by students.

Some children and adolescents have a very poor sense of the components of a story. They do not know approximately when something is likely to occur or what is likely to happen. Estimating childrens' knowledge of story grammar appears to be useful in estimating the likelihood of comprehension occurring (Bachor 1983). In assessing story grammar, children are asked, for example, "What is the beginning?" Students may also be asked to make a series of predictions about what will occur next in the text. In addition, students may be presented with scrambled stories to see if they are able to determine logical order.

Summary We have suggested a variety of assessment techniques that may be used to test formulated hypotheses about a student's reading abilities. The choice of the assessment tool will depend on how complete and current the available background data is.

After collecting information about how students read from background data and further assessment, the next step is to prepare an IEP in which specific instructional goals are stated. Once prepared, the IEP is implemented. The next sections present a number of instructional techniques and materials for this purpose.

APPROACHES TO READING INSTRUCTION

In the remainder of this chapter the focus is on reading instruction. Depending on the assessment of the student's needs, any single technique or combination of techniques may be selected. Before describing some of the methods, strategies and materials used to teach reading, we will briefly deal with two general considerations: first, the necessary ingredients of a reading program; and second, selection of material for students of different ages.

Necessary Ingredients of a Reading Program

Johns (n.d.) suggests there are three general, essential components of a reading program. The first requirement is knowledge about reading. The second is a good teacher. The last is an effective school.

The first and second requirements for an effective program are linked. The more obvious requirement is that teachers be knowledgeable about methods of teaching reading. Further, it is important that teachers be capable of integrating appropriate methods into their daily instruction.

The importance of a good teacher has been well established (e.g., Gates 1937). Factors like the amount of time spent teaching (Stevens and Rosenshine 1981), having a clear sense of purpose (Durkin 1975), and having the skills necessary to deliver instruction (Joyce and Weil 1972) are components of good teaching. Johns (n.d.) adds that having students complete pages in a workbook or other skill package is not teaching; rather, teaching requires some direct action on the part of the teacher to help the student to be better able to complete something he or she could not do before assistance was provided.

There is evidence from Great Britain (Great Britain Department of Education and Science 1975) and the United States (National Assessment of Educational Progress 1982; Venezky and Winifield 1979) that leads us to suggest schools themselves are very important to a reading program. They key components of an effective school may be summarized as 1) all people having a clear sense of purpose, developed, in part, by promoting staff development; 2) parental participation in the school; and 3) clear educational leadership.

Teaching Students of Different Ages

As we have seen, different assessment strategies may be selected, for example, to estimate knowledge of specialized vocabulary or to test skill in reading technical information. The same principle holds when choosing instructional procedures and materials. Two examples may serve to illustrate the importance of using age-appropriate materials and techniques.

Students who are having difficulty reading may not understand the predictability of some components of the reading task. The form of this problem may vary with age. For example, younger children, aged six or seven, may not understand that a word is the same each time it appears. Therefore, they may decode a word each time they see it, despite the close proximity of some of the repetitions. The following strategy has helped several children to learn that a word remains the same each time it appears. First, a repeated word is demonstrated to be the same word each time it occurs. This demonstration is done by finding repetitions of a word both on the same page and on several different pages throughout the story. The child reads each repetition and is told that it is the same word each time it is read. The consistency of the word is then reinforced in a game-like manner by playing "find and read the word." Repetitions should be found on various pages throughout the material. In this manner the child acquires "sight vocabulary." In contrast, however, we have found drilling high frequency word list to be of little use with young children. Children may learn the word list but they do not seem to learn the connection between words on the list and words in the reading text.

A second general example illustrates materials selection. Older students may react very negatively to reading materials that look as if they are meant for younger students. Materials may be rejected out of hand in such cases. While there is an increasing awareness of the need to adapt materials, there still is a shortage of interesting, well written, easy reading material that appeals to young adolescents. Such material as the adapted classics series (published by Globe Book Company, New York) or *Stars* magazine (published by Turman Publishing Company, Seattle, Washington) are helping to fill this void. However, this material, will not lead to the development of an age-appropriate vocabulary. If the intention is to have the adolescent acquire and use a better vocabulary, reading materials must be selected accordingly. In addition, we recommend that a writing program be incorporated in which the student uses his or her newly acquired vocabulary.

Commonly Used Methods and Materials

From the viewpoint of one working with students with special needs there is no obvious benefit in selecting one commonly used program over another, assuming that program is taught well. Again assuming that program is taught well, there likely will be some students who do not appear to benefit. They will continue to have difficulty learning to read. For example, some children in grade one who are taught to read using the reading experience or language experience approach learn the bad habit of guessing at words based on the first letter only. These children usually have an excellent memory and are somewhat impulsive. When they hear a story they acquire the main story line and much of the detail. When they read the story, they could be likened to oral story tellers who use the first letters as clues and guess the remainder of some of the words. If caught early this habit can be corrected without too much difficulty. If left, however, the guessing habit is extremely difficult to correct. Other children, also in grade one, who are learning to read by phonics may have trouble blending sounds into words. In the extreme case they may decode so many words on the page that their reading resembles someone with speech disfluency. Again if not addressed early this habit is difficult to change.

There are four methods or combinations of methods and materials commonly used in reading instruction: basal, language experience, linguistic, and individualized programs. Each one is described briefly below.

Basal Format The basal reader format consists of four main elements: readers, workbooks, teacher's manuals and tests (Swaby 1984).

In a basal reader format, readers are ordered to form a hierarchy which is graduated according to the degree of difficulty of the skills acquired and content read. Basal formats were not designed necessarily to reflect any one approach to reading instruction. Usually there are three or four readers in the first grade and one or two readers in each of the following grades for the remainder of the series.

The use of basal readers has been criticized extensively. Moe and Johnson (1980) examine two traditional criticisms of basal readers. First, the composition did not reflect all groups of people in a variety of roles. However, they feel that this criticism is no longer justified, with one exception: women continue to be stereotyped. Second, the syntactic structures found in basal readers did not match language used by children. Moe and Johnson suggest that syntax may have been sacrificed in favour of vocabulary control. Again, they also suggest that recent basals better reflect children's language. However, there still is room for much improvement in the readability of basals.

The misuse of basal readers and associated material also has met with its fair share of criticism. Durkin (1978) points to several concerns in using basal materials. She suggests that in the early grades there is too much emphasis on word learning. She also adds that in some cases children in early grades are over-questioned. And, still with regard to the early grades, she suggests that reading may become like a performing art. In the later grades, insufficient attention may be paid to building new vocabulary. There may be a heavy reliance on workbooks leading to an over-emphasis on filling in the blanks. Finally, she points out that some children lose any sense of the purpose of reading that they may have had.

Language Experience Language experience is used more often as a tool for beginning reading instruction than for later grades. The basis of the language experience approach is that learning to read is easier if initial language is supplied by the child.

Credit for the basic tenets of language experience often is attributed to Sylvia Ashton-Warner (e.g., Applebee 1978; Johnson and Moe 1980). It should be noted in passing, however, that Gray (1922) had pointed to the importance of meaningful text. Ashton-Warner (1963) calls her method "Organic Reading," and suggests it is a bridge to literacy. Central to her approach is the concept of "key vocabulary." Key vocabulary consists of words that are very meaningful to the child; they must be "made of the stuff of the child himself, whatever and wherever the child" (Ashton-Warner 1963, p. 32).

Using this notion of key vocabulary, children are encouraged to dictate their thoughts, observations, beliefs and questions. Herein lies the object of one of the main criticisms of language experience approaches. The success of this type of program is dependent on the teacher's ability to provide appropriate motivating experiences and to ensure continuity in both language and reading. This is doubly difficult when a language experience approach is used with children who have special needs because many of them lack the experience essential to provide content for this method. In addition, Johnson (1977) points out that some children will require assistance in preparing oral monologues. This is especially true for children and adolescents for whom English is a second language. On the other hand, this technique often is very much of interest to individuals with special needs and may be used as a starting point for both reading and writing. For example, oral monologues may be transcribed. This material may be read to help the student learn the relationship between spoken and written language. Later, the teacher and student may reread the stories and the teacher could help the student to substitute more advanced vocabulary. These new words then may be incorporated into stories that the student writes.

This method has also been criticized for the way in which skills are taught. For example, Petty and Jensen (1980) argue that skill teaching is incidental when the language experience technique is used. They suggest that the omission of important skills may have repercussions in later grades when students have to read in specific content areas. This criticism is an important one for special educators to consider because they will need to be alert to the possibility that referred students may lack certain skills.

Linguistic Approaches Bloomfield has been credited with the development of the linguistic approach to reading (Johnson and Moe 1980; Swaby 1984). The basic idea behind the linguistic approach is that when children start to read they already possess sufficient spoken language. Interestingly, this assumption is similar to the one made in language experience approaches. Bloomfield, however, made a different assumption as to the key to reading: children needed to learn that written symbols represent spoken words; that is, they needed to learn the code of which words are made (Johnson and Moe 1980). Learning this code in turn permitted them to make use of their spoken language in understanding new material that they might wish to read.

This approach has been criticized on the grounds that the language used in the stories is too restricted. While this criticism is warranted, linguistic readers are useful as supplements to language experience approaches. The use of linguistic readers at the Learning Assistance Center at the University of Victoria—in this case the *Miami Linguistic Reading Series* (Robinett, Bell and Rojas 1970)—has been of benefit to children in grade one who were referred out of a language experience program. These children appeared to need the extra structure and the highly predictable patterns found in linguistic readers.

Individualized Reading Programs Individualized reading programs usually are introduced after children have acquired some independent reading skills. Johnson and Moe (1980) point out that individualized reading programs are a logical extension of language experience; although no one approach to reading is a necessary prerequisite to individualized reading.

There are two common patterns used in setting up individualized reading programs. The programs may be either quite teacher-prescribed or may be student-controlled. Students may be encouraged to select and read a variety of materials or they may be provided with teacher-controlled selections from which they choose reading material. By selecting, either from a limited range or a wider range of materials, it is hoped that students will learn to enjoy reading.

The major criticism of this method is that it is student, and in part teacher, dependent. If students are not already interested in reading it may be very difficult to encourage them to begin an active reading program. In addition, instruction in skills may be lacking. Some students will not learn how to apply reading skills to the various content areas.

Summary The programs described above represent the more commonly used approaches to reading instruction. For each described approach there is considerable choice in commercially prepared materials. (Space does not permit us to describe the full range of commercially prepared materials that are presently available.) Given this wide range of choice both of methods and materials, many teachers are able to develop instructional programs for individuals with special needs.

In the next section, some additional programs and approaches are described in which further modifications have been suggested for individuals with special needs.

Special Programs and Approaches

Eight programs and approaches will be described in this section. They include *Distar Reading*, the initial teaching alphabet, *Words in Color, Color Phonics System*, Stott's reading programs, the Fernald technique, the Gillingham-Stillman method, and the *Individualized Blended Sound-sight Program*.

Distar Reading *Distar Reading* (Engelmann and Bruner 1969, 1974; Engelmann and Bruner 1969, 1975; Engelmann and Stearns 1972) consists of three highly structured instructional kits. Readers who are introduced to this program are assumed to need extensive prompting and practice in mastering the code of reading. In addition, orthographic modifications are made to some of the letters of the alphabet. There appears to be several purposes of such modifications; for example, some letters are changed in appearance to prevent common confusions, such as between "b" and "d." The various orthographic alterations are eliminated slowly over the course of the three levels of *Distar Reading*.

Distar Reading has been criticized for several reasons. Two of the more common criticisms are a) that there is too much scripting in the program and b) that there is too much emphasis on the code. The essence of the first criticism is that the amount of scripting prevents teacher creativity. The essence of the second criticism is that there is too much emphasis on the code used—synthetic phonics—and not enough on meaning.

The decision to use *Distar*, or components of it, should be based on such practical criteria as the constraints and goals of the program. Successful completion of *Distar* takes approximately two full years. This means that the teacher (or teachers) operating the program must be relatively certain of following through for two years, especially in view of the alphabet modifications and other prompting techniques. In some cases such program continuity may be in direct opposition to the goals inherent in a special education program. These goals may be to maintain or integrate the child or adolescent in a regular class. Such is the case with most resource teacher programs; therefore *Distar* is less appropriate for use in resource programs than in other special education placements. Also, in a situation in which *Distar* is being used by a resource room teacher, there may be a great deal of discrepancy between the requirements of it and those of the classroom reading program. Again this would suggest that *Distar* is less appropriate as a first choice in a resource room than some other program might be. However, when used in special class placements for handicapped learners as the main component of the reading program, *Distar* appears to be effective (e.g., Stennet and Isaacs 1979).

The Initial Teaching Alphabet The initial teaching alphabet (i.t.a.) is a 44-character alphabet. According to Downing (1979), it is appropriate for use with special needs students because it is consistent and eliminates the confusion in English spelling.

Downing (1965) attempts to clarify what i.t.a. is by describing what it is not:

a) i.t.a. is not a phonetic alphabet (p. xii);
b) i.t.a. is not a perfectly regular code for English phonemes (p. xii);
c) i.t.a. does not have only one single way of representing each phoneme (p. xiii);
d) i.t.a. is not a synthetic phonics method of instruction (p. xv);
e) i.t.a. is not associated with the demand for early reading (p. xvii); and
f) i.t.a. is not limited to a single series of basal readers (p. xvii).

According to Moe and Johnson (1980), there is little use at present of the i.t.a. However, there is some data to suggest it may be effective with special populations

(Downing 1979). The key factors in the decision to use this approach to reading would seem to be two considerations. First, is there enough material using i.t.a. to even consider implementing this program? Second, and more importantly, how smoothly will special needs children and adolescents transfer from the i.t.a. to regular orthography. Although Downing (1965) presents some evidence that this transition can be made, this question has not been answered satisfactorily.

Words in Color *Words in Color* (Gattegno 1968) is similar to i.t.a. in that the author's intention was to overcome the inconsistencies in the letter-sound relationships of the English language. The key to *Words in Color* is the use of eight colour wall charts in which a systematic organization of vowels and consonants is presented (Moe and Johnson 1980).

Bentley (1966) states that the main advantage of using this system is that reversals are eliminated. Moe and Johnson (1980), however, point out that the key disadvantage is that some children have difficulty discriminating between colour-code charts.

Color Phonics System The *Color Phonics System* (Bannatyne 1966) is another system in which the irregular orthography of English is systematized. It differs from *Words in Color* in that the number of colour codings is not as great. The colour code system used in this program seems more logical in that the colour used is the one found in the word. For example, the phoneme *ee* is coloured green to yield a colour-sound correspondence.

Bartel, Boyd and Tobia (1984) suggest that this program is best used as an aid in one-to-one instruction. In addition, this program is better used only as a learning aid rather than as a reading program.

Stott's Reading Program Stott (1970, 1971) developed two reading kits: the *Programmed Reading* kit and the *Flying Start: Learning to Learn* kit. Each one was developed with a slightly different purpose in mind but they both have a common feature as well.

In the first kit (Stott 1970), reading proper is addressed. Stott has incorporated two basic principles into this kit. First, he believes that the basis of reading is using "the phonic code as a fluent and near-instantaneous procedure (p. 11). His first goal is to get the non-fluent reader to realize that words are made up of sounds. Next, he suggests automaticity (not having to stop and sound out each word but being able to read the word as a sight word) is essential. He argues that these two goals are achieved by practicing sound-symbol associations in a simulating environment. The simulating environment suggested is the use of learning games or game-like learning activities. For example, picture bingo is used in teaching what is called "phonic fluency."

The *Flying Start* kit (Stott 1971) is based on the principle that children learn by playing. However, it was written for a slightly different purpose than *Programmed Reading*. The intention of the *Flying Start* kit is to change ineffective behavioural or learning styles (see Chapter 4 for a description of Stott's categorization of behavioural styles). In one game, for example, the two children playing cannot complete the puzzle unless they take turns. The objective in forcing cooperation is to reduce their impulsivity.

When considering whether to use Stott's program, it must be remembered that the *Programmed Reading* kit and the *Flying Start* kit were intended as supplements to a reading program. Students do very little actual reading when using either program; rather, the emphasis is on mastery of a code, phonics. Mastery of phonics is supposed to result in fluent reading. In addition, Allington and Strange (1977), in a general review of reading games, suggest that game-like formats for instruction must be evaluated carefully. Children do become caught up in the spirit of the game with the result that any instructional benefits are lost.

Fernald's Technique Fernald's technique (1943) may be seen as similar to the language experience approach to reading, but as having an additional component. The addition to the language experience approach is a modification in the manner in which the child or adolescent practices remembering new words. The student makes use of all four modalities—visual, auditory, kinesthetic and tactile—and thus it was called VAKT.

There are four stages involved in the use of the VAKT method. The first stage is multi-sensory. In each stage thereafter the use of one modality is discontinued. Throughout this process children do their own reading, never does the teacher read to them.

In stage one, the student is told she or he will be taught a new method of learning words. In addition, the child is encouraged to try to write as many new words as desired. In implementing this step, the teacher, at the child's request, writes a word on a large card or a piece of heavy paper. The child watches while the teacher writes the word in either manuscript or cursive, depending on what is used in the classroom. Next, the child traces the word until it is reproduced correctly without looking at the copy. Tracing is done with the first and second fingers. As tracing is completed, the child says the complete word by syllables (the method used for word attack). The criterion is that the whole word must be written twice without error and without looking at the copy. In addition, as in language experience, words are incorporated into stories so that the child or adolescent will learn to use words in context. After the child completes telling the story, it is recorded by the teacher before the next lesson so that it can be read.

In the second stage, tracing is discontinued. Stage one is stopped when the child no longer needs tracing in order to learn new words. To this end, tracing is reduced gradually until it is no longer necessary. At this point the child learns new words by looking at each word, saying it, and writing it. This is the same procedure as in stage one, but this time used without tracing.

The child has reached stage three when he or she can write new words by looking at the word in print. That is, the teacher is no longer required to write out the words. All the rest of the steps already established are followed.

The last stage has been reached when it is no longer necessary for the new word to be written. Fernald notes that at this point, as the student "looks at the word, the simultaneous association by similarity with words he already knows, together with meanings inferred from the context, gives him an instant perception of the word" (1943, p. 52).

The strength of this method would seem to be the systematic way in which prompts are reduced while interest in reading materials is maintained. The drawback may be the method of word attack. Tierney, Readance and Dishner (1980) suggest that the division of words by syllables may result in the distortion of "pronunciation of certain words, e.g., fat/her" (p. 175).

Gillingham-Stillman The Gillingham-Stillman method has been described as a phonics approach to reading (Bartel, Boyd and Tobia 1984); however, Gillingham and Stillman (1960) object to having their method classified as a phonics method *per se*. They refer to much "phonics" teaching as inappropriate and as being a "grunt and groan method."

The Gillingham-Stillman method is similar in some ways to Fernald's techniques in that children are taught by making use of various methods of presentation. The simplest way of describing this method is to say that the children progress through three sets of associations.

Through the first association, the students learn a phonogram; for example, a apple/a/. To begin, the teacher says the name of the letter and the child then repeats the letter name. This is the most basic requirement and it may be necessary to practice "echo speech" before progressing. In echo speech the child learns to correctly imitate the sounds to be produced. In the next stage of the first set of associations, the sound of the phonogram is made by the teacher and reproduced by the student while he or she looks at the phonogram. In the second association, the teacher makes the sound without allowing the student to see the phonogram. Then the student is asked to name the letter. In the third association, the teacher makes the letter and explains its form and orientation in space. The pupil then traces the letter, copies it and, finally, writes it from memory.

The above description is only an outline of the method; anyone wanting to use this technique is directed to Gillingham and Stillman (1960). The usefulness of the Gillingham-Stillman method may come not so much from the style of presentation of letters but from the systematic way in which new material is introduced. The order of introducing phonograms is planned carefully. Further, considerable emphasis is placed on students using what they learn. Writing, spelling and reading are integrated into this procedure. In addition, the importance of graphing student progress is emphasized so that improvements can be seen.

While these authors argue that they do not use phonics, in fact, a very systematic letter-by-letter, structured, name-sound-symbol association is taught. This set of associations is one of the key differences between the Fernald technique and this approach. The major drawback to the name-sound-symbol approach is one associated with all phonic-type approaches. Some children do not transfer from such detailed letter learning and word perfect reading to fluent, purposeful reading.

The Individualized Blended Sound-Sight Program (IBSSP) Brunsell, in the preface to Ingham's text on reading (1978), states that children in grade one become independent readers in two to four months after starting the IBSSP. Further, she asserts that an average grade one reader independently reads 70 to 100 books; better readers may read 200 or more books; and finally, poorer readers may read from 10 to 25 books. If these claims are accurate, this method holds considerable promise for reading instruction in general.

There are three steps in implementing IBSSP: 1) formal teaching, 2) interaction learning activities and 3) integration of the library reading-sharing period and the work period.

Four key aspects of the IBSSP should be taken into account when deciding whether or not to implement it. First, books selected from the library are the core of the program; the basal reader is still used but only to teach basic concepts orally.

Second, this is an in-classroom procedure in which the goal is as much individualization as possible. In order to meet this goal, reading groups are replaced with pairs of readers. These pairs spend one-half hour per day reading to each other from books of their own choice at their own level.

Third, this method was not built on the basis of adult understanding of reading; rather, Ingham asked children what sounds they heard and built words around those sounds. The sounds are incorporated in the following way. First, letters are presented individually. This presentation is brief, "just long enough to see it, hear it, say it, print it in the air and dramatize it" (Ingham 1978, p. 4). In this way sound-symbol associations are introduced. Next, letters are incorporated back into words. Also, several rules are introduced to assist children in learning to read. For example, rule 29 and 30 are about "or."

> 29. or says "or" and usually comes in the middle of words as in corn.
> 30. ore says "or" and comes at the end of words as in more. (p. 30)

Fourth and finally, a considerable amount of importance is attached to transfer and application. Ingham believes that the way to attain transfer is to have children spend time in learning centres. She argues that in using learning centres the

> teacher is able 1) to teach specific skills through the use of manipulative and written activities, 2) to set each skill to incorporate the area of practice, application and enrichment for that skill, 3) to increase his academic language skills as based on a structural set up. (1978, p. 119)

Summary The first set of programs or program supplements presented all involved some modification in presentation (e.g., Gattegno's use of colour codes) or in the orthography of the English language (e.g., *Distar*). This common feature means that, in each case, some extra prompts are provided for students. Given these program modifications, the resource room or regular classroom teacher who is deciding whether to implement one of these programs must consider the program merits in terms of two factors: a) the probability of transfer back to the regular program and b) the probability of students becoming kit-dependent.

In the latter group of techniques, attention is turned more towards the learner. It is focused on the way instruction is delivered to the student, not on the way the alphabet is presented. Thus, for example, Stott makes use of game-like instruction, while Fernald alternates the style of presentation, incorporating extensive practice while using several sensory modalities.

In looking at all of these techniques, shortfalls are found readily. All of the above techniques have been criticized by someone at some time. The important point is, however, that all of the above programs allow for a variety of ways of practicing reading, at least some of which children with special needs will enjoy. This is the key to their success. Student enjoyment of the instructional technique used by the teacher seems essential to sparking students' desire to learn to read.

Recognition, Fluency and Comprehension Strategies

In this last section, attention is turned to three issues that have been indirectly addressed throughout the chapter: a) word recognition and word attack, b) rate and fluency in reading and c) comprehension.

Before looking at each of the above issues separately, one caution seems necessary. Although a poor reader may need assistance in only one of the above areas, it is more likely that assistance will be necessary in all areas simultaneously. Dolch (1945) suggested five steps to follow in carrying out a remedial reading procedure. These are still relevant today, with some revisions and addition. They are as follows:

1. Assess to determine how and what the child (or adolescent) is reading. Having the student read what he or she can helps to build confidence.
2. Build up a sizeable sight vocabulary that the student recognizes quickly. Sight vocabulary is built by a) oral reading in class or by listening to others read, b) pairing good and poor readers so the fluent reader may provide assistance with words, and c) by studying sight words.
3. Practice some method of word attack.
4. Develop comprehension, starting with summarizing the meaning of passages. Begin with easier material and work towards regular classroom work. Retelling or summarizing, however, should not be overemphasized. The student should be encouraged to tell the main ideas.

> Just open a book, and ask him to read a paragraph, either silently or aloud. Use a paragraph rather than a sentence so that the words of the book cannot be remembered in parrot fashion. (Dolch 1945, p. 39)

5. Develop interest in reading by providing the child or adolescent with story ladders. Story ladders, as described by Dolch, are lists of books on the same topic which are ordered from easy to hard. In addition, encourage students to read widely, while following the story ladder theme in terms of difficulty. Try to suggest reading across subject areas, if possible, so that individuals with special needs learn that their ability to read is not confined to "reading story books."

Word Recognition and Word Attack

Four techniques for younger children Johnson and Louis (1985) describe several strategies that were designed for use with younger children. These are a) matching phrases, b) spoonerisms, c) progressive cloze and d) story ladders.

Matching phrases is a technique which takes a familiar text and cuts it into phrases. Phrases are selected and the child demonstrates where they fit into the text. Johnson and Louis suggest using nursery rhymes to start.

Using spoonerisms is also suggested as a technique to teach phonics. Again, the child starts with familiar language. Two examples are given, the first for vowels and the second for consonant/vowel combinations. In each case, the teacher models the needed change.

Old King Cole was a serry old moul.
Passy cut, passy cut, where have you been?

In progressive cloze, again familiar language in the form of rhymes is used. Single words are put on a card and the cards are placed in the correct order. As the rhyme is read aloud several times, the teacher removes individual words on a cumulative basis.

Finally, in Johnson and Louis' use of story ladders students are presented with rhymes in which the last part of the sentence is removed. The student fills in the text as well as he or she can.

Synthetic word families "Word families are [groups of] word elements that contain both vowel and consonants elements, to which can be synthetically blended an initial consonant element" (Tierney, Readance and Dishner 1980, p. 148). To use word families sentences may be written in which a pattern is repeated, as in the following series. Dan is a man. Is Dan a man? Dan ran. Is the man running? The man named Dan ran after a can.

The pattern described in the last paragraph is typical of that found in linguistic readers. Aukerman (1971) reports that linguistic readers are designed so that a word element ("an" in the example given) is found in all positions in a word: beginning, middle and end.

Syllabaries This technique is used to teach word recognition using the syllable as the unit of pronunciation. Cunningham (1975-76) describes the procedures used as consisting of seven steps. First, the number of syllables are determined "by counting the vowels (remembering that the *y* at the end of a word is a vowel): four vowels = four syllables" (1975-76, p. 129). Second, the word is divided into syllables using the following, ordered rules:

a) Divide between a root word and a prefix: re certify.
b) Divide between two intervocalic consonants (except consonant diagraphs, of course): re cer tify.
c) Divide before a single intervocalic consonant: re cer ti fy. (1975-76, p. 129)

Third, pronunciation rules are applied to each syllable. Fourth, the syllables are blended trying different acceptable sounds for each one. Fifth, if the word still is not recognized, different sounds are tried for some or all symbols. Sixth, if still unknown, other rules for pronunciation are applied, such as the *schwa* for unaccented syllables. Seven, if the student pronounces the word correctly and recognizes the meaning, reading is continued.

Reading strategy lessons Reading strategy lessons (Goodman 1975a) arrange instructional material in such a way that the reader uses language cues to make predictions, to verify them, and to try again if the first predictions prove incorrect. Basically, these lessons are designed as a variation on the cloze procedure, but include some additional rules. The material should incorporate familiar language; it should be unambiguous; it should contain natural redundant content; and it should be interesting to the student. The omission of words should be planned so that redundant words are eliminated. In this way, according to Tierney, Readance and Dishner (1980) the student becomes aware of both grammatical and meaning clues.

Reading Rate and Reading Fluency

Increasing reading rate Earlier in this chapter, the importance of varying reading rate for different purposes was discussed. This should be kept in mind when considering techniques to increase reading rates. Reading rapidly is not always desirable or useful.

Spache (1963) describes two ways to improve reading rate. First, he recommends prereading difficult materials. He suggests that students practice reading headings, topic sentences, and introductory and concluding paragraphs. Second, he recommends practice in scanning. His argument is that such techniques may not result in direct improvements in reading speed but students will learn "organized comprehension which is, in turn, the foundation upon which increased speed must be built" (Spache 1963, p. 251).

Various machines may be used to improve reading rate. Spache (1981) notes that while mechanical methods are useful, gains in rate after training usually are not maintained. Devices used include controlled readers and tachistoscopes. Spache (1981) warns that the use of such machines may result in eye strain or visual fatigue.

Neurological impress method This method was devised by Heckelman (1969) to improve reading fluency. Tierney, Readance and Dishner (1980) note that this method originally was termed echo reading.

Armitage (1978), in describing echo or neurological impress reading, suggests the following revised guidelines and steps:

> 1. Before being considered for Neurological Impress remediation, the student should be proficient in basic phonics, including the ability to blend sounds smoothly. Some students still read word by word, substitute words, leave off endings, and/or generally have great difficulty in handling a reading passage even after gaining proficiency in phonics.
> 2. The person reading with the child should be thoroughly familiar with the desired goals and the qualities considered important to fluent reading. Most success has been achieved with the *same* person working with the child during the entire period, preferably the teacher.
> 3. The pupil chooses a book which he or she would like to read. It is usually a very easy book because of the pupil's aversion to reading. High interest, low vocabulary books are very suitable because of the interest level and a continuity of sight vocabulary which will be reinforced daily. It is wise for the teacher to discourage any book that is obviously too difficult.

> 4. The teacher explains to the student that they are going to read together for 15 minutes each day for a week. . . .
> 5. The student and the teacher sit comfortably together with the student on the teacher's left and holding the book.
> 6. Both begin to read aloud. The teacher reads at the same speed as the student, being careful not to read ahead and have the child stumbling over the words. The teacher is providing a model of fluency for the pupil to imitate. The teacher uses a pen to point to the words as they read, thereby focusing the child's attention on the exact word being read. The pen should move smoothly to reinforce the smooth flow of reading.
> 7. The teacher should be aware constantly of any potential problem words that may be about to be read on the page. As they reach them, the teacher should stop, hold the pen in position, and give the pupil a chance to sound out the word, perhaps blending softly with him. If it is a sight word, the teacher should supply the word immediately, have the pupil pronounce it, then print the word on a separate paper for review later. (This item shows the importance of the suitable choice of material. Too many problem words may give the pupil a feeling of being defeated.)
> 8. Reading continues in this manner for an uninterrupted 15-minute period.
> 9. As the sessions progress and the pupil becomes more proficient, the teacher may gradually let the child take the lead in oral reading with the teacher's voice very slightly behind the child's as reinforcement.
> 10. Before each session begins, the pupil reviews the problem words from the previous day, receiving a check mark for the words he recognizes immediately. After 3 or 4 checks, the word may be removed from the list. (p. 23)

Progress should be checked on a frequent basis. This may be done by having students read on their own a short passage of the difficulty that they have been reading. As progress is noted the difficulty of the material should be increased gradually. Armitage (1978) stated that it may be seven weeks before any real progress is noted; however, some changes may be seen in three to four weeks.

Repeated readings Samuels (1979) devised a technique which he called repeated readings. This strategy is based on the notion that automaticity of reading (LaBerge and Samuels 1974) is essential.

Repeated readings is a very simple technique. Short passages are reread until fluency is reached and then a new passage is read. The problem with this technique and other similar strategies is lack of transfer (Moyer 1983). Each time a new passage is started, there is a reduction in reading rate and fluency as a function of the difficulty of the text and the reader's familiarity with it.

Comprehension

Some factors influencing comprehension Perhaps the most difficult issue to address adequately is poor reading comprehension. It is doubtful that a clear picture of the elements involved in comprehension will be gained by having students read short passages (10 to 300 words) only. Our clinical experience leads us to suggest that different strategies are necessary to read longer (1000 to 5000 words) passages. To comprehend longer passages most people can no longer rely on memory alone;

they require different techniques in order to extract the main points or to précis a lengthier text. Length is not the only variable that appears to have been downplayed by teachers, researchers and by those who prepare commercial reading tests. Factors such as text structure (Bos and Filip 1982), the students' facility with language (Samuels 1982), or the clarity of the writing style (Samuels 1982) are among the others that seem to affect text comprehension. Another important consideration is that at least some students with special needs seem either not to use their comprehension strategies independently or to use such strategies ineffectively (DiVesta, Hayward and Orlando 1979; Wong 1980, 1982). Furthermore, more attention seems to have been paid in the past to assessing comprehension than to teaching it. For example, students have been required to read a paragraph and answer a series of questions. While having children or adolescents answer questions may be a useful assessment procedure, asking questions will not teach students how to answer them. In other cases, children or adolescents have been required to fill in the blanks in order to estimate how well they understand. This technique may provide some general hints about completing that worksheet; however, little assessment information will be gathered about comprehension of text in a more general sense.

Many of the techniques already described will have an influence on comprehension. For example, if the student cannot read fluently, comprehension likely will be difficult. Similarly, prereading has an effect on reading comprehension (Sachs 1984). In the following section, some other strategies to increase reading comprehension are presented.

Beginning exercises for comprehension Comprehension appears to be, at least in part, tied to children's sense of words. Ingham (1978) suggests three strategies to address comprehension that may be started in grade one: logical comprehension, structural comprehension, and appreciative comprehension.

In logical comprehension, children are taught three steps. First, they are taught to read sentences as they might say them. The argument here is that the meaning of the sentence may change depending on the words emphasized. Next, children are taught to classify the key words they read. For example, if a child is reading a story about pigs, horses, cows and chickens on a farm, the teacher will ask the child to identify and classify the material. Pigs are not just pigs but animals; horses are animals; cows are animals; therefore, the story is about animals. In the final step the conclusion, the generalization, is summarized in a sentence, "This story is about animals on a farm."

In structural comprehension, children are taught to analyze the structure of a sentence (from left to right) in order to begin to comprehend it. Grammar, then, becomes part of comprehension. For example, in the simple sentence "Bill is walking to town," the teacher may introduce grammar in the following manner:

1. Who is walking? (Bill) "Bill" is the subject of the sentence.
2. What is Bill doing? (walking) "Walking" is the action word (verb) part of the sentence.
3. Where is Bill going? (to town) "To town" is the object of the sentence.

In addition, in structural comprehension the teacher may attend to the way the sentence is read. In this case the emphasis is on word order. For example, "Walking to town was Bill" or "To town, Bill walked" are variations of the basic sentence in which the meaning of the sentence is altered slightly. Sometimes sentence transpositions have no effect on meaning; however, sometimes significant (and incongruous or humourous) meaning shifts result, as in "The town walked Bill." By analyzing sentence structure in this way, reading and writing are both incorporated.

Appreciative comprehension involves children starting to read stories other than those found in basals and "reporting back." In this reporting back, students describe what they read and how they felt about what they read.

Semantic webbing This technique was developed by Freedman and Reynolds (1980) as a way of helping children organize and integrate information to make it meaningful. There are four basic elements to a semantic web: a core question, web strands, strand supports and strand ties.

The core question is selected by the teacher and written in a central circle. It must be worded in such a way that several answers are possible. This question is the key to semantic webbing as the remainder of the steps relate to the key question in some manner. The answers to the question, usually one word, are called web strands and are placed in smaller circles a short distance from the core question. Inferences, generalizations and facts are found in the story to support the web strands. These details are termed strand supports and are drawn in small circles and connected to the answer being supported. Any possible relationships among the strands are called strand ties. Strand ties are drawn in dotted lines to illustrate any connection between answers. In Figure 6-5, the components of semantic webbing are illustrated.

Freedman and Reynolds suggest that semantic webbing may be used to help children understand story plots in basal readers. To use semantic webbing, however, students must know why they are reading. For example, the reason for reading may be to answer specific questions or it may be to prepare a report or a composition. In addition, they must be familiar with the components of a story (see story grammar).

Schema theory or story grammar Stein and Trabasso (1982) state that the basis of schema theory can be traced back to Barlett's 1932 interpretation of how people remember. The main idea behind schema theory is that individuals use their prior experience to construct an idea of what will occur in any story. Story schema, then, are the ways a person represents the components of a story and the ways those components fit together (Whaley 1981). The other important element of this theory is that stories have a structure, or grammar, which can be learned. Practically, the purpose is to help children and adolescents make better sense of a story by teaching them the story grammar. Knowing what is about to happen or what likely will happen makes understanding easier. As a result, comprehension improves.

There are several different versions of story grammars (see Whaley 1981 for a readable review). We have selected the version given by Stein and Trabasso (1982) to present. According to these authors, there are two basic components in a narrative: setting and episode. Stories vary in the number of components found in them and the degree to which these components are embedded in one another. In Table 6-5 the

Figure 6-5 **The Components of a Semantic Web**

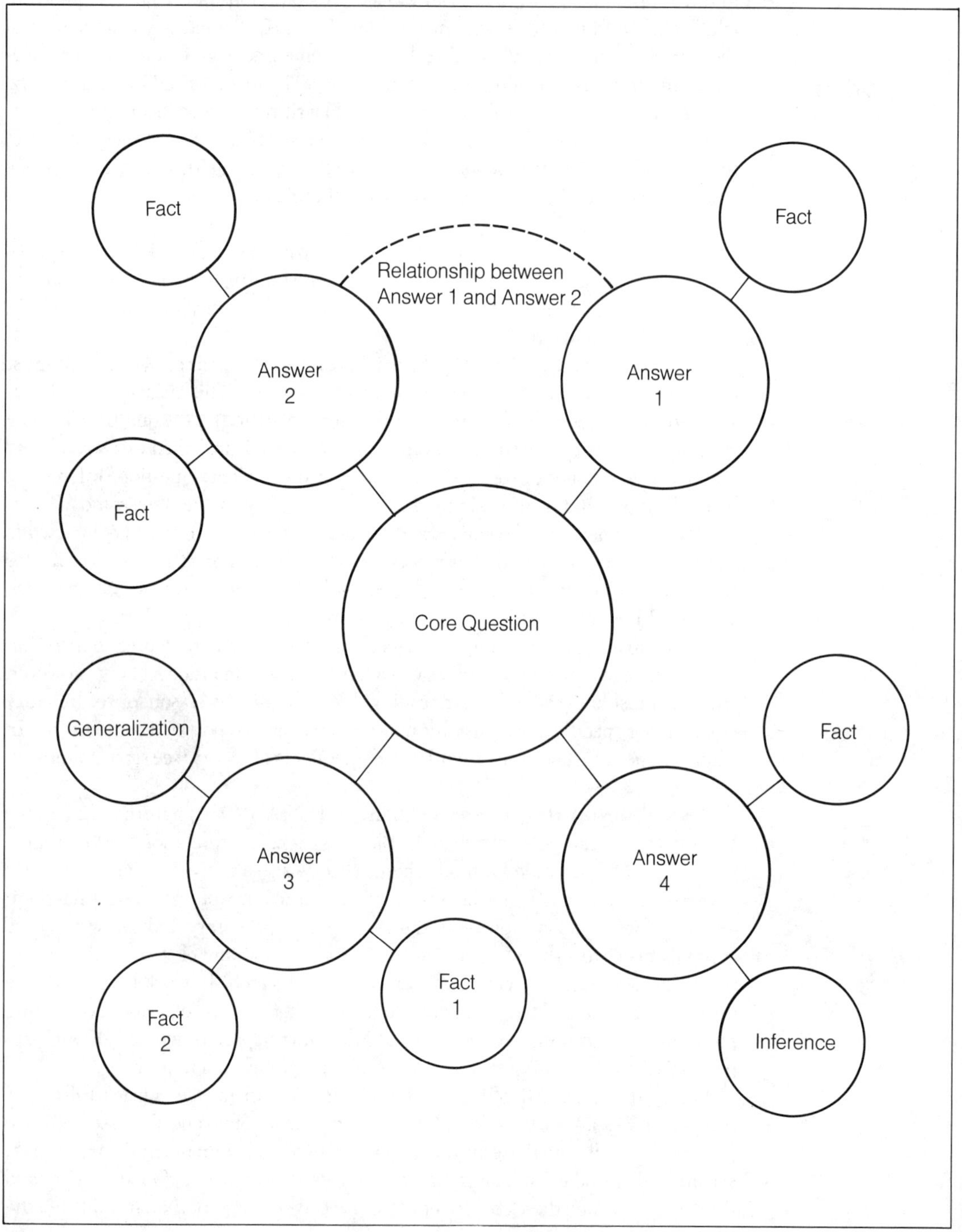

Stein-Trabasso story grammar is presented along with a general example of its use.

Any set of instructional materials may be used to teach story grammars. Different

Table 6-5 **Categories and Types of Causal Relations Occurring in a Simple Story**

1. **Setting**	Introduction of the protagonist; contains information about the social, physical or temporal context in which the story events occur.
Allow Episode	
2. **Initiating Event**	An action, an internal event or a physical event that serves to initiate the storyline or cause the protagonist to respond emotionally and to formulate a goal.
Causes	
3. **Internal Response**	An emotional reaction and a goal, often incorporating the thoughts of the protagonist that cause him to initiate action.
Causes	
4. **Attempt**	An overt action or series of actions, carried out in the service of attaining a goal.
Causes or Enables	
5. **Consequence**	An event, action or endstate, marking the attainment or nonattainment of the protagonist's goal.
Causes	
6. **Reaction**	An internal response expressing the protagonist's feelings about the outcome of his actions or the occurrence of broader, general consequences resulting from the goal attainment of the protagonist.

Example of a Well Formed Story

Setting	1. Once there was a big grey fish named Albert. 2. He lived in a big icy pond near the edge of a forest.
Initiating Event	3. One day, Albert was swimming around the pond. 4. Then he spotted a big juicy worm on the top of the water.
Internal Response	5. Albert knew how delicious worms tasted. 6. He wanted to eat that one for his dinner.
Attempt	7. So he swam very close to the worm. 8. Then he bit into him.
Consequences	9. Suddenly, Albert was pulled through the water into a boat. 10. He had been caught by a fisherman.
Reaction	11. Albert felt sad. 12. He wished he had been more careful.

Reprinted, with permission from Nancy Stein and Thomas Trabasso and the publisher, from Robert Glaser (Ed.) *Advances in Instruction: Volume 2* (Hillsdale, New Jersey: Lawrence Erlbaum Associates, 1982), 219.

texts, however, vary in their story structures. Therefore students' expectations about story structures need to be adjusted accordingly. One set of instructional materials is particularly well suited to helping students with special needs begin to learn story structures. The *Choose Your Own Adventure Series* published by Bantam and the *Choose Your Own Adventure, Junior Series* by Bantam-Skylark are excellent materials which vary considerably in difficulty level. Children enjoy reading them and deciding what should happen next in a story. In order to use these materials to teach story grammar, however, more attention must be paid to the questions. In the text as given, students are asked to make predictions by answering one of the questions. To teach story grammar, only one addition is necessary: children should make two or three predictions about what will happen next and explain why they think it will happen.

Whaley (1981) describes three techniques that may be used to teach story grammar. In a prediction task, children read incomplete stories and then write what they think comes next. After writing, the predictions are discussed in terms of the components of the story grammar. In a macro-cloze task, which is similar to a normal cloze procedure except that this time whole story categories (for example, setting) are deleted, students read the story and describe the missing element in either a written or oral form. In a task involving scrambled stories, a story is separated into various story grammar components, these are mixed up, and the students are then required to put them back together in a logical order.

Other techniques In addition to the strategies discussed to increase comprehension in this chapter, some other techniques described elsewhere are helpful. The majority of the strategies that are discussed in Chapter 10 may be used to aid comprehension. For example, the keyword method for remembering essential details may be used to help students review curriculum. In addition, the questioning strategies of Sitko, Slemon and Bachor (1983) may be used to help children and adolescents discuss and comprehend texts. In Chapter 11, organizational skills are discussed, the absence of which may have a bearing on comprehension. If, for example, a student does not know how to study, comprehension of material will be affected. Or, if a student's test-taking skills are weak, the impression may be obtained that poor comprehension of material has occurred. There are a number of other examples that could be selected from Chapters 10 and 11. In addition, some of the suggestions made in Chapter 12 for secondary students may be helpful.

Case Study

The final section in this chapter is a case study of a student with a reading problem. This case was selected because it is a difficult one. To start with, there is almost no information available. What is known is that Samantha, a grade six student, is having trouble understanding classroom reading material.

Steps 1 and 2: Background information and preliminary hypotheses

Samantha's background information was limited, to say the least. She was new to the school at the beginning of the grade six year. All that was available in the cumulative files where marks from grades one, two and three. Although of little use, these grades were in the passing range (C to B). Thus, at this point no preliminary hypotheses could be stated.

Step 3: Assessment of the individual

Since so little was known about Samantha, assessment was begun by observing in the regular classroom. The objective was to gain some idea about how Samantha acted in the classroom. Was she not understanding because she was not attending to instruction? This tactic proved to be less than helpful. Samantha, over the course of three half-hour observations, spent 89 to 98 percent of classroom time working. There was some evidence to suggest that this data was not totally accurate, however. Her classroom teacher reports that Samantha spends much of her time giving the appearance of working when she does not understand much of anything.

Thus, having found few helpful leads, the decision was made to administer a global test to get some sense of Samantha's reading abilities. A standardized reading test was selected and administered. The resultant data was interpreted and was found to be uninformative. It was learned that by the standards of this particular test Samantha was performing about average, her scores were at stanine four and five. Again, little progress had been made. At this point questions were beginning to form about the appropriateness of the referral. However, looking at Samantha's assignments suggested that something was amiss.

Next, Samantha was interviewed about her reading habits and her attitude towards reading. She reported that she didn't have any books at home. She never read, preferring to watch TV instead. Further, she didn't take books out during library period or at least attempted to avoid signing them out. She reported that "Oral reading was OK," but that she hated silent reading. Further, it was learned during the interview that Samantha read all books at the same speed, had difficulty with the "meaning of big words," and didn't look over chapters before reading them.

As a next step, some teacher-prepared tests were administered. It was learned that Samantha had a great deal of difficulty understanding technical terms and using references to answer questions in a science lesson. Three passages from the same story were selected to learn about her reading speed. These passages were selected to be of the same difficulty with about the same number of "big" words in each one. The results were confusing as her reading speeds varied considerably: for a passage of 300 words, 93 wpm; for a passage 200 words, 34.8 wpm; and for another passage of 220 words, 220 wpm. Further, Samantha retold the main ideas in the first and last case but not in the second. Also, her word attack skills were very weak.

Step 4: Reformulate hypotheses

As no hypotheses had been possible earlier, it was necessary to state them at this point.

a) Samantha's poor word attack skills seem to be interfering with reading speed and comprehension.
b) Samantha's vocabulary may be preventing her from understanding some passages.
c) Her reading speed is variable. It is influenced by her attitude to the material

more than the complexity of the reading material.

d) Samantha's attitude is poor, resulting in her spending as little time as possible reading.

Step 5: Assessment of instructional methods and materials

In order to assess instructional methods and materials, observations and trial lessons were completed. It was learned that Samantha worked reasonably well in small groups when assignments were carefully structured. She responded well to positive reinforcement—just a record of progress was enough to encourage her. It was discovered that Samantha loved to be the teacher.

Step 6: IEP goals

Six IEP goals were set for Samantha:

a) Increase reading speed for independent reading material to 240 wpm.
b) Increase ability to use word attack skills for unfamiliar words in reading material, from 50 percent correct to 80 percent correct.
c) Increase accuracy of questions answered in all instructional materials, from 75 percent accurate to 95 percent accurate.
d) Increase ability to answer text-explicit and text-implicit questions, from 60 percent correct to 85 percent correct.
e) Increase on-task behaviour during teaching sessions to 90 percent from 60 percent.
f) Increase amount of silent reading done.

Steps 7 and 8: Implementing and evaluating the IEP

The major goal, although stated last, was to increase the amount of time spent reading. Over the period of six months, 1200 pages were read in several different books. More pleasing was that by the end of the year Samantha had read part of a book at the grade six level.

Reading speed was improved in part by the above procedure. For each 200 pages that Samantha read independently and for which she summarized the main ideas, she was allowed to teach a lesson for ten minutes. Throughout the year, reading speeds were sampled and increased to 260 wpm. This rate was maintained even with shifts in difficulty.

Three word attack methods were taught: using context clues, using structural strategies and using phonic strategies. A teacher-made post-test indicated that Samantha could find the meaning of words 75 percent of the time—less than was hoped. While she was much better at using context, Samantha still had considerable difficulty when required to decode some words.

A teacher-made test was used to assess the ability to answer questions. Samantha improved to a 73 percent level. She continued to have difficulty answering text-explicit questions.

Lastly, periodic observations were used to check Samantha's behaviour. Although considerable support and encouragement was necessary over the course of the year, the goal was met.

7 MATHEMATICS

Mathematics has tended to be seen as the third "R" in special education (Cawley et al. 1979). This, to some extent, reflects its place in the curriculum generally. The effect of this perception has been slow detection and minimal support for many special students in the mathematics area in the past. There are, however, indications of some change as educators become more aware of the importance of mathematics and science in the curriculum for all students (McLeod and Armstrong 1984; Smith 1984; Bachor 1985). Such changes include the development of specialized math clinics (Robitaille 1976), the inclusion of cognitive skills in remedial curricula (Sharma 1981) and the addition of computer-based curriculum supplements (Leidtke and Muir 1984; Saper 1983).

CAUSES OF PROBLEMS IN MATHEMATICS

There have been several general explanations given concerning why students have problems in mathematics. The most frequent general hypotheses include the following. 1) The emphasis on literacy in the curriculum results in less time being given to mathematics. This is seen as especially relevant in the primary grades and results in problems not being detected as early as are problems in the language area. 2) The acceptance in the larger community that mathematics is not essential for adult survival (especially for girls) results in time and resources not being allocated to it as readily by either the family or the school. 3) The poor preparation for teaching mathematics of many elementary school teachers results in poor instruction by and poor attitudes of the teacher. 4) A general lack of student readiness exists in the visual-motor and concrete experience areas that relate to mathematics. 5) Students may have problems in abstract and conceptual thinking. 6) The possibility exists of generally poor attitudes and motivation for competence among all pertinent persons. 7) Reading problems have a pervasive effect. 8) There is a high frequency of students who come to a new curriculum without having mastered the prerequisite skills. Each of these hypotheses may contribute to an understanding of why a certain student is failing mathematics and so each needs to be addressed.

Hypothesis one can be empirically tested at the local school level. If less than an hour per day is allocated to mathematics in the early grades, then it will be impossible to cover the curriculum and ensure that the students have enough practice time to internalize the basic concepts being taught. In addition, if the mathematics period is often the one taken away for special events and/or is used to make up for language classes that are missed, then one can soon appreciate why students will have mathematics difficulties.

Hypothesis two can also be examined empirically by the classroom teacher, the special education support teacher and the principal. Does mathematics get its fair share of the budget in terms of materials and opportunities for staff inservice? Is there a frequent input of new and exciting materials to motivate the students and teachers to spend more time on mathematics? Do the parents feel that mathematics are important to their child's educational experience? In a recent study (Crealock and Park 1984) parents of children from 4 to 12 were asked to list activities they did at home to support their child's academic experiences. Mathematics related activities were reported considerably less frequently at each age level than were language related activities. In addition, only 40 percent of the mothers indicated that they liked mathematics as contrasted with 95 percent who reported liking reading. These findings related positively to statements concerning their children's attitudes and competencies in mathematics. This, in turn, supports a general hypothesis about mathematics being undervalued in terms of the actual behaviours and attitudes of those concerned.

The third hypothesis concerns teacher preparation and adequacy of instruction in mathematics. In a recent study, Fitzmaurice (1980) asked 62 learning disability teachers to rate themselves on competency in teaching 35 mathematics skills. Over 90 percent felt themselves competent to teach numeration, counting, place value, common symbols, computational processes of addition, subtraction, multiplication and division, regrouping and money. These skills reflect those competencies taught in the primary grades and suggest that within this limited repertoire teachers function well. In the higher level skills, that is, those which involve conceptual understanding, teachers were not as confident about their abilities. Less than 50 percent of the teachers felt able to teach number theory as it relates to various bases or to prime and composite number; the processes used to find area and volume of common geometric figures; or an understanding of the concepts of time, linear planes, weight, liquids and temperature. They also felt unable to assess student attitude to math, to use error analysis, to assess verbal problem solving or to use non-traditional conceptual approaches to mathematics instruction. These findings are supported by the recent report on the Canadian component of the *International Education* study reported by Walstrom et al. (1984). In this study the authors found that many higher order concepts were not taught and that teachers generally relied on the methods suggested by the textbook rather than expanding their repertoire by using alternate or additional methods.

Hypothesis four concerns the extent to which basic readiness skills affect mathematical skills. This has been an area of interest of many authors (Richardson 1984;

Whyte 1980; Johnson and Myklebust 1967; Hammill, Bartel and Bunch 1984). The areas of problems most frequently examined include awareness of spatial concepts such as over, under, in front of, beside, middle, left, right, etc.; awareness of the concept of reversibility; the ability to classify objects along single or multiple dimensions, the ability to seriate or order events; an understanding of object permanence and of conservation of number, area, mass and volume; and the possession of adequate gross and fine motor coordination.

The fifth hypothesis concerns the influence of perceptual and conceptual skills for the student. In the perceptual area, a teacher must be aware of how well a student attends to the appropriate stimuli; how well he or she discriminates both auditorially and visually between relevant and irrelevant material; how appropriate the auditory and visual memory skills are for both simple and sequential memory tasks; and how able the student is to retrieve necessary information from long-term memory. In the conceptual area, one must be sensitive to students' needs for concrete experiences before they can handle abstract concepts and symbols (Nesbit 1979) and to their sense of mathematics as a subject involving rote learning as opposed to logical thinking (Crealock 1981).

Hypothesis six involves the role of attitude and motivation. To some extent this has already been addressed through hypothesis two; but it does bear reexamination, this time focusing on the student. Interviews and observation of behaviour give the teacher a good sense of the extent to which the student sees himself or herself as competent in mathematics and of how accepting he or she is of a poor mathematical self-concept. This is particularly relevant for the female student who may not have any good female models (Crealock and Park 1984) and who may feel that competence in mathematics is not necessary to her sense of the accepted female role (contemporary feminism notwithstanding).

In hypothesis seven we examine the pervasive role of reading and language skills in mathematics competence. Some authors have differentiated mathematics difficulties in terms of those due primarily to language and reading problems versus those due primarily to problems in quantitative thinking (Johnson and Myklebust 1967; Kosc 1974). The specific language problems associated with number include reversals of digits, ignoring zeros, inability to write numbers, and confusing symbols; while reading problems generally interfere seriously with all problem solving and even an understanding of basic instructions past the early grades. This is especially problematic because reading competence in mathematics demands that the student read thoroughly since all the information is relevant and must be comprehended as a whole.

The final hypothesis involves the competencies that students bring to their new experiences. Over 50 percent of the variance in attaining a new skill is accounted for by the acquisition of the necessary subskills (Bloom 1971;Gagne 1977). Most learners with special needs come to a mathematics concept with spotty subskill competence at best. Such lack of competence must be diagnosed and then compensated for by review and additional teaching if these students are to succeed in mastering new material. This issue will be addressed in detail later in the chapter.

GOALS FOR INSTRUCTION IN MATHEMATICS

As one begins to assess problems in mathematics, it is important to be aware of the general expectancy for mathematics competency in the kindergarten to grade eight curriculum. Otherwise, it is too easy to over- or underestimate the degree of mathematical competence a student has. Tables 7-1, 7-2 and 7-3 give levels and competencies for set and number operations; geometry, graphing and measurement; and problem solving, vocabulary and symbols respectively in grades one through eight. In examining these tables, one is quick to appreciate how spatial awareness problems, language difficulties, memory deficits or reading problems would begin to manifest themselves in poor mathematical achievement. An example of this is easily seen in Table 7-3. The language demands at the oral, written and visual levels are quite specific and require that the student be able to handle abstract concepts and symbols if she or he is to achieve competency of all the subskills at even the primary level.

While it is conceded that some students may not attain all the competencies expected of the average learner, there have been several attempts to determine which competencies are most essential for the special needs student to master. In general terms these areas of competence include numerical and computational skills, time and measurement, basic geometry, application of mathematical concepts to life situations, a positive attitude towards mathematics and towards self as a functional user of mathematics, and an ability to solve problems that have a mathematical component. More specifically, the Ministry of Education of the Province of British Columbia has detailed these competencies to include the use of comparisons and sets; the ability to read and write numerals and to relate them to numbers and counting; the ability to solve problems involving addition, subtraction, multiplication, division, decimals, fractions and place value; the ability to use calculators and estimation; the ability to translate word problems into number problems; the ability to use common measuring instruments, to tell time, to use maps and graphs, and to use all denominations of money; the ability to use spatial concepts in understanding shape, size, area, volume and mass; the ability to use money appropriately, to prepare a monthly budget, and to use banking services and contracts advantageously (*Special Education Core Curriculum Supplement* 1983). While other provinces would emphasize some of these competencies to a greater or lesser degree, most seem to agree on them as the essential ones for the lower achiever in mathematics.

McLeod and Armstrong (1982) found that students with mathematics deficits are generally functioning at a grade three to four level and thus need fairly basic remediation. They also found that over 60 percent of respondents at the intermediate and secondary level indicated they had major deficits in division, fractions, decimals, terminology in fractions and general mathematics vocabulary, multiplication of whole numbers, percent, place value and measurement. McLeod and Armstrong suggest that, with such seriously deficient skill levels, remedial teachers probably should abandon basic skill instruction and focus on life skill mathematics for the special needs student—a concept that is consistent with the core curricula suggested above.

The decision concerning what to teach will reflect individual provincial and district mathematics guidelines. For students who are receiving mathematics instruc-

Table 7.1 **Set and Number Operations**

Level	Competencies
Grades 1, 2, 3	Understands and uses associative property, communicative property and identity property
	Addition and subtraction facts to four digits with regrouping
	Multiplication facts to 50
	Division facts to 50 with remainders
	Counting to 9999
	Place value to 9999
Grades 4, 5, 6	All multiplication tables
	Add, multiply and divide using decimals, fractions and mixed numbers
	Can use factoring, prime numbers in multiplication
	Can work with decimals to the 10 000th
	Counting to billions
	Count by 1s, 2s, 3s, 4s, 5s, 6s, 7s, 8s, 9s, 10s
	Count using Roman Numerals to 1000
	Place value to billions
	Can express data as a percentage
	Can find ratio of two sets
Grades 7, 8	List factors and multiples of numbers
	Can convert between decimals and fractions and percentages
	Can write decimal numbers using powers to 10
	Can round off numbers to specific place value
	Write sets of equivalent numbers
	Perform calculations using a series of operations

Based on the math guidelines used in British Columbia (1982), London, Ontario, Board of Education (n.d.) and the Ontario Ministry of Education (1980).

tion primarily within the special education setting, the goals will reflect where the student is at this point in time and should relate to a modified core curriculum. For students who receive their mathematics instruction in the regular class, the teacher will have to choose those components of the full curriculum that are most essential and give the student as much support with them as possible.

Much of the above material will be included in the background information that is

Table 7.2 **Geometry, Graphing and Measurement**

	Competencies		
Level	**Geometry**	**Graphing**	**Measurement (time, money)**
Grades 1, 2, 3	Can name, describe and draw a circle, square, rectangle and triangle	Can use graphs to record measurement	Can estimate and measure length, area, mass, volume, capacity Can tell time to minutes Knows time units from minutes to years Can solve money problems using all coins
Grades 4, 5, 6	Can name and describe a cube, prism, cylinder, cone, tetrahedron, pyramid, sphere, quadrilateral, parallelogram, pentagon, hexagon and octagon	Can interpret and construct concrete graphs, pictographs, bar graphs, line graphs, circle graphs	Can estimate and measure distance Can use appropriate instruments to measure Can use decimal notation with metric measure
Grades 7, 8	Can draw all geometric shapes Can use concept of congruency for segments, angles and triangles Can use a compass appropriately Understands Pythagorean Theorum	Can graph on a number line	Can use degrees to describe angular measure

Based on math guidelines used in British Columbia (1982), London, Ontario, Board of Education (n.d.) and the Ontario Ministry of Education (1980).

Table 7-3 **Problem Solving, Vocabulary and Symbols**

	Competencies		
Level	**Problem solving**	**Vocabulary**	**Symbols**
Grades 1, 2, 3	Translates word problems into numerical problems and solves	Uses faces, edges and corners to describe geometric shapes Knows names of circle, square, triangle, rectangle Uses expanded notation to 999 (units, tens, hundreds)	$=$ $\neq$ $>$ $<$ $+$ $-$ $\times$ $\square$ $(\ )$ $\div$ $\overline{)\quad}$ ¢ $
Grades 4, 5, 6	Uses alternate methods to solve problems Writes open sentences for word and situation problems Estimates answers to problems and checks answer	Knows all language pertaining to decimals, fractions, sets and numerical operations	
Grades 7, 8	Analyzes a verbal problem Constructs a visual model to aid problem solving	Understands the terms whole, rational, irrational, real number, integers Uses connectives such as "and" and "or" in open sentences	a, b, c ε ϕ e n v

Based on math guidelines used in British Columbia (1982), London, Ontario, Board of Education (n.d.) and the Ontario Ministry of Education (1980).

available concerning the student who is having mathematics problems, although it may be necessary to seek further information. Additional information will be found in student records, interviews with parents and referral forms themselves. In some cases there will not be enough background information available to form any hypothesis and therefore it will be necessary to do more global assessment in order to focus in on problem areas. By the time the student is referred there should be sufficient data to isolate major problem areas and possible causes. Examples of the general problem area(s) would include computational skills involving regrouping, word problems, etc. Examples of causes or tentative hypotheses would include reading skills at a low enough level to undermine mathematics problem-solving skills; limited skills within the family and/or limited acceptance of the son's or daughter's inability to do mathematics; student capability of mastery of relevant subskills being undermined by the use of an inappropriate method (e.g., use of abstract rather than concrete strategies); or student mastery of concepts requiring more time than is generally available to him or her, etc.

ASSESSMENT IN MATHEMATICS

The next step in the decision-making model involves testing to determine the student's individual strengths and weaknesses. The hypotheses generated above should direct much of the testing since information must be attained that speaks to each hypothesis. Such testing can be done using both commercially prepared and teacher-prepared assessment instruments.

Commercially Prepared Assessment Instruments

There are several commercially prepared tests on the market that are helpful in providing an initial understanding of where the student is in relation to others or to the competencies expected at his or her grade level. These are summarized in Table 7-4. Three of the tests (CTBS, CAT, and *Key Math*) have been normed on both Canadian and American students, one has been normed on British Columbia students and the remainder only provide American norms.

Key Math (Canadian edition, Connolly, Nachtman and Pritchett 1979) has been used widely in Canada to determine the individual skills and deficits of students who are having mathematics problems. It is criterion-referenced, administered in a one-to-one situation and nicely formatted to maximize interest and motivation. The original normative sample consisted of 1222 American students, in grades kindergarten to seven (Salvia and Ysseldyke 1981) with the Canadian version being standardized on an additional 769 students from 19 school districts across four provinces—British Columbia, Manitoba, Ontario and Nova Scotia. The subtests measure numeration (24 items), fractions (11 items), geometry and symbols (20 items), addition (15), subtraction (14), multiplication (11), division (10), mental computation (10), numerical reasoning (12), word problems (14), missing elements (7), money (15), measurement (27) and time (19). Because of the Canadian standardization the test can be used as either a norm-referenced or criterion-referenced test, although it is the latter that is of most value to the educator.

Another test that has been normed and standardized on a Canadian population is the *Canadian Test of Basic Skills*. It was administered to 32 137 students from a representative sample of schools across all ten provinces and the Yukon and Northwest Territories. The split-half reliabilities in mathematics range from 0.79 to 0.86 in the six grades tested but the confidence band for grade equivalents ranges over several grades. There is one mathematics test for the primary levels, three for the grade 3 to 8 level and one for the grade 9 to 12 level. Each test has a specified time limit. Items measure number, money, time, problem solving and computation. Raw scores, grade equivalents, standard scores and percentile ranks are provided. Although the test purports to be both norm-referenced and criterion-referenced, only

Table 7-4 **Standardized Tests in Mathematics**

Test	Category	Grade level	Layout	Content	Technical adequacy	Administration
Key Math Diagnostic Arithmetic Test. Canadian Edition (Connolly, Nachtman and Pritchett 1979)	Diagnostic; criterion-referenced; norm-referenced	K-6	Easel format	Operations, applications, metric supplement	Normed on small American and Canadian samples; reliability range is 0.28-90; content validity adequate	One-to-one; power test; easy to administer
Canadian Achievement Test (CTC, McGraw-Hill Ryerson 1977, 1978, 1981)	Norm-referenced; criterion-referenced	1-12	Paper and pencil	Some matching to Canadian curricula	Canadian norms; reliability and validity adequate	Group test; has time restriction; some practice items
Canadian Test of Basic Skills (Hieronymus, Linguist, Hoover and King 1978, 1979, 1980, 1981)	Norm-referenced; criterion-referenced	K-12	Paper and pencil	Partially based on Canadian curricula	Canadian norms; confidence bands wide; reliability and validity adequate	Group test; restrictions; no practice items past primary level
Stanford Diagnostic Arithmetic Test (Beatty, Madden, and Gardner, 1976)	Diagnostic; criterion and norm-referenced	1-10	Paper and pencil	American content, most of which parrallels Canadian; no metric component	American norms; validity adequate	Group administered
Metropolitan Achievement Test Mathematics Instructional Battery (Hagan, Farr, Prescott and Barlow 1978)	Criterion-referenced	K-10	Paper and pencil	American curricula	Good standardization; reliability and validity adequate	Group test, teacher can go from test data to instructional objective

Test	Category	Grade level	Layout	Content	Technical adequacy	Administration
Wide Range Achievement Test, Mathematics Subtest (Jastak and Jastak 1978)	Norm-referenced	1-10	Paper and pencil; some oral items at level 1	Standard computational skills tested; no metric items	Standardization, reliability and validity data are inadequate	Individually administered; easy to score, quick
Peabody Individual Achievement Test, Mathematics Subtest (Dunn and Markwardt 1970)	Norm-referenced; criterion-referenced	K-12	Easel format	Standard curricula; no metric component	Good standardization; poor reliability and validity	Individually administered; easy to use screening measure
WISC-R Arithmetic Subtest (Wechsler 1974)	Norm-referenced	K-10	Oral question and answer; written questions for later items	Very sparse sampling of numerical and problem-solving skills	Good standardization; reliability and validity limited	Individually administered
B.C. QUIET (Wormelli 1983)	Norm-referenced	1-7	Easel format, paper and pencil or oral response	Based completely on British Columbia curricula	B.C. norms; reliability marginal; validity adequate	Individually timed test; easy to use

norm-referenced scores are supplied; the user must determine the criterion-referenced scores by matching student performance on individual items to the mathematics skill the item tests. In summary, this test has good reliability and was standardized on a Canadian group; however, it is problematic for use with special education students because of the difficulty in determining the criterion-referenced scores and because of the time limitations it imposes. This latter issue is one that affects all students to some degree. Nagy (1984) found that in an analysis of 12 000 students who took the CTBS at grade four, 20 percent of them did not reach 75 percent of the items on the mathematics subtests. This suggests that the CTBS is biased against the slower student, who must get his or her marks from only that portion of the test that is completed while the faster student gets them from the entire test.

A second group achievement test that has been normed on the Canadian population is the *Canadian Achievement Test* (CTC, McGraw-Hill Ryerson 1977, 1978, 1981). It is a multi-levelled group-administered test that has been based in part on Canadian curricula (it is an adaptation of the *California Achievement Test*) and standardized on a representative sample of 76 000 students from across Canada. The reliability coefficients range from 0.80 to 0.97 and content validity is acceptable, although, again, the confidence intervals for percentiles and grade equivalent scores vary. There is a 20-item locator test to help the user determine the appropriate level for further testing. This is followed by two subtests that measure basic number facts and concepts and applications. There are more practice items provided in the CAT than there are with the CTBS; plus, the test booklets have more space in which students may respond. These two features make the test attractive to the user who is testing students with problems in mathematics. In a recent comparison of these two tests (Bachor and Summers 1984), the authors felt that the *Canadian Achievement Test* was a more useful instrument for assessing low achievers than was the *Canadian Test of Basic Skills*. Their reasons included the shorter administration time, easier to use booklet format, availability of practice items and greater technical adequacy. While the CAT is not used as widely as is the CTBS, is is gaining support in Western Canada and in pockets in Central and Eastern Canada.

The *Wide Range Achievement Test* (WRAT) is a widely used assessment instrument in Canada. While it was developed using American norms, there have been some attempts to develop local norms. Pope, Warrington and Adams (1982) standardized the WRAT on all the students in grades one, three and five in three schools in Victoria. These three schools were selected as a cross sectional representative sample. Students that had failed a grade or who were in special education were not included, which does bias the results. At the grade one level, the mean grade equivalent was 2.36, at grade three it was 3.86 and at grade five it was 5.24. The results suggest that the WRAT may overestimate at the grade one level; but results are close to expectancy at grades three and five. It should be remembered however that the WRAT is not generally felt to be a reliable diagnostic instrument.

The *Peabody Individual Achievement Test* (PIAT) is generally considered to be a norm-referenced test. However Algozzine and McGraw (1980) demonstrated how the mathematics subtest could be used as a criterion-referenced test. They have identified 14 major subgroups of arithmetic behaviour and the corresponding test items. These include number discrimination (items 1, 2, 3, 4, 12, 15, 17, 23, 33), size discrimination (items 5, 6), shape discrimination (items 11, 22, 41), general information (items 19, 21, 28, 35, 40, 53, 69), counting and addition (items 7, 9, 10, 23, 26), subtraction (items 8, 14, 24, 27), money items (items 16, 18, 29, 42, 43), multiplication (items 30, 24, 26, 46), division (items 31, 32, 38, 39, 52), fractions (items 13, 20, 47, 55), numerical relationships (items 37, 44, 48, 51, 54, 60, 62, 65, 72), word problems (items 45, 49, 50, 58, 59, 61, 63, 66, 78), geometry (items 56, 57, 64, 68, 73, 74, 75, 79, 81, 84), and algebra (items 67, 70, 71, 75, 77, 80, 82, 83). The PIAT cannot be used to define precise problem areas, but does provide the user with a method of sorting out the responses the student makes and, therefore, to better understand his or her strengths and weaknesses. As Algozzine and McGraw state:

> ...the use of a diagnostic testing approach to the PIAT mathematics subtest, within the range of the behaviors sampled, provides more information from which to begin to plan meaningful educational programs than simply recording global performance scores. (p. 77)

The B.C. QUIET (*British Columbia Quick Individual Test*) was developed through the Education Clinic at the University of British Columbia (Wornelli 1983). The arithmetic test consists of 5 oral and 60 written questions ranging over the B.C. Core curriculum from kindergarten through grade ten. The author states that it should be used with a child who has been referred for academic achievement problems in order to determine that a deficit relative to the student's peers exists. If the deficit shows on this screen test, they recommend further testing to determine the specific problem areas. The test was normed on 150 students at each grade level. The students were randomly chosen according to school size, community size and gender. Physically handicapped, language impaired or mentally handicapped students were excluded. Reliability ranges from 0.75 to 0.86 which falls below the 0.9 level recommended by Salvia and Ysseldyke (1981) for a diagnostic test. The standard error of measurement is low. Content validity appears adequate.

The WISC-R mathematics subtest has been included in Table 7-4 because of the widespread use made of this test as a diagnostic indicator of achievement in mathematics. Schenck and Levy (1981) found that 75 percent of IEP goals were developed using this as their prime measure of mathematics competence. Salvia and Ysseldyke (1981), however, point out poor reliability and validity for this subtest and suggest that it is not technically adequate to be used in this way. It also tends to be confounded by its heavy dependence on auditory sequential memory, which makes it even more inappropriate for use with many handicapped students. To support its use diagnostically, Levy (1981) gave the WISC-R to sixth graders who had been classified as either disabled or not disabled in mathematics according to scores obtained by using the WRAT. He presented the questions in four formats: verbal presentation-verbal response, verbal presentation-graphic response, graphic presentation-verbal response, and graphic presentation-graphic response. For both groups, the most prevalent testing mode (verbal presentation-verbal response) yielded the worst performance. When this was combined with grade six vocabulary (versus grade three level), performance was even poorer. While the authors interpret the overall findings as support for use of the WISC-R as a diagnostically valid measure, agreement with their conclusion depends on one's acceptance of the WRAT as a valid diagnostic tool—that validity is questionable. It also suggests that the reader ignore their other findings concerning the inadequacy of the presentation and response format for good results.

In summary, there are several tests available that have been standardized on Canadian populations. They include both individual and group administered tests and so yield adequate normative data to the user. Diagnostic data can be obtained but only by careful persèverance on the user's part. None of the tests listed in Table 7-4 have enough items to be truly diagnostic, although they can provide data on problem areas that should be probed further through teacher-prepared assessment devices.

Teacher Prepared Assessment Instruments

Once the student has been identified as having a possible mathematics deficit, further assessment is required to pinpoint the specific error patterns and subskill gaps that exist. This can be done through a variety of informal methods.

One of the advantages of a teacher-prepared instrument is the relaxed format that it allows. Plata and Calvin (1984) encourage testers to give the students time to get used to them, to administer tests over several days, to evaluate performance as soon as possible, to develop their own tests based upon local curricula and to get feedback on how to improve the assessment instrument. Thorough evaluation should flow naturally from a careful task analysis of the final objective. Items should be included from each objective in the analysis and should be presented so that the student moves from easy to difficult ones. In Figure 7-1, a task analysis for addition of two digits with carrying is presented.

An informal test developed from this would include 3 to 4 items for each of the 11 frames to determine the student's present mastery level of the relevant subskills. This approach gives the user a sense of where the student is to go (final objective), which prerequisite skills are necessary to accomplish the objective, and where the student currently is in obtaining mastery of those prerequisites.

A major area for informal testing in mathematics involves readiness skills. The essential readiness skills include auditory sequential memory, attending to material, directionality, spatial relationships, vocabulary and symbol usage, one-to-one correspondence, counting, conservation of quantity, classification, and seriation (Smith 1976; Johnson 1979; Kaliski 1967; Wallace and Kaufman 1978; Nesbit 1979; Hammill, Barlet and Bunch 1984). In Table 7-5, these skills and suggested methods for evaluating them are listed. It is important not to take these skills for granted since they underlie all subsequent understanding and competence in mathematics. A final readiness skill that needs to be assessed is reading. This can be done through any of the techniques described in Chapter 6. Failure to appreciate the pervasive influence of reading in mathematics may seriously undermine any remedial approach.

Attitude towards mathematics is also important to assess. Students who have serious negative attitudes are much more difficult to remediate than those who feel positive or at least neutral about the subject. One quick, non-threatening way to get

Figure 7.1 **Task Analysis in Mathematics**

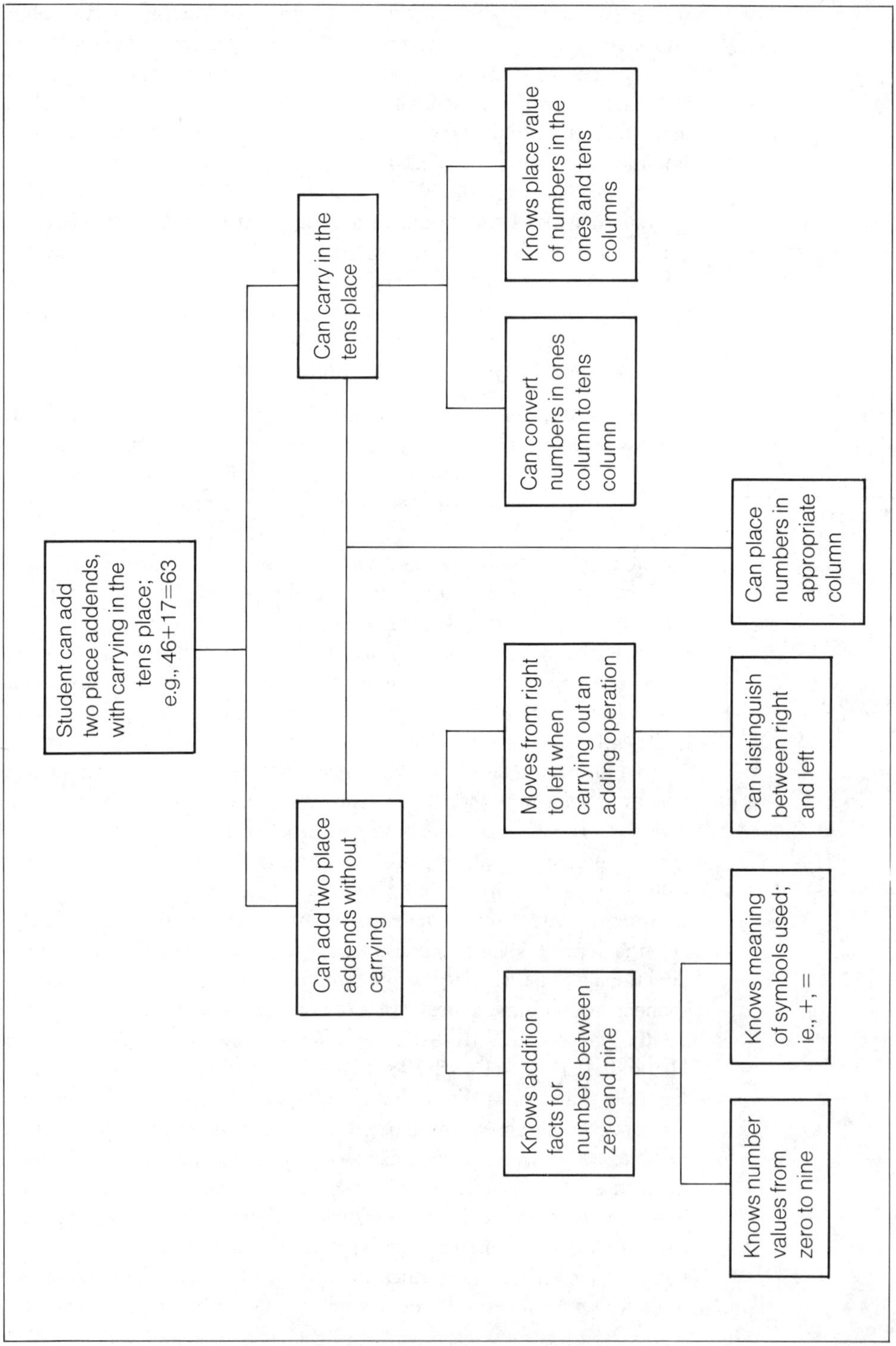

Table 7-5 **Informal Assessment of Readiness Skills**

Skill	**Informal assessment**
Auditory sequential memory	Have students repeat orally administered mathematics instructions. Start with short instructions and increase the number and complexity of the instructions until ceiling level reached.
Attending to material	Observe how much time a child spends attending to math work, systematically scanning for information.
Directionality	Observe tendency to go from left to right in solving numerical problems.
Spatial relationships	Have student copy problems to check for correct vertical alignment; do a multiplication or division problem to check for alignment. Observe tendency to transpose numbers.
Vocabulary	Have student name geometric shapes and descriptions, numerical terms and concepts.
Symbols	Have student identify symbols used at his or her level and explain their meaning.
One-to-one correspondence and matching	Ask student to pass out materials to each student in class. Have student copy a row of articles.
Counting	Have student count as far as possible. Have student count and point to 1, count and point to 2, etc.
Conservation of quantity	Vary content and spatial configuration of number sets or sets of objects.
Classification	Ask student to identify the object that does not belong with the others. Have student sort objects along various single dimensions and multiple dimensions.
Seriation (ordering)	Have students order a series of objects that are graduated for size, colour, density, weight, etc.

information concerning attitude involves the use of the semantic differential, i.e., bipolar adjectives, (Osgood 1957) with a three-, five- or seven-point scale such as that shown in Figure 7-2. The concept to be reacted to is presented at the top of the sheet, followed by six to eight adjective pairs presented in random order to avoid the student responding to a positive or negative set. The user can choose any bipolar adjectives that appeal to him or her and can also use the same technique to test reactions to other general or specific events such as mathematics, exams, division, fractions, etc.

This same information can be obtained by having students respond in a more traditional test-like manner. For example, students may complete items like the following: "Having to work out my own solution to fraction questions makes me feel uncomfortable: 1) strongly disagree 2) disagree 3) undecided 4) agree 5) strongly agree" (Alberta 1980).

Observation of student behaviour in mathematics also yields good information about attitudes. Schminke (1978) suggests that teachers rate students on a five-point scale ranging from always to never on the following behaviours.

1. listens carefully to instructions
2. asks questions in class
3. sticks at seatwork until finished
4. groans, fidgets, drops pencil, etc., during lesson
5. joins math club
6. spends time at math centre
7. likes math games
8. uses math spontaneously in shopping, calculating sports averages, etc.
9. works ahead in math book
10. asks to leave the room during math
11. likes math class
12. likes to help others in math

While readiness skills and attitudes are important to assess, most low achievers have serious deficits in the subskills that support more advanced work. Approaches to testing include task analyses, check lists, error analysis, oral interviews and observation.

A task analysis of each meaningful task will highlight the lower level skills so they can be assessed. In Figure 7-1 a task analysis in addition was presented with subskill test items. Similar analyses in problem areas provide specific assessment and lead directly into remedial objectives and strategies.

Checklists are also useful ways of assessing skills, especially if they include related skills that are logically sequenced. In Figure 7-3 a checklist for fractions that could be used to check competence with elementary level students is given. The essential hierarchy of tasks is defined, which allows the teacher to check off the competencies

Figure 7-2

Bipolar Adjectives Scale to Assess Mathematics Attitudes

Directions:
Check the number for each adjective pair that best says how you feel about mathematics. Check only once for each pair of adjectives. You will have seven checks when you are finished.

Mathematics

A	Very A	Somewhat A	Neither A nor B	Somewhat B	Very B	B
	1	2	3	4	5	
Great	_	_	_	_	_	The Pits
Cruel	_	_	_	_	_	Kind
Interesting	_	_	_	_	_	Dull
Insignificant	_	_	_	_	_	Important
Easy	_	_	_	_	_	Hard
Dull	_	_	_	_	_	Shiny

that the child comes to the class with and to write in the date of mastery of the newly taught concepts for each student in the mathematics group.

There is a growing trend towards assessing students' work through error analysis (Bachor 1978, 1983, 1985; Hopkins n.d.; Crealock 1981). Few students approach mathematics problems randomly; they may make errors but there usually is a pattern to those errors. Some of the more common error patterns for the low achiever include

1. saving face and putting down anything in the answer space,
2. literal interpretations of instructions,
3. simplification, omitting questions or steps in the procedure,
4. faulty computation,
5. faulty algorithm,
6. poor alignment of numbers,

Figure 7-3

Checklist to Assess Understanding of Fractions

Skill	**Student**					
	Tom	Carol	Martha	Pete	Geoff	Molly
Can identify fractions ½, ⅓, ¼, ⅙, ⅛ of a concrete object						
Can identify fractions ½, ⅓, ¼, ⅙, ⅛ of a pictorial representation of an object						
Can identify fraction symbols						
Adds simple fractions with same denominators						
Adds simple fractions with different denominators						
Adds mixed fractions with same denominator						
Subtracts simple fractions with common dominator						
Subtracts mixed fractions with common denominator						
Subtracts mixed fractions with different denominator						
Multiplies simple fractions						
Multiplies mixed fractions						
Divides simple fractions						
Divides mixed fractions						

7. inappropriate notation in regrouping,
8. problems with place value and regrouping.

Saving face is a frequent strategy and shows up as bizarre responses that don't seem to follow a pattern. Literal interpretations are involved when students view subtraction as taking the smallest number from the larger number no matter where it occurs in the problem, e.g.,

$$\begin{array}{r} 438 \\ -\,289 \\ \hline 251 \end{array}$$

This may also be the error pattern in addition when students view it as adding all the numbers together, e.g., $39 + 75 = 24$.

Simplification occurs most often as an omission of a step, or steps, in complex problems such as the following: 1) omission of the middle number in the multiplier, e.g.,

$$\begin{array}{r} 396 \\ \times\ 487 \\ \hline 2772 \\ 1584 \\ \hline 18612 \end{array}$$

2) ignoring the number to be carried in the tens column, e.g.,

$$\begin{array}{r} 86 \\ +47 \\ \hline 123 \end{array}$$

Faulty computation occurs when the student doesn't know his or her basic facts. In the problem below, the student did not know 9 x 6, 9 x 7, 9 x 8, or 9 x 9; therefore, any problem containing these factors was wrong even though the student could handle all other aspects, e.g.,

$$\begin{array}{r} 97 \\ \times\ 46 \\ \hline 942 \\ 388 \\ \hline 4822 \end{array}$$

Faulty algorithm is the most frequent error pattern found in students' answers to multiplication and division problems. It is evident when students omit steps or do them in the wrong order because they cannot remember the complete operation.

Poor number alignment is a frequent problem with younger low achievers. It is compounded when students have to copy work from texts to their own workbooks since there is a tendency when copying to misalign numbers as the following example shows:

```
 837
- 21
----
 606
```

In copying the question, the student placed the numbers in the subtrahend in the middle of the numbers of the minuend, subtracted from the units and hundreds columns where the numbers seemed to fit and then added a zero to make the number look right.

Place value as a concept is very difficult for many students and they may handle it through simplification (forgetting the notation indicator) or through an inappropriate manner of notation such as 1) forgetting the indicator, e.g.,

```
 67
+29
---
 86
```

2) making the notation so large he or she adds it instead of the tens value, e.g.,

```
 ¹56
+95
----
 101
```

or 3) placing the notation in the hundreds column and adding it, e.g.,

```
 188
+ 15
----
 193
```

At times it is not easy to find the error pattern in a student's work. A technique that can save the teacher time is the oral interview (Brownell 1928). The oral interview involves the following steps.

1. Observe the child's performance in mathematics.
2. Choose a problem type that the child is having difficulty with and ask the child to solve a representative problem.
3. Ask the child to talk out loud and tell you each step he or she is thinking about as he or she does the problem.
4. Record the child's entire response. Do *not* try to teach the child. Note posture, attitude, time required, etc.
5. Analyze the oral response for error pattern.

Leidtke (1983) recommends the oral interview as a way to discover computational errors or incorrect procedures. He states that students like them because they get individual attention and teachers like them because they help detect students who respond too literally, who have good defenses or face-saving mechanisms, or who interpret the problem in an unexpected manner. An example of the latter is a child's response to the teacher's request to make up a story for 5 and 2. Answer: "Once there was a flower who was five years old. She had a brother who was two years old and the mother and father were really old, around 25 and 52" (p. 34). An original response but not what the teacher expected.

In summary, readiness skills, attitudes and subskill competence each need to be assessed through informal teacher-developed methods. The methods to be used include task analysis, check lists, error analysis, oral interviews and behavioural observation. Once the teacher has pinpointed the problem areas, the next step is to decide how to remediate the problems that have been identified during assessment.

TEACHING STRATEGIES

There are several general considerations that must be taken into account in any specific teaching strategy for students with special needs.

Reading level Bachor (1985) and Levy (1984) have suggested that the syntax and vocabulary level of many instructions and word problems are too difficult for the mildly handicapped student. A rewriting of these in a less complex form or a careful review of the advanced vocabulary is often necessary to ensure that the student can comprehend the materials and work independently with them. This is especially true since reading in mathematics is more difficult for most students than is prose reading. Such difficulty occurs because reading mathematics requires a very thorough deliberate search for detail rather than a general comprehension of the intent of the author.

Procedure aids Some students have difficulty remembering general instructions, assignments or details about mathematics. Additional reminders may be given, such as having the student repeat instructions or problems back to the teacher, having instructions on the chalkboard for easy reference, or having the student and/or teacher ensure that all necessary instructions are written down in the student's workbook and encouraging the student to refer to the workbook.

Concrete versus abstract presentation Many students need to have mathematics concepts presented in a concrete format long past the point where textbooks advocate that methodology. Piaget (Flavel 1963) suggests that students can't handle concepts at the abstract level until they are 12 to 15 years of age generally and even then only with subject material with which they have had much experience. It would seem, therefore, that the low-achieving student would need concrete presentations well into his or her secondary school program (Case 1982). In a study involving 435 grade seven students in Alberta, Bye, Harrison and Brindley (1980) found that 56 percent of students were at a concrete level of functioning in fractions and ratio but that the textbooks and teachers were teaching to that level only 4 percent of the time. This mismatch makes these topics difficult for all students to learn but they are additionally problematic for the low-achieving ones.

Learning hierarchies These should be developed from a knowledge of general curriculum guidelines and specific problem areas. Both testing and remediation should focus on the subskill training that many students require.

Spatial problem Many students cannot handle the large number of problems often presented on a worksheet. Reducing the number of items from five to ten per page allows the student to complete a page and experience success. It also avoids any problems that students with poor motor coordination have with large writing. Some students also have difficulty lining up their numbers correctly and may need vertical lines drawn on their worksheets to ensure appropriate number placement. Another method to aid with proper alignment is to turn a normal page sideways and then do the problem. For students with right-left confusion, green lines on the right side of their worksheets will remind them of where to start when doing numerical problems. Use of coloured dittoes make this a fairly easy modification to put into practice.

Self-reliance Because few teachers have time to be constantly on call to a student and because low-achieving students need to develop a sense of their own adequacy to solve problems, it is important to develop mechanisms that allow the students to check their own work independently of the teacher.

There are many commercially produced workbooks and other materials available that reinforce basic principles. A guiding consideration in choosing them should be the degree to which their use allows the student to be independent from the teacher at both the directions and instruction stage and at the feedback stage. Few students can afford to wait for someone else to explain how they are to do remedial worksheets, nor can they wait for feedback on the correctness of their efforts. Few teachers have the necessary time immediately available to instruct and correct the work of the many students who need to do either remedial or general review sheets. Some of the most useful materials that do allow the students to do their work on their own are the *Veri-tek* (1977) materials. These involve cases of 12 or 24 plastic tiles and a series of workbooks in the basic math skills for kindergarten to grade 12. The student places the tiles number side up in the top half of the case, answers the questions given in the workbook by number, finds the answer in the answer box, and places his or her tile on the corresponding letter in the bottom of the answer case. When all the problems are solved, the student closes the case, turns it over and opens it. Each exercise page in the workbook contains a geometric design that the tiles must match. If some of the tiles don't match, the answers were incorrect and the student removes the mismatched tiles, returns the case to its original position and resolves the problems that were incorrect. These materials involve both concrete and abstract manipulations, allow the student to work independently of the teacher, reinforce basic skills, and motivate through their use of visual designs and immediate feedback.

Success Most low-achieving students have not had success with mathematics. Remedial work should be structured so that students can obtain 95 percent success on all their efforts. This will require a very careful sequencing of material. The payoff comes in a developing self-confidence and independence in the mathematics area.

Practice Most low achievers can attain mastery of essential skills if they have enough time and practice to thoroughly learn them. This includes massed practice opportunities while they are acquiring the skill and spaced practices to reinforce the skill throughout the year.

Maintenance and generalization If newly learned skills are to be maintained and to be transferred to other related tasks, they must be reviewed frequently and the transfer to other tasks needs to be taught. Blankenship and Baumgartner (1982) describe a program for low achievers in mathematics in grades three, four and five that includes reinforcing students for generalizing skills, training several response exemplars, developing and using a verbal rule, staying with one skill until it is mastered and plenty of modelling and demonstrating of both the original and transfer tasks.

Two researchers based in Canada have suggested broad guidelines for remedial mathematics programs. Bereiter (1979) feels the following objectives are most useful.

1. Emphasis should be on finding out the answer rather than on the answer itself.
2. What is learned should be generalized to many problems with different ways of approaching them given to the student.
3. Beginning math should be made coherent rather than presented in isolated experiences.
4. Instruction must be thorough so the individual will have enough opportunities for practice to reach mastery.
5. The math program should be taught so the individual gains confidence in his or her math ability.

These are each compatible with the general strategies suggested earlier.

Juliano (1982) has translated the general strategies into a six-stage program designed to reach the reluctant math student. His model includes assessment; coping tools (manipulative or pictorial materials that help the student become independent of the teacher); concept base (concrete to abstract); automatic process (having basic facts and concepts in immediate recall); student's learning style; and interests and reinforcement techniques (meaningful, shaping).

In mathematics, as in most disciplines, remedial instruction may involve fully developed programs which are specifically designed for the low achiever, modifications of programs used in regular classrooms and/or materials developed specifically to teach an identified skill area. McLeod and Armstrong (1982) found that most special education teachers use commercial remedial materials, text adaptation, teacher-made materials and the current text in descending order of frequency. Raphael and Wahlstrom (1984) found that regular class teachers use the textbook and its accompanying manual on a regular basis but only rarely incorporate additional commercially prepared or teacher-prepared materials into their mathematics lessons. These findings suggest the importance of choosing appropriate commercial materials since they will often comprise the essential aspects of the remedial program.

Commercially Prepared Materials

Five mathematics programs will be described in this section. They include *Project Math* (Cawley 1979), *Precision Teaching* (Great Falls, Montana, n.d.; Neitherart and Csapo 1978), *Mathematics Their Way* (Baratta-Lorton 1976), *Cuisenaire Rods* (Davidson 1969) and the *Distar Arithmetic Program* (Engleman and Carnine 1972).

Project Math was developed specifically for the low-achieving student. The units encourage interaction between teachers and students and allow for considerable individualization (Cawley 1971, 1976; Cawley et al 1979). The teacher may construct something, present something, say something or write something. The student has basically the same options. In this way there are 16 different ways in which the teacher and student can interact. Further, Cawley emphasizes that students must clearly understand that what they are doing makes sense. In *Project Math* (Cawley et al 1979) "making sense" is accomplished by the authors providing practical examples of situations where math instruction does make sense. For example, in the administrative guide, the issue of measurement is introduced by a job advertisement for a waiter/waitress. The authors point out that the job has less than desirable conditions, however, without knowledge of measurement "the applicant might receive a totally distorted picture of the job. And this is only the beginning! Most aspects of living require a knowledge of measurement" (p. 32).

Precision Teaching is a highly structured approach to teaching computation that incorporates continual experiences with materials in a format that provides repeated measures of learning progress and feedback to the student. Its goals are to increase rate and fluency of math computation for each individual student. It can be used for average and exceptional children from kindergarten to high school in large group, small group or individual settings. There are four basic steps to the program: screening and identification of skill level, direct daily measurement of progress, charting, and data decisions. At step one, the student is tested for one minute on speed of copying numbers (grade 1+), addition facts (grade 2+), subtraction facts (grade 3+), multiplication facts (grade 4+), and mixed facts (grade 5+). Scores are compared to a predetermined standard, and students are classified as being at an acquisition level, fluency level or proficiency level. Only students in this last level are seen as able to generalize and maintain the skills already learned. The others move into step two and are either specifically taught the skill (those at the acquisition stage) or given one-minute drills on a daily basis to increase rate and accuracy (fluency level). The third step involves charting. To do this, the students are taught to score their drill sheets and record their scores on a sheet that may span up to a three-month period. The teacher reviews these charts frequently in order to make decisions about instructional intervention (step four). If students are at goal, the teacher could raise the goal level, change the response format to a more difficult one (oral to written), move on to a more difficult skill or maintain the activity and timing to reinforce the skill further. If the student is making little or no progress, the teacher could use flash cards or alter the one-minute drills. If the student is regressing (corrects decelerating), the teacher could do some direct teaching, go back to an easier level, have the student count, introduce contingency management (make a contract), introduce paired timings where student can do two drills and choose the best for charting or maintain the practice with the timed sheets. All of these contingencies are outlined in the materials.

Neithercut and Csapo (1978) have developed materials based upon the precision teaching model that they call *Learning Math Precisely*. These materials can be used with large or small groups in both a remedial and review capacity. They work because they help the teacher implement the following instructional variables:

> "1. sequential steps for individualizing instruction; 2. systematic procedures,

> contracts, to reinforce the learner's correct responses; 3. daily measurement of the learner's performance." (p. 1)

Students are motivated to maintain their efforts through charting their progress and self-competition, through individual contracts that are drawn up between teacher and student with meaningful reinforcers that are chosen by the student, and through as much social praise and encouragement as the student seems to need.

There are many good features of the precision teaching approach. For the teacher, it makes teaching decisions more straightforward, it allows her or him to make precise educational judgments, it provides a structure for collecting information, it shows how to analyze and interpret the information, and it makes the information easy to communicate to students, parents, and/or colleagues. For the students, it ensures that they know exactly where they are in terms of skill acquisition, and it allows for the development of independent and metacognitive skills since they have control over their performance through attending to the task and monitoring their speed, which, in turn, actualizes their sense of competition in a fair manner.

A third approach to teaching mathematics to all students in a manner that reflects individual strengths and weaknesses is the *Mathematics Their Way* program developed by Baratta-Lorton (1976). The basic premise of this program is that children are helped to develop understanding of and insight into the patterns in mathematics through the use of concrete materials that are carefully organized to maintain interest and active involvement with learning experiences. The skills and concepts taught include linear and geometric patterning; sorting and classification; counting forward and backward; comparisons and conservation of length, quantity, weight, volume and duration; graphing; number concepts; mathematical labels and symbols; place value; and relationships within the number system. It also involves applications of concepts learned. This is an excellent basic level program that ensures that the student can both manipulate and understand essential concepts in mathematics. It involves a non-traditional approach to the teaching of mathematics that is based on sound cognitive, developmental and motivational principles. The pace may need to be modified for the low achiever but the sequence should remain intact.

A remedial approach that also emphasizes exploration of mathematical concepts through the use of concrete materials is that invented by Cuisenaire and developed for use in elementary schools by Gattegno (Davidson 1969). This approach is based on the philosophy that mathematics is a process of observation and a discovery of relationships. The materials are 291 wooden Cuisenaire rods of varying colour and length. The method involves a four-stage process that can be used at any level of skill. Stage one involves independent exploration of the rods. Stage two involves independent exploration and activities directed towards equivalence, sequences, patterns, transformations, concepts of greater than and less than, and odds and evens. Stage three involves directed activities using mathematical notation. Stage four expands the directed activities to include math notation with the number equivalence that is given to the rods. Initially the rods are used in a discovery mode and later for verification when the student has progressed to paper and pencil answers. While empirical research does not indicate that this approach is better than others, it is a useful component of a remedial program because of its use of concrete materials in a manner that is interesting and motivating to students.

The *Distar Arithmetic Program* (Engelmann and Carnine 1972) is a highly structured approach to teaching basic skills in mathematics. The lessons are based upon a thorough analysis of the concepts and skills required for subsequent higher order problems. They introduce new concepts in small incremental steps that prevent difficulties before they occur. Linguistic and perceptual cues are specifically programmed into the instructional activities. Both diagnostic and achievement testing are built into the program. The program has been found successful in helping children acquire basic skills in mathematics (Osborne and Nibbelink 1975).

As mentioned earlier, the backbone of most teachers' math programs are the text and workbooks used in their school district. These reflect provincial curriculum guidelines and though they are often written by a local team, they are generally field-tested across the country. For the teachers of special needs pupils, they provide a good overview of the scope and sequence expected in various grades and divisions plus many basic strategies for teaching new concepts. An example of this is the *Its Mathematics* series (Cathcart et al. 1982) which features a low reading level to allow easy access to concepts; problem solving plus computational strands from grade one up; frequent formative evaluation checks; a variety of materials and activities to accommodate different ability levels; attention to variety, transfer and motivation; and supplementary materials for students who need extra review. For some students, there may be too many items per page in the workbooks and review pages but these can be halved or altered to fit the child as necessary.

A text series specifically developed to teach consumer math is *Mathbase: Practical Skills and Applications* (Pogue, Walpole and Cooper 1981). Its table of contents includes topics like Wheels: The Cost of Driving a Car, Sports Math, Earning a Living, Meals: The Mathematics of Food Preparation, Saving and Borrowing, Planning a Journey, etc. It also includes two extensive appendices on an introduction to algebra and a review of basic skills. The format is varied and interesting and the main thrust is both practical and motivational for the older low achiever in math.

Strategies for Teaching Specific Deficits

While many low achievers can experience success with a program that has been modified as needed, many students will also need specific remediation for skills that they have not mastered elsewhere in their school program. The following remedial suggestions relate specifically to such an approach (Nesbit 1979; Wallace and Kaufman 1978; Hammill, Bartel and Bunch 1984; Crealock 1981;Bruner 1973; Richardson 1984).

Readiness skills

One-to-one correspondence Have the student duplicate pegboard designs, match buttons with button holes, etc. Following concrete matching, have the child match on worksheets (1 flower, 1 leaf, etc.). After single concept matching, increase to multiple concept matching (1 fork, knife, spoon, glass or sets with four, five, six, etc. things in them). Do class or home jobs that involve one-to-one matching such as passing out paper, pencils, etc., setting the table, giving out drinks, or sharing coins equally among family members. Present worksheets with objects having one missing feature (dog with no tail, sweater with one arm) which need to be matched to single features.

Seriation and ordering Present materials randomly and ask that they be ordered according to one specific dimension (height, width, colour). Have student reproduce an order of objects backwards. Start with concrete materials and later have child renumber pictorial representations on a worksheet in correct order.

Number concept Have student listen to a clap, a block being dropped into a pail, a finger snap, etc., and say the correct number. Reverse the procedure by saying or showing a number and having the child give an auditory response. Have the student count objects in sets or count beads or other concrete objects. Prepare a set of boxes that are numbered and ask the student to put the correct number of objects and/or the appropriate number symbols in each box. Have several balls with numbers on them that children roll or toss to each other; when the teacher calls out a number, the student with that ball (e.g., number three) must bounce the ball three times. Have students develop number scrapbooks where numbers from one to ten each have a page and students paste in the correct number of pictures to represent the number.

Classification Sort anything and everything available according to single dimensions initially and, later on, according to several dimensions. Use attribute blocks to sort. Present objects with common attributes and ask the student to identify them (things with grooves, green things, things with tails). Sort decks of cards by suit, number, colour, face versus number. Make scrapbooks of things sorted on one dimension (Christmas things, red things, family things) and/or on several dimensions (smelly stickers, fuzzy stickers, smelly fuzzies, etc.). Have students identify the one thing that doesn't belong in a set of related objects or concepts and tell why it doesn't belong.

Visual spatial ability Play games such as Simon Says, Follow the Leader, and Obstacle Course to give student real experiences of over, under, beside, up, down, in front of, middle, sideways, backwards, etc. Practice similar concepts on pegboards, flannel boards and, later, on worksheets that ask students to indicate where objects are to be placed according to specific directional or spatial instructions. Have students select objects on rotational concepts such as "those above the line, to the left, on the bottom of the page," etc.

Number operations

Addition Present concrete materials of several types (bottle caps, pennies, shells, buttons) and have the child count out answers to number problems. Pass out cards, each with a number on it, to each person in the class; then call out an addition problem, and have the student who has the answer stand up and shows it. Vary this by using bingo cards and having those who have the answer put a marker on it. Roll die and have children add up the total of the roll. Prepare worksheets for number stories of any sum and have students put in as many as they can think of.

A cognitive strategy that encourages students to manipulate numbers involves their understanding how many addition problems can be reduced by doubling or converting to ten. Many children have mastered the doubling facts (5+5, 8+8, 9+9) and when they are shown that 8+6 can be reduced to 7+7 if both numbers share a 1, they have reduced the difficulty level of the number even though they have added a step. Zeller (1983) has suggested giving the doubles a concrete representation such as 2+2 wheels on a car, 5+5 on a foot, 9+9 players on a baseball team to make acquisition of the original doubling fact easier. Another way of reducing the cognitive difficulty of an addition problem is to convert any problem using 9 to 10 so that 9+5 becomes 10+4 which is easier for most students to handle.

Another strategy for motivating students to both match and add involves matching letters of the alphabet to numbers so that a = 1, b = 2, . . . z = 26. Then students can be asked to match the letters in their names to the appropriate numbers and add them up to find the most costly person (MCP) or least costly person (LCP). This can be varied and used with names of towns, sentences, spelling words, hard reading words, etc. Another variation involves choosing a number and finding words that fit it; for example, if 100 is chosen, "useless" is worth 100 points. The basic task involves matching and addition while the variation gets the students thinking about numbers and how they can be manipulated for various purposes.

Students also respond well to addition tricks. Figure 7-4 involves the addition of five-digit numbers (Polak 1975). Students will do this many times to try and beat the teacher. Once the teacher tells them how he or she does it, they then have a trick to show off to parents, peers, or siblings, etc., that has positive motivational effects.

Palindromes (numbers or words that are the same when read backwards or forwards) can also be used to reinforce addition skills. Any two-digit number, when added to its converse, will turn into a palindrome in six or fewer steps.

Example:

$$\begin{array}{rl} 15 & \\ +\ \underline{51} & \text{Step 1} \\ 66 & \\ +\ \underline{66} & \text{Step 2} \\ 132 & \\ \underline{+\ 231} & \text{Step 3} \\ 363 & \end{array}$$

The one exception is 98 which needs 24 steps to turn into a palindrome, which would keep some students productively challenged for a considerable length of time.

Subtraction The major problem experienced by students in subtraction is regrouping. One interesting approach to this, developed by Schminke et al. (1978), is the low stress algorithm. In general terms it involves breaking the learning experience down into small steps, providing the individual with a written record of each step, eliminating the abstract aspect of working with unseen digits, facilitating the analysis of an individual's work using error analysis, and offering the student the advantage of immediate feedback and reinforcement of success after a brief exposure to the strategy. More specifically, this helps many children with the regrouping in reverse aspect of subtraction. It differs from conventional approaches in three respects.

1. The necessary regrouping of the minuend is recorded between the minuend and subtrahend.
2. It reduces the conceptual demand from one involving multiple digits to one involving one- or two-digit numbers.
3. All renaming of place values is completed before any subtraction takes place.

Example:

$$\begin{array}{rrr} \not{6} & \not{2} & \not{4} \\ 5 & 11 & 14 \\ -2 & 7 & 5 \\ \hline 3 & 4 & 9 \end{array}$$

In this example, the student is regrouping 624 as 5 hundreds, 11 tens, and 14 ones before he or she actually subtracts the two numbers. The principle advantages of this approach are that it reduces the memory load for students and reteaches them how to group numbers. It is especially useful for students who have failed using more traditional approaches.

Figure 7-4

Five-digit Number Addition

1. Student: Gives a 5-digit number	8 1 4 6 3
2. Teacher: Gives number to total 9 in each place when added to student's number	1 8 5 3 6
3. Student: Repeats step #1	7 4 8 0 9
4. Teacher: Repeats step #2	2 5 1 9 0
5. Student: Repeats step #1	3 9 4 6 7
6. Teacher: Gives answer immediately and students are intrigued by how he or she could determine the answer so quickly	2 3 9 4 6 5

Solution:

a) Subtract 2 from last digit for number in ones column of answer, e.g., 7-2=5

b) Copy down numbers from the next four places of the 5th number (i.e., the tens to ten thousands) into the appropriate answer column, e.g., 3 9 4 6 5

c) Put a 2 in front of the other 5 numbers, e.g., 2 3 9 4 6 5

Figure 7-5 shows two fun activities that are useful for motivating students to continue practicing basic subtraction facts. And in Figure 7-6, there are several activities presented that encourage both addition and subtraction. These types of activities can be used many times over the year but the introduction of them is best spaced so that the teacher has something new with which to excite the students. New activities may be introduced monthly or more frequently depending on how many the teacher feels are useful.

Figure 7-5

Subtraction Activities

A. 1089

Directions:

1. Choose any three digit number, making sure the first and last digits are at least two numbers apart (i.e. 417 but not 415).
2. Reverse the digits and subtract the smaller from the larger.
3. Reverse the answer obtained in step 2 and add the two numbers.
4. Record your answer. It will always be 1089.

Example:

Step 1.	469	1.	981
Step 2.	964 -469 495	2.	981 -189 792
Step 3.	594 +495 1089	3.	792 297 1089
Step 4.	1089	4.	1089

B. Subtraction by Pairs

Directions:

1. Write down any 4 digit number.	6 8 3 0
2. Moving from right to left, write down the difference between the first two numbers, then the second and third, third and fourth, fourth and first.	6 2 5 3 4 3 2 3 1 1 1 1 0 0 0 0
3. Repeat step 2 with the subsequent answer until you reach 2222, or 0000, whichever comes first.	

Figure 7-6

Addition and Subtraction Activities

A. Adding by 1s and 2s

Directions:

1. The point of the game is to be the first to reach 20 counting by ones and twos.
2. Two students play together or the teacher may play with a student.
3. The first player starts by saying one or two, the second player adds one or two and so on reversing turns until one player reaches twenty.

Examples:

player 1 : 1	player 2 : 3
player 1 : 5	player 2 : 6
player 1 : 8	player 2 : 10
player 1 : 11	player 2 : 13
player 1 : 14	player 2 : 16
player 1 : 17	player 2 : 18
player 1 : 20	

B. Subracting by 1s, 2s or 3s

Directions:

This is very similar to A above except that the first player starts at 21 and subsequent responses decrease the total by 1, 2, or 3. The winner is the first person to get to zero.

Multiplication Have students use graph paper to create a 10x10 number grid. One to ten is marked in across to top row horizontally and down the left side vertically. Then the student fills in the rest by counting by 2s, 3s... 10s. The grid can then be taped to the desk or the front of the math workbook. Show the students the patterns that emerge: the numbers across and down are the same; counting 1s, 2s, 3s, etc. is the same as multiplying by those numbers. Also show them what the patterns of the nine times table are. This is also a good way to introduce graphing and charting since the students become familiar with the use of graph paper and can see how to plot points that change along the vertical and horizontal axes by regular intervals.

Bruner (1973) has described an interesting way of helping students with the nine times table. If they read it from the multiplication grid, they can see that as the number in the tens column increases by one, the number in the ones column decreases by one. This then leads to an awareness that the sum of the two numbers in any nine times product is always nine. They can also see that the answer to a nine times problem is the answer to the equivalent ten times problem minus the multiplier, i.e., $9x7 = (10x7) - 7 = 70-7 = 63$. This reinforces the idea that numbers are to be thought about and understood, not simply memorized.

Another way to help students master the nine times table is to encourage them to use their fingers. Although one does not want students to depend on concrete aids forever, such aids can help many students in the transition period from skill acquisition to skill mastery. The most dependable concrete aids are one's fingers. Have students number their fingers one through ten. Then ask them to locate the number of the multiplier (i.e., the number of groups of nine) on their fingers, put the indicated finger down and read the rest. All fingers to the left of the down finger count as tens and all to the right count as ones.

Example: 7x9 = 6 tens and 3 ones, or 63.

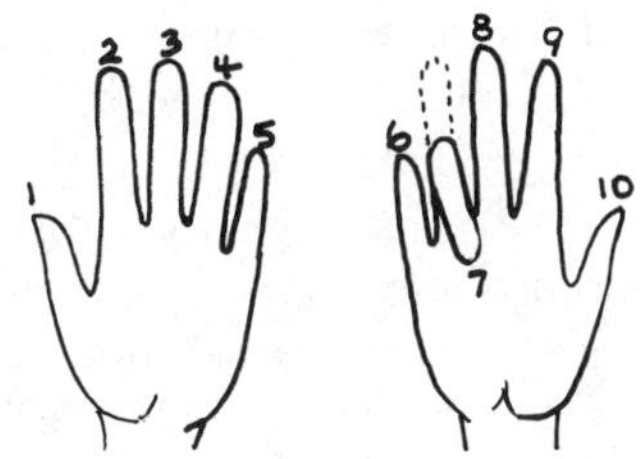

This reinforces the concepts involved in place value also.

A somewhat more complex form of finger multiplication involves the multiplication of numbers six through ten by six through ten. The fingers are numbered on each hand (six to ten), and then the multiplier and multiplicand are joined. The fingers above the joined fingers plus the two that are joined are read as tens; the numbers on each hand left below are multiplied together to give the units value. This seems complicated initially but students are intrigued by the technique sufficiently to want to master it, and then they have a dependable concrete aid.

Example: 8x7 = 5 tens plus 2x3 ones = 56.

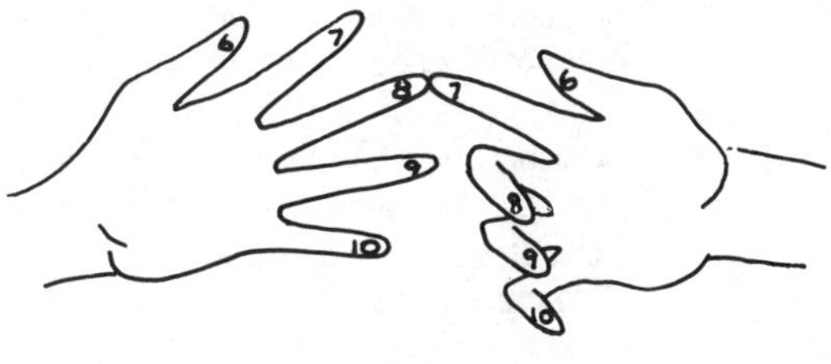

This technique is also useful for reinforcing the concept of multiplication by zero. The student can see readily that no matter how many fingers may be left on the one hand, if there are none left on the other, the answer is zero.

One of the most useful and motivating ways to reinforce basic mathematics skills is through the use of card games and card tricks. Gollick (1973) describes many of these which have been used successfully in her special education clinic in the Montreal Neurological Institute. Children as young as four can handle the concepts involved in card games. Depending on the maturity of the students, both rote and cognitive approaches to understanding and mastering mathematics are involved. For example, in a game called "Up the River" (or "Oh Hell" depending on where you are from) which involves the use of a trump suit to take tricks and the necessity for players to estimate the number of tricks they expect to take in each hand, at least seven distinct skills are reinforced. These include categorizing, adding, counting, calculating probabilities, using more-than-less-than concepts, developing strategies, and modifying strategies in response to external situations.

Gollick (1973) also describes several card tricks that students enjoy practicing on each other and showing off to friends and relatives. One such card trick involves the procedure illustrated in Figure 7-7. This card trick supports adding, subtracting, multiplying, following directions and memory skills. There are many more such ideas available that support the students' interests and competency in mathematics for a very minimal cost (i.e., several packs of cards).

Many of the more popular games that students play can also be used directly or modified in minor ways to reinforce skills. Such games as Monopoly, Snakes and Ladders, Yahtze are but a few that can be part of a mathematics centre in any class. These games have the additional advantages of being known to parents who can be encouraged to play them with their children and of being available for little or no cost since parents of older children who no longer are interested in the games are often quite willing to donate them to the class if teachers make their needs known. This becomes increasingly more important as budgets become tighter.

Other basic skills There are several other basic skills with which low-performing students often need additional help. These include categorization, counting, place

Figure 7-7

Card Trick Used in Mathematics Review

A. The audience (students) will

1. Draw a card, but not show it to the person doing the trick.
2. Multiply the denomination by two (cards 2-10 have face value, the ace = 1, jack = 11, queen = 12, king = 13).
3. Add three to answer in step 2.
4. Multiply answer in step 3 by five.
5. Add to the answer in step 4 the following values: 1 if a club, 2 if a diamond, 3 if a heart, and 4 if a spade.
6. Say the answer in step 5 aloud.

B. The trickster (teacher) will

7. Subtract 15 from the answer given in step 5.
8. In the remaining number, the first number (or first two in three digit number) is the denomination and the last number indicates the suit.

An example would be as follows:

Mike chooses the Jack of hearts	step 5: 125+3 = 128
step 2: 11x2 = 22	step 6: "128!"
step 3: 22+3 = 25	step 7: 128-15 = 113
step 4: 25x5 = 125	step 8: = "The card drawn was the Jack of hearts" (11 = a jack and 3 = a heart)

value, money, fractions and telling time.

Categorization Students should be encouraged to group anything and everything from very basic concepts such as green things to more elaborate ones such as "trees that don't lose their leaves in winter." They need opportunities to do this at both the concrete and abstract levels. Concrete materials such as Cuisenaire rods, attribute blocks or bottle caps, buttons, sticks and stones can be used in a variety of ways. Wallace and Kaufman (1978) suggest having students cut out pictures of similar things (for example, hats) and paste them on cards. They also suggest having students match the cards for the numbers of items on them. Cards can be developed with different ways of stating math problems and students must find all the equivalent ones. For example there could be at least four cards developed to illustrate a math problem like 4+3 = ______. Those four cards could include a word problem, a concrete problem with four red discs and three blue ones, a semi-concrete problem such as

or a semi-abstract problem such as 0000+0000 = ______. Students would get practice with 4+3 = ______ and would have practice sorting out each problem presentation from others simultaneously presented.

Counting Counting is one of the most basic skills students must master if they are to proceed to more complex tasks and if they are to develop essential consumer math competency. Children should be able both to count simply by rote and to count meaningfully. The former can be done through frequent recitation of counting by 1s, 2s, 5s, 10s, etc., while the latter takes much more effort and understanding. It has been the practice to ensure that students understand math first, on the assumption that then correct performance would follow. With low-achieving students, however, the reverse seems to be true at times, with understanding following rote performance. For example, once the student can count automatically from one to ten, he or she can then be asked to move beads along a string or to move blocks from *a* to *b* while counting. He or she can then count the other students in the group, the books on a shelf, or the number of problems on a page, for example. Worksheets can contain sets of different numbers of objects that the student must count and record; items with spaces before or after the number to be filled in (e.g., ______ 19 ______); duplicated numbers in a set from one to ten that the student must find; etc. Dot-to-dot puzzles where the dots are in sequence can be used since these range from easy to difficult and are generally motivating to students. A number line on which the numbers are matched to centimetres can be attached to the top of the child's desk to reinforce both counting and measurement.

Place value This is a very difficult concept for students to understand. It is an abstract one and often students need many more concrete experiences with it before understanding comes. Some students may perform better when taught in rote fashion first and in a meaningful fashion either simultaneously or afterwards. For example, in handling a place value question involving a three-digit number, e.g., 328, students need to repeat concepts—e.g., if there are three numbers, the first is in the hundreds place, the second is in the tens place, and the third is in the ones place—so they can answer questions involving the number. In addition they should have concrete materials that aid their understanding of this. Cuisenaire rods and other similar materials provide this sort of aid, as do pennies, beans, markers, etc. For low-functioning students, counting out the three in the hundreds place (that is, 300 beans) is more meaningful than using three Cuisenaire rods in reinforcing what a number in that place really stands for. Once students have had lots of concrete experiences with place value equivalents, they can do similar things in a semi-abstract manner by drawing in the numbers of items in each place on a worksheet. Worksheets are divided into two columns (or three, four, etc.) and students make a circle to indicate how many items are actually in each place. For the number 39, the sheet would look like this:

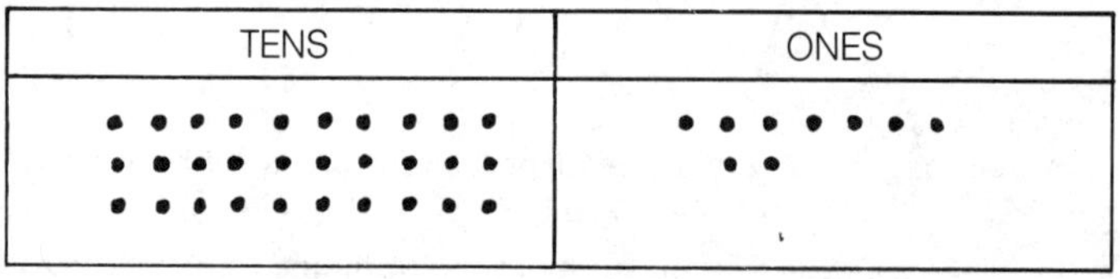

Later they can use similar worksheets but only to indicate in which column each digit in the number should be placed. Cards can be made with the amounts in each place drawn in on one side and the actual number on the back.

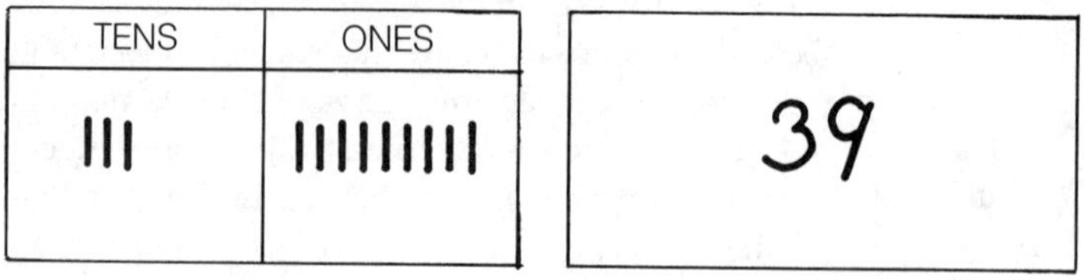

Front

Back

The teacher may present the front of the card or the student may go through the pack and, having figured out an answer, may then check it on the back of the card.

Teachers can work with the entire class in several ways. One way involves questions like "I am thinking of a number with two hundreds, six tens, and four ones, what is it?" Students can write down the answer at their desk or on the blackboard, or they can answer orally. Other teacher-led questions include: "What place does the three represent in 23?" or "Can you write a number with nine in the hundreds place?" Another large group lesson that combines both an abstract and semi-concrete approach with a motor response involves passing out 30 cards to the class. The cards range in denomination from zero to nine and in place value from ones to hundreds; for example:

3
TENS

The teacher says will number 403 please come forward and the three students with those numbers come up to the front to make the number.

4	0	3
HUNDREDS	TENS	ONES

If there are less than 30 in the class, some students will have more than one number to keep track of; if there are more than 30, the teacher can make a double set and then two groups of students would come forward for each number.

Money Understanding the value of money is an essential consumer skill that all students must master. It is more easily understood if actual money or a good replica is used in teaching and if the problems presented are real to the student. Have students identify actual coins and paper money, use worksheets with coin pictures on one side and the money equivalent on the other for matching. Paste objects of interest on cards that also show their cost and have students count out the money each object costs. Show students the many different ways one can make up a sum of money and have students decide which is the most efficient way. A grocery store can be set up in the classroom with actual things or play things for sale so students can see what these items cost and get practice making change.

Figure 7-8 presents a task analysis for teaching students how to make change which has been used with students from grades two through ten. It has been successful remedially because it includes only the essential skills and does not attempt to teach related ones such as subtraction and addition.

Older students can receive essential practice in handling money through the actual planning of a classroom event such as a party or field trip that includes fund raising, budgeting, getting best value for money, etc. Comparison shopping is another useful activity. This can involve comparing grocery items from the newspaper grocery ads; checking on articles of clothing or toys through several catalogue listings; obtaining menus from local restaurants and calculating the price of a nutritious meal; etc. Many basic number operations can also be reinforced through teaching money since this topic has intrinsic motivational appeal for many students.

Figure 7-8

Task Analysis for 'Making Change'

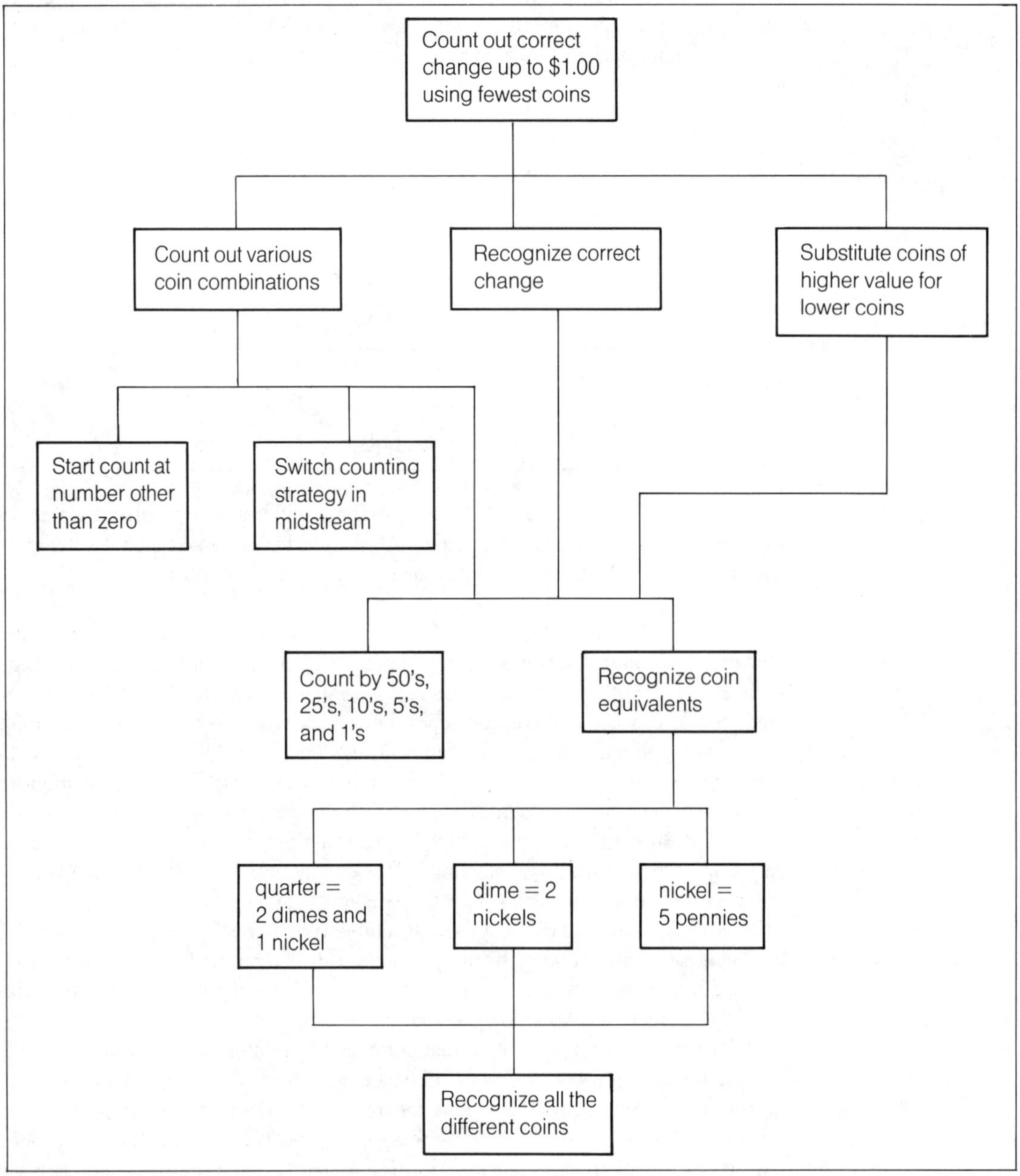

Adapted from M. Semmel, S. Thiagarajan, H. Stolovitch, *Tips for Teachers: Task Analysis* (Bloomington: Centre for Innovation in Teaching the Handicapped, Indiana University, 1974), 3.

Fractions This too is a concept that students need to work with concretely for a long time before it is introduced in its abstract form. There are many opportunities to do this with younger children, especially around activities that involve food. For example, a cake can be cut into halves, quarters, eighths, etc.; apples can be cut up and shared among small groups. Cards can be dealt out in Fish or Concentration so that each partner has half, a third, a quarter, etc., depending upon how many are playing the game. Flannel boards and pie-shaped pieces of cardboard can also be divided easily to show students what a fraction looks like in relation to the whole. After students have a solid understanding of fractions from concrete experiences, they can then work on worksheets that initially show the same fractions (i.e., half an apple, one-sixth of a pie) in a two-dimensional manner and then later in a more abstract form. Prepare sheets that show the transition, for example:

Piece of Pie

Worksheets can help students with equivalent fractions like this one in which students circle the equivalent fractions in each line and match them:

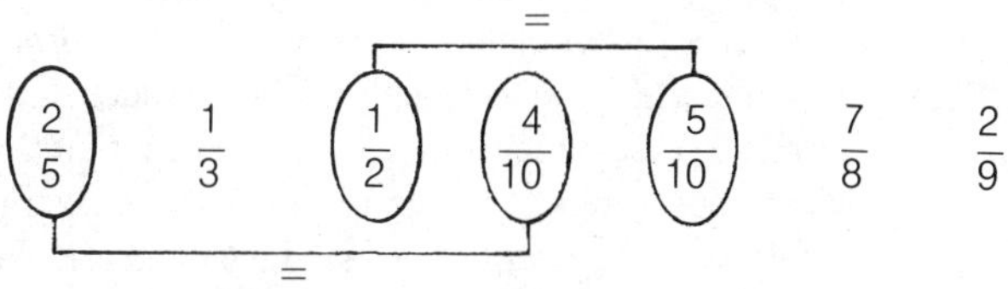

Telling time This involves reading the face of a time-piece and understanding the meaning of units of time. With the advent of digital clocks, reading the face of a traditional clock is not as essential as it once was, but it is still a useful skill for students to master. It is helpful to begin teaching the hours, half hours, quarter hours, five minutes intervals, one minute intervals, seconds, and their meaning before and after the hour. Then one should teach the space equivalents of each of these concepts for clocks that don't have full number markers. Each of these items should be mastered before the next, more complex feature is added. Understanding is increased when each time measure is taught in relation to the activities it applies to. For example, when teaching the hour, match nine o'clock with school starting, twelve noon with lunch time, five o'clock with dinner time, eight o'clock with bedtime, etc. The half hours can be paired with the beginning or end of specific activities and with the quantity of work or fun that can be accomplished in that time period. To give students a kinesthetic sense of time markers and of time passing, one student can be a clock while another arranges his or her arms to match the time called out by the teacher. This can also be done with paper plate clocks that the students have made in art. During initial teaching of how to tell time, the teacher should reinforce the students learning by asking them what time it is or by saying "it is now nine forty-five," or "eleven sixteen," etc., whenever the opportunity presents itself. Students can become aware of time by keeping a personal log that includes the exact time of an activity, the activity itself and its duration. This can be especially helpful for students who have difficulty appreciating where time goes and how long they spend on-task and off-task. They can also be helped to understand time better by performing tasks with a specific time limit such as the one-minute worksheets in precision teaching, running on the spot for three minutes, being given ten minutes to finish lunch or 30 minutes to complete their math homework, etc.

Problem solving Ideally students have had many opportunities to experience mathematical problems in sentence form and so the transition from numbers to words should be easier than it seems to be for many students. Hammill and Bartel (1978) suggest that the major causes of student difficulty in this area are lack of practice and inadequate prerequisite skills. They indicate that the essential prerequisite skills for effective problem solving include reading, computational mastery, estimation, ability to organize sequences of operations and ability to write complete sentences. They suggest the following steps should be followed in problem solving: read the problem through quickly, reread it carefully to determine what is given and what is required, refine your hypothesis, write out the numerical sentence, perform the operations, check your answer and state the result in correct sentence form.

Crealock (1981) has developed a technique for problem solving that requires students to determine the relevant facts given in a problem, to brainstorm ways of solving the problem, to choose the most appropriate method and, finally, to solve the problem. This is especially useful for older students who have failed math earlier and can be motivated through an awareness that thinking helps in math. It also provides an opportunity to integrate concrete and abstract approaches to problem solving. In Figure 7-9 students suggested trial and error methods, visual semi-concrete methods and abstract methods that allowed the teacher to explore algebraic relationships with a group that had already solved the problem through simpler, albeit less efficient, means. In this approach, students always use the same technique whether they work individually or in groups, even though they are encouraged to see how many different ways can be employed to solve the problem.

Other aspects of helping students with problem solving include simplifying the syntax, explaining the key vocabulary and cue words, using analogue problems with easier numbers and then showing students how to use them with larger numbers and more difficult concepts, illustrating problems pictorially, and having students write or dictate problems for themselves and the class.

Record keeping There are many skills that need remediating and many strategies that can be used effectively. The final general guideline concerning both skills and strategies, however, involves adequate record keeping. Students can do much of their own record keeping through the use of charts and graphs that record their progress. But teachers must also keep records in some form that is consistent and reliable. None of us have sufficient memory skills to ensure that a student's progress (or lack of progress) will be readily appreciated unless we have kept track of it and can access the pertinent files easily. Time spent organizing record keeping is time well spent (Connolly 1980; Zigmond 1983).

Figure 7-9

Using Creative Problem Solving in Mathematics

Original Problem:
Mike went to visit a farm. He saw lots of animals running around. He counted 60 heads and 140 legs. The farmer had only chickens and rabbits. How many of each kind of animal did he have?

Fact Finding

2 legs—chicken

4 legs—rabbit

140 legs altogether

60 animals altogether

Idea Finding

1. 70 of each animal
2. kill while counting
3. count them as they go by
4. divide them
5. count the eggs
6. stamp the chickens
7. put a clamp on each animal
8. tag them
9. chicken = 2 legs, 1 head
10. 90 chickens
11. 40 of each kind
12. 30 of each kind
13. 40 of one, 20 of other
14. cut off heads & legs & put in bag
15. draw 60 circles (heads)
    —put 2 legs for each head
    —20 legs left
    —put 2 more legs on as many animals as possible
16. get more rabbits
17. kill some chickens
18. half are chickens and half are rabbits
19. divide 60 heads into 140 feet
20. try different numbers until one works
21. ask the farmer how many he had
22. ask the chickens how many of them there are
23. look at the rabbit-chicken foot ratio 4:2

Note: Number 15 and some form of trial and error attempt to solution were selected as the best solutions.

Solution Finding

1. 

   10 rabbits

   50 chickens
2. a form of trial and error
3. total animals = 60

   chickens = x

   rabbits = 60-x

   2x + 4 (60-x) = 140 legs

   2x + 240 - 4x = 140

   100 = 2x

   50 = x

   ∴there are 50 chickens and 10 rabbits

Adapted from R. Noller, P. Heintz and D. Bleuler, *Creative Problem Solving in Mathematics* (Buffalo, New York: DOK Publishers, Inc., 1978), 14.

Case Study

The final component of this chapter involves looking at a case study in mathematics and analyzing it through the decision-making model described in Chapter 2. This case involves John S., a grade seven boy who was having difficulty coping with the regular curriculum.

Step 1: Background information

John is a grade seven student who has always been in the lowest math group. This year he is falling further behind, his attitude is deteriorating, his CTBS scores are at a mid grade five level in mathematical computation and at an early grade six level in problem solving, and his teacher has requested help in assessing his specific weaknesses and setting up a program for him. His intelligence is in the average range on group tests. He receives adequate marks on the rest of his subjects. Also, he seems to work best with teachers who actively involve him in lessons. His family is supportive but not too worried about his progress.

Step 2: Preliminary hypotheses

1. John has had chronic math problems and probably has not mastered several of the prerequisite skills necessary for grade seven work.
2. This year he is expected to do more work on his own and he finds this difficult to do.
3. As he fails more frequently in math, his motivation and interest in the subject are becoming more negative.

Step 3: Assessment

Individual testing was done with John to determine the extent to which the three hypotheses developed in step 2 could explain his problems.

A board-developed math test that taps into math skills in the local curriculum was administered and scored. It indicated adequate computational skills in basic operations but some problems with fractions, ratio and algebra. The resource teacher probed this with a test he had developed himself and found that John had problems reducing fractions to their lowest terms and in finding equivalent fractions, both of which are necessary subskills for calculating ratio.

A personal interview to probe John's attitude towards math was conducted which revealed that he is increasingly negative about the subject. In addition, he finds it hard to sit through the whole math period, can't follow the sequences of topics and how they relate to each other, doesn't understand the terminology used in algebra and can't see any real value in studying math. He also hates to do all the homework that is assigned.

Step 4: Reformulate hypotheses

1. John needs remediation in fractions, ratio and basic algebraic concepts.
2. John needs to have opportunities to use mathematics in areas that interest him so he can see how math can be meaningful.

3. Further probing needs to be done to determine John's interests and learning style preferences.

Step 5: Trial lessons

An interest survey was administered to John. Results suggested that his main interests are cars, hockey, adventure movies and computers.

Observations, interviews and trial lessons indicated that he attends best in small group situations with regular feedback from his teacher and interaction with his teacher and peers. He attends least well when working on his own at home or at school. The reinforcement that seems to maintain his attention best is the promise of free time to pursue his own interests.

Step 6: Development of IEP goals

1. John will be able to understand and apply main concepts in algebra.
2. John will develop an understanding of the practical aspects of math.
3. John will attain mastery of the following math concepts:
   a) reduction of fractions to their lowest terms,
   b) calculation of equivalent fractions, and
   c) comparison of two selected groups as a ratio.
4. John's attitude to math will improve.

Step 7: Implementation of IEP

Remediation will be handled by both the resource teacher and the grade seven teacher. Goal 1: John will construct a personal dictionary of terms. He will review concepts on a computer with a remedial drill program. Goal 2: John will be presented with problem-solving situations using newspapers, comparison shopping, etc., based on his interests in sports and computers. Goal 3: John will review fraction concepts using concrete mathematical, then pictorial, representations and, finally, abstract symbols. This same progression will be applied to the teaching of ratio. Goal 4: This will be attained through successful mastery of goals 1 to 3. The above activities will be carried out in the resource room. The grade seven teacher will involve John more in the classroom discussion by asking him to explain his answers and by checking on his understanding. John will be given more opportunities to work in small groups with practical problems the resource teacher will set up for the regular class. New lessons will begin with a review of the previous one. New lessons will include a written outline of the new work and an application to meaningful problems where possible. Assignment dates and procedures will be written in his work book. John will earn free time when his regular in-class work is finished early and his homework is done correctly.

The grade seven and resource teachers will share responsibility for John's progress. All course outlines, test schedules and project dates will be given to the resource teacher so he can review work with John. The resource teacher will sit in on the regular class to see how John is transferring skills taught in the remedial class to the regular class. Both teachers will meet every two weeks for ten minutes to discuss progress.

Step 8: Evaluation

Evaluation will be done in two months, at the end of the first term, and at the end of the year. The fraction probe and interview will be used again to check specific progress. Both the resource and regular teacher will observe John's homework, time on-task in class, and willingness to participate in math class as a second check on increasing competence and positive attitude development.

8 ASSESSING AND TEACHING COMPOSITION SKILLS

Many adults, whether they have learning problems or not, seek ways to minimize the kinds and amount of writing they do. Such avoidance may, in part, be due to the way they were taught to write. Graves (1978) points out two factors which may have resulted in such avoidance. First, in some classrooms writing was "seldom encouraged and sometimes not permitted" (p. 30). Second, Graves suggested that too frequently students wrote as a form of punishment. Such punishments are familiar to everyone and include memorable experiences like writing 1000 times, "I won't chew gum." While writing lines is no longer a frequently used form of punishment and while there has been a recent concentrated effort to highlight the value of writing well (cf. Britton 1970; Graves 1978; Gundlach 1982), some children and adolescents continue to write poorly.

One reason for the continued poor quality of compositions may be that children are not clear as to why they are writing. For example, Shook (1982) surveyed 108 children in grades one and two, asking them why they write. Six percent of these children did not know and 56 percent answered: "It's fun," "The teacher says to," or "To get good at it." To some extent such answers may be typical of responses given to any question asked of grade one and two children about school-related concerns. Students' concerns about composition, however, do not stop at uncertainty about the purpose of writing.

As students get older, they report further discontentment about writing. In a relatively recent survey of writing achievement completed in the United States, approximately 25 percent of the surveyed nine year olds reported they did not enjoy writing (National Assessment of Educational Progress 1980). It was observed 1) that enjoyment of writing decreases with age and 2) that an increasing number of students believed they could not write. In another case, Graves (1982) surveyed grade seven students. These individuals found it difficult to even select a topic about which to write. They believed that writing is an artificial act.

Since writing is perceived as problematic by a sizable group of normally achieving students, it should not be surprising that it is seen as an obstacle by students with special needs. The composition skills of the learning-disabled have received considerable attention and the majority of examples will be drawn from this group of special needs students. For example, Lerner (1976) suggests that poor writing skills are the most prevalent communicative disability among learning-disabled students.

There is ample documentation that learning-disabled students do not write as well as normally achieving students (Poplin et al. 1980; Morris and Crump 1982). There is some evidence, however, that learning-disabled students may write better than one would expect. For example, using the *Test of Written Language* (Hammill and Larsen 1978) as a criterion, Poplin et al. (1980) reported that learning-disabled students stayed within one standard deviation of the mean on the vocabulary and thematic maturity tests. In another case, Morris and Crump (1982) found no difference in T-units (a measure which estimates complexity and length) among groups.

Speculation as to why writing is problematic for learning-disabled students ranges from the hypothesis that poor writing is a function of deficits in oral language (Wiig and Semel 1980) to the suggestion that the inability to write is a consequence of poor reading (Myklebust 1973). Both points of view may be partially correct. Some students may have difficulty writing because they adopt speech conventions that are inappropriate when incorporated in compositions. Difficulties may also be apparent when writers cannot read appropriate background information or edit (reread) what they have written. We suspect, however, that other factors besides deficits in oral language and poor reading skills contribute to inappropriate compositions among the learning-disabled. We have already suggested that some students do not see writing as important; therefore, they may not devote time to it. Or, students may lack the technical skills necessary to write (see Chapter 9 for suggestions). Like any other skill, good writing does not happen by magic; it must be practiced.

In the remainder of this chapter we will cover four areas. First, a summary of some of the current viewpoints about composition is presented. Next, we will discuss some of the roadblocks to written expression, many of which are found as products in the writings of students with special needs. Third, we will examine techniques that may be used to assess composition skills. Finally, suggestions are offered that may result in better written expression.

HYPOTHESES CONCERNING INADEQUATE COMPOSITION

Children and adolescents with learning problems may not compose well for two major sets of reasons. First, a student may compose poorly due to the lack of, or weaknesses in, other related skills needed at some stage of the writing process—reading is one example. Second, poor composition may be the result of weaknesses in skills associated with composing itself. In other words, the impact that reading, listening, speaking, or all three taken together have on composing must be kept separate from how composition itself takes place. Each of these influences is explored in turn in the next two sections of the chapter.

Indirect Influences on Composition

Reading How well we read and what we read may be hypothesized as having, at the least, an indirect influence on how we write. In other words, as we read different styles of writing we may attend, for example, to the way in which sentences are written. Secondly, especially for expository writing, difficulty in reading will interfere with library research at the very minimum. Myklebust (1973) takes the position that poor writing is a consequence of poor reading. This argument would suggest that reading and writing form a hierarachy. The relationship between reading and writing, however, may not be as straightforward as Myklebust suggests. Britton (1980), for example, argues that learning to read and write must be seen as part of the process of mastering written language. This latter viewpoint would suggest that while there is some interrelationship between reading and writing, it may not be hierarchical. For instance, Gillingham and Stillman (1960) introduce techniques to associate reading and writing earlier in their program. Reading and writing may need to be taught in parallel, rather than in sequence; that is, composing may be learned best by introduction earlier rather than later. Learning to compose and to read simultaneously may have an added benefit in the intermediate grades. Graves (1978) reports that more emphasis on writing results in better reading comprehension and improvements in math problem solving.

In suggesting that reading and writing be presented simultaneously, we are assuming that a child or adolescent has an adequate oral language. In such cases where oral language is inadequate, initial instruction may have to take the form of a language experience approach. Such an approach may be necessary where cultural expectations are different as well, as in the case of language arts for Native students (Ministry of Education, British Columbia 1982).

Oral language Oral language, or conversation or speech, is, on the surface, more obviously related to composition than is reading, as both oral language and composition are ways of conveying messages. Oral language appears to be one of the keys to understanding difficulties in composition (Wiig and Semel 1980).

Many learning-disabled students have trouble expressing themselves clearly and succinctly in oral discussions. While at first glance they may appear quite glib and chatty, careful observation of their speech often reveals several weaknesses. Many do not have the full range of nouns and adjectives in their vocabulary and so either overuse those they are familiar with or substitute pronouns (often inappropriately) in much of their discourse. This results in less precise descriptions of events which can be both misleading and time consuming. Myklebust (1964) suggests that learning-disabled students often retain narrow word meanings, are able to generate fewer examples of words in general categories and tend to understand and use vocabulary in a restricted, concrete manner.

A second area of weakness in oral language is the general inability of learning-disabled students to organize their discourse effectively. They may forget cogent aspects of their argument; flit from point to point without finishing a thought; expect their listener to have the same background experience they do and so leave out essential aspects; garble the sequence and logic of their description or thesis; and run on with little awareness of the effect of their speech on their audience.

Many are unable to remember several instructions when these are presented simultaneously. Also, many are unable to communicate material in the sequence required. This can be aggravated by the tendency of these students to be distracted by events around them such as the presence of several classmates when they are trying to talk to the teacher, noises in the hallways, sounds on the street, or attractive visual displays in the classroom.

Learning-disabled students will often produce more agrammatical sentences and sentences of shorter length than do normal children (Wiig and Semel 1975). In addition they will show greater response lags in producing sentences, a problem which can result in frustration and intolerance on the part of the listener who is trying to understand them.

Since learning-disabled students incorporate agrammatical forms into their speech, it should not be surprising, to the extent that speech appears in student writing, that their compositions are flawed. De Beaugrande (1982) has suggested that two characteristics of speech (which he terms disturbances and types of redundancy) may interfere with written production. There are eight different types of disturbances and four different types of redundancy listed by de Beaugrande.

Disturbances

1. Fillers: Fillers maintain or support the continuity of a discussion. They result in such things as run-on sentences when found in writing. Examples of fillers include: non-words (eh, um), feedback signals (yes, right, ok) and connectors (and, however, but, nevertheless).
2. Lexical variants: These are words which are more typical of speech than they are of writing (yeah, all-right, cause).
3. Simple restarts: The author begins writing again, restating a word or a phrase (Once upon a time, once upon a time).
4. Revised restarts: In this case the writer changes part of the phrase while repeating it (Once upon a time, once long ago).
5. Shifts: In the middle of a statement, the writer takes a new direction (There are three new boys in town and a baseball game was held on Saturday night).
6. Post-completions: Another word or phrase is added after a sentence has been completed. (John had a new car, a red one.)
7. Non-agreements: Use of grammatical inconsistencies (If I would have knowed you were coming, I would have sawed you off).
8. Vagueness: The meaning of the sentence is not clear (The thing is that.... My experience would suggest....); or the structure of the sentence is confusing (Whenever I go fishing I catch one).

Types of Redundancy

1. Simple recurrences: The speaker repeats what was initially stated following a digression to start the conversation again.
2. Revised recurrences: Again, repetition occurs but with slightly changed wording or revised expressions.
3. Self-paraphrases: Concepts are restated using different expressions.
4. Needless wordage.

Looking at de Beaugrande's list, several common writing errors can be imagined. Caution must be observed, however, as some good writers intentionally use such techniques to highlight a point or to ensure an argument is not likely to be misinterpreted. The main point, however, is that such speech conventions do, if used continually, disrupt compositions.

While speech and writing are similar in some senses, differences do become apparent when conversation is compared to composition (Bereiter and Scardamalia 1981). In conversation, there is an opportunity to clarify what is being said. When you read a completed composition, there is no such opportunity. Writing, then, is different and perhaps more difficult than conversation (Bereiter and Scardamalia 1981). The opportunity to interact with the audience is eliminated when you write, it must be imagined. Yet, many students do not appear to know why they are writing or for whom. Perhaps this partially explains why the compositions of younger students and of some learning-disabled students resemble some part of a conversation. They may be waiting for an imaginary audience to help them with what comes next. There is some evidence to suggest that children are awaiting for a signal to start again, a prompt to say more. Scardamalia, Bereiter and Goelman (1982) report that asking children to write more resulted in about double the number of words written (from 57 words to 112 in grade 4, and from 90 to 194 in grade 6).

The transition from speech to composition for young children may be made through drawing. Vygotsky (1978) argues that his analysis of children's drawings suggest that they are "a particular kind of child speech" (p. 112). Britton (1982) argues further that spoken language meets the majority of young children's needs. It is only when children value "written language as a vehicle for stories that they are likely to form an intention to write" (p. 163). Therefore, drawing and trying to write down stories may be important steps in beginning to learn to write.

Suggesting that drawing may be an important transition from speech to composition does not imply that students with special needs should be taught to draw. It suggests only that for young children drawing may be a form of graphic speech. Seeing drawing as important may help us to understand the obvious labour a child puts into making letters and the pride he or she feels when successful. Indeed, children may be drawing the letters rather than writing them. Scardamalia, Bereiter and Goelman (1982) report that some children in grade three are still drawing letters.

Direct Factors Interfering With Written Production

Student and teacher attitudes toward composition If students believe that they can't write, they are less likely to want to try; however, if their teacher believes that students can't write, they won't get the opportunity to try. Attitude towards composition, then, will directly influence the amount written. Students, for example, may be reticent to write if they know that their teacher places emphasis on recopying as opposed to rewriting. Attitude, however, may be changed. We suggest that the most direct way to modify attitude is by modifying instruction. The suggestions offered later in this chapter have been found effective, over time, in creating a more positive attitude towards composition.

Handwriting There are a few occasions in which handwriting and concern about layout (spacing between letters, words, headings, paragraphs, etc.) may interfere with written production. The key to such interference is that the student is so worried about handwriting or layout that he or she loses sight of content or even changes content to accommodate the technical problem he or she is having.

In looking at handwriting, the concern is not legibility per se but rather the consequences of the way a writer moves a pen or pencil. An example of a left-handed student will illustrate this point. Bradley, a young boy, was sent for help in writing. He never wrote more than two sentences in a period, no matter what the assignment was. His writing was also illegible. Bradley was quite uncomfortable when he arrived and was asked after he was observed writing: "Can you see what you are writing?" His answer was "No, can anyone?" Bradley held his pen the way a right-handed person might and, as a result, masked every word he wrote. A few weeks of appropriate modelling and practice eliminated both poor handwriting and increased written production.

In addition to those suggestions offered in the next chapter on handwriting, Harrison (1981) and Milone, Jr. and Wasylyk (1981) offer some good advice for handwriting instruction for left-handed students. Harrison (1981) includes both manuscript and cursive models that are helpful for left-handers.

Another factor which may interfere with written production among students with learning problems is the lack of a well developed sense of layout. Letters, projects, presentations are each improved if attention has been paid to layout. The students who have difficulty knowing where to place components of a composition may not produce what they had intended.

A major educational implication which might be drawn from the above, is that more time should be devoted to writing instruction for students with learning problems. Writing means more, however, than copying or other such tasks that emphasize only legibility. Legibility is a legitimate concern but one that should not be the primary focus of the lesson. There is some indication, however, that copying (transcribing work, redoing assignments to make them neater, etc.) is a major component of writing time. Leinhardt, Zigmond and Cooley (cited in Silverman et al. 1981) report that 75 percent of the time (19 of 25 minutes) learning-disabled students devoted to writing was spent copying. Beginning readers spent less than five minutes per day completing writing of any kind. Most of this time was spent filing in the blanks in worksheets or in ditto sheets. Students with second to fourth grade reading skills averaged only 7.5 minutes per day completing creative writing activities. Our clinical experience leads us to suggest that writing time devoted to copying and willingness to write are often negatively correlated. In convincing children that they can learn to write, several obstacles must be overcome. The attitude that "writing equals copying, so don't write too much" is only one of these roadblocks to writing.

Spelling The extent to which spelling may interfere with writing production likely has been underestimated. Scardamalia, Bereiter and Goelman (1982) suggest that spelling may interfere with sentence construction, although only in certain circumstances.

> If the meaning of the sentence being written has already been fully constructed ...attention to a spelling problem cannot interfere with construction of *that*

> unit of meaning—although it could interfere with construction of some other unit of meaning if the writer was thinking about it at the time. (p. 182)

Difficulty in spelling may not modify the meaning of a sentence that has already been thought out; however, difficulty in spelling may have two other influences on the sentence being constructed. First, the original words chosen may be modified to accommodate spelling level. As a result, both the form and length of the sentence may be modified. Second, since the words selected affect meaning, the clarity of the intended meaning may be influenced.

Inability to spell in an extreme form may have even more serious repercussions. Students who have become convinced they cannot spell sometimes believe they cannot write at all. In such cases, poor spelling is usually accompanied by problems with other aspects of technical form such as punctuation, capitalization and syntax. Such students have been observed writing nothing at all, attempting to copy from others or from other sources, or trying to convince someone else to write down what they want to say, or some combination of all of the above activities.

Content generation and organization One of the major difficulties faced by students with learning problems, and others too for that matter, is deciding what to write about and how to organize the resulting content. Many written products consist of a set of scattered thoughts (not necessarily in sentences) which leap from one idea to another.

Two kinds of long-term memory search may take place when someone wants to write about a topic (Bereiter and Scardamalia 1981). The first search is for general knowledge: what does the writer know about the topic. The second search follows the more general search. It is used as the writer decides what specific goals (to be funny, to prove a point, to elaborate on an idea, etc.) he or she wants to accomplish as writing continues. Bereiter and Scardamalia (1981) argue that normally achieving students have difficulty with both of these searches. Inability to complete the first type of search will result in statements like: "I can't think of anything to say" or "Who wants to write on that dumb topic, anyway." Successful or partially successful completion of a general topic search that is followed by failure to address specific goals will likely result in a series of disconnected thoughts. Readers of such writing will have difficulty following along as the writer typically has no awareness of audience needs.

A third point is relevant for the poor writer. In addition to having difficulty with the general and specific memory searches required, he or she may have problems with the instructions of an assignment. For example, a teacher may ask a class to write a short summary of the basic issues involved in some topic that will reflect both their own reading from the text and classroom discussion. Some students will have difficulty remembering and isolating each aspect of the assignment; they may not remember exactly how to write a summary, or how many words or sentences the teacher means by "short"; and, they may have difficulty integrating classroom discussion and text materials, especially if there is any disagreement between the two. For such students, even if they know the content, their knowledge of the meaning of specific instructions may be too poor to allow them to complete the assignment well.

In this example the instruction concerning length ("How much do I write?") was made fairly specific by the words "short." In cases where the length requirement is left open or is much longer, the question of "how much" can cause considerable anxiety. Such anxiety may not effect writing production; however, concern for length has resulted in a series of connected quotes and needless word counting exercises, among other things.

Another difficulty faced by learning-disabled students is the inability to organize (See Chapter 11). This inability manifests itself in terms of organizing time, space, planning and thinking. Often the student's lockers and desks are so disorganized that they have trouble finding the right books and locating actual assignments—both within their texts and in any number of notebooks they might have used for an assignment. This sort of problem is often accompanied by difficulty organizing their time so that they can sit down and do the work required. They are late, miss periods, and can't judge the amount of time needed to complete assignments (Kronick 1979). The inability to think clearly often is reflected in all of the above activities. Caution is necessary in arriving at the conclusion that poor planning or lack of organization is indicative of poor thinking—such may not be the case. Poor thinking may be in evidence when sentences in a student's composition are not linked to one another well. For example, a student may state three discrete ideas believing that each of the three represents a fully developed sequence of thoughts.

ASSESSING COMPOSITION SKILLS

In general, written language may be seen as having three aspects. These are: a) one's attitude toward written language, b) the content of the written product, and c) the technical form required to express the content properly (Alley and Deshler 1979). Suggesting assessment and teaching strategies that focus on the first of these two issues are the next two topics to be addressed in this chapter. The latter issue is treated in Chapter 9 on technical issues in writing.

Salvia and Ysseldyke (1981) suggest that there are three important considerations when assessing language skills. First, they suggest that cultural background may influence the obtained results. This will be especially true when one is asking a non-native English speaker to write a composition. Second, they point out that the age of the child is important. This is obviously true when evaluating written product: the typical expectations for younger children emphasize acquisition; whereas for older students different types of writing are expected depending on what is to be written. Third, they suggest that intellectual competency should be considered. Intellectual competency will influence such things as the vocabulary used in a composition.

It is important to take the three factors discussed by Salvia and Ysseldyke into account when assessing a student's writing ability; however, a) these factors are *indirect* influences on actual composition skills, and b) these factors are all "givens," i.e., they are not focuses of assessment or remediation. By contrast, the next consideration, attitude towards writing, is important in two senses: it can be changed and it has a direct influence on whether composition occurs.

Assessing Attitude Towards Writing

If a student does not want to write or believes that he or she cannot write, his or her feelings will be translated into actions that are easily observed. Students having such attitudes likely will write very little when requested to do so, or, if they do write, their efforts usually are accompanied by complaints such as "I can't write" or "I don't like writing."

Questions to guide observation Attitude towards writing, then, may be determined by careful observation. The following three questions may serve to guide observation, though it must be understood that no one indicator necessarily means a person does not want to write.

1. How long does a student take to start to write? In some cases the problem may not be in not wanting to write but in not knowing what to write. If, however, a student typically spends the majority of his or her time sitting, as opposed to writing, attitude towards writing in general, or towards starting writing, or towards deciding what to say may be the problem. In general, the more negative the student's feelings are towards writing, the more time will be taken to start composing.
2. How much does the student write? The amount a student writes may be another indicator of his or her attitude towards composition. Students with writing problems often will refuse to write more than a sentence or two. They often feel that they will have to rewrite or recopy sentences they have already written after the teacher has read their work.
3. How much does a student protest when asked to write a story? Some students who do not like to write are very vocal about their feelings. They very quickly will express how they feel. They tend to voice these feelings more readily when only the teacher (or other person requesting them to begin writing) is present.

Teacher-prepared attitude measures In addition to observing students as they begin to compose, the techniques described in Chapter 4 may be used to learn about students' attitudes towards composition. For example, individual students may be interviewed or checklists or rating scales may be prepared and administered. Rating scales are especially useful when a relatively independent, objective measure of change in attitude is needed. This might be the case when change in attitude has been proposed as an annual goal for a student's IEP.

Assessing Composition

There are two possible options when the objective is to sample either how well a composition has been done or what progress is being made towards a final draft. First, commercially prepared tests may be administered. Second, teacher-prepared measures may be used. Both possibilities will provide useful data that will be helpful in preparing an IEP if one is necessary.

The probability that the obtained estimates are representative of any student's present composition abilities will be increased by following the guidelines specified in Chapter 3. To review, first the layout of the test booklet needs to be evaluated. Second, the test needs to be evaluated as to its technical adequacy. Depending on whether the text is criterion-referenced or norm-referenced, slightly different questions should be asked about technical adequacy. Third, the test should be evaluated for administrative considerations such as "How much time does the test take to administer?" As much as possible these criteria should be applied to both commercially prepared tests and teacher-prepared tests.

In addition to the suggestions offered in Chapter 3, the following specific suggestions may be useful in assessing written language. First, the probability of obtaining a reliable estimate of composition skills will be increased if a minimum of two samples of written language are obtained. Further, these samples likely are best obtained over time (a minimum of one day apart) and written on different topics (students' degree of competence may vary depending on the topic). Finally, both the reliability and the validity of any sample of written language may be affected by a variety of extraneous factors such as the way in which the material to be written about is introduced.

Commercially prepared tests Authors in the United States (e.g., Hammill and Poplin 1982; Wallace and Larsen 1978) suggest that written language tests may be inadequate for diagnostic purposes. Two major reasons may be offered as to why such tests are inadequate. First, some commercially prepared tests do not meet all the criteria specified for technical adequacy. For example, Salvia and Ysseldyke (1981) report that the validity data of the *Test of Written Language* are mixed. They especially question the content validity data of this measure. The second reason offered for suggesting that tests of written language are inadequate for diagnostic purposes is a general one, based on the decision-making model presented in Chapter 2. All such tests provide only a single estimate of written language in which performance independent of instruction is reviewed. We have suggested that, when writing an IEP, performance should be evaluated in the context of instruction as well.

In Canada, the choice, if choice is an appropriate word in this case, is limited even more than in the United States. At this time there is no commercially prepared test designed to provide an estimate of composition skills that is normed in English-speaking Canada. Thus, selecting any commercially prepared test involves compromise, either in the type of skills assessed or in the appropriateness of the standardization procedures.

One possible compromise is to choose subtests from an achievement test that has been normed in Canada. However, neither the *Canadian Test of Basic Skills* nor the *Canadian Achievement Test* (described in Chapter 3) require students to compose, so the data obtained from these two achievement tests is not helpful in providing information about students' ability to prepare initial drafts. Both tests do yield some data about editing skills, however. The *Language Skills* or *Written Language* (depending on the level) subtest of the CTBS does require that students recognize errors, which is one of the necessary steps of editing or rewriting. The comparable subtest in the CAT provides even more information about editing skills. In this case, the student is required to complete tasks that better demonstrate editing skills such as reorganizing sentences into a paragraph.

A second possible compromise is to use tests devised in the United States. The problem in this case is that the results likely will have to be analyzed as if the test were teacher-prepared. In other words, the completed standardization procedures may mean that any such test is inappropriate for use as a norm-referenced test in Canada.

The *Test of Written Language* (Hammill and Larsen 1978, 1983) falls into the category of having been standardized and normed in the United States. The TOWL consists of two techniques designed to assess different aspects of written language. In the first section of the test, three pictures are presented and students are asked to write a story about them. The authors claim that having a student write about a series of three pictures results in a spontaneous language sample. However, it is spontaneous only in the sense that students are preparing an original composition. In the second section, students are given three subtests in which they are to spell words, identify appropriate word usage, and add correct punctuation and capitalization to provided sentences. The data collected in the second section is more appropriate for analyzing technical skills in writing (see Chapter 9).

Three scoring procedures are described to analyze the story written about the three-picture sequence in the TOWL. There is a scoring procedure for the vocabulary used in the story, the thematic maturity of the composition, and the student's standard of handwriting. The first two are the most useful for the present purpose of determining the degree of attained composition skill.

The vocabulary score is to be calculated only when the story is comprised of 50 or more words. The main criterion used to estimate vocabulary is the length of the word, those of more than 7 letters receiving one point. The authors report no internal consistency data for this scoring procedure. They do report three studies in which test-retest reliability was 0.38, 0.56 and 0.90. The higher reliability was obtained from a sample of high school students who were asked to write the same story. The lower reliabilities were obtained when the instructions were to write another story about the same set of pictures. The authors argue that the former technique is the better one; however, we suggest that the latter is a better way of estimating the vocabulary used in stories about the theme, rather than calling for story vocabulary recall. In any case, the authors themselves conclude that the reliability of the vocabulary subtest is questionable.

Thematic maturity is intended to provide an indication of the student's ability to write in a logical fashion. Twenty criteria are provided that are used to obtain the raw score. Some of these criteria are very subjective; for example, "Writes story as part of a dream sequence?" (Hammill and Larsen 1978, 1983, p. 22). In our opinion, however, it is possible for a student to write a good story without using such elements and without influencing the thematic maturity of the story. Unfortunately for students who choose not to include such material, omitting "dream sequences" and other similar subjective criteria will result in a lower total raw score.

Again no internal inconsistency data is reported for this scoring procedure. Test-retest reliability is better for this scoring procedure than for the one for vocabulary. On the same three studies, the test-retest reliability is 0.84, 0.68 and 0.72. These data are acknowledged by the authors as being borderline. This concern with reliability is especially true when this test is being used for diagnostic purposes.

The problem with using either the vocabulary or the thematic maturity scores alone or taken together is that the test-retest reliabilities are somewhat low. The only option in this case, then, is to use the total score as a composite estimate of the student's ability to compose. If it is decided that the standardization and normative samples are similar enough to Canadian conditions, a total score may be calculated. This score has a mean of 100 and a standard deviation of 15. The total score does have a reasonable test-retest reliability; however, some of the reported intercorrelations among the scoring procedures and the other three subtests are on the low side.

The authors are helpful in suggesting several informal techniques for evaluating compositions. Most of these fall under what we have termed the technical components of writing. One score that may be helpful in estimating composition skills is called the thought unit, or T-unit. A T-unit is defined as subject and predicate clauses that, taken together, stand alone. Sentence fragments, subordinate clauses, or interjections are not counted as thought units.

Teacher-prepared estimates of writing skills Given the lack of appropriate commercially prepared composition tests for use in Canada, at the present time the best choice is for the teacher to prepare his or her own tests. There are two main ways that data may be collected about students' strengths and weaknesses in composition. One, the teacher may select to present a stimulus to a group of students or to an individual student in order to obtain a sample of written language. Two, spontaneously written compositions or compositions that have been prepared for classroom assignments may be analyzed. In terms of IEP preparation, the latter option has the added advantage of being representative of what the student has completed in instructional conditions.

Designing tests to measure composition In assessing composition skills, most typically students are presented with some stimulus designed to encourage them to begin writing. This stimulus material may be presented in several different forms; for example, as a series of pictures similar to a cartoon strip. If this choice were made it would be important not to include captions. The cartoon-like pictures then provide the material that the student responds to and provides him or her with a sequence of ideas that may be used in developing the story. This was the format employed in the TOWL, in which three pictures were used. The number of pictures does not have to be limited to three, however.

The most common stimulus is a picture or series of pictures. Myklebust (1965) recommends that only one picture be used, however, a series of pictures makes it possible to add time as potential story element. Myklebust has proposed several criteria that make considerable sense whether using one or more pictures as a stimulus. When administering a writing test to an individual, the following three of Myklebust's criteria are useful in selecting a picture or pictures:

1. Action must be portrayed so that the stimulus material is as interesting as possible.
2. The selected stimulus must lend itself to imaginative thinking and be conducive to allowing students to write about more than is portrayed in it.
3. A definite background and foreground should be included in the stimulus.

Many of the criteria suggested by Myklebust will apply whether pictures (or a single picture) or some other stimuli have been chosen. Pictures, however, have the added advantage of always being present when the story is being written, whereas other stimuli may have to be remembered and written about at the same time.

Scoring criteria for teacher-prepared tests Since the purpose in analyzing the resultant composition is to prepare instructional objectives, there are no clear-cut scoring criteria. One may count words per sentence, or analyze the syntax used, or count the number of thought-units. Such scores are useful in setting the criteria for whether or not an instructional objective should be included in an IEP. They will not, however, be very useful in determining what composition skills should have been scored in the first place. In short, in evaluating compositions it is impossible to avoid subjective judgments about what skills are important in a student's learning how to compose.

In analyzing any composition, then, the criteria necessarily will be subjective. In general, however, answering the following questions may be helpful in evaluating students' compositions:

1. Does the story have a beginning, a middle and an end?
2. Are the ideas in the story connected to each other in some way? Or, does the story skip from one idea to another leaving large logical gaps to be filled in by the reader?
3. Do the ideas presented in the story relate to the stimulus materials?
4. Is the presentation of the story appropriate for the age and grade of the student writing it?

These criteria may be applied to compositions that are written under test conditions or under instructional conditions. The purpose is not to attach any score but to determine, relatively speaking, the student's strengths and weaknesses in composition so that instructional goals may be set; that is, some normative criteria may still be useful in planning instructions.

Once a first draft has been written, attention may be turned to how well the student edits his or her written product. Does he or she know how to proofread? Can he or she rewrite sentences if necessary? Does he or she know how to combine two or more short sentences into a longer one? Are there problems with grammar, or word usage, or with both grammar and word usage? Are words misspelled? Is the vocabulary used appropriate for the student's age and grade? For suggestions in addressing these technical aspects of writing, the reader is directed to Chapter 9.

IMPROVING WRITTEN LANGUAGE

Written language problems usually are identified in students who are enrolled in secondary school. For this reason, the majority of instructional suggestions made in this chapter are more appropriate for older students. Many of the instructional strategies, however, are applicable to (and have been used with) students in the late elementary grades and in junior high school or middle school.

General Issues

There are some general issues which affect remedial programs addressing written language. First, if speech patterns are so disturbed that they nullify any written language instructional strategies, oral language may have to be addressed first. Next the goals of the instructional program must be clear. Furthermore, the method of program evaluation must be well established. Finally, the amount of time available for remediation must be well defined.

The first general issue involves the relationship between speech and writing. It was suggested earlier that speech patterns can interfere with written production. If a student has difficulty with fluency, logical presentation and vocabulary at the oral level, their remedial program will have to involve oral experiences. Alley (1977) suggests the following four steps in developing speaking skills:

1. Building a speaking vocabulary
2. Learning linguistic patterns
3. Formulating sentences
4. Developing ability for listener to comprehend the speaker. (p. 39)

Another very simple strategy is to point out to the student that he or she is using inappropriate speech patterns in compositions. Sometimes simple awareness followed by self-monitored or teacher-monitored practice of alternative phrasing will reduce or eliminate the use of an inappropriate non-word, word or phrase.

The second general issue to be addressed concerns the goals of the remedial program. If the intent is to prepare the students for university, the goal will include content and formal training in all areas of academic written expression. If the primary concern, however, is to prepare the student for a career in the commercial or trade areas, the remedial focus will be on highly specific practical goals such as those suggested by Edgington (1973). When formulating the goals of a writing program, it is also necessary to keep in mind the separation between the mechanical components (form) and the productive components (content) (Wiederholt 1978). If the goals centre on expanding the vocabulary and length of written production, then emphasis on form should be a secondary concern. Goodman (1975) refers to the child who realizes after the third rewrite (when the composition is half the length it was initially) that the less you write, the fewer mistakes you make. If mistakes are "punished" (by loss of marks, large x's on the page, etc.), students very quickly reduce their amount of writing.

Another basic issue is evaluation. How do you define success in a remedial program? Traditionally, success has been measured in terms of higher scores on standardized achievement tests. However, if readiness for functional living in society and a healthy self-concept are the primary goals, then a better match between goal statements and evaluation items must be found. Touzel (1978) worked with a variety of experts in secondary learning disabilities on a project that involved evaluation alternatives. The methods that were universally considered appropriate included: 1) following up both graduates and drop-outs of programs, 2) using a variety of sources of information, and 3) doing ecological observation to determine whether graduates were using skills taught in the program outside of the school environment. Bachor (1979) further argues that low-achieving adolescents must be assessed through tests that yield information about their information processing skills if that, in fact, is what we are interested in improving, rather than being assessed through traditional ability and achievement tests that often bear little relationship to what is being taught and merely confirm the deficits already known to exist.

The final issue in deciding on a remedial program involves making realistic decisions about what can be accomplished in the time available. While everyone hopes that the adolescent with learning problems will be functionally competent in all the essential writing skills, the reality of most students' situation demands that the most necessary skills be carefully chosen for instruction and that a systematic plan for teaching such skills be implemented. It is useful in this regard to remember that education does not end when students leave the secondary school. A growing number of adult literacy programs, such as the Adult Performance Level Program offered in Kansas, are being offered across North America; thus, the poor writer has more than one chance at mastery of written language skills.

Changing Attitudes

Because of their long history of failure, special needs students need to experience early success in remedial programs if they are to succeed. It is essential in the initial periods to break new material down into many small steps that can be mastered in a short period of time. Listing each new competency and checking it off as the student masters it provides visible reinforcement. For many, success itself may be sufficient reinforcement for actual learning. For others, more tangible rewards may be necessary. Special privileges, attention, time with teachers, opportunities to do highly valued work, praise from "prestige" adults (i.e., the principal, vice principal, family friend, father or mother), valued articles such as records, points toward motorbikes, holidays, etc., can be used judiciously to get students motivated sufficiently well to stay with new programs long enough for those programs to have a chance to work. It is not always easy to determine which reinforcers will work with individual students, so great care must be given to choosing successful ones.

In the initial stages, it is important that students recognize that their ideas are valued and that they will not be penalized for technical errors in their writing. Encouraging them to write several times a week about something (anything) that interests them will give them practice and get them into the habit of writing. Having them write for 5 to 15 minutes at a specific time each day encourages them to write without censoring their product first. Many of them have been told to "think first, then write," an approach which has tended to inhibit their writing rather than produce a neater product. To write first and evaluate later is a pleasant change that will give them a chance to evaluate themselves as writers after practice. Encouraging them to keep a daily journal or diary also increases their writing experience and is less threatening since it is not seen (and therefore not evaluated) by others.

Another aid to improving the attitude of the student is to present the material in a different format than those with which he or she is familiar. This can include the use of audio-visual material such as tape recorders or videotapes, where possible. Once the student has outlined a short composition or paragraph, it can be taped, listened to, transcribed, revised if necessary, and then the final draft may be transcribed again. (The transcription process can be made less cumbersome by the use of a microcomputer and a word processing program.) Using this technique, the student will have written an outline and translated it into a composition. He or she still may require assistance with the technical aspects of composition, that is, in translating speech to composition; however, the student will have written the outline and organized his or her thoughts into a logical sequence. Two out of three is a great improvement for many poor writers.

Changing what students use to write with is often a useful strategy. One idea is to encourage the use of a typewriter. Although the initial learning may take time, typing may be a very useful skill in and of itself. Sometimes learning to type has long-term and unanticipated benefits, such as greater facility when using a computer. In other cases, using the typewriter is compensatory in the sense that the students themselves are better able to read their own work and their attitude towards writing sometimes changes. On at least one occasion, learning to type has resulted in more attention being paid to spelling and, over time, resulted in better spelling. However, even if spelling remains atrocious, it helps to be able to read the work. Taking time out to learn to type may also be useful in itself because it focuses on learning something new rather than on trying for the ninth time to master something old.

Another highly motivating approach involves using a microcomputer (see Chapter 14). Now that this technology has become increasingly affordable, it is not unusual for a school to have microcomputers at the students' disposal. There are at least four ways that using microcomputers may help the student with special needs. First, they may be used as a reinforcer: although this is a minor role of the computer, students will work very hard for the opportunity to use one. Second, the computer can be used as an aid in managing a remedial program: IEPs can be formulated and records maintained on one. Third, software to be used in remedial programs has been developed. Good computer programs are not yet in great abundance in this area; however, Hannaford and Sloane (1981) provide an excellent checklist to evaluate the ones currently available. With the rapid changes in technology, the range of available programs will soon cover most topics. Fourth, the computer may be used as a tool. Students may learn to program the computer themselves, but this will have little pay-off in terms of written language. The use of the word processor, however, may have a large pay-off. Students can write and edit their work without having the time consuming task of recopying. As a result, making changes is no longer a task to be dreaded. The use of the word processor may be one of the better tools to help effect a positive change in attitude towards writing.

Every successful writer needs a large amount of experience on which to draw. Providing opportunities that enrich the poor writer's repertoire will likely help him or her write with greater confidence. Watch a film, go on a field trip, invite an interesting person to class and then have the student with special needs write about the experience.

Although the last few paragraphs have stressed the need to improve the student's attitude towards writing, it is useful to keep in mind that most students with learning problems want to be able to write well. What holds them back is poor history.

Once they have confidence that you know what you are doing and can structure a program where they will be successful, they are usually willing to try anything at any level. Don't belabour the fact that they should have mastered a skill at grade four; however, don't be afraid to go back to that skill level if that is where the deficit occurs. In representing skills, it usually is necessary to modify the content which is selected to illustrate the skill. Students with learning problems are affected very much by what they consider to be age- and grade-inappropriate materials.

Facilitating Content Generation

Bereiter and Scardamalia (1981) note that there are two difficulties associated with content: a) getting access to knowledge and b) giving order to it. In discussing priorities for a written language program, we suggested that the first one, having something to write about, came before the second one, being able to write it correctly. In many cases the skills will develop simultaneously. As students begin to lose their inhibitions about writing, their vocabulary, sentence length, and syntax will often increase and improve. In all cases, concern about the other form elements should wait until a self-image as "one who can write" is well established. In other words, having something to say should precede trying to say it well.

There are several techniques that may be used in order to encourage students to write better than they do at present. Some of these techniques are "contentless" in that they are ideas that may be incorporated into most composition lessons. Others are designed to aid in thinking about specific types of content. Each type of technique is presented below.

Contentless prompts The use of "contentless prompts" is suggested by Bereiter and Scardamalia as a technique to start student's writing. Contentless prompts are of two types. The first type is simply a prompt that will encourage as much writing as possible. It may be as straightforward as "See if you can write six more sentences." The second type is a prompt that will get a student to start writing; for example, providing a beginning sentence. It should be kept in mind however that, as Bereiter and Scardamalia point out, contentless prompts are useful only for initiating writing, not for improving it.

Techniques to keep writing As simple as it sounds, the suggestion to keep writing sometimes is sufficient to keep students composing once they have started. This little extra support may act as a confidence builder.

Another more formal strategy to keep students writing is suggested by Graves (1978). He encourages teachers to use a "process-conference." The process-conference is a brief meeting of about five minutes held while the student is writing. These conferences may be used to prompt content, to expand it, or to review a draft. Graves stresses that teachers remember that a composition is never wrong, only unfinished. The conferences are used to provide guidance along the path to a completed product.

Concrete prompts to aid composition Concrete aids to composition may range from helping a student write a sentence to helping the student organize the way better.

Providing beginning sentences is one suggestion that provides an opening for the student who may otherwise be unable to start. Ending sentences may be tried as well, however, they were proved not to be as effective as beginning sentences in helping writers to start composing (Bereiter and Scardamalia 1981).

A variation on the above idea is to use a sentence shell. This provides a format for students to use to express their personal point of view. Many students with learning problems do not know how to go beyond the simple declarative statement of like or dislike. Providing them with specific sentence structures can expand their repertoire of response forms. It also shows the need for support statements after an expression of personal opinion. An example of this type of support statement is "I believe _______________ is the correct position to take because _______________."

Students with learning problems need to feel their opinion is acceptable and valuable but that it must be stated clearly and supported by relevant information. Teachers can supply a variety of these form statements. They may have students use them as opening sentences for short answers to tests or for paragraph writing, and then suggest a format for further development. Bereiter and Scardamalia (1981) termed this type of prompt "abstract planning elements." Here the student is provided with a format for what to say next. For example, a format like the following one may be taught:

1. Opening statement of opinion
2. Three statements of fact that support the opinion
3. Statement of a contrasting opinion
4. Reconciliation of contrasting opinion and the student's own. (This is optional but can be used after the student is comfortable with the basic process.)
5. A concluding statement that reaffirms the initial statement of opinion.

One variation on using the idea of "abstract planning elements" is suggested by Scardamalia, Bereiter and Fillion (1979). The aim is to use the reader's point of view in persuasive writing. Since immature writers are rather self-centred when they present arguments, they usually state only why they hold a certain opinion, not why others should. This task provides a first step towards the much harder task of convincing another person. The procedure is as follows:

1. The teacher discusses with the class what constitutes a "peeve" or specific point of view about a certain situation. The teacher can introduce a topic that is somewhat controversial and give reasons to support one side of the issue. The students then give arguments to refute the teacher's points. The important thing is that the students give arguments against the specific points the teacher raises, not against independent points. This demands careful listening on their part.
2. Students pair off. Each student lists a peeve and at least two reasons for it. Topics should be ones where objective data is available rather than being simply ones that involve personal taste.
3. The students exchange lists and each tries to write a short passage trying to persuade his or her partner to adopt a more favourable opinion. The teacher may have to remind students that they don't have to present real reasons for

supporting it.

4. When the pairs are finished, the lists and passages can be read aloud with everyone discussing how well both sides were argued and suggesting other points that could be raised.

The following is an example of this exercise:

> *Peeve:* I dislike people who dye their hair.
> *The responding passage might be:* Although properly dyed hair looks completely natural, people don't dye their hair to fool others. They dye their hair to look and feel better. Others dye their hair for a change, for excitement. This is no different from getting a new haircut or buying a new piece of clothing. It is not dishonest to want to look and feel one's best. (Scardamalia, Bereiter and Fillon 1979, p. 20)

Prompting content Contentless prompts may help in ordering content, getting started writing, in saying more, or in all of above; however, students must have something to say first. The most basic method of expanding students' ideas involves giving greater credence to the things they do. This will include the TV programs and films they watch, the concerts they attend, the sports they enjoy, the hobbies they have, the people they know. Any of these can become suitable topics for written prose. Once students are writing, providing guidelines for critical evaluation of such experiences can be included. Critical writing may serve as an introduction into essay or journalistic writing, which may give students a different perspective on media editorializing. Encourage them to use bigger words, ones they may not be able to spell without help. Give them the spelling if they need it and be glad their interest in words is at a high level even if their ability to write them is not.

Cramer (1978) suggests two strategies that may help students to begin to write better. One is story dictation. A second is following writing models. With these strategies, as with others in this section, both figuring out what to say and how to say it sometimes are addressed in the same technique.

Story dictation This technique takes three days to complete. On the first day, the teacher records the story thought up by the child. Next, first the teacher reads the story, then the child reads it. On the second day, the story is read again with or without teacher help. Finally on the third day the story is reread again and word cards are given for those words which were read correctly. The advantage of such a method is that the teacher can work with several students due to the three day cycle and the fact that reading and writing are integrated.

Using models to compose Providing a student with a model from which to write may provide him or her with assistance in two ways: a) models may help the student in thinking of what to say, and b) models may help the student organize what he or she wants to say. Cramer (1978) suggests two methods of using models. In close imitation, the student reproduces verbatim or summarizes something heard or read. In loose imitation the student uses part of the content or the structure of a story (or poem) as a basis for writing. For example, the content of a familiar tale such as "Little Red Riding Hood" may be used as a basis for a story in which the fates of the characters are different than in the original. Taking another approach, just the pattern of the story may be used as the basis of a new one. In this case, changes are made in the names of the characters, the point in time, or the place—even the planet—on which they live; that is, in any or all the content dimensions of the story. Furthermore, loose imitation is not restricted to prose. Students may write in a prescribed form in poetry as well. One example is the five line poem described in the section below.

Writing poetry Students often respond favourably to new or different activities. Writing poetry may be one such novel activity, although there may be some initial resistance to the idea.

A first easy step is to write poetry in groups in which each person only has to write one line. The result is a group poem that often turns out to be quite delightful. To begin, a starting line might be provided to a group of students, each of whom would then think of one line that relates to this first line. It is surprising how little redundancy results. The lines may be grouped in some manner, for example, by categories of thought, and then organized into stanzas.

Patterned poetry is another beginning point. One should be careful in selecting patterns, as some are very difficult to follow. For example, Haiku, the Japanese lyric poem has a fixed seventeen-syllable form that simply points to an event that has moved the poet. While in its truest form the event usually involves nature, it can be otherwise, as this example illustrates.

On the infinite stairway
To my heart
She has climbed
But two rungs

Haiku can be fun to write; however, attaining the final form of seventeen syllables often is difficult to accomplish.

There are other, much simpler, forms of patterned poetry than Haiku. One of these types should be selected when beginning this type of poetry. For example, one type of patterned poetry has a structure which uses the same opening and closing line along with a conjunction. The following poem is an example of this type of poetry:

  Their love thrived
because
  they wanted it too
because
  they were beginning anew
because
  the day was too
because
  their love thrived.

The idea for this type of poem came from a book by Cramer (1978). It is an excellent source of ideas for teaching patterned poetry (and other types of poetry and prose as well).

An identity as a novice poet can be motivating and encouraging to the student who has rarely accomplished anything creative in written language. Another poetry form that a student with learning problems might attempt is the form of the cinquain or five-line poem (Caldwell, Schooley and Thomas 1976). For this, a student needs only pencils and imagination. The directions are as follows:

Write verses on any subject by following this pattern:
First line: one word title
Second line: two word description of title
Third line: three word action description of the title
Fourth line: four words expressing a feeling about the title
Fifth line: a one-word synonym

An example is:
Learning
Hard work
Reviewing many times
Satisfying when finally accomplished
Progress

Writing prose A very common question asked by all students about any prose they write is "How long must it be?" Perhaps one of the most frightening components about beginning to write prose is the idea that it is long; that is, a composition is by comparison longer than anything else a student with learning problems has been required to write. When first asking for compositions from students with learning problems, length should be de-emphasized.

When they worry about length, students often are wondering what they are going to say. Thinking up content for creative prose or expository writing may be cause for legitimate concern for them. Providing students with techniques that help them think about what they are going to write about will reduce the concerns they have.

Two parallel strategies help students with special needs think about content for prose. The first is the story grid format (Crealock 1980). The second, although not originally so termed, might be called story charting (Hennings 1982). In each case "brainstorming" or "factstorming" is utilized to generate the initial content. The results of this process are then organized through the use of a grid or a chart.

In using the story grid format, students are first asked to name several different content categories for a story. They can include hero, heroine, villain, minor character, theme, climax, mood, setting, time, moral, denouement, etc. Initially, the teacher may want to supply the categories to be used; however, as soon as possible the student should become involved in the category generation. In addition, other categories may be included that make it easier to start writing; for example, one of the categories might be an opening sentence.

The number of categories developed will depend upon how sophisticated the writer is. Reluctant novices might be best with only four elements while a more experienced group may be able to handle ten or more. A familiar form is the "telephone format" which uses seven categories. The directions are as follows:

1. Choose seven major content categories of a story. Write these across the top of the grid.
2. List ten specific items under each. (Brainstorm for these with class.)
3. Have students write down their telephone number vertically. This is their story formula.
4. Move across each category in the grid; find the item that corresponds to each number in the formula and write it down.
5. Write a story that will include each item in a logical manner.

When the students have gotten this far, you may have them share the story with you or with the others in the group. Story grids may be completed with a single student as well; however, the amount of time taken to think up the content is increased dramatically. In addition, at this point students may be encouraged to do some editing and rewriting. Figure 8-1 presents an example of a seven-category grid developed by a class of nine 14 year old learning-disabled students.

Initially, learning-disabled students tend to lump as many elements as possible into as few sentences as possible. With practice, however, they can expand their compositions. In addition to giving the students specific aspects for their story writing, this format also forces them to use their imaginations to make all the elements work together in some reasonable fashion.

If the composition doesn't turn out as well as they would have liked, they can place part of the blame on a poor formula rather than entirely on themselves. Having a story not turn out is an excellent opportunity to introduce rewriting and to emphasize that "It is not incorrect; it is not finished yet." In rewriting, students may be introduced to technical components that help make compositions better. Sentences may be combined or separated to make the meaning clearer. Misspellings may be addressed so the reader will know what the intended word was. The rule of thumb we are suggesting in introducing editing and rewriting is that the student needs to see that there is a reason for making a change: it makes the story read better.

Two examples of compositions based on the story grid format are given in Figure 8-2. The first is written by a 15-year old student who still was struggling with composition. As this first example is an initial draft, it lacks any semblance of polish. It contains some technical problems as well. The second is written by a younger student who had considerable trouble writing anything when he first was referred. He believed that he was not capable of writing. This example is a second draft, which still needs further rewriting. In both cases, we have included the categories with which the student started.

Story charting is a technique that may be used in conjunction with story grids; the skills developed in each are compatible. The format for story charting is slightly different, and more extensive use is made of group sharing (Hennings 1982). Students start by thinking of content; subsequently, they are required to categorize the brainstormed information.

Figure 8-1 **Seven Category Story Grid**

Plot-Action	Hero	Heroine	Feeling	Place-Location	Thing	Time
1 fighting	1 Superman	1 Wonder Woman	1 bored	1 Canada	1 monsters	1 10:00 a.m.
2 floor hockey game	2 Spiderman	2 Miss Piggy	2 sad	2 New York	2 ivory ape	2 noon
3 swinging on vine	3 Tarzan	3 Spider Woman	3 happy	3 London	3 Black Beard	3 Christmas
4 fly	4 Schmoo	4 Angie	4 miserable	4 Pakistan	4 ghost	4 summer holidays
5 swim	5 Super Samurai	5 Karen	5 excited	5 Florida	5 hamster	5 Halloween
6 stretch	6 Batman	6 Super Girl	6 mad	6 California	6 cougar	6 spring
7 learning to drive	7 Robin	7 Shipiro	7 scared	7 Venus	7 Creature from the Black Lagoon	7 winter
8 runs	8 The Hulk	8 Bat Girl	8 frightened	8 Milwaukee	8 Jeremy	8 fall
9 saving people	9 Aquaman	9 Isis	9 stunned	9 Moon	9 Jolly Green Giant	9 April
0 acting on a stage	0 Fred	0 Mom	0 shocked	0 Hamburg, Germany	0 green forest	0 June

Figure 8-2 **Examples of Stories Written Using a Story Grid**

Example A:
The Hulk
great legends
The bear attacked the man
penguin
pain
kill all good guys
present

The Great Legends
This is the story about The Great Legends. This is a story about when the incredible Hulk tried to kill all the good guys.

One day last century, some kind of war was going on. They say that it was the penguin. He took over the incredible Hulk and told him to kill all the good guys. The battle was on. The Hulk kept fighting and fighting, but I didn't believe it until at the present time I was watching the television set when a man said that he saw the incredible Hulk saved a man from a lot of pain by pulling a bear from him. Just before the bear bit off his head. But he already had an arm bitten off by the shoulder and all you could see was blood all over and veins and a bone.

Example B:
shoot fire
Green Lantern
Spider Woman
unhappy
Beaconhill Park
bike
3:00 p.m.

Spider woman was trapped on a rugged cliff holding on the roots of a tree for the sake of her life. She had been riding along a bumpy path when she gazed up to watch Green lantern shooting fire in the air. Suddenly she was falling over the edge of the cliff.

Unhappily eyeing the rocks below, she reached out and grabbed at the protruding roots. Thoughts were storming through her head as to she was at Beaconhill Park at 3:00 p.m. Her pleasurable ride had now become a frightening experience. Suddenly she got an idea as to how to save herself. She would let go of the roots and try spinning around in the air before she hit the rocks. "1-2-3 go!" she said and she went around once, twice but then she was crushed to death on the rocks below and the seagulls got a free diner.

Hennings suggested that having students write stories on topics in science and social studies (themes should not be based on personal experience) may develop the ability to organize expository content. She argues that in allowing students to learn to organize "informational content for writing, students gain insight into how authors handle complex ideas on paper; in so doing, they are refining their schemata for comprehending this kind of content" (p. 8). In this way writing is interrelated with reading comprehension. The eight, slightly modified, steps for using story charting are as follows:

1. Have the students gather material by factstorming. (This procedure is the same as brainstorming.)
2. Have the class categorize the obtained facts into related groups.
3. Divide the class into writing groups. Have each group write approximately one paragraph based on the devised classifications. Have each group then edit its paragraph.
4. Have the class as a whole order the written paragraphs. Products may be typed or copied on chart paper.
5. Have the students move back into writing groups and prepare an introduction and conclusion.
6. You may have students compare and contrast beginnings and endings, for example by following the steps of creative problem solving (see Chapter 10). In this way a single composite introduction and conclusion may be generated.
7. Next, have the class find similar types of written products in which the topic or structures are parallel. These compositions may be examined to answer such questions as the following:

   a) What are the big categories with which this writer started? How do we know?

   b) What system of heads and subheads is this writer using? What do they tell us about the way the topic will be developed in this section?

   c) What is the main—or most important—topic of the section? How do we know?

   d) What kind of information has the writer put into the introduction of the section? What clues have been given in the introduction as to the organization of the material to follow?

   e) What kind of information has the writer put into the concluding section? Are any clues given as to the most important points included in the section? (Hennings 1982, p. 13)
8. Go back to developing data charts and writing. Tasks can be based on completed readings, which may be used to reinforce the process of reading text-explicit information (Pearson and Johnson 1978). On the other hand, tasks may be designed that require students to go beyond the information given, which may include synthesizing information and making judgments based on obtained facts.

By using combinations of the above ideas and materials, students with learning problems should be able to generate and order content. After working with a variety of prompted content generation strategies, they will need to practice thinking about and organizing content with less assistance.

A Case Study

The final component of this chapter is a case study of a student who is having difficulty composing. We will use the decision-making model presented in Chapter 2 to review the case of Clinton. Clinton is in grade six at the time of this referral.

Step 1: Review background information

Clinton was doing very well in all subjects, except when he was required to compose. He was admitted to the gifted program in his school based on standardized test scores and his verbal abilities. He had to withdraw, however, because he could not write well. This withdrawal further affected his willingness to even attempt to write so that when the referral was made he had reached the point where he refused to write.

Step 2: Preliminary hypothesis

Clinton's removal from the gifted program has reinforced his opinion that he is not capable of writing.

Step 3: Assessment of individual strengths and weaknesses

In assessing Clinton, his attitude to writing was found to be a key concern. In an interview, Clinton suggested that he never wanted to write again. He refused to let anyone read what he wrote other than the resource teacher who was helping him.

Work samples of his writing over the period of the beginning of the school year to the time of referral revealed two things: a) his compositions were getting progressively shorter; and b) the vocabulary he was using in his stories was becoming simpler.

Step 4: Reformulate hypothesis (develop preliminary IEP goals)

a) Clinton needs to be convinced that composition does not have to be painful. Trial lessons might emphasize supportive writing strategies.

b) Clinton needs to be encouraged to use his vocabulary when he writes. Perhaps sentence combination exercises may help.

Step 5: Assessment of lessons and materials

Trial lessons were used to convince Clinton he could write when provided with assistance. Effective support included brainstorming ideas which he could choose to include in anything he wrote; exploring at least three ways that he might begin composing and having him select the "best" beginning; and having him write about his ideas on paper or on the computer.

Step 6: Development of IEP goals

a) Clinton will prepare a first draft of material that will be a minimum of 50 words in length.

b) Clinton will incorporate age- and grade-appropriate vocabulary into his composition.

c) Clinton's attitude towards inability to compose will become more positive as evidenced by his willingness to let others (regular classroom teacher, gifted program teacher and principal) outside the remedial setting, read his work.

Step 7: Program implementation and evaluation

Over the period of six months, Clinton became convinced he could write. His compositions began to lengthen considerably. The goal of 50 words was reached and passed within three months. It took, however, the remainder of the period of the IEP before Clinton was sharing his writing. He started by allowing his composition to be shared with others in the resource room. Next, he surreptitiously stole a few stories home to share with his younger sibling. Finally, near the end of the six-month period, he allowed two things to happen: a) other teachers could read his writing, and b) his "best" compositions could be posted in his regular classroom.

9 TECHNICAL ASPECTS OF WRITING

Students write for many reasons, often as a response to external demands from teachers to complete projects, copy notes, write tests, or compose texts, and sometimes as a response to internal needs to express their knowledge and feelings to themselves and others in a way that can be understood. Ideally, much of the reason for writing will come from the latter motivation where the content to be expressed is of interest to the student and consists of information he or she wants to explore and communicate to others (Appleyard, Leitenberg and George 1984). One of the prerequisite skills for writing is the ability and willingness to talk to others as a first means of communicating thoughts and feelings. Once students appreciate the importance of such communication, they easily accept the discipline involved in learning to both speak and write effectively.

According to Kohl (1974) "vocabulary, spelling and grammar become the means of achieving more precise and sophisticated forms of expression and not merely empty ends in themselves" (p. 24). The process involved in this form of communication, however, is much more complex than that involved in oral language communication. Bereiter and Scardamalia (1982) have been engaged in an extensive writing project in central Canada designed to help educators understand the problems involved when a student attempts to make the transition from conversation to composition. The main areas of difficulty they have identified include writing without an immediate respondent, problems in active memory search for content, problems in shifting from local or small components to whole text planning, and problems in knowing when and how to re-write and edit a written text. Imbedded within these four areas is the issue of competency in the mechanics of writing.

Hannan and Hamilton (1984) have developed the following list of benchmarks in the development of writing in elementary students.

1. scribble
* 2. beginning spelling
* 3. labelling of illustrations
4. wanting to write and show work to others

* 5. wanting assistance with the mechanics of writing
* 6. over-using rules
* 7. increasing length of story
* 8. increasing awareness of writing conventions
9. becoming discouraged, getting in a slump
10. writing for a variety of reasons
*11. improving style through greater variation
12. becoming more independent

In this list, the seven asterisked benchmarks refer to technical aspects while the other five refer to motivational issues. Once again, the importance of basic skills is highlighted.

This chapter will discuss the basics specifically as they relate to spelling, handwriting and syntax. General issues also will be discussed and, finally, a case study will be presented illustrating the process and techniques developed in the chapter.

GENERAL ISSUES

Instructional time In the past few decades, there has been considerably less time spent in formal teaching of writing skills. Largely, this stems from the competition for scarce educational time between the core curriculum and either newer subjects such as French at the elementary level and values education or expanded programs in the areas of art, music and physical education. More recently, however, there has been a move back to scheduling time in the curriculum for writing, spelling and grammar. It is being recommended that 15 minutes of spelling and 15 minutes of handwriting be given daily at the grade one to three level; that 15-minute periods in each be given three times weekly in grades four to six; and that reviews and checks be given as necessary in the older grades.

Writing and literacy The role of writing in a general approach to language and literacy is also an issue of debate. Some authors would prefer to see each skill identified and given its own time and curriculum. Others see reading as the most essential language skill and, therefore, they subordinate others to it. More recently, writing is emerging as the subject that best allows the child to be taught in an integrated manner. The work of Graves (1978) is a familiar example of this approach. In his work, students write for much of their school day and through this practice they generate their own stories to read, at the same time developing spelling and grammatical and handwriting skills as necessary. A more specific offshoot of this has been the development of scribble writing programs (Errey 1984; Heald-Taylor 1984). This picks up on the natural tendency of preschoolers to scribble and to intend that their scribbles have meaning. This tendency is capitalized upon the first day they begin school. Heald-Taylor (1984) describes a scribble writing program in grade one that started children writing and used this as their core language experience. Children were encouraged to write in every possible form—notes to each other, lists, letters, labels, poetry and stories. They were also encouraged to take risks in spelling, to illustrate all their work, to confer with peers and adults, and to use their writing as material to be read. At the end of the year, all students were well prepared for grade two language demands in the following ways: they could communicate a message in writing, they had developed left to right directionality, had gained manipulative control of their writing instrument, were aware of sentences and spacing between words, and had developed sound-symbol relationships. In addition, they had developed positive concepts of themselves as writers and readers.

Studies like the above point to the necessity of reevaluating the role of writing in the total curriculum. More specific research has looked at the relationship between reading and writing directly (Belanger and Martin 1984; Blackburn 1984). There is considerable evidence supporting the correlation between reading and writing but very little that looks at the causal relationship. Belanger and Martin's (1984) study looked at grade nine and ten students in an Alberta community who had shown reading improvement after training in the *S.O.S. Reading Technique* to see what effect this had on subsequent changes in writing skill. Writing samples were obtained in three 30-minutes sessions and scored for major sentence errors, productivity, T-unit length and overall quality. The overall findings showed no significant improvement in writing skills over a six-month period for either those students who had shown reading gains or control groups who hadn't. While this study joins the ranks of those which demonstrate correlational rather than casual rélationships between reading and writing, the authors suggest that one of the problems with much of the writing research is that it fails to deal adequately with the motivational aspect. Most educational and psychological inquiries into writing look at student product that has been artificially obtained as opposed to looking at writing that has been internally motivated. Writing resulting from students' desire to express themselves might well show a different relationship to reading. The present findings, however, do lend support to the importance of teaching writing skills either independently or as the overall structure for language development rather than relying on improved reading skills to automatically produce improvement in writing skills.

Reader- versus writer-based reading While most children's early writing is expressive (i.e., they write to express their own feelings about an event), gradually, as writing becomes more narrative, students begin to be more aware of the potential readers of their work. Flower (1980) has defined reader-based writing as being much more explicit and detailed than writer-based in that it presents text in such a way that anyone reading the material can understand it. This is in contrast to the writer-based work that is more typical of the immature student and that assumes without warrant a shared experience among all potential readers. One of the ways our own work has attempted to address this issue is to compare the ratio of pronouns to nouns in a given text (Crealock and Sitko 1985). Students whose work is writer-based tend to use more pronouns which are not as explicit nor as elaborated, while students whose writing is reader-based use more nouns and elaborate them more extensively. Blackburn (1984) has shown with young writers that while first drafts are often writer-based, part of the process of revision is to make them more reader-based. In this process the child is asked to take a second look at the text, seeing it not just through his or her own eyes, but through those of their classmates. It would appear, then, that a legitimate objective in a writing program would be to make the student primarily reader-based in his or her writing.

Mainstreaming One of the critical issues in any text that attempts to show educators how to individualize programs to accommodate the exceptional learner is to decide which aspects basically involve good teaching to all students and which require small-group or individual time to be effective. For the average regular class teacher, the majority of changes suggested need to be ones that can be adapted for the whole class. Thus the guidelines listed, timing required, assessment and remediation described are each applicable to large groups. Where the remediation is limited to one-to-one interaction, it is suggested that this be done by resource teachers or volunteers who have been well trained. Therefore, programs like the Gillingham Stillman, which require an individual remedial setting, are mentioned but not elaborated upon.

Approaches to assessment and remediation Assessment can be done through commercially prepared instruments, teacher-prepared instruments or some combination of both. Our experience has been to rely primarily on teacher-prepared tests, observations and checklists; therefore, these will be described in more detail than commercially prepared tests. Remediation also can be done through commercial programs or teacher-constructed materials. Again our preference is for teacher-constructed materials or teacher-adapted commercial materials.

Another issue in remediation is whether to approach the student holistically or analytically. Several authors have recommended that low-achieving students should be taught from a specific problem perspective (Gajraj 1983; Blau and Loveless 1982) since their subskill development is often spotty and needs careful analysis. Others have recommended the more global approach described earlier in the writing based approaches to the total language curriculum. A combination program that encourages the student to write as much as possible in the regular class but which also focuses on specific weaknesses identified through a careful analysis of error patterns and subskill inadequacies, would appear to best serve the poor writer.

SPELLING

Spelling involves the act of forming words by putting letters together in an appropriate sequence. There is general agreement about what an appropriate sequence is, although Canadian and American spellings can differ slightly in some cases. Children who are good spellers have internalized correct spelling to such an extent that they rarely have to direct any mental energy to the actual writing of a word. Children who are poor spellers, however, often put so much effort into the form of the word that they have little energy left over to focus on how well they are communicating the ideas themselves.

Simon (1975) has developed a task analysis that identifies the stages an average speller goes through in spelling a word. The first stage is to search for the word in long-term memory and, if it is there, to write it down. If it is not there, the student moves into the second stage which generally involves a phoneme-grapheme match strategy in which the student a) calls on past knowledge of phonetic analysis and syllabication, b) checks on context to determine the correct form if a homophone is involved and c) makes a final selection of a spelling pattern from the list of options available for that phoneme. The third stage involves writing the word out and then checking it for accuracy against a visual recognition imprint of the word. This may mean writing out several possible forms of the word and choosing the one that "looks right." Batstone (1981), in an attempt to test this theory with fourth grade students in London, Ontario, found that students who had spelling difficulties fell down in the last stage—that of checking their answer against a visual imprint already in memory. This suggests that poor spellers either have not seen or have not attended to words correctly spelled often enough to have a solid visual imprint against which to check. Anecdotal comments from teachers concerning the accuracy of their own checking mechanism after working with poor spelling models for some time would support this.

Other studies that speak to the differences between average and poor spellers are those of Boder (1971) and Carpenter (1983). Boder has identified three types of poor spellers—the dysphonetic, dyseidetic and alexic. The largest group are those who are dysphonetic. She reports that this accounts for 63 percent of poor spellers. These are students whose primary deficit is in analytic or phonic skills. Their spelling errors are generally non-phonetic, often bizarre and may include many word or syllable substitutions. Dyseidetic students have a primary deficit in perceiving words as visual gestalts. Their phonic skills, however, are usually intact. Their spelling errors are usually good phonic equivalents. The last group, the alexics, have deficits in both analytical and gestalt processes. Boder's findings suggest that normal spellers spell 70 percent or more words correctly in any spontaneous writing passage. Dysphonetics spell about 50 percent of sight words correctly; but they spell less than 50 percent of non-sight words either correctly or as good phonic equivalents. Dyseidetics spell less than 50 percent of sight words correctly; but they spell 70 percent or more non-sight words correctly or as good phonic equivalents. Generally, students who lack analytic or phonic skills are both more frequently identified as poor spellers and have a more serious spelling problem. This is consistent with Simons (1975) analysis of the importance of the grapheme-phoneme matching approach in spelling.

Carpenter's (1983) work supports the particular importance of a phonemic strategy in spelling for older students. He found that poor spellers in the grade four to six level show error patterns like those of average students in grades one to three. In both cases, one error pattern emanates from a poor sight spelling repertoire. Average spellers in grades four to six, however, use phonic strategies more frequently and effectively than do poor spellers at that grade level, which indicates the increasing reliance on this strategy over the years.

In this section we will attempt to show how children's spelling difficulties can be accurately identified and remediated in such a way that children with learning difficulties can be freed up, at least in part, from their concern over correct form so that they are able to expend more effort in developing good content.

Before we look at specific assessment and remediation, however, one general caution seems appropriate at this point. In most cases, there are no lifetime cures for poor spelling. In our experience, children who have serious difficulties in the primary level continue to have problems throughout their writing lifetimes. These problems manifest themselves when the speller is under some stress (emotional, physical or social) and/or when academic material of a more difficult nature is introduced. I hope this statement does not discourage you from teaching your pupil to spell. Remember that although we said there are no permanent cures for spelling problems, there are ways of relieving the disability that can bring a level of competency to the child that will enable him or her to cope more easily with most of the writing demands encountered in school and later in the work environment. This is especially true in later years when the pressure for accurate spelling is lessened. It is ironic that while spelling can be a monkey on the back of many students during their school career, adults who do finally complete their schooling are often excused much more readily. Perhaps this happens because so many of us remember our youthful trials with spelling.

Assessment

Background factors There are several skills we often take for granted when teaching children to spell. These include left to right progression in visual scanning, spatial skills, auditory discrimination, auditory memory, fine motor coordination and visual-motor skills. We also assume that children are motivated to spell correctly and that they can organize themselves well enough to perform competently. For the poor speller any or all of these backup skills may be weak. You can obtain initial information about the child's skills and attitudes while he spells, analyzing his errors and talking to him about his view of himself as a speller. In addition, there are formal and informal means of assessing both these underlying problems and more specific academic difficulties.

Commercial Tests Formal assessment is generally carried out through commercially prepared tests that are norm referenced and have been tested with enough children to provide guidelines on normal development.

You can test for visually and auditorily related problems with the *Illinois Test of Psycholinguistics* (ITPA), the *Marianne Frostig Developmental Test of Visual Perception*, the *Monroe-Sherman Tests*, the *Woodcock-Johnson Psycho-Educational Battery*, the *Slingerland Tests*, the *Impact Testing Materials* and/or the *Wepman Test*. These tests will indicate how well the child does compared to a normative group on a variety of underlying skills that relate to spelling (Fulmer and Fulmer 1983; Chernoff 1980). They will also give you a profile of the relative strengths and weaknesses the child exhibits that can provide some useful guidelines for developing remedial programs. However, in our experience these tests generally assess skills that are too globally defined to provide sufficient direction for individual children with specific problems.

Other commercially prepared tests that provide direct information about writing skills include relevant subtests of the *Canadian Tests of Basic Skills* (CTBS), the *Canadian Achievement Test* (CAT), the *Wide Range Achievement Test* (WRAT), the

Test of Written Language (TOWL) (Hammill and Larson 1978) and the *Test of Written Spelling* (TWS) (Larson and Hammill 1976). The first three tests provide overall scores on general competence that are useful in providing an overall estimate of the student's level of skill, while the latter two tests provide some additional information of a criterion-referenced nature. The TWS includes both words that are "predictable" and which, therefore, can be spelled through a knowledge of phonics rules or generalizations and words that are "unpredictable" in the above sense and which rely on the student's sight word knowledge. The user can analyze the student's performance on these two types of words to plan for remedial work. However, a more in-depth analysis of error patterns is still required in order to develop a carefully sequenced individual program. Devising such a program requires further assessment through the use of teacher-prepared and/or other informal materials.

Teacher-prepared instruments Informal assessment refers to the use of testing materials prepared locally and done at the school by the consultant and by the classroom teacher. It is usually criterion-referenced and is related directly to the skills and competency expectations of that classroom. Informal assessment can be done through the use of checklists, error analysis, interviews, observation strategies, or some combination of each.

Checklists These are usually created by drawing up a list of competencies expected at the relevant grade level and then observing the child's daily work and/or devising a test containing the competencies to be assessed in order to determine what he or she knows. For example, you might convert a competency such as "The child will be able to spell one-syllable words containing short vowel sounds" to the following check list item:

Can spell one syllable words with short vowel sounds
YES NO

Error Analysis This approach differs from using checklists in one basic way. When using checklists, the teacher knows what competencies the child is expected to have acquired and goes about testing whether or not the student has acquired them. When conducting error analysis, the teacher looks at the child's work in order to determine patterns in the errors he or she makes. The teacher has no specific expectations of competencies before analyzing the child's efforts.

Some errors cannot be classified but many can. Often children use an incorrect spelling strategy that can be decoded. Examples of incorrect strategies include phonetic spelling, a tendency to omit medial sounds, and difficulty with transpositions. Figure 9-1 presents an error analysis used to make sense of a child's spelling mistakes. It includes ideas from Spache (1973), Thomas (1979) and our own experience.

We find using an error analysis helpful in several ways. First, it provides us with more information about the child. We no longer simply think, "Oh no, he can't spell half the words he should." Instead, we are challenged to find out what he or she is doing and to find some pattern in the errors. Second, we find that we have a much better grasp of what the child's problems are and of how to translate the information

Figure 9-1

Error Analysis for Spelling		
Type of Error	**Error Example**	**Correct Form**
Omission of silent letter	mak	make
Omission of sounded letter(s)	chalging	challenging
Addition of letter(s)	boey	boy
Substitution of letter(s)	ghosp	ghost
Phonic equivalent	goste	ghost
Transposition, reversal	frist	first
Doubling error	grabing	grabbing
Apostrophe error	theres	there's
Handwriting error	fefth	fifth
No response or bizarre error	soce	pound

into remedial work. At times an error pattern can almost jump out at you. For example, a primary source of errors for some children may be difficulty with a particular rule, such as the silent "e" rule. Remediation in such a case focuses on specific teaching and practice with many instances of the rule. Other children show several kinds of errors. And there are some who show all types at once. In these cases, the type of error that accounts for most of the difficulties can be given priority in remedial planning. It is helpful here to remember that many children may be able to spell a word correctly on a dictated test but not in free writing or in their notebooks. An error analysis ideally includes writing from several sources so you can assess errors that are typical of all of the student's work.

Both error analysis and checklists allow you to keep constant records of the child's progress. Depending upon the maturity of the child, you may ask him or her to keep a list of errors either at the back of the spelling book or in a separate work book. An entry with the date, error and the corrected spelling beside it helps the child to focus on words he or she has difficulty with and helps the teacher keep a running check on errors and progress. Then, at regular intervals you can run the errors through an analysis. An added advantage to having children record their own errors is providing them with an opportunity to be independent. Making correct word lists available so they can check on spellings and having them keep records allows students to exert some control over their own program. In addition, recording the correct spelling may be sufficient for correcting the error. If a student has serious difficulty with visual memory, it may be more useful to have him or her record only the date and the correct spelling of the word so that he or she does not confuse the correct with the incorrect spelling.

You may want an informal check on whether or not your pupil prefers a phonetic approach to a visual approach to spelling. Children whose auditory skills are adequate will often substitute phonic equivalents such as "ones" for "once." This is especially noticeable when they are trying to spell an unknown word phonetically. If the word is spelled according to simple phonetic rules, they may often get it right. On the other hand, children who are dysphonetic (Boder) and prefer a visual teaching style often succeed on basic sight words and words which are the exceptions to phonic rules since these words have usually been drilled through visual exercises (Camp and Dolecourt 1977).

Figure 9-2, which presents an example of a story written by Dan, an eight year old boy, illustrates how error analysis can be used in evaluating a child's strengths and weaknesses. From this type of analysis we can begin to build on the strengths while correcting the weaknesses. For example, Dan needs more practice with silent *e*. This practice will include presenting him with many examples of the rule and then with some non-examples to ensure that he can discriminate between the correct and incorrect forms. He should be encouraged to listen for each discrete sound in multisyllable words and to discriminate more finely between similar phonetic sounds. Much of this can be done through using the tape recorder and can be integrated into stories he himself creates.

Interview A short but well thought out interview with a student can add considerable information concerning his or her spelling problems. Specific areas to be covered include general attitude to spelling, cognitive awareness of strengths and weaknesses, the duration of a spelling problem and its effect on the student's work generally. Cognitive awareness involves asking the child to tell you how he or she goes about trying to spell a hard word. This will tell you if the child can articulate his or her strategy and whether it is an effective one. Also, an interview at the assessment stage prepares the way for later interviews and discussions through which you can explain the remedial approach and try to enlist the students support for it.

Observation Observation can take many forms, all of which include recording the date and behaviour involved. This seems obvious, but one of the most frustrating experiences an educator has in trying to understand a student's problems is finding data in the school record that is undated and/or not clearly specified. A frequent type of observation form is one that records the number of correct spellings on a weekly test. Others look at more general behaviours such as absence on spelling test days, time on-task during spelling study periods, washroom visits during spelling, and verbal or non-verbal reactions (e.g., groans, grimaces) to spelling period or tests. These are useful in understanding the depth of the problem and in checking later after successful remediation.

Having completed your assessment of the student's spelling profile, what do you do?

Remediation

Remedial approaches can include helping the child with underlying problems or with specific academic problems. And in some cases it is possible to modify a commercial program to meet the student's needs. As mentioned earlier, the underlying problems that may be interfering with the child's spelling performance include problems in handwriting skills, spatial orientation, organizational skills, left to right progression, auditory discrimination and memory. Giving a child help in these areas often makes the spelling of specific words much easier. However, most children will also need very specific instruction and practice with individual words. There are some children who can work with a commercial program if it is adapted to their individual needs. We will discuss each of these approaches in turn.

Figure 9-2
Error Analysis of a Writing Sample

Hulk

The Hulk is	The Hulk is
green. He is strong.	green. He is stonge.
He is bigger than a door.	He is biger then a door.
He can pick up 11 000	he can pick up 11000
pounds. He can break steel	paunes. He can stelle
doors. The Hulk is	doors. he Hulk is
Mister Universe. He eats	mister unoves. he eat
bones. He can kill a	bons. he can kill a
bear. He can kill an	bear. he can kill a
elephant. I am The	ealephet. I am The
Hulk. He has a friend	Hulk. he has a feand
Brickman.	Brick Man.
The Hulk's shirt	The hulks shirt
rips. He can kill	Ripes. He can kill
15 gorillas at once.	15 Grillas at ouns.
He can bust two	He can pust to
brick walls. He	Brick walls. he
loses his temper. He	Luses his temper. He
has green eyes. He can	has geen eys. He can
bend steel bars in	bened stells bors in
jail.	Jall.

Weaknesses

1. Phonetic spelling. About half of Dan's errors involve spelling according to phonetic equivalents rather than the correct visual form.
2. Overgeneralization of the silent e rule. He seems to be aware of the rule but not clear about using it, e.g., "stelle" for steel, "ripes" for rips.
3. Omission of consonant sounds. This occurs in sequences of two or more consonants in both initial and final positions.

Since these three categories account for most of his errors, they would be the areas one would try to remediate first.

Strengths

1. The most obvious strength is Dan's willingness to write fairly lengthy and involved stories. Although he has many errors, he spells about 75 percent of the words correctly, which makes his story quite readable for the most part.
2. He has fairly good auditory skills. Although he may rely too much on phonetic spelling in some cases, applying this technique logically to unknown words allows him to write with greater variation and creativity than he could otherwise, and yet be understood.
3. He can use some appropriate consonant blends as in "strange" and "Brick," and although he overgeneralizes the silent e rule he is aware of its basic function.

Underlying problems

Handwriting skills For many children, poor handwriting and poor spelling go hand in hand. Since ways to improve handwriting will be discussed in detail later in this chapter, I will just remind you of this relationship. A child who loops both *e* and *i* can appear to be making a spelling error when he writes a word like "friend." On the other hand, a bright child may deliberately write a letter wrongly if he or she isn't sure of the spelling, in hopes that the error will be blamed on poor handwriting rather than on inaccurate knowledge.

Spatial orientation If a child has difficulty with proper spacing within and between words, markers can be helpful. Putting a toothpick between letters and a finger between words is one example of such an aid. Once children start into cursive writing, some of the spatial difficulties disappear, especially those involved with spacing between letters. Some children benefit from using a marker to stay on the line. It is also helpful to distinctly number pages and allow a fair bit of space between each word and each line on a dictation test. We have also found that it is useful to many children to have them write on every second line in their notebooks. When they do that, minor deviances from the lie don't add extra confusion to the page since there still is sufficient space between the line of work above and the one below to make each line stand out in a distinctive pattern. Concrete aids such as these can help many children, but it is important to remember that they are *only* aids and that the purpose is to help the child improve to the point where he or she can function without them.

Organizational skills Many children with learning disabilities have very poor organizational skills. Time spent helping them organize their desks both inside and out can be useful. Timing how long it takes a child to get from your instruction, "It is time now for spelling dictation," to actually writing the first word can be helpful to both you and the student. You have an objective measure of how long the child does take that can be used as a baseline in charting the student's progress. The child has a concrete demonstration of how he or she gets behind right at the start. Timing yourself performing the same task (or timing another child if this would not create a peer problem) can give extra information to the child about his or her relative speed. Once the child's time has been established, it should be recorded. Then you can show the child how to reduce that time factor. At this stage it is useful to help the child organize his or her desk by discussing the best places to put things and emphasizing the value of having a specific place for each item. Initially, frequent checks (several times daily) are usually necessary, but these can be decreased as the child becomes used to organizing his or her desk and begins to internalize where things belong. This is one of those tasks that may take a long time to initiate but which yields great payoffs for both child and teacher once a routine is established.

Left to right progression Lack of this skill will make both reading and writing difficult for the child. We have used the task analysis approach to present a learning hierarchy that will help you teach this skill most effectively. Figure 9-3 shows the discreet objectives a child must master if he or she is to scan visual material properly.

Figure 9-3

Task Analysis for Directionality in Reading and Writing

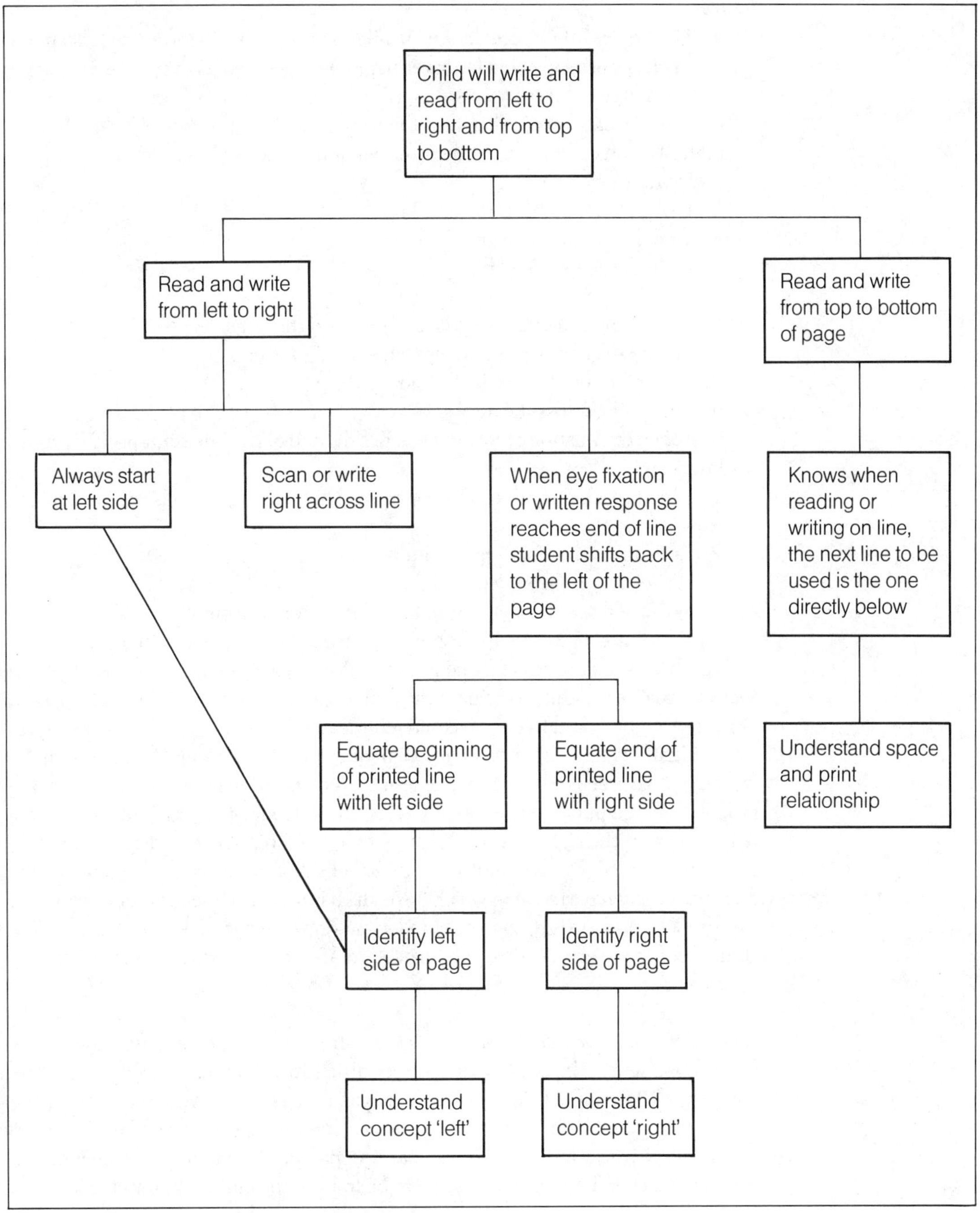

While a task analysis is developed from the top down to ensure that all the necessary prerequisites are included but that irrelevant ones are omitted, the teaching sequence is from the bottom up. Therefore the logical sequence would be approximately as follows:

1. Understand the space and print relationship. This involves understanding the very basic principle that letters either joined together (as in cursive writing) or used with very small spaces between them (as in manuscript writing) form words and that the larger spaces between them separate the discrete words. While most children can recognize the space-print relationship in reading, some do have difficulty with it when writing.
2. Understand the concept "left."
3. Understand the concept "right."
4. Identify the left side of the page.
5. Identify the right side of the page.
6. Equate the beginning of printed lines with the left side.
7. Equate the end of printed lines with the right side.
8. Start reading a line at the left side of the page.
9. Scan right across the line.
10. When eye fixation or writing reaches the end of the line, shift back to the left of the page.
11. When reading or writing along a line, know the next line is the one directly below.
12. Read and write from top to bottom.
13. Write and read from left to right.
14. Write and read from left to right and from top to bottom.

At this point you know what you want to teach, and the sequence in which you will teach your objectives. The next things to develop are strategies (lesson plans) that will enable the child to achieve these objectives. For example, in order to teach objectives eight through ten you could tell the child verbally "your eye moves from left to right," demonstrate this with your finger, and encourage the child to use a marker with a green line on the left side and a red line on the right side as a scanning prop. You could put a green strip of tape on the left side of the student's desk and a red strip down the right side as permanent concrete reminders of left to right progression. You might also want to have the child involved in a program like the *Michigan Tracking System* in which he or she has to scan a line of letters looking for specific letters or words and then underline or write them. You could time the student for speed and accuracy and have him or her chart or graph his or her own progress.

Auditory Discrimination Since many of our approaches to reading and spelling involve phonetics, the child with auditory difficulties is at a serious disadvantage. For some children it will be necessary to begin at quite basic levels in order to shape their auditory discrimination skills to the level required for good language functioning. Initially, you can work with global sounds like the mooing of a cow or the ringing of a bell. The second step would be to have the child discriminate between unlike single sounds like *b* and *w* or *a* and *o*. This would be followed by having the child discriminate between like sounds such as *b* and *p* or *e* and *a*. The next step involves putting those single sounds into words that must be discriminated, for example, words such as "pat" and "bat." For younger children, using puppets can add fun to this exercise. In addition, if the child is answering through the use of a puppet, we have found that she or he is often more outgoing and less afraid to risk an answer than when answering the teacher without such an aid. If the puppet makes a mistake, it is not as humiliating for the child as if he or she had made the mistake. When the child can discriminate between simple like and unlike single-syllable word pairs, the next step is to begin to make the words more difficult. Expose him or her to vowel and consonant blends, two-syllable simple words, complex two-syllable words, three-syllable words, and so on.

The above suggestions have focused primarily on activities that support the training of the back-up skills we often assume children have. Now we would like to describe some activities that help a child develop actual spelling competence.

Specific remedial strategies

Choosing the words to teach One of the first things to decide when you are developing an individual spelling program is what words to teach. We have used the following three criteria in making up appropriate word lists.

1. Teach the words that are most frequently used in general communication. We have been guided here by Thomas (1979) who determined lists of the 3000 words that account for approximately 95 percent of all words used in written communication from writing samples of students across Canada. Figure 9-4 lists the 100 most frequently used words. We have drilled these words in formal spelling periods and used them as frequently as possible in other subject areas.
2. Teach the words that will facilitate the child's functioning in the classroom. This can include vocabulary words in textbooks, directional words the child needs to be able to work independently at activity centres, and labelling and instructional words that are found frequently throughout the class on work-sheets and bulletin boards.
3. Teach the words that the child wants to be able to spell. Motivation for learning these words is usually high. Also they are often more difficult than the words chosen according to the previous two criteria and therefore will contain word parts that can be useful in teaching more mundane words.

While these three criteria are the most useful in making up an individual spelling program, there are several other factors you may want to consider. They include the frequency of use of particular words in the specific locality you are in, the importance you and your school administration place upon accurate spelling for communication, and the difficulty or phonetic complexity of the word. These points will be especially important in considering when to teach certain words.

Once you have decided what words to teach, you must decide how many words to teach at once and in what sequence to present them. A group of three to six words can usually be taught in one week. However, some children will only be able to learn

Figure 9-4

One Hundred Words Most Frequently Written By Elementary School Children in Grades One through Six

the	had	very	see
and	said	if	will
I	like	not	an
a	but	be	now
to	have	little	dog
was	got	could	people
in	were	home	over
it	are	this	by
he	at	down	man
we	me	can	big
my	go	back	into
of	up	as	come
is	all	about	has
on	with	them	did
you	day	saw	play
they	his	house	once
that	out	two	too
went	would	mother	their
when	get	from	off
then	him	our	or
for	her	what	your
she	came	do	know
there	some	time	dad
one	because	just	around
so	going	after	us

(These 100 words and their repetitions account for 58% of the total word count.)

From V. Thomas, *Teaching Thinking* (Toronto: Gage Publishing Co., 1979). Reprinted with permission of the publisher.

one word per week while others may be able to handle 10 to 20 (Hill and Martinis 1973). Poor spellers usually require short but frequent spelling periods. We have found daily periods of from 10 to 15 minutes are most useful. We have also made many worksheets that the children can work on at any time. These were made available at a spelling centre to give the children extra spelling experiences. In this way children spend 15 minutes daily in structured spelling and another 5 to 15 in spelling-related activities. Spelling games, riddles and puzzles are also available at the spelling centre.

There are several additional guidelines for establishing spelling programs in classrooms (Gajraj 1983; Tremblay 1982). They include teaching through all modalities, starting at the beginning rather than assuming the student can spell at supposedly easier levels, teaching in a careful sequence, modelling the spelling task, pacing the instruction to the individual or small group, continual evaluation of progress, and development of positive teacher and student attitude towards spelling.

Spelling rules The role of rule-teaching in spelling is somewhat controversial. Some people argue that the majority of words do conform to a pattern and therefore recommend teaching the pattern and how to use it. On the other side of the fence, however, are those who suggest that the learning of rules is just one more demand on a child that has limited pay-off. They point out that rules which can help a child with many words are often quite complicated and confusing. In addition, children with auditory deficiencies have a great deal of difficulty with words that conform to an auditory pattern and so they do not benefit from a heavy emphasis on rule learning. We have also found that children need a good sight vocabulary first before they can use rules. Many poor spellers do not have this.

It has been our practice to teach rules when they do seem to aid a child in correct spelling but not to place undue emphasis upon them. Two rules that often do seem to help children with the addition of suffixes are a) the rule for silent *e* and b) the rule for doubling (Lutz 1973). The rule for silent *e* states that with words ending in silent *e*, drop the *e* before adding a suffix that begins with a vowel; for example, "state" becomes "stated" when *ed* is added. However, retain the *e* before adding a suffix that begins with a consonant; for example, "state" becomes "statement" when *ment* is added. The doubling rule states that when words end in a vowel-consonant sequence as does "hit," you should double the final consonant before adding a suffix that begins with a vowel but not before adding a suffix that begins with a consonant; thus, "hit" becomes "hitting" when *ing* is added, but becomes "hitless" when *less* is added. Older students can often make better use of a rule approach than younger students. The rules listed in Figure 9-5 can be helpful to them.

Visual teaching A correct visual image seems essential to spelling success. This belief is reflected in the visual approach recommended in many programs. The steps involved in teaching spelling visually usually include the following:

1. See the word in isolation.
2. Say the word.
3. Think about the word; form a visual image of it in your mind.

4. Write the word.
5. Check your own spelling.
6. Repeat steps when necessary.

For many learning-disabled children, the visual method has to be augmented through the support of other senses. The *Fernald Multisensory Approach* (1943) developed by Grace Fernald and Helen Keller has been successfully used to teach spelling for decades. With this approach, the child traces letters (kinesthetic), sees his tracing (visual), says the letters aloud (vocal), feels the letter (tactile), and hears what he says (auditory). The following eight steps outline the suggested procedure.

1. The student asks for a word and the teacher will write it on the chalkboard or on a piece of paper in large script (visual).
2. The child traces the word and says it aloud as he or she traces (kinesthetic, auditory).
3. The child hears himself say the word and sees it (auditory, visual).
4. The child studies the word.
5. When the child is ready, the teacher rubs the word out and the child writes the word.
6. The child turns the paper over and writes the word a second time.
7. The teacher plans to have the child use the word in written expression.
8. The teacher makes sure the child can get the correct spelling of the word any time he or she is unsure of it.
9. When competitive spelling occurs (such as a spelling bee) the words are written, not given in oral form.
10. As the child is more confident in learning new words, the kinesthetic aspect is dropped, then the tactile until finally the student learns the new words through a more traditional look-say method.

Because some students have serious difficulty with the visual modality (Boder 1971), Blau and Loveless (1982) have challenged both the Gillingham (1979) and Fernald methods' use of visual teaching. They suggest that the visual should be placed last and that remediation should focus on the other modalities. Their remedial approach—TAK/v—combines older techniques with two unique features. They blindfold their students with opaque goggles, tell the students the word, and then place three dimensional letters (5 cm high, 1 cm thick, 2-3 cm wide) in their left hand one at a time in proper sequence. They find that both immediate and delayed recall of correct spelling is at the 80 percent level. Left-handed manipulation is significantly superior to right-handed. They suggest this occurs because the left hand taps into the right hemisphere strengths in spatial arrangement and manual patterns.

Reversals Reversals of letters or transpositions between letters constitute a frequent type of spelling error. While this type of error is very common among young children, when it persists with much frequency into the late primary and junior grades it needs some remedial intervention.

If the reversals are of the sort in which one letter is confused with another, such as

Figure 9-5

Spelling Rules Worth Teaching

1. q is always written as qu. It never stands by itself; e.g., quick, queen, quarrel.
2. No English word ends in v or j.
3. Regular plurals are made by adding s; e.g., animals, horses, monkeys, cliffs.
4. To form plurals of words with a hissing ending, add es; i.e., after s, x, z, sh, ch. E.g., buses, foxes, buzzes, wishes, churches.
5. Words ending in an o preceded by a consonant usually add es to form the plural; e.g., potato-es, volcano-es. (Exceptions: pianos, solos, Eskimos.)
6. Nouns ending in a single f change the f to a v before adding es to form the plural; e.g., leaf, leaves, wolf, wolves. (Exceptions: dwarfs, roofs, chiefs.)
7. ck may only be used after a single vowel that does not say its name, and at the end of a syllable or root word; i.e., it follows a short vowel. E.g., track, pick, rocket, wreckage.
8. One-syllable words ending in both a single vowel and a single consonant always double the last consonant before adding an ending.

E.g.	stop	stopped	stopping
	flat	flatter	flattest
	swim	swimmer	swimming

(Exceptions: fix, box, fox, mix—x is same as ch, that is, it counts as a double consonant ending.)

9. Words ending in a single l after a single vowel double the l before adding a suffix regardless of accent; e.g., cancel-l-ed, travel-l-er, signal-l-ing, metal-l-ic.
10. If a word of more than one syllable ends in a t preceded by a single vowel and has the accent on the last syllable, double the final consonant.

E.g.	permit	permitted	
	admit	admitted	
	regret	regretted	
but	visit	visited	Don't double the t.
	benefit	benefited	

11. Drop the final e from a root word before adding an ending beginning with a vowel, but keep it before a consonant.

E.g.	love	loving	lovely
	drive	driving	driver
	settle	settling	settled
	grace	gracing	graceful

12. "All" and "well" followed by another syllable only have one l; e.g., also, already, although, welcome, welfare.

13 "Full" and "till" joined to another root syllable drop one l; e.g., useful, cheerful, until.

14. The sound ee on the end of a word is nearly always y. (Exceptions: committee, coffee.) y and not i is used at the end of an English word. (Excpetions: macaroni, spaghetti, vermicelli (Italian) and taxi (short for taxicab).)
15. If a word ends in a consonant plus y, change the y to i, before adding any ending except ing.

E.g.	party	parties	heavy	heaviness
	marry	married	funny	funnily
	carry	carried	pretty	prettier
but	cry	crying	hurry	hurrying

16. When c is followed by e, i or y, it says s; otherwise it says k.

E.g.	centre	ceiling	circle	cycle	
	cottage	cave	cream	curious	clever

17. ti, ci, and si, are three spellings most frequently used to say sh at the beginning of all syllables except the first

E.g.	national	patient	palatial	infectious
	gracious	ancient	musician	financial
	session	admission	mansion	division

(Exceptions: ship as a suffix, e.g., worship.)

18. i comes before e when it is pronounced ee, except when it follows c, or when sounded like a as in neighbour or weigh; e.g., brief, field, priest, receive, deceive, ceiling.
19. A silent e on the end of a word makes the vowel in front say its own alphabetic name. e.g. hate, ride, cube, bake, shire, mere, lobe. (Exceptions: done, come, some, give, have.)
20. When w comes before or it often says wer as in "worm"; e.g., worship, worst, worth, work, worried.

Adapted from L. Wells, paper prepared for inservice special education course, Toronto Board of Education 1981.

a *b-d* reversal, the child can benefit from a multisensory approach and lots of practice. The multisensory approach involves forming the letter in sand, finger paint, plasticene, Play Doh, crayon, pipe cleaners or sandpaper. An old piece of window screening can be used in several ways. One way is to print the letter on a piece of paper, place the screen over top and have the child trace the letter. A second way is to put the screen down first, lay a piece of paper on top and have the child draw the letter with a wax crayon. This produces a coloured visual imprint with a rough surface that the child can trace over later. In both cases, a tactile impression is made along with the visual one.

Many children confuse *b* and *d* because these letters are often taught concurrently. We find problems occur much less frequently when the *b* is presented singly. After introducing the letter, its distinguishing features are identified (line, circle, circle to right of line) and the child is given lots of practice in tracing the letter while saying it aloud. The letter can then be incorporated into a word which also is traced and said aloud to reinforce the tactile-auditory connection. Once you are sure the child can recognize and print the *b* correctly whenever it occurs, teach the letter *d* in much the same way. The third step is to present them simultaneously to the child and to help her or him define the discriminating features if need be. Incorporating them both into a word like "bed" which has a logical pictorial structure to it

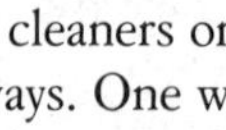

can also provide the child with a helpful concrete image for further reference. *p* and *g* can also be reinforced pictorially.

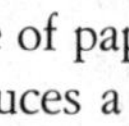

If the reversals are within the word (transpositions of letters) some of the same techniques can be used. All of the tactile, kinesthetic ideas can be applied to words as well as to individual letters. Saying the letters aloud as they are being traced or written adds an auditory dimension to the experience. Tracing can also be done on the blackboard or in the air quite effectively. Another variation of the tracing method involves projecting an image on a bare wall using a transparency on an overhead projector.

It helps to stress the beginning letter of each word and left to right progression. Teaching writing using cursive script also seems to correct the reversals problems with some children. Using a typewriter can also help since the letters are placed far enough apart that they do not lend themselves to reversal difficulties. It helps to type out troublesome words and even to colour code letters that are commonly confused.

As was the case with letters which are reversed, letters within words and whole words that are commonly reversed should be taught first one at a time, then paired together so that the child learns to discriminate between them, and, finally, they should be tested. An example would be to teach "was" first; give lots of practice until the child has reached 100 percent mastery in dictation and within short sentences; then teach "saw" in a similar fashion; and follow this by discrimination exercises in which the child must identify and use each word correctly.

A final strategy to help reversals is to write on the child's back and have him identify the word and/or to have several children use their bodies to write simple words on the floor. The children usually enjoy these activities and they further reinforce correct spelling.

Method for teaching individual words Whatever the spelling error pattern may be, teaching words individually makes sense with poor spellers. Mullen (1971) suggested a method based on the *Michigan Spelling Program* (Smith 1970) which we have used in a modified manner to develop individual word work sheets. The steps involved once you have chosen your target word are illustrated in Figures 9-6 and 9-7.

After using a worksheet such as this, the younger student has written, visualized and traced the word 19 times, while the older student has had 32 experiences with the word. He or she has had considerable exposure without any conflict or confusion from any other word. The children have shown 100 percent mastery after working through a sheet. Although some of them do forget the correct spelling when tested later or when using the word in other work, our success rate with this approach is better than without it. Hill and Martinis (1973) used a similar method with a small group of learning-disabled children and compared their accuracy rate after using this approach to their accuracy rate after using a more traditional spelling program. Three weeks after completion of the trial program, the students could correctly spell significantly more words than they had been able to spell after having completed the traditional program. In fact, after a six-week program, they were able to spell 1580 words. This finding is quite significant since one of the frustrations a teacher experiences in working with poor spellers is their inability to retain knowledge and demonstrate it once direct teaching is over.

The worksheets used in this approach always follow a similar pattern and can be made up by a small group of teachers to share or by a single teacher who needs to individualize a program. They allow a student to work independently at school or at home with words that he or she needs to master.

Figure 9-6 **Spelling worksheet for younger student**

friend

Peter and his friend saw a monkey jump.

1. Make a yellow banana around each word that spells "friend".

frend	friend	friend	
friend	fred	freind	
fried	friend	frind	
friend	friend	friend	friend

2. Fill in the missing letters

fr__d	fri___	fr___
fri__	fr_nd	_ie__
	f____	___d

3. Print the word friend four times on the back of this paper.

4. On the back make a picture of you with your friend. Tell what you are doing.

Adapted from spelling worksheets developed by H. Homeniuk, St. Leo's School, Brantford, Ontario, 1984.

Recording spelling errors One of the most effective ways to improve spelling performance is to show children how their spelling differs from a correct version. Kauffman et al. (1978) compared spelling performance under two conditions: 1) model only—the child was praised for correct spelling; the teacher would write out the correct spelling of a misspelled word and then the child would write the word correctly; and 2) imitation plus model—the child was praised for correct spelling. However, under the second condition, if a child spelled a word incorrectly, the teacher would say "this is how you spelled the word" and write what the child had written; then the teacher would say "This is the right way to spell ______________" and

Figure 9-7

Spelling Worksheet for Older Student

Name ______________________________

Answer

1. My *answer* is correct.

2. *Look at* the word "answer." Close your eyes and picture it. Does the picture in your mind match the word?

   Answer

3. *Circle all* the correct spellings of "answer."

answer	plans	answer
answered	answer	sweat
answer	swear	answer

4. *Fill in* the missing letters.

ans_er	an__er	___wer
a_swer	answ_	ans____
an_wer	_nsw_	an____

5. *Circle only* the letters in each row that spell "answer."

a	m	n	z	s	w	o	e	1	r
v	a	k	n	s	v	w	e	p	r
a	m	n	z	s	w	1	e	o	r

6. *Circle all* the correct spellings of "answer."

answer	ansver	ansyer
amswer	ansure	amswer
aswer	answer	answer
answer	answer	answer

7. *Trace over* each word pattern.

   answer answer answer

8. *Write* the word "answer" *three times.*

   1. ____________________
   2. ____________________
   3. ____________________

9. Turn the page over and *write out* the word "answer." Check it.

would write it correctly; and, finally, the child would write the corrected word. The authors found that the children who were taught following the second method learned the correct spelling faster and showed a higher level of performance on subsequent tests. It was felt that this method worked because it focused the child's attention on the aspects of the word that were "non-instances" of the correct form, a process which helped the child discriminate more accurately. This is probably also helpful in raising the child's spelling consciousness—helping her or him to internalize a sense of correct spelling. Nulman and Gerber (1984) replicated this finding but stated that their students found being reminded of their errors aversive and cautioned educators about using this technique extensively. In both the Kauffman and Nulman and Gerber studies, post-testing occurred within days of final training and so neither could describe the long-term effects of this approach. This is especially significant since Rogers (1981) found that she could train poor spellers to 100 percent accuracy using three different methods, but in a three month post-test all were back to pretraining levels for words that had not been reviewed during the interim period.

We have found it useful to encourage children to keep a record of their own spelling performance through the use of charts and graphs. These can be simple graphs that ask the student to plot correct spelling words against dated spelling tests. More complex records such as the chart shown in Figure 9-8 have been used by Miller (1984) on the Gulf Islands of British Columbia. This chart includes the words the student is working on plus a running record of the number of times words are spelled correctly in sequence and on a scheduled review (tenth day). Words are dropped after two consecutive correct responses, retested on the tenth day and then added or dropped from the active list until a three-month review. This chart can be used either by teachers or students, and, although it appears difficult to the students initially, Miller reports no difficulty in student use. The chart is meant to be used as a component of a flexible program designed to maximize the effectiveness of instructional and learning time through a) inclusion of only words that need to be learned, b) a continuously evolving word list, c) daily testing, d) practice feedback and reinforcement, e) self-correction and analysis of errors and f) routine review and recycling of misspelled words. The child receives concrete reinforcement as his spelling improves; he or she practices charting techniques and both child and teacher have a permanent record of progress.

Evaluating spelling performance Spelling can only be evaluated on actual spelling tests. Pointing out and correcting errors in other work is essential since you don't want the child to continue seeing an incorrect visual image, but the child's work should not be down-graded in any way. In the process of actual evaluation, whether you use a test-study-test method or a study-test-study method, it is often helpful to have the child make up a personal dictionary of troublesome words. I find index cards (7.5 cm x 12.5 cm) that can be filed in recipe-size boxes are most useful. Students can make the boxes in art class and decorate them in any way. Writing the correct spelling of a misspelled word and underlining the difficult part gives each child his or her own reference source. Either writing out the meaning of the word and/or drawing or pasting on a picture to represent the word is an extra aid. Filing

Figure 9-8

Daily Spelling Work Record Sheet

Name: Mike   Time Period: Nov. 19-30, 1985

	Number of words correctly spelled on each day									
Words	**M**	**T**	**W**	**T**	**F**	**M**	**T**	**W**	**T**	**F**
	4	5	2	3	4	3	4	4	3	14
green	✓	✓								✓
hopscotch	✓	✓								✓
mouse	✓	✓								✓
Canada	✗	✓	✓							✓
hockey	✓	✓								✓
funny			✗	✓	✓					✓
Halifax			✗	✓	✓					✗
together			✓	✗	✓	✓				✓
everybody			✗	✓	✓					✗
watch				✗	✗	✓	✓			✓
their						✗	✓			✓
island						✓	✓			✓
government						✗	✗	✓	✗	✓
nineteen						✓	✓	✓		✓
asleep								✓	✓	✓
shark								✓	✓	✓

Adapted from R.D. Miller, "Organizing an individualized, evolving word list program," unpublished, University of Victoria, 1984.

these words alphabetically and showing how they are related to others using the same root gives students further personalized language enrichment. Encouraging children to use their dictionary boxes frequently will help them review words. Most LD children need frequent review tests to ensure retention of accurate spelling, especially of high frequency words.

Spelling programs We know of no commercial spelling program that can be used with LD children without some modification. However, if you have determined specific spelling goals for your pupil and are able to adapt a program to meet them, many prepared materials can be used successfully. The overall goals of a good spelling program should be the following:

1. To teach children to spell the words they will use as adults. Therefore, the bulk

of the words should come from the list of approximately 3000 words that both Dolch and Thomas have found account for approximately 95 percent of the words in everyday usage.

2. To help children learn and internalize rules and spelling generalizations so they can attack new words with accuracy.
3. To create a spelling consciousness in children and a desire to spell properly for good communication.
4. To give children ample review and rehearsal experiences.
5. To integrate spelling into reading and writing.
6. To develop an interest in words and in creative writing.

If the program used in your school addresses these general goals, then you should be able to modify the timing and sequencing to meet the individual needs of your pupil. Some spelling programs introduce writing experiences that your pupil may find confusing. You should omit these initially, if not completely, from the program. A basic program thoroughly learned will serve you and the pupil best in the long run.

If you would like to try a rather different approach, the "stick reading" program developed by Kavorkian (1977) is a good example of an interesting, concrete method of learning to spell. In this program, stick reading substitutes symbols for letters. There is a simple code for spelling monosyllabic words with short and long vowel sounds. The code can also be expanded to show how the spelling of words changes when endings are added. The method involves three sets of sticks, each a different size and colour. The consonants are one colour and are long and rectangular. The vowels are a second colour and are small and square. The endings are of a third colour and are rectangular and grooved, the number of grooves representing the number of letters in the ending. For initial consonant blends you place one stick on top of another. This procedure involves the following steps: 1) Start with "bat," "milk" and "pet." The teacher demonstrates the method to the child by writing the word and spelling it using the corresponding stick figure under it.

As children work with the words of this type, they see and feel that one-syllable short-vowel words have one vowel followed by at least one consonant. Through this process a visual-tactile impression is formed. The process becomes a way of translating the abstract letter symbols and rules into a semi-concrete form for words that *do* obey rules. 2) To add endings to short-vowel words, children are told that "milk" is ready for an ending but "bat" and "pet" are not. The rule is that there must be two consonants after the vowel in a one-syllable word in order that the vowel stay short. Therefore they have to double the final consonant in "bat" and "pet" before adding an ending.

3) To add endings on long-vowel words, the silent *e* must be removed to make the word ready for the ending. Any root word must end in a consonant to be ready for an ending.

When children remove the final *e*, they do so physically as well as mentally, and this integration of the concrete and the abstract form of the words seems to make the spelling task easier for the learning-disabled child.

The authors of this program feel stick reading works because it reduces a complicated area of learning to two basic formulas—one for short-vowel words and one for long-vowel words. Although the author uses actual sticks, a classroom teacher could make the figures out of cardboard so that each child could have a set at a very low cost. In these days of reduced budgets, this would make the program possible in most classrooms.

Another program of a more traditional type that can be easily modified for use with children who are having spelling difficulties is the *I Can Spell: Super Guide to Teaching Spelling* program developed by Rogers, Covell and Slade (1978).

The authors have chosen words from the "most useful" word lists. They sequence the words so that the children learn the phonetically regular words and then lead them carefully into more difficult spellings. There is an emphasis on transfer of spelling skills to words outside the program. The *I Can Spell* texts are designed so that the child can work through them on his or her own. Although the authors don't advocate the teaching of spelling rules, they strongly stress the importance of arranging words according to a pattern. When words are grouped together because they contain a certain group of sounds represented by a common group of letters, they are not complying with a rule so much as illustrating a general principle. Therefore the program insists upon a careful grouping of words according to their auditory and visual consistencies. This approach is based on grouping principles used by Schonell many years ago (1942). The program combines discussion, games, review exercises, creative writing activities, nonsense poems and riddles to foster an interest in both creative and everyday writing activities.

New words are generally learned over a five-day period, although this can be extended for children who are having difficulties. The authors recommend 30- to 40-minute periods daily. We would modify this to two 15- to 20-minute periods daily rather than the one longer one. They also recommend administering a pre-test on the first day and a post-test on the final day. Words spelled incorrectly on the pre-tests are then practiced for the middle two to four days, depending on the length of time in the sequence. The practice procedure is a fairly strict one and involves the following five stages: a) the initial writing stage in which the child says each letter and word to himself while writing out the entire word three times; b) the tracing stage in which the child traces the word in the air; c) the visualizing stage in which the child covers the practice words and visualizes the word on the next line; d) the second writing stage in which the child writes the word on the next line; and e) the final, checking stage in which the child checks the word. This process is repeated for each word and the method is stressed and internalized.

For children with difficulties in learning spelling, the authors suggest a maximum of three new words per week. They also suggest giving the children more practice opportunities, insisting on legible handwriting, giving the child plenty of time for tests that are quickly corrected for immediate feedback, and encouraging the child to keep a record of his or her own hard words.

The *I Can Spell* series comes with two workbooks for each child. These notebooks are used together, which requires some organizational skill on the part of the child and may initially be somewhat difficult for poor spellers. However, the ideas are good and there is plenty of variation, review and practice of a creative nature with which the children can become involved. After every two new word lists, there are break periods built into the word workbooks that include poems such as "Alligator Pie," rhymes, riddles and creative writing suggestions that encourage the child to use the words he or she has been learning. At the higher levels, the program becomes quite integrated with history and geography, mathematics and general language skills. The pictures are fun and can help children make sense out of the words used.

Another program that was initially developed for poor spellers but is now used in regular classrooms is the morphographic spelling program (Dixon and Englemann 1976). It is a direct instruction program developed for use in grades 4 to 12. There are 140 lessons in the program that are carefully sequenced to teach single-root morphographs and the rules by which they are combined into increasingly more complex words. The technique combines active participation of all students, attention getting and maintaining activities, a fast pace and continual feedback through worksheets and summary charts. Both teachers and students sign a contract that outlines the point and grading system (Benham 1984). This program has been used extensively across Canada with positive results (Earl, Wood and Stennett 1981). A second version, "Spelling Mastery," developed for younger students (Dixon, Englemann and Meier 1980; Dixon and Englemann 1981) has also been shown to result in greater spelling gains than a more traditional language arts program (Lum and Morton 1984).

Precision teaching materials have also been used successfully with both regular classes and special needs students (Great Falls n.d.). The words to be learned are generally drawn from high frequency lists with additional emphasis on "spelling demons," i.e., words most often misspelled (Thomas 1979). All work is done in one-minute timed exercises, but the student needs can vary from "see to say" in which the student sees the word on a page and repeats it for one minute to "hear and write" in which the words are dictated at a faster rate than the student can write which both forces the student to decide which word to write and gives room for personal improvement.

The programs described above can easily be used with poor spellers. As mentioned earlier, however, most commercial programs can be used most effectively if they are adapted for the specific child or group of children you are working with.

Games Another approach to spelling remediation involves using games (Forte and

Pangle 1976; Anderson and Groff 1968). Games have long been used as fun reviews in spelling. While spelling bees and other group games of a competitive nature are not recommended for low-achieving students in their usual form, they can be modified in the following manner: the class is divided into teams; the teacher asks each child a word that he or she knows is within that child's own competence, rather than a word chosen at random; then the child writes the answer at the blackboard or at his or her desk and a point is recorded for the team if the word is spelled correctly. The game is scheduled to last for a certain length of time and no child is eliminated for a wrong spelling. In this way both the strong and the weak spellers get equal time and the weak spellers can see how others spell words as well as having practice themselves.

We often use Scrabble (children's or regular) to give children extra spelling practice; but again, we modify the game. For immature spellers, we include only the vowel and consonant sounds we have been working on in class and expand the variety of letters as the child demonstrates greater competence.

Crossword puzzles of varying degrees of difficulty can be constructed. Figure 9-9 demonstrates how a child can practice spelling easy words while being exposed to more difficult language principles. These again offer children a novel way to review words. They also show patterns in words. As well, they can be used to help children spell words they need to know in other disciplines.

Figure 9-9

A Simple Crossword Puzzle

Across

1. A small rodent
2. The past tense of eat
3. The sum of 7+3

Down

1. A small rodent
2. The past tense of eat
3. The sum of 8+2

1. R	2. A	3. T
2. A	T	E
3. T	E	N

Several game formats encourage children to build words that are as long as possible; for example, building "stairs" such as the following:

```
STEP
   A
   T
   HELP
      O
      TOWN
         O
         WAS
```

You can have the children build word lists by changing one letter at a time. In this case "foot" becomes "boot," which becomes "boat," which becomes "coat," and so on. You can vary this by adding an additional rule such as only the first, last or middle letter can change. In this way, the children build word families of their own and you can see how well they have understood the structure. A further variation of this kind of word game is to encourage the child to increase the length of the word by one letter at a time for as far as he or she (or the group) can go; for example, "a" becomes "an," which becomes "ant," then "rant," then "rants."

It is not unusual in many classes for teachers to have an activity centre primarily devoted to games children can play. Because it is our feeling that children with learning problems need considerably more time to rehearse and practice both spelling and reading, we have found that it is most useful to develop games that are similar to commercial fun games but that employ activities that strengthen the language experiences of the classroom. An example of such a game that we have developed is Super Speller. Super Speller is like many other games in which a child rolls a die and moves several spaces on a board. The major difference, however, is that children have to choose the length of word they would like to spell and if they spell it correctly they can move the number of spaces corresponding to the word's length. In this way, the individual has to think about her or his own spelling competency in order to make the best decision. In addition, both players are involved in a spelling experience—one is actively attempting to correctly spell a word while the other is checking the opponent's accuracy from a master card. Teachers can develop different master cards that can be useful for the child at several different difficulty levels in a spelling program. Ideally, teachers should try to ensure that spellers of approximately equivalent competency play this or any other game, whether teacher-made or commercially prepared. Otherwise, the weaker speller could become quite frustrated. Teachers themselves might want to play the game with the child. It could also be one of the activities sent home with the child. In this case, either an older sister or brother or the child's parents can play with her or him. This provides the child with another opportunity to practice as much as he or she needs. The important thing is to be as creative as possible in the means through which you help your pupil maintain the interest and motivation necessary to attain spelling competency.

We have now listed remedial activities that include programs, games, back-up skill development and methods of teaching for specific content or problem areas in spelling. The second aspect of language-learning that we will discuss in detail is handwriting.

Figure 9-10

Super Speller

Super Speller: Instructions

1. Each player chooses a marker.
2. Player 1 asks player 2 for a word that has from two to seven letters in it.
3. Player 2 then chooses a word from the appropriate list and asks player 1 to spell it.
4. If player 1's response is correct, he or she can move as many spaces as there were letters in the word. If the response is incorrect player 1 does not move.
5. Players alternate between spelling and dictating words.

Materials Required

1. Two—four markers
2. Word list
3. Super Speller Game card

Word Lists for Super Speller

2	3	4	5	6	7
to	the	they	there	little	because
in	and	that	would	mother	speller
it	was	went	going	people	catfish
he	you	when	could	around	firefly
we	for	then	about	wizard	chimney
my	she	said	house	author	forever
of	one	like	after	bigger	perhaps
is	had	have	their	reader	clothes
on	but	were	green	purple	balloon
so	got	with	apple	celery	indoors

Figure 9-10

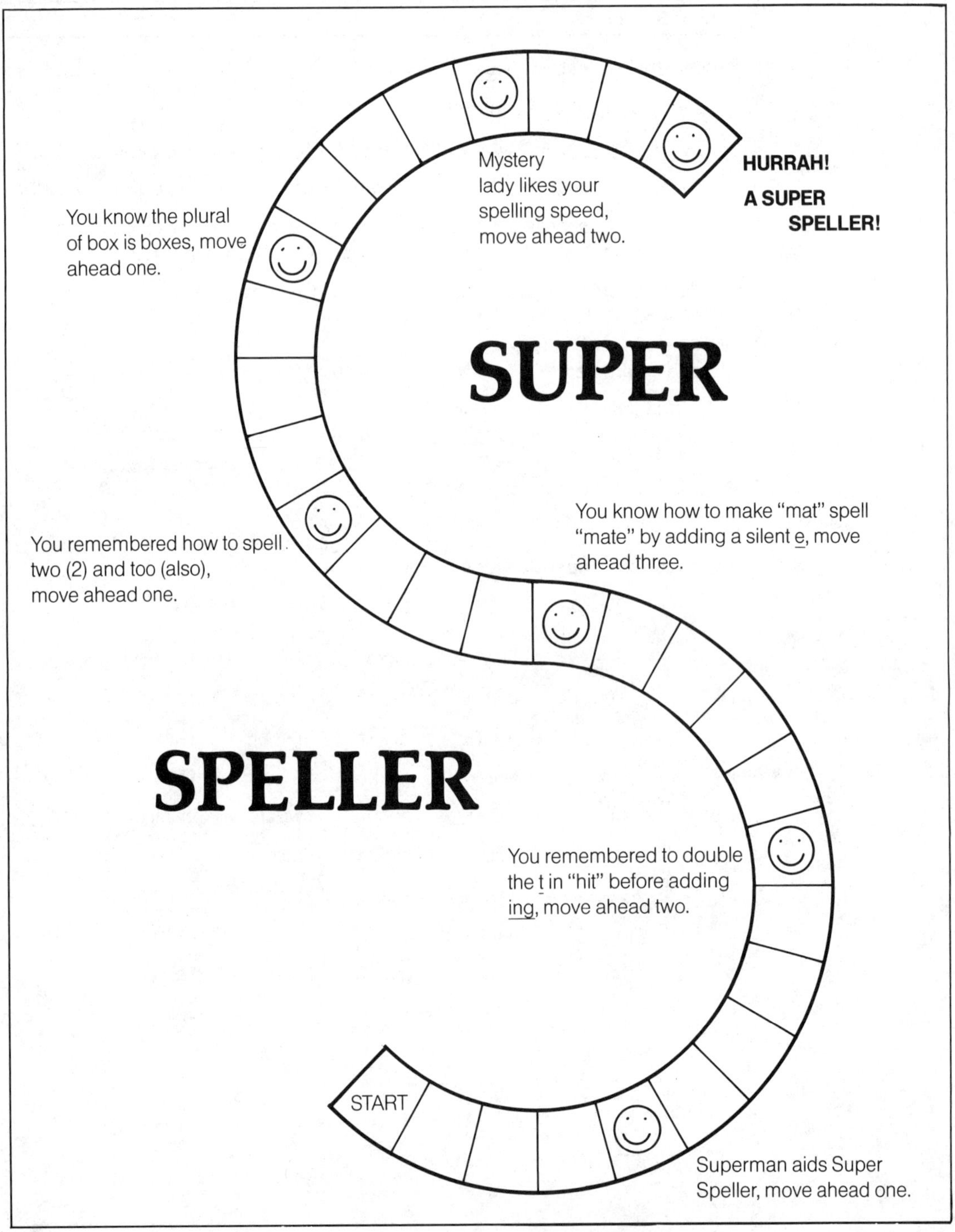
Mystery lady likes your spelling speed, move ahead two.
HURRAH!
A SUPER SPELLER!
You know the plural of box is boxes, move ahead one.
SUPER
You know how to make "mat" spell "mate" by adding a silent e, move ahead three.
You remembered how to spell two (2) and too (also), move ahead one.
SPELLER
You remembered to double the t in "hit" before adding ing, move ahead two.
START
Superman aids Super Speller, move ahead one.

HANDWRITING

For many children with learning difficulties, putting their ideas down on paper presents them with an almost impossible task. Orally, they communicate effectively; but on paper, many are almost illiterate. Part of the problem is often difficulties in spelling. However, for many students the fine motor coordination and spatial skills required for legible handwriting are very poorly developed. Others have difficulties in handwriting that are primarily the result of ineffectual initial teaching and lack of rehearsal time. There is also a controversy over teaching manuscript versus cursive to young children. Hagin (1983) and Neimen (1971) have reviewed the findings in the literature and stated that manuscript had been found to be easier to write as an initial script and had been found to be an aid in spelling. Teaching manuscript has been found to be consistent with the motor development of young children since it is composed essentially of circles and lines, which are within the motor capabilities of beginning writers. Furthermore, the independence of letters and good spacing both between words and between lines makes manuscript more legible than cursive. On the other hand, Neimen also found that cursive writing has large support from both home and community, it offers the writer more opportunities for individuality and children are generally able to write faster in cursive than in manuscript. In addition, children with poor coordination often find the pencil easier to control with cursive writing than with manuscript since it allows them to write in continuous motions and thereby avoid the stop and start and frequent reversals that are evident in manuscript writing (Early et al. 1976). A Canadian study tried to evaluate the relative advantages of cursive over manuscript in terms of reading and writing speed, reading comprehension and detection of spelling error; but this study found no significant difference between the two writing forms in any of these dimensions (Currie 1981). While there is no definitive data to support teaching one or the other script, we have found that most children are more motivated to write than to print and that they also tend to have fewer reversal problems with a cursive script.

Objectives There tends to be general agreement about the broad aims of a handwriting program for low-average students. They usually include the following:

1. To write in a legible manner.
2. To write with sufficient speed and fluency to handle normal writing expectancy.
3. To write with ease.
4. To appreciate that handwriting is an important tool enabling the student to communicate his or her ideas effectively. (Currie 1981)

Examples of a more specific scope and sequence are those used in the London, Ontario Board of Education (1981).

Grade 1: Development of preprinting skills.
Teaching printed form of letters.

Grade 2: Development of increased skill level in printing.
Reducing size of printing to normal expectancy.
Introduction of cursive writing (lower case only).

Grade 3: Completion of transition to cursive (lower and upper case letters).
Maintenance of print skills for use in maps and diagrams.
Grade 4: Consistent improvement in cursive writing skills.
Developing an awareness that both speed and accuracy are important factors for handwriting.
Grade 5: Continued improvement in handwriting.
Increased facility and speed.
Grade 6: Maintenance of all cursive and manuscript skills previously taught.
Emphasis on self-improvement.
Grade 7: Adherence to acceptable handwriting standards.
Appreciation of individuality of handwriting skill without compromising legibility.
Grade 8: Development of a desire to write at acceptable levels in all written work.

It becomes obvious in looking at this sequence of objectives that students are generally expected to know and be able to demonstrate an acceptable level of handwriting by grade five. Instruction beyond that point is primarily practice to achieve greater proficiency and commitment to good handwriting.

Assessment

General problems There are several underlying factors that may contribute to poor writing. These include mild spastic conditions, reactions to medication, perceptual problems and emotional difficulties. Another general type of problem that may be present involves the writing materials themselves. This type of problem includes how a pencil is held, the posture of the pupil, whether the child is right- or left-handed, how the paper is positioned, the thickness and length of the pencil, sharpness of the lead, and so on. For some children, helping them organize their desks and setting them up in a manner that facilitates writing may result in much improvement.

Commercially prepared testing Few of the achievement tests used in Canada give the teacher any specific information on handwriting because they involve computer-scored tests that only ask a child to check the right answer. One individual test that has been developed by Hammill and Leigh (1983), the *Basic School Skill Inventory*, has a handwriting subtest that can provide considerable useful information to the teacher. After having watched the student write in either manuscript or cursive, the teacher answers the following questions:

1. Does the child write from left to right?
2. Does the child demonstrate an easy three-finger grasp of the writing instrument?
3. Can the child write his or her first name?
4. Is the child's sitting posture appropriate?
5. Can the child copy a word from a model at close range?
6. Can the child copy a word from a model on the chalkboard (far range)?
7. Can the child stay on the line?
8. Can the child write his or her last name?
9. Can the child write letters on request from the teacher?

10. Can the child write simple words dictated by the teacher?
11. Can the child correctly copy instructions from the chalkboard?
12. Can the child write simple sentences dictated by the teacher?
13. Does the child use correct punctuation and capitalization when writing?
14. Can the child spell simple words correctly?
15. Can the child write a complete sentence consisting of at least four words?
16. Can the child write a simple story consisting of at least three sentences?
17. Can the child share information with others in written form?

These questions are ordered hierarchically and provide the teacher with a good check on current skill development. The authors have norms available, but the more valuable use for the results is as a criterion-referenced test which the teacher can use in planning remediation. A teacher-made checklist that includes these 17 questions might provide similar information to the practicing teacher, especially since this test is not widely available in Canada (Doree and Kennedy 1984).

Informal or teacher-based assessment One approach to assessing handwriting involves the teacher writing a sentence such as "The quick brown fox jumps over the lazy dog" on the chalkboard. Then the child is asked to copy it five times in his or her usual handwriting, three times in his or her best handwriting, and as many times as possible in his or her fastest handwriting within an approximately three-minute period (Towle 1978). This gives the teacher opportunities to look at the child's writing under various conditions that simulate different kinds of classroom tasks. In addition to this approach to analyzing handwriting, samples of classroom work should also be gathered and assessed. Some of the general problems that can be identified from both these types of writing samples include the following:

1. Does the child have spacing difficulties between letters, words or sentences?
2. Can the child keep on the line?
3. Are there excessive indications of crossing out and erasing words?
4. Are lower case letters evenly or unevenly made?
5. Are capitals made correctly; is the punctuation correct?
6. Is there a mixture of cursive and manuscript?
7. Is the slant consistent or is there an erratic variation within small writing samples?
8. Is there a malformation of specific letters?
9. Can the child produce a reasonable number of letters in the time period? (Bailey 1977)

It is sometimes difficult to determine just how bad a child's writing skills are, especially if he or she is in the early stages of writing. Figure 9-11 shows a handwriting sample from an average grade four pupil and from a grade four pupil who has serious difficulties in the language area. Although both boys have been writing for nearly two years, Bobby's development has been very slow while Matthew has made good progress in size, spacing, correct form, fluency and slant. Bobby will often trace over material and erase in his attempts to write properly. He still has difficulty with

Figure 9-11

Writing Samples from Two Grade 4 Boys

1. Who lived in the old castle?

sir charls and sir albert and o o olaf

2. What did the wizard give them?

the wizard gave theme a goskam

3. What was the one thing that bothered them about the castle?

then Bothered alot the dragon

4. What did they do to the dragon?

they killed the ugley drag om.

Average Work Sample from Bobby, Grade 4.

March 30, 1979

Dear Mr. Lott,

Thank-you for lending us the PET computer. We enjoyed playing with it. My favorite educational game was "Snoopy" and my favorite fun game was "Escape".

Yours truly,

Matthew

Average Work Sample from Matthew, Grade 4.

correct spacing, the size of his letters often varies and he labors over written work for a much longer time than do most of his classmates.

Figure 9-12 presents a checklist we have found useful in determining problems with specific letter formations. If we apply this checklist to Bobby's work in Figure 9-11, we find his most common errors occur with looped *t* (8), confusion of *m* and *n* (6) and with unconnected and poorly connected letters (1). Matthew's work has no errors of this kind although he does crowd his letters somewhat and has some erasures (due to spelling rather than writing corrections, however).

Remediation

In developing a remedial approach, the following guidelines are recommended:

1. Teach handwriting directly, following a carefully devised sequence.
2. Teach it for 15 minutes daily at the same time, as far as possible.
3. Let students know you feel handwriting is important.
4. Give immediate feedback and reinforcement.
5. Use consistent language to describe aspects of the program.
6. Insist on good sitting posture, proper paper placement and good pen grip (see Figure 9-14).
7. Use good quality lined paper.
8. Use the correct writing instrument for the individual student.
9. Use a published writing series.
10. Surround children with models of good writing.
11. Teach the children with models of good writing.
12. Determine acceptable criteria and stick to them.
13. Once the student has learned to write, encourage lots of writing. (Gajraj 1982; Milone and Wasylyk 1981)

Once you have determined the sequence and timing of a program there may be specific problems to address. The following techniques may be helpful in overcoming such problems.

Eye-hand coordination and fine motor coordination These techniques should precede and accompany actual writing exercises for many pupils. They include having children do puzzles, pegboard designs, sorting, matching, cutting, pasting, colouring, and so on. Encouraging geometric form copying, doing coding exercises and many tracing exercises will support writing exercises with other children. The former activities are more general and can be used with younger children (grades one to three), while those in the second group are more specific and therefore more appropriate with older children (grades four and above). While these activities and other more specific programs that improve eye-hand coordination are useful, they do not spontaneously insure improvement in handwriting skills. Legible, fluent handwriting requires specific teaching and much practice. There may be a royal road to Rome, but there is no easy road.

Prompting Some children can be helped in their writing skills by actual prompts by the teacher. Physical prompts can include direct manipulation of the child's hand. Gestural prompts would involve showing the child how a letter or word is formed by

Figure 9-12

Checklist for Determining Problems in Letter Formation

Type of Error	Frequency
1. Unconnected or poorly connected letters	
2. Open *g, d, a, o, p, q, s*	
3. Closed *c, h, y, u, b*	
4. Undotted *i, j*	
5. Uncrossed *t*	
6. Confusion of *m* and *n* or *u, v* and *w*	
7. Unlooped *e, l*	
8. Looped *i, t*	
9. Bottom loop reversed on *f, g, j, p, q*	

exaggerated gestures with the arm and hand. Verbal prompts would include telling the child the letter *d* starts out like the letter *c*.

Sound-symbol training Lebrun and Van de Craen (1975) found that if children were prevented from mouthing their words as they write them, they tended to make more spelling mistakes. They found that children needed to be trained to isolate speech sounds before they could be trained to write them. Strategies they used included rhyming games, drawling out their words, and feeling the differences between words that resemble each other such as "van" and "fan." They suggest that alphabetic script consists of rendering phonemes to graphemes. Therefore, children in kindergarten should be trained to make this type of connection. For example, children should be encouraged to draw a circle, square, cross or dots to the sounds of a whistle, a tinkle, a rattle, a clap, etc. They suggest also that children learning to write should be trained in grammatical analysis since they are trying to acquire the ability to express themselves in writing.

Tracing Many children respond well to tracing. This technique can be used in several ways. One is simply to trace over shapes that the teacher has copied onto a ditto sheet. This can be varied by placing a small piece of screen (approximately 21 x 30 cm) over the dittoed sheet and having the child trace over the shapes with his or her finger, thus getting a tactile impression of how the word is correctly formed. You may also have the child copy a letter onto a sheet of paper that has been placed on top of the piece of screen. The crayon marks leave a raised imprint that again strengthens the tactile experience of the word when the child traces over the letter with his or her finger. Once a child has had considerable experience tracing completely formed words, you can try presenting him or her with stimuli that are partially faded out. Next, you could present word shapes formed of lines that are not joined and, finally, of dots that the child must connect.

Other suggestions Other remedial approaches to handwriting include using materials that give the child good visual support; for example, heavy coloured chalk on chalkboard, crayons on plain paper, finger paint and clay. Some children need to be given concrete starting points on their paper such as coloured arrows for directional cues or red and green dots for start and stop. When teaching manuscript, it is useful to teach the child to print with a continuous stroke rather than a stop and start. Encourage oval form and slight slant to the right. Have children write numbers in a vertical rather than a horizontal array. For some children it helps to associate a letter with something of interest, for example, an *s* with a worm. As mentioned previously, it often helps to encourage the child to say the letter as he or she writes it. Introduce only one form of alphabet letter at a time. We have found that children who confuse *b* and *d* often do so when they have been taught those two letters simultaneously. Teaching the child one letter at a time and giving the child enough practice to attain mastery avoids much of the confusion and tendency to reverse these two letters. At the chalkboard have a child use half a piece of chalk since it is often easier for him or her to manipulate. Encourage self evaluation of handwriting by the child himself or by peers if you have a cooperative group of children.

Sequencing handwriting instruction Since many children are not adequately taught how to write, we will describe a cursive writing program that was developed for exceptional children by Heyman (1977) and modified slightly for our use.

The program begins with pre-writing exercises which are used for approximately ten days. These exercises last for 10 to 15 minutes a day and involve the child and teacher working at the chalkboard. The sequence of pre-writing exercises is as follows:

1. The teacher draws two circles that touch side-to-side which are a comfortable size for the child to trace. The child starts at the 12:00 o'clock position and traces each circle outwards simultaneously with each hand. The teacher initially should guide the child's hand until he or she is comfortable with the tracing exercise and then have the child repeat the exercise with ten additional double circles.
2. Repeat the process with the child tracing the circles inwardly.
3. The next exercise has the child tracing race tracks which cross at the child's midline, e.g.,

The child should use his or her dominant hand for this exercise.

4. Have the child trace circles in the same direction with each hand simultaneously, that is, trace to the right with both hands, trace to the left with both hands.
5. Child draws straight lines with continual in and out movements with each hand simultaneously, e.g.,

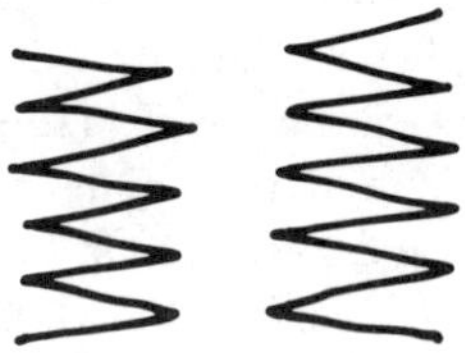

Once the child has completed all the pre-writing exercises, he or she is ready for instruction in specific letters at the chalkboard.

1. The teacher makes a row of cursive *e*'s and guides the child's hand in tracing them. Then the child traces the row of *e*'s alone five times. He or she then traces a second row of *e*'s five times. Next the child makes a row of *e*'s on his or her own, and, finally, covers the chalkboard with *e*'s. At this stage, the teacher should watch for correct form and for left to right progression. If the child is writing the *e* correctly, he or she is ready for the next letter, *l*. The teachers points out that the only difference between the *e* and the *l* is size, and so now they have two letters in their repertoire. The child then practices the *l* by tracing a line and writing his or her own line, and so on.
2. The teacher introduces *i, t* and *u*. The teacher makes a row of *u*'s which the child traces and then writes. The teacher points out that the cursive *i* and *t* have the same form with the addition of a dot and a line. These are written as continuous rows and then the dot or the line is added depending on whether the child is writing *i*'s or *t*'s. At this point the child does four minutes of warm-up such as he did previously (circles, race tracks, lines). He or she reviews the previous day's lesson and then does a new one.
3. The teacher introduces *m* and *n*. The teacher makes a row of each, points out the difference between them, and emphasizes that each letter begins and ends on the line. Now the child is ready to begin writing words. He or she has the letters *e, l, i, t, u, m,* and *n* in his or her repertoire and so can combine them to write words like it, let, net, men, ten, nut, mitt, tin, tell, mitten. The teacher dictates several words and the child practices the letters within words.
4. The teacher introduces the letter *c*. For the first time, the child must reverse the motion; therefore, this letter requires more guidance and tracing than the previous ones. Competency with the letter may take two to three days. Dictating words like cute, ice, mice and cut will reinforce the skill.
5. Introduce the letter *a*. This is like the *c* since it only involves an additional joining stroke and is usually mastered quickly. Again, the child traces the letter and writes it himself until mastery is attained.
6. The *d* is introduced as an *a* with the joining stroke extended upwards.

7. The *g* is introduced as an *a* that goes down below the line. In each case, the similarities between the new letter and others the child has already learned are pointed out. Also, words that contain combinations of the new and old letters are constantly dictated to the child to give him or her practice with the letters in word form. This also reinforces spelling skills, although this is a secondary concern during writing exercises.
8. The *o* is introduced. While this letter is not difficult in itself (show the similarity to *c*), the child will need help learning to connect it to others.
9. *w* is introduced as a *u* that ends up in the air like an *o* and so is connected to other letters like an *o*. Several examples such as won, was, want, wet, etc. are given for practice.
10. The *j* and *y* are introduced as similar to the *i* and *u* except that they go below the line.
11. The *p* is introduced as a *j* that curves back over and then closes at the line.
12. *q* is introduced as a reversed *g*.
13. The *b* is then introduced. It follows the form of *l* and ends like a *w*.
14. The *f* is introduced. It is like a *l* with a loop above and below the line.
15. The *h* is introduced as a combination of *l* and *n*.
16. The *r* is introduced. This letter may be a bit difficult for the child because of the slight point at the top and therefore will need more rehearsal than some of the others.
17. The *s* is introduced. Again this letter is different from the others and generally requires a great deal of tracing, copying and practice within words.
18. The *x* is introduced.
19. The *v* is introduced as like an *x* that ends in the air.
20. The *z* and *k* are introduced and again require much practice.

At this stage, the child has traced, copied and used all the letters in words. To ensure mastery, blindfold the child and have him or her write the letter on the chalkboard. This will enable you to check which letters need further work since writing letters should be automatic at this stage. The reason why you are spending the time to teach handwriting properly is so that the child is able to write independently and fluently; therefore, you should check this now before the child begins to write on paper.

Now that the child can write all the letters correctly at the chalkboard, he or she is ready to begin writing on paper. The teacher should have dittoes prepared that contain one letter per ditto. On the top half of the page the letter should be written out several times for the child to trace; the bottom half of the page should be lined so that the child can write out the letter herself. It usually takes about two weeks before children are able to write everything in cursive in the correct size. For some children it may take a little longer to reach an appropriate fluency rate (Towle 1978).

Learning to write the capital letters is the final step. Start with *I* since it is used so frequently. After the child has learned *I*, begin with *A* and go through the alphabet. Use the same procedure as with lower case letters. Once the child has completed the exercises on the chalkboard, he or she is ready to begin work with paper. Construct ditto sheets that focus on one letter per sheet. On the top half of the sheet, the child traces the letter while on the bottom half he or she will write it out without the model. It usually takes about two weeks for children to be writing everything in cursive in the appropriate size.

This method has been used successfully with seven to ten year old children with handwriting problems. Because the sequence and materials are well structured, you can use faster children to help slower ones. You may also encourage parents to work through the letters on a chalkboard or on paper at home in those cases in which extra rehearsal would be helpful to the child.

Another handwriting program that has been used extensively in Canada is that developed by Schieven-Parr (1978). She identifies five groups of letters that are similar in shape, much like the grouping described above. The letters are taught in groupings because she finds the skill gained in learning to write one letter transfers to other letters of the same group.

Group 1: c o a d g q
Group 2: e l b h k f
Group 3: n m v x y z
Group 4: i t u w j p
Group 5: r s

There are work sheets for each letter that include a lot of tracing of the single cursive form (*c*) and of the double (*cc*) to give students practice in joining the letters. Both on the letters which the student is to trace and in each space in which the student is to write the letter freehand, there is a marker which shows where to start forming the letter. The program is straightforward, well sequenced and useful for both an entire class and/or for individuals who need extra practice since each worksheet is self-contained and structured in a manner that allows the student to work on it independently.

The precision materials developed to teach cursive writing by Pelander and Willis (n.d.) are also well constructed both to teach beginning penmanship to large groups and to remediate problems since they contain branching components that allow for extra review. The materials deal with small cursive letters, connecting strokes, words, phrases, capital cursive letters, sentences and paragraphs. Worksheets are timed and evaluated continually, and progress is graphed by students.

Self-instruction Meichenbaum and Goodman (1971) have developed a program incorporating metacognitive features. There are five steps involved that move the student from dependent to independent functioning (Ashford 1984).

1. The adult models a task while the child observes.
2. The child performs the task while the adult verbalizes it.
3. The child again performs the task but verbalizes the instructions himself.
4. The child performs the task while whispering the self-instructions.
5. The child performs the task and self-instructs covertly (silently).

The following is an example of a learning sequence:

> Okay, what is it I have to do? You want me to copy the picture with the different lines. I have to go slow and be careful. Okay, draw the line down, down, good; then to the right, that's it; now down some more and to the left. Good, I'm doing fine so far. Remember go slow. Now back up again. No, I was supposed to go down. That's okay. Just erase the line carefully.... Good. Even if I make an error I can go on slowly and carefully. Okay, I have to go down now. Finished. I did it. (Meichenbaum and Goodman 1971, p. 121)

Self Evaluation Throughout this chapter, from the guidelines through descriptions of specific remedial approaches, we have talked about the value of self-charting. An example of the type of chart the student can use to do this is found in Figure 9-13.

Figure 9-13

Student Self-Evaluation Penmanship Chart

My Handwriting Chart	Name:				
Check (✓) if yes	**Sept.**	**Nov.**	**Jan.**	**Mar.**	**May**
1. Capital letters are twice as large as small letters.					
2. The differences among *o*'s, *c*'s, *a*'s are clear.					
3. The round parts of *m*'s, *n*'s and *u*'s are clear.					
4. Loops on *f*'s, *j*'s, *g*'s, *y*'s are clearly drawn below the line.					
5. Loops do not intercept writing on the lines above and below.					
6. *T*'s are crossed.					
7. I do not use a circle to dot my *i*'s.					
8. Letters are not crowded.					
9. Letters all slant in the same direction.					
10. Words are separated by clearly defined spaces.					
11. There are good margins.					
12. My writing is neat to look at.					
13. There are no smudges on the page.					
14.					
15.					

Adapted from A.B. Currie, *Instruction in Handwriting in Ontario Schools* (Toronto: Ministry of Education, 1981).

In this example, students are asked to evaluate their own letter forms, spacing, sizing and the overall quality of their writing.

Posture and pencil grip Good posture makes good penmanship easier. Figure 9-14 shows the best posture for right-handed students with recommendations for changes if the student is left-handed.

Figure 9-14

Paper Position and Correct Posture for Handwriting

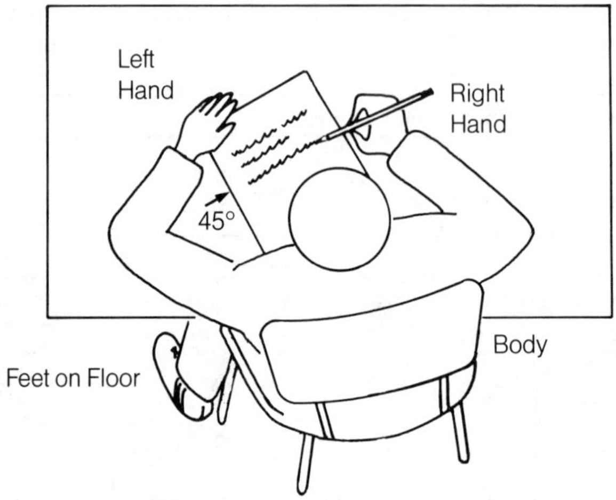

Note: This paper position would be reversed for left handers who write up* but maintained for left handers who hook.†

*Bottom-up position
†Book position

Pencil grip is important too. The most frequently recommended pencil grip is a loose three-finger grip in which the pencil is held by the thumb, first and second finger. Commercial three-sided hard rubber grippers are sometimes put on pencils to help the children remember the three-finger grip and to force them to hold the pencil approximately one centimetre above the lead. An inexpensive variation is to cut a piece of sponge to fit around the pencil in the same three-sided shape. This focuses the student's attention on proper grip and makes the pencil easier to hold.

Speed There are guidelines as to how fast a student should be able to write such as, for example, the following standards given by Currie (1981, p. 38):

grade 3: 40 letters per minute
grade 4: 50 letters per minute
grade 5: 60 letters per minute
grade 6: 67 letters per minute

grade 7: 74 letters per minute
grade 8: 80 letters per minute

For students who have difficulty in handwriting, however, these levels of expectation would be reduced initially and perhaps never totally attained.

Left-handers Approximately 11 percent of all students are left-handed. And for many of this group handwriting is a problem. In the past, educators have tried to make them switch to using their right hands, but new research into the influence of brain hemisphere preference has made us aware that the left-handed individual should be allowed and supported in writing left-handed. Some left-handed writers use a "hook position" (in which the hand is at or above the space to be written in) which may interfere with their ability to read what they are writing and also may make it more difficult for them to form their letters correctly. Recent evidence, however, suggests that this position is not any more debilitating than a "bottom-up position" (in which the hand and arm is below the space that is being written in) which is more typical of the right-handed writer and does not visually interfere with other work on the page. Therefore, in this matter as well, educators are encouraged to support the basic writing approach of their students and to suggest ways of maximizing the students' skill development.

Left-handed students should be placed as far from the windows as possible so that they don't cast a shadow on their own work. They should be given left-handed equipment to work with as much as possible (e.g., scissors, desks). Also, left-handed adults or older students should be encouraged to come in and model certain skills for the student whenever possible; e.g., crocheting, using a handsaw. Having dealt with handwriting problems, we will turn now in the next section, to look briefly at related technical aspects of writing.

GRAMMAR AND USAGE

Grammar refers to the system of rules which governs the way words are arranged in sentences. Usage refers a) to the specific choices students make among those rules when speaking and writing and b) to the acceptability of those choices (Crowhurst 1980). For example, "Mary lamb a little had" is an example of inaccurate grammar, whereas "Mary done good on her test" is grammatically acceptable but of questionable usage in responding to a test item. Students who have problems in writing have been found to demonstrate more problems in appropriate usage than average students (Anderson 1982).

There are some commercially prepared tests that deal with grammar and usage. The TOWL (Hammill and Larsen 1978) has two subtests—word usage and style—which can yield useful information. Again, however, careful analysis of a variety of student work can provide the most comprehensive assessment of grammar, capitalization and punctuation. Weiner (1980) has listed ten points that a teacher should check in determining the level of grammatical competence a student demonstrates in his or her written work. These ten points can easily be put into checklist form as in

Figure 9-15. Such checklists take time to complete; but for individual students they can provide useful guidelines for subsequent program objectives.

In a more detailed look at acceptable criteria for punctuation and capitalization; sentence structure; subject-verb agreement; verb tense, sequence and principal parts, Evans, Brown and Marsh (1977) have suggested the following criteria for grades seven and eight.

1. Punctuation and capitalization: Initial capitalization and final punctuation errors should be few and internal punctuation should be used sufficiently well to not confuse the reader.
2. Sentence structure: None or very few errors in sentence fragments or run-on sentences with comma faults. No sentences in which unrelated ideas are lumped together with conjunctions. No dangling clauses.
3. Subject-verb agreement: None to a few errors of agreement but with evidence that student knows rules in straightforward sentence situations.
4. Verb tense, sequence, principal parts: Maximum of three percent errors even in poor writers in principal parts identification or shifts in tense sequence.

Figure 9-15

Diagnostic Evaluation of Grammar Skills

Skill	Yes	Sometimes	No
1. Subject and predicate agreement			
2. Correct tense			
3. Plurals formed correctly			
4. Possessive endings correct			
5. Appropriate word order, no omissions			
6. Complete sentences			
7. Appropriate punctuation			
8. Appropriate capitalization			
9. Variety in sentence structure			
10. Complex sentences used correctly			
11. Appropriate amount of information per sentence			

Adapted from E. Weiner, "Diagnostic evaluation of writing skills," *Journal of Learning Disabilities* 13 (1980).

Remediation

Remediation generally is approached in the larger context of writing for other purposes (note-taking, projects, compositions); however, some students do need exercises that relate specifically to their chronic grammatical problems and/or that help them appreciate appropriate usage.

Capitalization and punctuation These skills are usually best taught in conjunction with actual writing exercises. It is often the case that they were taught long ago; moreover, they usually will not have been reviewed in recent years. Some students will have forgotten or will continue to forget the various forms. Therefore, a single page on which different types of sentences or parts of sentences are clearly labelled and accompanied by examples and on which correct punctuation is listed can be useful.

Building sentences and paragraphs Oral language is often composed of sentence fragments rather than complete sentences. This makes direct translation from the oral to the written form very difficult for many learning-disabled students. They require sentence-writing practice that moves them from simple to complex forms in a systematic manner. A logical sequence should include the following.

1. Simple declarative sentences involving a noun and verb; e.g., "The dog barked."
2. Declarative sentences with subject and predicate modifiers; e.g., "The large black dog barks loudly."
3. Declarative sentences involving subject, object and verb; e.g., "The dog barked at the cat."
4. Declarative sentence with an object and with modifiers; e.g., "The ferocious dog barked continually at the neighbours' cat."
5. Simple compound sentences; e.g., "The dog barked and the cat ran away."
6. Complex sentences; e.g., "The dog, who had been sleeping, barked at the cat."
7. Simple compound sentences and complex sentences with more extensive modifiers.
8. Other sentence forms such as the question and exclamatory sentence, without and then with modifiers.

Sentences should be incorporated into larger structures such as paragraphs and compositions for further practice. A variation of this involves having students write or select sentences from comic books, magazines, newspapers or texts and categorize them.

Scardamalia, Bereiter and Fillion (1979) suggest the "Three Sentences" activity to help students write a smooth succession of sentences. The procedure for this is as follows:

1. The teacher brings to class a simple sequence of three sentences and demonstrates the task procedure with the help of the students.
2. For homework, students are assigned to find and copy an appropriate sequence of three sentences (from books, magazines or newspapers) to use in class the following day.
3. Students begin with three professionally-written sentences that follow one another and convey a complete thought unit. The sentences are labelled A, B and C. (For example: "(A) Children squabble with each other. (B) It's a fact of life and part of the growing-up process. (C) But sometimes it gets out of hand.") The students must write new A and C sentences, retaining—and re-copying—the original middle sentence, B. The new succession of senten-

ces need not add up to the same idea as the original, but [it] should make sense and convey a complete thought.

4. The two sets of sentences are then read aloud, and the class or some smaller group of students guesses which is the original set. The class might then discuss the reasons for their guesses, and what makes up a smooth, coherent piece of writing. (p. 55)

Principal parts of speech Having students identify these in specific sentences or passages is a useful and necessary review for all students some of the time and for some students on a regular basis. Another way to do this that is more fun is to use a concept like "Crazy Compositions." An example is given in Figure 9-16. The form can be made longer or shorter depending on grade level; the major topic can be varied; the parts of speech can be varied to suit the class's needs. These can be run off and made available at a writing table so that students can do them individually. Alternatively, the class can do them together with the teacher asking for each part of speech in sequence and then reading the story or having a student read it to the class. This exercise can be done in the five minutes the class is waiting for something else (e.g., lunch, a bus trip) and reviews material in an interesting way.

A final example of a technique for reinforcing sentence structure is the creation of pattern books such as the one illustrated in Figure 9-17. Pattern books are usually used to encourage creative writing in students, but they can also be used to draw students' attention to sentence structure. In Figure 9-17, the repeated parts are "If its my ______________," and "Why can't I ______________?"

Children fill in the blank left in each clause and illustrate the completed clauses with a picture. The pages are then put together with the first clause on one page and the second clause on another. Teachers can reinforce punctuation, capitalization, correct choice of words and complex sentence development. These stories are both motivational (If it's my room, why can't I mess it up? If it's my brother's toy, why can't I hide it? If it's my homework, why can't I forget it? If it's my good dress, why can't I paint it?) and instructional.

Case Study

We would like now to share a case study with you that will demonstrate some of the approaches we have discussed in this chapter.

Step 1: Background information The following information was included on the referral form. Vince is a ten year old boy from an average middle class home. He has no serious behavioural problems but is becoming very frustrated with school work. His mathematics is marginal, his spelling is very weak and he finds any written work difficult. His oral language skills are good, he interacts well with his peers and has a supportive family.

Step 2: Preliminary hypotheses

1. Vince has average ability.
2. Vince has problems in mathematics, spelling and writing.
3. Vince is becoming frustrated due to his lack of academic success.

Figure 9-16

Crazy Compositions: A Fun Review of Parts of Speech

Vacations

Last year we went to ____________ (place or city) with ____________ (noun) and ____________ (noun). We had a lot of ____________ (time) and decided ____________ (infinitive) again this ____________ (time). So we all packed up our ____________ (noun), said ____________ (salutation) to ____________ (pronoun) and took off. The first thing we did was decide what we would see. Our list included a ____________ (type of music) concert, ____________ (play) and ____________ (activity). We also decided what we would buy. ____________ (girl's name) wanted to buy a ____________ (colour) ____________ (thing); ____________ (boy's name) wanted to buy a ____________ (adjective) ____________ (thing) and Mom and Dad wanted to buy ____________ (adjective) ____________ (plural noun). It took us ____________ (adverb) to do everything. ____________ (adverb) we went ____________ (place) ____________ (adverb). It had been a ____________ (adjective) holiday.

Adapted from placemat at Pizza Hut (1984).

Step 3: Individual testing Vince was tested on the WISC-R, WRAT, and *Wepman Auditory Discrimination Test.* A writing sample was analyzed for error patterns. Figure 9-18 presents a summary of his formal testing results and Figure 9-19 presents a sample of his uncorrected work from a creative writing exercise.

The results of formal testing show average intelligence. There is considerable variability in Vince's test profile. This is a pattern which is sometimes found with learning-disabled children. His strengths lie in the verbal area primarily and include good social and general comprehension skills and good reasoning ability. In fact, if

Figure 9-17

Use of Pattern Books to Reinforce Sentence Structure

Figure 9-18

Results of Psychometric Testing

Name: Vince Morash **Date of Birth:** July 16, 1966 **Date of Testing:** March 30, 1977			**Age:** 10 yr., 8 mon. **Grade:** 5	
WISC Results			**WRAT Result**	
Information		11	Reading	5.1
Similarities		12	Spelling	3.0
Arithmetic		7	Mathematics	4.4
Vocabulary		11	**WEPMAN Results**	
Comprehension		14	No serious auditory	
Digit span		7	Discrimination problems	
Picture completion		10		
Picture arrangement		11		
Block design		7		
Object assembly		8		
Coding		5		
Verbal IQ	113	103		
Performance IQ		87		
Full scale IQ		95		

you prorate his verbal intelligence score on the basis of the information, similarities, comprehension and vocabulary subtests, he has a verbal IQ of 113, which is above average. He shows serious weaknesses in arithmetic, digit span, block design and coding. The difficulties in arithmetic and coding are consistent with problems he has in mathematics, spelling and writing at school. The results of the *Wide Range Achievement Test* (WRAT) show adequate reading skills, but spelling and mathematics skills are one to three years below grade level.

The writing sample in Figure 9-19 shows good spacing, sizing and form but many problems with spelling, plus some with grammar and capitalization. He is also very slow in completing written work. The error analysis chart given in Figure 9-1 was used to code Vince's errors. The results indicated that most of his spelling errors involve substitution of letters, especially of vowel sounds. Vowel blends such as *ay, ar* and *ou* seem especially difficult for him. He seems to have some rules for capitalization but they are not appropriately applied. While his difficulties are obvious, he also demonstrates many strengths. He is able to stick to the topic and describe playing hockey with a minimum of words. Although he uses "is" rather than "are" in the first sentence, for the most part he shows a good grasp of sentence structure and variation. He can spell correctly most of the short supporting words that are essential to written communication, but he does have difficulty with the longer descriptors.

In summary, in the spelling area his primary weakness seems to be with letter substitutions in two-syllable words. In the writing area, his form is good but his fluency is poor. These, therefore, were the two aspects of his written work that we focused on initially. Because mathematics was not as weak, it was decided to concentrate on spelling and writing.

Figure 9-19

Uncorrected Writing Sample

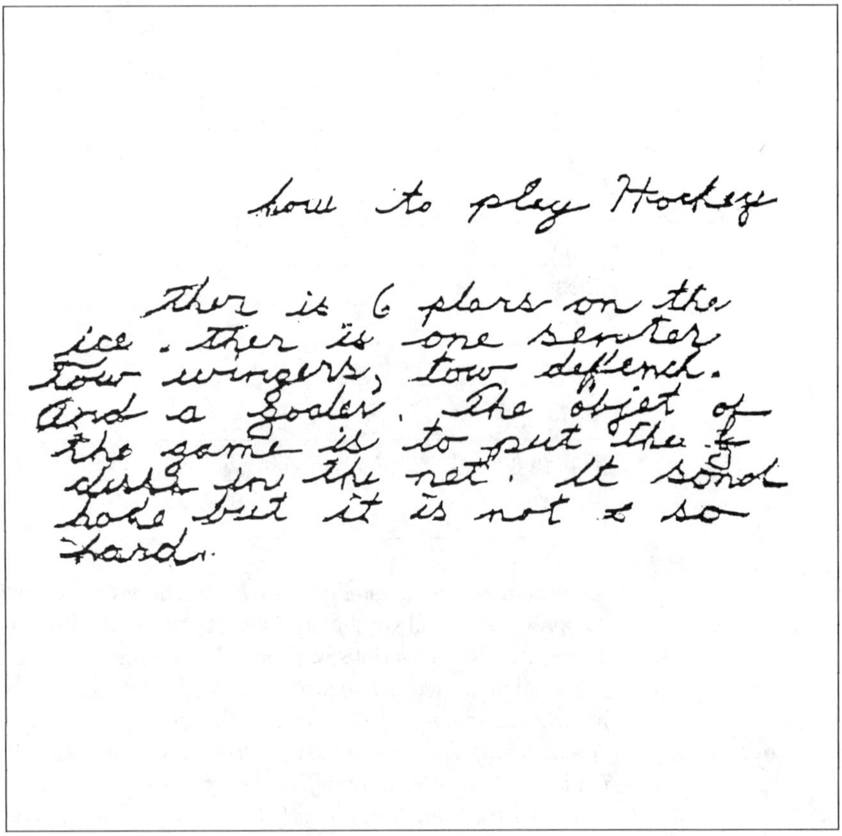

Step 4: Preliminary IEP *goals* From the testing and background information, the following IEP goals were developed:

1. To improve spelling skills
2. To achieve competence in mathematics at grade level
3. To improve writing fluency and speech
4. To decrease frustration with school work

Step 5: Further testing Vince was taught in the resource room with a small group of students for two weeks to see how he responded to small group instruction. His ability to retain in memory words or sentences that he himself wanted to write or that were dictated to him was also checked. A learning hierarchy was developed and he was tested on each objective.

Figure 9-20

A Learning Hierarchy for Spelling

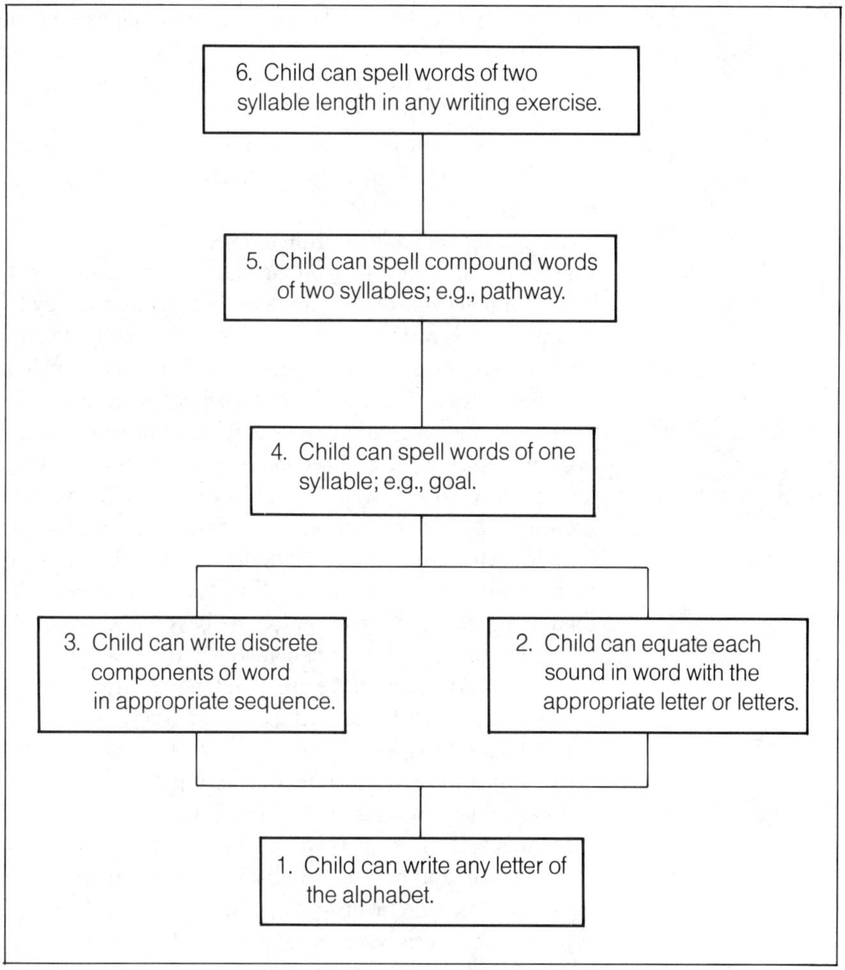

He learned well in the resource room so we scheduled him in for a regular period. Memory had seemed to be a weakness on the digit span subtest of the WISC-R; but, in fact, it did not appear to transfer to the verbal area since he had no difficulty with auditory memory on these tasks. In terms of the learning hierarchy, Vince had no difficulty with the first objective but had considerable problems with the others.

Step 6: Final* IEP *goals

1. Vince will be able to match phonemes with their appropriate graphemes.
2. Vince will be able to spell words of one syllable, compound words and any two-syllable word.
3. Vince will improve his penmanship skills to grade level in fluency and speed.
4. Vince will develop a positive attitude to writing and spelling.

Step 7: Remediation Because the school Vince attended used a resource centre model to give remedial help to children in need and because he had functioned well there during the short trial period, it was possible to schedule him for one hour per day of remedial help for a full school year. He shared this hour with three other children. During this hour he worked on mathematics, spelling and writing for approximately 20 minutes each. He was highly motivated to succeed and apart from one- to two-minute breaks between subject areas to exchange materials and do a few stride jumps, stretching exercises, toe touching, etc., he worked consistently through the hour.

In spelling he worked through objectives two to six on the task analysis. He had considerable difficulty with phoneme-grapheme matching. Frequently he would demonstrate knowledge of the sound-symbol match in one syllable words but not be able to transfer this to the two-syllable word. For example, he rarely misspelled "light" but had much difficulty with "lightning." He spent much time dictating sounds into the tape recorder from sheets the teacher had prepared. At times the sounds were discrete and at times they were embedded in words. He would play these sounds back, underline them on a second copy of the sheet and then compare the two sheets for accuracy. In this way he could pace himself, and his teacher could check the accuracy of his work periodically. He kept daily check sheets on the number of sounds correctly identified and then later on the number of words correctly identified. These were reviewed every Friday with his remedial teacher and the results on them suggested the direction for the next week's work. Because he had difficulty using the correct sound-symbol match with the longer words, he practiced words he wanted to learn through the use of individual ditto sheets, similar to those described earlier. Once he had mastered the sound in several two-syllable words, we would go back, isolate the troublesome sound and test it with a new word. This seesaw approach of learning the discrete sound-symbol match, memorizing it in a high-interest word and then testing it in other two-syllable words seemed quite effective for him.

He was in the remedial program for one year, at the end of which his spelling was at grade level and he was comfortable with written work. While he is able to feel reasonably competent in spelling, however, he still requires close supervision and extra practice. He is rarely penalized for incorrect spelling but when an error is noted in a new word, his teacher draws up a spelling sheet and he completes it on his own time. If the error occurs with a word he generally can spell correctly, it is pointed out to him and he corrects it. If it occurs more than once, he writes it out, sometimes traces it and occasionally will work through a complete sheet containing the word. He has also been helping a few younger children in the remedial class and one of his tasks is to make up work sheets for them. This gives him additional practice with easier words and reinforces his skills. As mentioned earlier, poor spellers are rarely cured but they can be helped with a carefully sequenced program, regular lessons, continual supervision and much rehearsal and review.

Vince also spent approximately 20 minutes a day in remedial writing. Initially he did exercises at the chalkboard similar to those described earlier (Hayman 1977). Since the form of his writing was not a problem, he did these exercises primarily as warm ups. They were timed however and by the end of the week he could quickly fill the board with circles, crosses, lines and race tracks. This was followed by rows of letters, again done primarily as warm ups. This stage lasted a week. He found these activities fun and responded well to changing positions (from sitting at his desk to standing at the board), to the larger muscle movements that board writing involved, and to the time checks. He was timed for two-minute blocks at the end of each period. He began doing this himself with a portable oven timer after day three. When the buzzer indicated the time was up, he would count the number of letters or figures he had made and record them. These data were recorded on a daily log and plotted on a graph as part of his remedial mathematics. He later summed them over varying time periods and found means and modes. He took great satisfaction in this activity. It became a concrete record to which both he and his teacher could refer.

Since fluency was Vince's major problem, we concentrated primarily on building up his speed after the first week. He continued to work at the blackboard at varying written tasks. Initially he would copy a piece of work that the teacher placed just above him on the board. Again he would time a two-minute sequence and chart the number of letters he could complete in that time period. Our general goal for copied material was 50 letters per minute. He was able to achieve this standard at the end of three weeks although his form degenerated somewhat. In two more weeks, however, this form improved to its former appropriate level and he had reached the objective. The next step was to have him write materials that had previously been dictated on a tape at a rate of approximately 50 letters per minute. This frustrated him initially but again within two and a half weeks he was handling this task well. The final stage at the chalkboard was to write spontaneously at a similar rate. He would take time at his desk to rough out what he wanted to say in point form and then go up to the board to write it out in appropriate paragraph form. This took an additional eight weeks. Again he recorded and graphed his times daily so his progress was visible, though slower, at this stage. Throughout this latter stage he would practice with written and oral copying in order to keep up his fluency.

As you have no doubt recognized, Vince was an excellent pupil to work with. He was well motivated, cooperative and although he was somewhat frustrated initially, he was not crushed. He did, however, have occasional off days on which he would regress. This occurred more frequently earlier on in the remedial period but he will still occasionally have poor days. The graphs and records helped a great deal in this regard. They gave us all actual data with which to check the frequency and duration of such slumps. They also helped Vince and his teacher understand them since they met weekly for a general review of his progress. Often something quite specific had occurred in either the academic or social area that had contributed to the poorer work. Seeing slumps in black and white seemed to make them easier to talk about and easier to handle in most cases. As in spelling, Vince's written fluency is at grade level.

Step 8: Evaluation Vince himself kept daily charts and graphs on his progress which were reviewed weekly with his remedial teacher and which provided the basis for the sequence of remediation both in the resource room and in his regular classroom. He, his regular class teacher and the resource teacher met every three months to review the progress in both classes. Finally, at the end of the year, the resource teacher tested Vince with a battery of standardized achievement tests. In all areas of remediation, he was functioning at grade level, his frustration was within normal limits and he was beginning to enjoy school again.

He is now in a regular grade seven class but has continued on a monthly basis to time his fluency, record it and discuss it with his regular class teacher, who has incorporated many of the strategies used in the remedial class into her regular program. At this stage Vince seems to have overcome his academic problems but he and his teachers continue to monitor his progress so that he has the feedback, support and review he needs in order to maintain average grade level functioning. It is a never ending struggle but a worthwhile one.

10 THINKING SKILLS

Thinking skills are rarely taught directly in the school curriculum. Rather, it is assumed that one of the by-products of basic skill learning and the acquisition of information will be the development of higher level cognition and thinking skills. In part this stems from the belief that creativity, originality and problem solving cannot be taught effectively; and in part it stems from the fact that traditionally thinking skills have not been a formal part of the curriculum. Because they have not been taught, there is a feeling that they should not be taught that is shared to some degree by both students and teachers. This was well illustrated to us during classroom discussion following a creative problem-solving exercise with primary level students: the class having solved the problem in a quite original manner through the use of group brainstorming, one grade one girl said that the exercise was fun but not what you learned in "real" school.

In this chapter we will explore how thinking can be taught in "real" school. First we will examine approaches to assessment. We will then look at remediation from the perspective of using the present curricula more effectively. And we will present descriptions of some techniques and programs that have been written specifically to enhance thinking skills.

A focus on the teaching of thinking skills has long been part of the enrichment curriculum for the gifted, but only occasionally has it been suggested for average and low achievers. However, there has been concern for the apparent lag and/or even lack of development of some students in those skills that do enhance general problem-solving. Some authors have suggested that many students lack the required language skills (Koopman 1983; Kronick 1983; Regan et al. 1978). This problem might include not understanding words and phrases like "because," "if *a*, then *b*," "prediction," "testing out your idea," etc. Others have pointed to problems a) in systematically searching for required data; b) in assuming a reflective approach to answering questions in uncertain areas (Gerber 1983); c) in amassing a sufficiently large repertoire of cognitive approaches to be able to use the most effective ones (Keogh 1977; Craik and Lockhart 1972); and d) in using sufficient mnemonic or memory aid strategies to both code and retrieve material from long-term memory stores (Pressley et al.

1980; Bereiter and Scardamalia 1981). One other area of concern has been the role of developmental lag in some students with learning difficulties. Several authors have suggested that low-achieving students are not different from normals in their basic thinking skills but that they do differ in the length of time it takes them to move from one level to another (Sinicrope and Mick 1982; Gerber 1983). This is especially evident in the move from concrete to abstract operational thinking.

ASSESSMENT

While it is agreed that students do differ in their ability to think things through and solve problems effectively, there is not a standardized approach to assessing difficulties in the field. Much of the work referred to above noting cognitive problems comes from educational research rather than from instructional practice. The testing on which this research has been based involves seeing how long it takes a student to handle different tasks; identifying the search strategy he or she uses to examine data or to search for information in his or her own memory; identifying cognitive style such as impulsivity-reflectivity; identifying locus of control; or using tests specifically designed to measure creativity (Jaben 1983). Koopman (1984) and Ross and Fabiano (1983), in extensive literature reviews of the area, have identified many commercially prepared tests. However, their use in the field is extremely limited. Some of the more frequently used ones are listed in Table 10-1.

Although use of testing per se has not been typical in Canadian classrooms, there are many ways of informally observing a student as he or she sets about solving problems. Some of the more useful behaviours to note are as follows:

1. Answers impulsively rather than reflectively.
2. Sporadic unplanned data search as opposed to systematic approach.
3. Tendency to focus and foreclose on one answer rather than explore other, possibly better ones.
4. Poor understanding of cause and effect relationships.
5. Limited use and understanding of problem-solving vocabulary or syntax.
6. Passive versus active approach to problem-solving.
7. Little or no evidence of a risk-taking mentality (that is, student is not willing to make a cognitive or intellectual leap) but basically is concrete and data bound in his or her problem solutions.
8. Inability to answer questions that go beyond factual knowledge and that ask students to think about the issue.

Whether the decision to teach thinking skills has been reached as a result of identifying problems through formal testing or informal observation, or whether that decision is the result of a philosophical commitment to the idea that "teaching thinking" has a place in the curriculum, the teacher can approach the issue either globally, by instructing the entire class, or specifically, by offering instruction to a few students. He or she must also decide whether to augment the materials already available and use them in a more effective manner, whether to design his or her own techniques or

Table 10-1

Commercially Prepared Assessment Instruments to Measure Cognitive Skills and Thinking Processes

Test	Skill measured	Comments
The AC Test of Creative Ability	Quantity and uniqueness of ideas	Reliability and validity adequate; group administration
Cognitive Abilities Test (CAT)	Ability to abstract and see relationships among symbols	Standardization and reliability good; validity data meager; group administration
Conceptual Level: Paragraph Completion Method	Capacity for conceptual complexity as indicated by discriminability, differentiation and integration.	Normed on Canadian students; reliability good; requires students to write out lengthy response
Feuerstein's Learning Potential Acquisition Device (LPAD)	Concrete and abstract reasoning skills	Limited reliability and validity data; lengthy individual administration
Wechsler Intelligence Scales Block Design, Object Assembly	Field dependency or field independency	Standardization good; reliability poor to moderate; individually administered
The Group Embedded Figures Test	Field dependency, field independency	Reliability and validity good; group administered
Watson-Glazer Critical Thinking Appraisal	Critical thinking	Reliability acceptable; group or individual administration

Adapted from R.R. Ross and E.A. Fabiano, *The Cognitive Model of Crime and Delinquency Prevention and Rehabilitation: Assessment procedures* (Toronto: Ontario Ministry of Correctional Services, 1983); and J. Salvia and J. Ysseldyke, *Assessment in Special and Remedial Education* (New York: Houghton Mifflin, 1981).

programs tailored to the needs of particular students, or whether to use some combination of the two approaches.

REMEDIATION

Modification of Materials and Instructional Approaches Currently Available

There are many general approaches to improving the thinking skills of students. They include improving their memory retrieval skills, questioning techniques that demand thinking, concept attainment strategies, metacognitive strategies, brainstorming, and perceptual and conceptual flexibility training.

Improving memory Strategies to help memory, known as mnemonics, have been around for centuries. Some techniques date back to the Greek philosophers, while others are constantly being explored and expanded. Those we will describe here include the key word method (Pressley et al. 1980; Miller, Galanter and Pribram 1960), acronyms (VanderZanden 1980; Salsberg 1982) and chunking. There are two important aspects to the use of mnemonics. One involves the original learning of the new material. Here the mnemonic provides the learner with a method of linking the new and relatively meaningless material to previously learned meaningful material. The second is the retrieval mechanism. Since the meaningful material linked with the new content is usually over-learned (e.g., the numbers 1 to 10), it is not hard to remember; it then, through association, stimulates memory of the new material that has been learned and so the student is able to retrieve the new material from long-term store.

The key word method involves identifying a well-known word and then having the student associate the word to be learned with it through a bizarre image which then connects the two. It has two major forms. One is that used by Pressley et al. (1980) to aid students in foreign language learning. In this case, the keyword is an English word that sounds like part of the foreign word to be learned. Examples for French would include associating "beurre" (butter) with the image of a stick of butter full of burrs, "haricots" (beans) with the image of beans wearing hairy coats, or "pamplemousse" (grapefruit) with the image of a moose with grapefruit pimples. This technique has been shown to help with foreign language learning and has been generalized to other tasks by both normal and slow-learning students (Pressley 1984).

A second form of the key word method is that proposed by Miller, Galanter and Pribram (1960). It also uses well-known words coupled with new words connected through bizarre visual images to facilitate memory and retrieval. In this case, the well-known words are the numbers one through *x* (as many as there are words on the list). These, in turn, are connected with short words that rhyme with the numbers

Figure 10-1

Using the Key Word Method to Review Aspects of Ancient Egyptian Civilization

Number	Rhyming Word	Concept	Visual Image
1	bun	hieroglyphics	Squibbles on a bun
2	shoe	Nile	blue laces that run through the shoe
3	tree	pyramids	pine tree in triangular form
4	door	sphinx	great human headed lion on door
5	hive	King Tutankhamen	King Tut and queen bee in hive
6	sticks	cat worship	Cat goddess carried on sticks
7	heaven	class structure	three doors to heaven
8	gate	food	lots of food hanging on a gate
9	line	clothing	clothesline full of clothes

and then with visual images. An example is shown in Figure 10-1 in which the key word method is used as a way of remembering the essential concepts of ancient Egyptian societies in grade five social studies. This method is also used successfully in remembering lists of many kinds.

While the key word method relies largely on the visual modality, acronyms make primary use of the auditory modality. Acronyms involve the creation of a single word or sentence in which either all the letters or the first letter of each word represent the first letter of the words to be memorized. Examples of single-word acronyms include HOMES to remember the Great Lakes, MRS. VANDERTRAMP to remember the thirteen irregular French verbs. Other familiar examples include the many acronyms we give to associations to which we belong (CEC, Council for Exceptional Children; KLCVI, Kirkland Lake Collegiate and Vocational Institute, TAS, Thorndale Agricultural Society).

An extension of the use of acronyms to provide a mnemonic for a larger amount of content is shown in the following one for remembering the countries of South America in order of size (Woolfolk and Nicolich 1984). In this more expanded form of the acronym approach there is a combination of a new word plus several phrases, each formed by the initial letter(s) of several words.

	Br	Brazil
BrAP	A	Argentina
	P	Peru
Col	Col	Columbia
Bol	Bol	Bolivia
of		
Ven	Ven	Venezuela
Chile	Chile	Chile
PeeU	P	Paraguay
	E	Equador
	U	Uruguay
+	Three ex	Guyana
	Guineas	Surinam
		French Guiana

Sentences have also been used as memory aids. You could help a student remember aspects of Egyptian civilization through the key word method described earlier or through the following sentence: "Healthy nice people see kings and cats coming from closets." In this sentence, the first letter of each word is the first letter of each of the nine topics. More familiar sentence acronyms are, "Men Very Easily Make Jugs Serve Useful Nocturnal Purposes," to remember the names and order of the planets, or, "On Old Olympus' Towering Top, a Finn and German Viewed a Hop," to remember the cranial nerves. There is no limit to the interesting acronyms both teachers and students can create as memory aids.

The auditory modality is also the most important one in the use of rhymes, verses and songs to remember things. Many young children will sing the alphabet to find the letter they want and most of us recite "Thirty days hath September..." to determine how many days there are in a month.

The final mnemonic to be discussed is the use of "chunking." In a very influential paper in 1956, George Miller suggested that the capacity of working memory is seven items, plus or minus two bits of information. In other words, we can only hold five to nine distinct pieces of information simultaneously in our minds while we think about a problem. Others have suggested that for young children or slow learners it may be even less. Given this finding, Miller and others have suggested that what we need to do is find the best ways of incorporating as much information as possible in each item. A straightforward example is the seven-digit telephone number (e.g., 439-1963). Theoretically, this should exhaust our working memory capacity because it has seven distinct features. However, if I can chunk the first three numbers (439) as the university area code and the last four (1963) as the year Kennedy died, then I just have to remember two bits of information—"university" and "Kennedy"—to be able to recall the number when I need it. This greatly reduces the difficulty of the task. In dealing with more complex sets of information, the classification systems we teach in many subjects are another form of chunking. If a student has a well-developed hierarchical coding system for the organization of plant life, and if he or she is given an unknown vegetable (artichoke) and told it is a plant, then that student has already got a wealth of information available to him or her about the artichoke that can be retrieved from memory and used to describe and discuss its features. He or she does not have to be taught anything except the few distinct features of the artichoke in order to have a full repertoire of meaningful information about that new vegetable that can be used when necessary. Teaching students how to chunk, or organize their information meaningfully, goes a long way towards improving memory retrieval and subsequent problem solving.

Questioning strategies One of the most important means of stimulating thought among students is through classroom questioning. Questions that require students to go beyond the information immediately presented force them to search through their memory to find additional information that will help them to make better sense of the present data and to put it into a broader context that may include application and problem solving. Asking questions that demand this is especially important since studies have shown that the quality of student response almost directly parallels the type of question asked (Taba 1966). In addition, it has been found that teachers generally ask factual questions (Sitko, Bachor and Slemon 1980) and that this is especially prevalent in classrooms for the low achiever (Brady 1977). Blanton, Sitko and Gillespie (1976) have reported that teachers of the handicapped tend to focus on questions that determine whether the student has decoded correctly in reading. This results in a serious tendency to under-emphasize asking questions that require higher-level language strategies.

Sitko and Slemon and Bachor (1983) have devised a questioning module designed to increase the spread of teacher questioning strategies. They categorize questions as either non-transformational or transformational. Non-transformational questions are of two types: recall and discrimination. Recall is involved when the information required can be retrieved directly from memory of the text; e.g., "What town did Anne of Green Gables live in?" Discrimination questions involve making a discrimination among two or more concrete stimuli; e.g., "Which of these numbers is the same as the model?" or "Which letter is clearer, the first or second?" Decoding questions in reading are generally considered to be discrimination questions. Transformational questions can involve either relating concepts or problem solving. Relating concepts may require the explanation of a concept or comparisons among concepts; e.g., "What is a tree?" or "How are a pine tree and a maple tree different?" Problem-solving questions involve the use of previously learned rules, principles or information to solve a problem or make an inference; e.g., "How could we get more time for sports in the school program?" or "What do you think about mainstreaming in Canadian schools?" Transformational questions require the student to transform the information in a new way in order to answer the question, while non-transformational questions only ask that the student recall the information correctly.

In response to the concern that most teachers ask primarily non-transformational questions, Sitko and Slemon (1982) devised a module that was taught to teachers, outlining the importance of a more diversified portfolio of questioning techniques and giving practice on the four types of questions. Teachers were observed before they worked through the module, during and after they read it, during feedback and after termination of feedback. Feedback involved observing teacher questioning in regular classroom settings and then showing and discussing the results with the teacher immediately after the session. Most teachers were not aware of their initial tendency to ask primarily recall type questions and responded to that information and to the subsequent feedback by asking an approximately equivalent number of questions in all four categories. As would be expected, recall and discrimination type decreased while conceptual and problem-solving questions increased. At final observation, there was approximately an equal number of transformational and non-transformational questions being asked. The major changes came after the teachers read the manual. This suggests that being aware of the necessity to ask a wide range of questions, plus having materials that provide lots of opportunity to see examples of types of questions and to practice them is sufficient to expand teachers' questioning approach. The correlation between student response and teacher question was 0.95, which further supports the assertion that students can and will answer questions that require them to think if that is what is demanded of them.

Concept attainment Jerome Bruner has developed a concept attainment model that involves the student in an active, dynamic, thinking manner that is not usual in more traditional ways of teaching concepts (Weil and Joyce 1978). This model assumes that the learner is active and interested, that the concept is meaningful and that the general approach is intrinsically motivating. There are six stages to a concept attainment lesson. These are illustrated in Figure 10-2.

Figure 10-2 **Example of a Concept Attainment Lesson**

Stage 1:	Teacher gives examples and non-examples of the concept	gas X (non example) milk ✓ (example) beer X apples X orange juice ✓	
Stage 2:	Students generate hypotheses about the identity of the concept	1. It is natural foods. 2. It is foods we drink. 3. It is foods our mothers want us to eat. 4. It is healthy drinks. 5. It is drinks I buy in the store.	
Stage 3:	Teacher and students accept or reject hypothesis on the basis of present information	1. Reject because apple is a natural food that is not an example of the concept. 2. Reject because beer is also a food we drink. 3. Reject because most parents want you to eat apples. 4. Accept because the two examples are both healthy and drinkable. 5. Reject because you can buy beer in the store.	
Stage 4:	Students give more examples to check hypothesis	apple juice ✓ coke X orange X water ✓	healthy drink not healthy not a drink healthy drink
Stage 5:	Define the concept	The concept is things that you drink that are healthy.	
Stage 6:	Discuss the attributes of the concept	Things you drink would be known from previous years. What makes a drink healthy would include being a natural product, containing vitamins and minerals, not containing an undue number of additives, etc.	

The lesson illustrated in Figure 10-2 could be part of a unit on nutrition; but rather than the teacher explaining why foods are nutritious or not, the students themselves are thinking about the issue, using information in their memory store and getting practice with the problem-solving strategies of hypothesis development and testing. We have used this approach successfully with low- and normal-achieving students in regular classroom settings and have found that they all can contribute to the lesson. Teachers must ensure, however, that all students get a chance to answer and not allow the class to be dominated by a few. Asking the less forthcoming students for the third or fourth in a series of responses seems ideal since they have the model of the first two but can still suggest an answer fairly readily; whereas waiting until later means that all the easy hypotheses will have been given already and those students may not be able to generate more complex ones.

Metacognition One of the major new approaches to teaching thinking is that described by Meichenbaum (1972, 1982; Meichenbaum and Goodman 1971) as cognitive behaviour modification. A brief description was presented in Chapter 9 in connection with writing skills, but the major thrust of the approach has been in teaching students how to think about solving problems in the academic and social areas. There are six major steps for the teacher to follow in setting up a program.

Step 1: Identify a problem area.
Step 2: Break down the skill into components.
Step 3: Explain why the skill needs to be taught.
Step 4: Teach the skill through a) a description of the steps, b) modelling by the teacher, c) practice by the student and d) much feedback and praise during the lesson.
Step 5: Student should demonstrate the acquired skill.
Step 6: Teach for transfer and generalization.

The following example shows how a teacher could use this approach:

Step 1: Searching systematically for the information presented in the problem and/or needed to solve the problem.
Step 2: Components of a systematic search strategy include:
1) determining the problem,
2) determining the information required to solve it,
3) determining how and where to get the information,
4) getting the information, and
5) recording the information in a useful manner.
Step 3: Explaining that the first step in any problem-solving exercise is knowing the factual data you have to work with. Without it, you may make many suggestions that have no reality base, which is a waste of everyone's time.
Step 4: Teach the skill. Teacher says aloud, "Our problem to-day is to plan for a Christmas class party. Before we start, though, we need to collect some important information. We need to know how many of us will be attending the party, we need to know whether we want food, games and gifts, we need to know how much money we can spend on the party, and we need to know where we will have it. Now is there anything else we need to know? Oh yes, we need to know when we will have it and for how long it will last. Is there anything else we need to know about? No, good, then lets figure out how we can get the information. We can find out how many are attending by asking everyone in the class if they are coming and deciding whether we will invite anyone extra. We can take a vote on what things we want to do at the party. Bill can check our class bank account to see how much money is available and we can decide among ourselves now what day and where we want to have it. Mary, you should then check with Mr. Boag about whether the space is available and what we have to do towards clean-up after, since you are our property person. Let's get all this information now and then put it on the chalkboard so we can all look at it and plan how we can use it for the best Christmas party yet."

Steps 5 and 6: If the teacher was in the initial stages of teaching information gathering procedures, she would say all this aloud as a model to the class of how she would go about data collection, and then have the students say it aloud as they copy down the steps they need to follow. If she were using this to review and reinforce good data collection, she could let them tell her aloud what data collection had to be done, or she could have them work on it quietly in small groups or individually as she checked on whether they really could systematically plan a procedure for data collection.

In implementing this strategy, Meichenbaum (1982) suggests that teachers ask themselves the following questions:

1. Have I identified an appropriate skill and determined all its components?
2. Do my students have these skills within their own repertoire. If they don't, can I teach them in the time available?
3. Have I asked the students whether they think this is a worthwhile skill to develop?
4. Have I provided enough experiences for my students to be able to generalize this skill to other areas? Have I planned for specific transfer to other areas that are appropriate?
5. Have I planned for sufficient review and follow-up practice so that they will maintain this skill and use it effectively?

This approach has been used successfully with individuals and small groups but has not been used extensively in large groups. The basic approach, however, is compatible for large group teaching, especially in introductory teaching sessions. It may also be effective in tutoring situations where good problem solvers are encouraged to talk out aloud and share their thinking strategy with others.

Deferred judgment and brainstorming Deferred judgment and brainstorming are the central components of many problem-solving activities. The former is one of the hardest things for students to learn and yet if they do not feel that any ideas they are coming up with are worth developing, they will dry up without giving their natural creativity a chance. Osborn (1963) stated that

> creativity is so delicate a flower that praise tends to make it bloom, while discouragement often nips it in the bud.... Every idea should elicit receptivity, if not praise. Even if no good, it should at least call for encouragement to keep trying. (p. 50)

We can't stress this often enough. Deferring judgment is difficult for children to do and at times equally difficult for teachers when they are exasperated with what appears to be verbal "fooling around." All children have natural creativity and what starts off as a ridiculous idea can often lead to a much more productive one and/or be pruned itself to be the best solution possible for a given problem.

There is no technique available to teach children the value of deferred judgment.

Common sense approaches such as frequently reminding them that every idea is worth recording and demonstrating this yourself are the most effective ways of teaching it. As you proceed with problem-solving sessions, each child will become more confident about suggesting ideas and you will see a steady increase in spontaneous idea production, especially among the quieter LD children who need the supportive environment that deferred judgment provides.

Creativity is often blocked by perceptual, emotional and cultural sanctions. One of the most devastating emotional experiences for anyone is looking foolish in front of his or her peers. If that fear is removed, it is much easier for a child to let himself or herself go. Consistently reassuring and reinforcing the child by recording his or her ideas drives home the point that everyone has the capability to produce ideas that are worthy of consideration.

While deferred judgment is the affective cornerstone of creative problem-solving, brainstorming is the cognitive cornerstone. Since quantity breeds quality, it is essential that the children be encouraged to produce as many ideas as possible. First, select a recorder. You may want to choose more than one, especially once you get off the ground, so that ideas are not lost due to slow recording. You may also want to rotate recorders, changing them every few minutes or at some useful interval. Initially, it may be easier for you to record, in such a way that the ideas are available for all the children to see—either at the blackboard or on experience chart paper (large pieces of paper, mounted on a stand, that can be flipped over as you would a coil notepad). Eventually, however, you may prefer to have someone else record, which will leave you free to encourage and support ideas from the floor.

Once you have your recorder and blackboard space; all you need are ideas. Start off by simply asking the students for as many ideas as they can think of. If you have some very verbose children, limit them to one idea at a time so that as many children as possible have a chance in the initial stages. When they seem to be petering out, coach them to stretch their imaginations by statements like "Let's fill up this board," "Everyone think up one more idea," "We'll work on this for one more minute," "Let's try for 25 ideas." If this gets them going, stay with them until all their ideas are exhausted—don't stop after one minute, for example, just because that was the original prompt. You may have some difficulty with this last suggestion—we certainly did at first. However, our own experience and that of others (Osborn 1963; Parnes, Noller and Biondi 1977) has shown that there are considerably more original ideas generated in the last half of a brainstorming session than in the first half. This seems to occur because many ordinary, culturally approved ideas tend to be offered initially, while the more unique creative ideas occur when students are really having to work and go beyond the obvious.

After you have prompted your students as suggested above, you may still want to encourage them to think further about possible solutions. There are several approaches to this.

Using cue words Osborn (1963) has developed a list of useful cue words to help students go beyond the ordinary. Those words and the techniques for which they stand include the following:

1. Minimizing—take an idea that you already have and see how you could change it by making it smaller in some way.
2. Maximizing—what happens to an idea if you make it bigger? Having thought about using pictures, why not expand on this? How about painting a mural that will show the development of the topic over the pertinent time period? Specific maximizing prompts can include "make it taller, fatter, heavier, thicker," et cetera. Any suggestion of this sort can trigger an idea.
3. Combining—are there ways that some of the ideas can be put together to yield a new solution? A mural in which each picture is designed to illustrate a very specific point might be more useful than a free-flowing descriptive mural.
4. Re-arranging—this can include substituting, modifying or adapting ideas in a novel manner. Are there ways of taking some of the initial suggestions and changing them in a manner that might be more useful? How can they substitute for a basic element in the original solution?

The matrix method A very specific approach to helping students generate new ideas is the matrix method. The grid used in the matrix method is illustrated in Figure 10-3. Categories that would yield useful ideas are defined across the horizontal and down the vertical. In the example, in Figure 10-3, we used the general categories "how ideas can be expressed" across the top and "who will express them" down the vertical. The students suggested "written, spoken, on tape, cartoons and graphs" as sub-categories of "how ideas can be expressed," and "students, teachers, and coaches" as subcategories of "who will express them." The students might next use "what, how, when, or where" as questions to which to respond in order to fill in each box. This would force the students to think about combinations of persons and modes of expression that they might not have otherwise considered.

At times it may seem that answers are moving into the ridiculous, but using such a grid does force students to try to think in unique ways. You may prefer to use the matrix approach after the students have had some experience with brainstorming; but at any stage it has a great deal of value in helping children come up with a greater variety of solutions. It can also help students clarify their thinking since it organizes their thoughts into very specific areas. This approach also gives the students the additional reinforcing effect of seeing the boxes filled up. It is essential to use simple matrices initially. Finally, since they provide so much structure to help the imagination, even slower students can usually complete them.

Go crazy A final prompt in a brainstorming session that can often elicit interesting ideas is to say "go crazy with ideas, anything goes, the wilder the better." If your children have not been used to having their ideas accepted, you may find this quite helpful in the loosening up stages. Somehow it is all right to suggest a "way out" idea if that is what the teacher has told you to do. It is perhaps unfortunate that the children depend so completely on the teacher's instruction, but it can be a useful aid to encourage them to let their imaginations roam as far as they can.

The above suggestions should help you get the children used to producing quantity in brainstorming sessions. The more they produce, the better chance there is of

Figure 10-3

A Matrix for Improving Communication in School

	How Ideas are expressed				
Who Expresses Ideas	**Spoken**	**On Tape**	**Written Test**	**Graphs**	**Cartoons**
Students					
Teachers					
Principal					
Coaches					

finding the best solution to their problem. Since brainstorming can be used in many activities, in time, the students will be as knowledgeable as you about the process. You can also post the problem at the back of the room for a week and encourage the children to add to the idea page daily. This might be easier for the shyer children who have difficulty expressing an idea orally in a group session. Then at the end of the week you can follow up with the selection of the best ideas.

Brainstorming is fun and cognitively stimulating for all class members. If students know that it is a regular part of their classroom experience, they spend time on their own thinking about unique solutions to classroom problems. While the original exercise should always be free from judgment, once you have a quantity of ideas, judgment is exercised in choosing the best ones. This process will be described in detail later in the chapter.

Perceptual flexibility It has been mentioned before that the way in which we perceive our environment limits our creativity. We experience the environment through our own eyes and ears rather than either from another's perspective or through the use of other senses. Therefore, activities that encourage a broader perceptual experience will enrich the potential problem-solving skills of children.

Methods of classification are one of the problems that can be worked upon. One way of classifying things that is rarely taught in any formal sense is the use of smell. However, thousands in the food, restaurant, cosmetic and perfume businesses use that sense daily. Therefore, in order to enrich the children's ability to classify, bring in materials that have to be classified by smell, texture or taste. You might want to brainstorm classification systems with your own class and then, as part of a class project, have them include as many items in each subsection of the system as possible. Specific guidelines for expanding children's understanding of how to classify can be found in Samson (1976).

Challenge their perception in other ways too. You can, for example, teach them to appreciate how context can distort our perception of the dimensions of any object. Figure 10-4 shows various ways to demonstrate this idea using lines. Ask the children if the lines are the same length or not; if they think the lines are different, ask them which seems longer, which shorter. Then ask them how they would check their answers? You can repeat this exercise with any combination of lines. You could also ask students to look at line figures such as those at the right of Figure 10-4 and to consider what the lines remind them of? How could they incorporate these lines into other drawings?

Another exercise is to have the children look at an object from the sides, top and bottom and record their observations. To get the students to think more about observation ask them to simulate a car accident or an argument between two children, or to consider a teacher's view of the classroom. By doing this you can stress how we are limited in our understanding of an event by the perception we have of it. That perception is governed both by how much of the event we actually experience (i.e., do we hear what Mary said to John or do we only hear John shouting back to Mary) and by how our previous experience has conditioned us to interpret the event. Experiences of this kind make it easier for the children to see things from the other person's perspective, which, in turn, benefits interpersonal interactions. They also facilitate idea generating since students learn to approach any problem from more than one point of view.

Figure 10-4

Perceptual Line Configurations

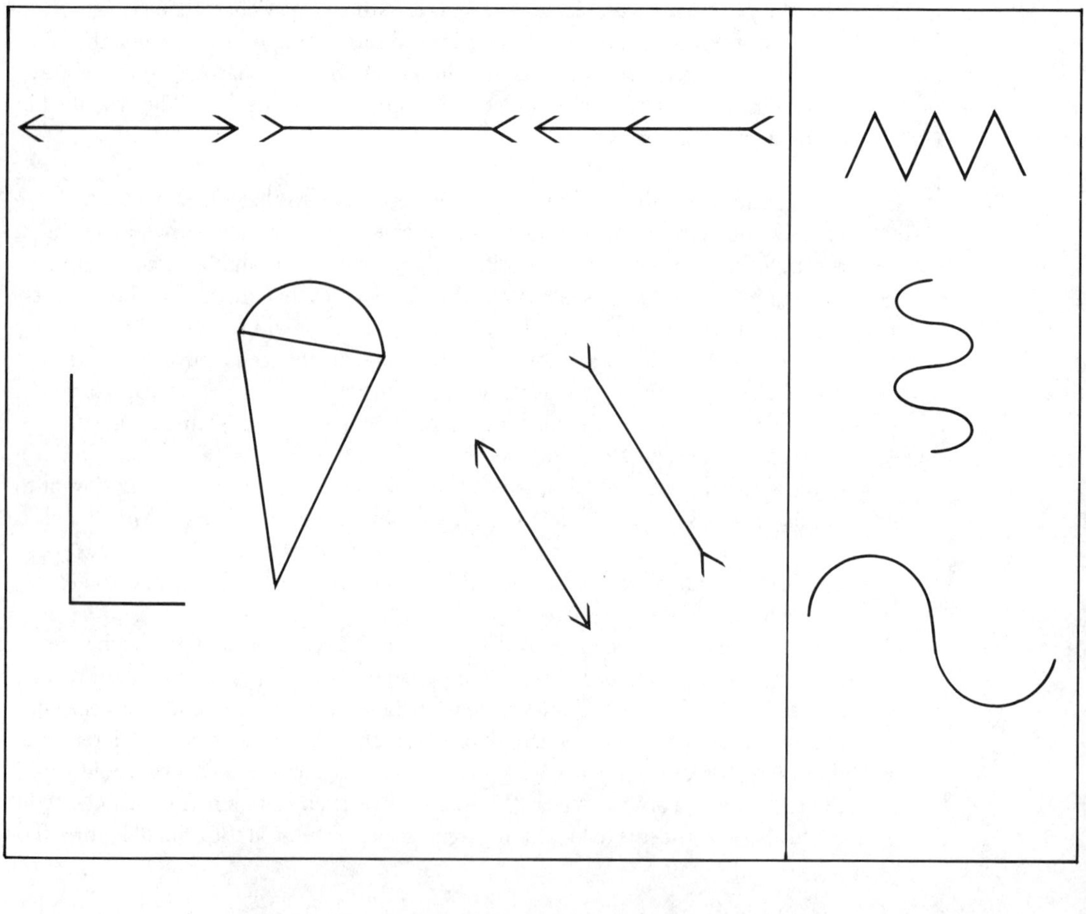

For children who enjoy playing with language, making up anagrams can be stimulating. Because these can vary in difficulty they can be used with children at any age level. Encouraging the children to think in terms of analogies will also broaden their verbal thinking skills. Stimulation for analogies can come from subjects they are currently studying; for example, "nut is to tree as *x* is to flower" could be used as part of a junior science lesson, or "tail is to sail as tub is to *x*" as part of a rhyming exercise in phonics. The important thing in any exercise involving analogies is to help students determine the relating factor in each example (Samson 1976). You can also help extend students' perception by giving them verbal experiences that involve spatial relations such as those illustrated in Figure 10-5.

Figure 10-5

Verbal Spatial Problems

Stimuli	**Answer**
1. standing mis	misunderstanding
2. 	hole in one
3. exxposure	double exposure
4. Rforkoad	fork in the road
5. Blouse	see through blouse
6. ICE	thin ice
7. U P	split up
8. S U N	sun down
9. W A T E R	water fall
10. lip lip	tulips

Adapted from *Owl* Vol. 9, No. 6 (1984).

Other verbal activities that are both fun and thought-provoking include "definition rhymes" and "hidden orchestra." To do definition rhymes, present a noun-modifier combination and ask the students to generate a new noun-modifier combination that means the same but also rhymes. Examples of this are

happy father = glad Dad
false pain = fake ache
mature complaint = antique critique
old scold
melancholy fellow = sad lad

With hidden orchestra, present sentences (without the underlining) and ask the students to find the instrument hidden in the passage. This task forces them to look at the letters in a new way, not as part of a group of letters forming a word separated by a space but as distinct in themselves. Examples include:

Look at the rich man doling out his money.	mandolin
Clap Ian on the back.	piano
Clean the tub after you have taken a bath.	tuba

Examples in both these activities can also be generated by the students to enrich the experience.

Specific Programs and Techniques

Instrumental enrichment The instrumental enrichment program developed by Feuerstein (1979) has been used in several major centres in Canada to promote a higher level of thinking skills in adolescents (Giblon 1981;Muttart 1984). The general goal of the program is to make the student, whose environment has not allowed him or her to function to potential, a better problem solver at school and in life generally. Specific subgoals include the following:

1. To correct deficit cognitive functions.
2. To acquire the vocabulary, labels, concepts, operations and relationships relevant to the thinking skills program.
3. To form habits and systematic approaches to problem solving.
4. To have the student become intrinsically motivated.
5. To create insight and reflective thinking.
6. To have the student shift from the role of passive recipient and reproducer of information to the role of active generator of new information.

Feuerstein sees the mediated learning experience that is provided by a knowledgeable teacher as the essential link between the environment and effective learning. Through the *Learning Potential Assessment Device* (LPAD) he determines the nature and locus of deficits as they occur in the input, elaborational or output phase. Input phase refers to the student's awareness and perception of the information in the environment; elaborational phase refers to the cognitive processes that go on in the student's mind as he or she combines an understanding of the new information with that already in memory; and output phase refers to the final response the student makes to the environmental demand. The testing approach is a test-teach-test one in which the testor is interested not only in what the student already knows but in how quickly and in what way that student can learn those components in which he or she fails. The materials in the instrumental enrichment program are identical in kind to those in the testing program although they take the learner through a much more extensive experience in each component of the program.

The program itself consists of 15 units that are graded in difficulty from simple to complex. They are paper and pencil exercises with a game-like approach that teach students to compare, analyze and organize ideas and to solve problems. Some of the exercises involve temporal sequences while others involve spatial activities. The exercises are essentially content-free but they are easily adapted to discussion in various problem areas. For example, one set of exercises requires that students find dot patterns that originate from different angles and spaces on the page without actually turning the page. They have to instead "turn their heads" internally to see where the pattern is. This leads easily to discussions about how rooms look from different perspectives and, later, to discussions of how certain problem situations are seen from the point of view of different people involved. The materials themselves are designed to be intrinsically motivating. In addition, there is considerable extrinsic reinforcement given through peer and teacher interaction directed at creating student awareness and understanding of his or her achievements. There is further reinforcement in some exercises through self-corrective feedback devices. In all cases, feedback comes quickly from either the materials or those involved in the teaching experience.

This program has been developed for use with a wide variety of students, ranging from the low-achiever to the gifted. It has been used with individuals, small groups and entire classes. Teachers are required to have taken the training courses and to use the specific materials developed by Feuerstein and his colleagues. These materials and training are readily available in major Canadian settings. Even for those who cannot avail themselves of the specific program however, the basic philosophy can be implemented in all classrooms. This philosophy implies that students should be assessed in the most positive manner possible and that the results of assessment should not be expressed in terms of what the student knows or doesn't know, but rather in terms of how quickly he or she can learn and what learning strategy suits him or her best. Remediation involves interesting materials that he or she is intrinsically motivated to work through—materials that demand an active, thinking approach with lots of opportunity for interaction and feedback from others.

Narrol, Silverman and Waksman (1982) have taught the cognitive skills program developed by Feuerstein (1979) to vocational high school students. These students were taught cognitive and thinking skills daily for an entire school year. Post-testing showed significant gains in cognitive thinking plus anecdotal reporting of an enhanced self-concept but no direct transfer to academic achievement. It may be, however, that the transfer gains are there but take time to become evident. Or it may mean that transfer has to be specifically taught if it is to occur. Teaching students how to think more effectively is an end in itself, however, and should not have to show direct payoff in achievement to warrant its place in the Canadian classroom.

Stimulating cognitive growth This program was developed by Regan, Leithwood,

Sullivan and McConnell (1978) to develop a child's ability to think. It was designed for the primary grades, but could also be adapted for use in junior grades. The program contains five units composed of several learning activities that could be taught on their own in 10- to 15-minute lessons, could be integrated into other program activities or could be used to capitalize on situations that arise more or less spontaneously. Unit I develops concepts and skills such as observing, describing and "finding out" that students need in order to do the tasks presented in later units. Unit II introduces tasks that involve inferencing, determining relevance, classifying, sequencing and determining which effects make a difference. In Unit III the skills developed in Unit II are extended into use in solving simple problems. Unit IV outlines experiments that further use the problem-solving principles. And Unit V outlines experiences that focus on identifying, solving problems and applying or extending classification skills. These units have all been devised for use in the regular curriculum and can augment programs in reading, mathematics, language or science. Teacher behaviours include modelling, checking and probing for student understanding and remediating as necessary. Each of the lessons is well developed and ready for use. In addition they involve active thinking behaviour that focuses on information that is available to most students. This allows for participation by all rather than just the more competent students. An example on classifying from Unit II illustrates this.

> "Let's look at this group of numerals (e.g., 5, 15, 25, 55, 45). How are these numerals the same? (All have a 5.) I want to put some more numerals in this 5 group. Can I put in this one, 35? Why? Can I put in this one, 22? Why not? How about 50, 200, 10, 72, 85, etc?" (p. 63)

This lesson goes on to apply the same principles to a group of words and to a group of children who ride the bus to school. In each case students are asked to determine what the class is and what attributes allow entrance into the class and they are given practice with the rules for class inclusion and exclusion.

Teaching problem solving This program has been developed by Ross and Maynes (1982) for use with elementary and secondary school students. It includes many techniques that allow the teacher to teach the whole group but to follow up the lessons with individuals or small groups. The program uses problem organizer sheets which include spaces for how and when decisions are made concerning the following elements: focus (what is the problem), framework (how will it be reported), data sources, data itself, predictions, observations, conclusions and final report. Aspects of the problem-solving model can be taught individually or the model can be taught as a whole. Aspects of the model can also be used to facilitate teaching in various subject areas. In Table 10-2, Browne (1984) has shown how four of the steps (identifying the task, defining the problem, isolating the problem elements and formulating an hypothesis) can be used in English, social studies, math and science, respectively.

Social cognition programs Spivak and Shure (1981) have developed an interpersonal cognitive problem-solving (ICPS) skills program that includes 72 lessons sequenced for maximum effect. They have taught this to children and adolescents from various socio-economic and intelligence levels and report positive findings. The IPCS skills include

1. the capacity to generate alternative solutions to problems;
2. the ability to consider the consequences of one's social acts in terms of their impact both on other people and oneself;
3. the ability to articulate mentally the sequence of steps necessary to carry out a specific solution to an interpersonal problem;
4. social causal thinking; i.e., having an awareness of the existence of motive in human behaviour;
5. sensitivity to problems, an awareness of the variety of interpersonal problems that can beset people and an ability to articulate them if asked;
6. dynamic orientation, the ability to look beyond the surface of human behaviour and to appreciate that it may be caused by underlying motives.

Ross and Fabiano (1981, 1983), on the basis of an extensive review of the literature, describe a cognitive program which they feel would be appropriate for teaching adolescents who have difficulties associated with developmental delay in the acquisition of the cognitive skills which are essential for effectively coping with social situations. Their program includes social perspective-taking, means-ends reasoning, interpersonal problem-solving and impulsivity control. They suggest that the person who implements such programs should be able to motivate the adolescent to

Table 10-2

Adaptation of Ross and Maynes' Problem-Solving Model in the Classroom

Problem Type	Subject	Activity	Problem	Follow-up: Remedial
Comparative "Identifying the task"	English	Critical reading	"Which English format is best for conveying humour?"	Riddles
Select alternatives (decision making) "define the problem"	Social Studies	Using legends	"How can natural resources best be shown on maps?"	Simulation games
Statistical relationship (correlational) "isolating the elements"	Match	Calculating area	"How can knowing the area of a triangle help solve the area of a quadrilateral?"	Kit: *Think Tank* (SRA)
Casual relationship (manipulating variables) "formulating hypotheses"	Science	Soil profile	"What is the best soil upon which to build your house?"	Kit: *Productive Thinking* (Bomar)

Adapted from J. Ross and F. Maynes, *Teaching Problem-solving* (Toronto: OISE Press, 1982).

be involved in the program, should be empathic to the student's thinking and feeling, should be able to model effective reasoning and problem-solving, and should be eclectic in his or her approach to interaction.

The province of Alberta's decision-making curricula (1971) provide a series of units for grades one to six that includes skill and knowledge experiences in decision-making in social areas. The curricula are designed to enhance social, inquiry, physical and language skills. Examples of social skills involve handling face-to-face relationships, talking about feelings and reactions, and role playing. Examples of inquiry skills include data collection, data organization, data recording and data interpretation. Language skills include extending vocabulary in order to express ideas effectively. The specific content varies according to the unit. For example, the grade one unit, entitled "What do I need to grow up," has as its objectives the development of a realistic and positive self-concept; an awareness of love, loyalty and responsibility; an appreciation of the individuality of people and family groups; and a sense of appropriate personal responsibility as a member of a family. It allows for many discussion opportunities and experiences that reinforce the development of inquiry skills.

Creative problem-solving technique Creative problem solving (CPS) is a technique designed to improve a child's thinking abilities and to help him or her to handle both academic and social problems with greater skill and efficiency. Many students with learning difficulties will have difficulties throughout their lifetime both in organizing their thinking and in communicating effectively through writing. They may have serious doubts about their own ability to think productively and may, therefore, be more comfortable letting others do their thinking for them. While remediating the areas of difficulty is usually the most frequent route followed in determining a curriculum for children with learning difficulties, providing them with compensatory skills also has a place in the program. By learning how to solve their own problems through applying the CPS technique, they will learn a skill that will aid them in their everyday functioning and will foster a more positive self-image.

The creative problem-solving (CPS) technique developed by Osborn (1963) and Parnes (1977) offers school-aged children an opportunity to learn a skill that will supplement and support their basic academic ability. It is the problem-solving aspect rather than the creative aspect that is the most useful skill for the children to master. Figure 10-6 shows the structure and sequence of the CPS technique. Selection of an appropriate problem, determination of relevant facts, generation of ideas, selection of the best solutions and finally, construction of an action plan are each components of the problem-solving approach. These components can be best remembered using the mnemonic "pray for inspiration as soon as possible" (PFISAP = Problem, Facts, Ideas, Solutions, Action Plan). In fact, hoping for inspiration is often the first choice strategy for many exceptional children but when that doesn't work, hopefully they will adopt CPS for better success. As Figure 10-6 shows, both divergent and convergent thinking are required at different stages for successful implementation of the model. This opportunity to explore many ideas coupled with the discipline required to choose the best ones is a unique feature of CPS.

Figure 10-6

Creative Problem-Solving Sequence

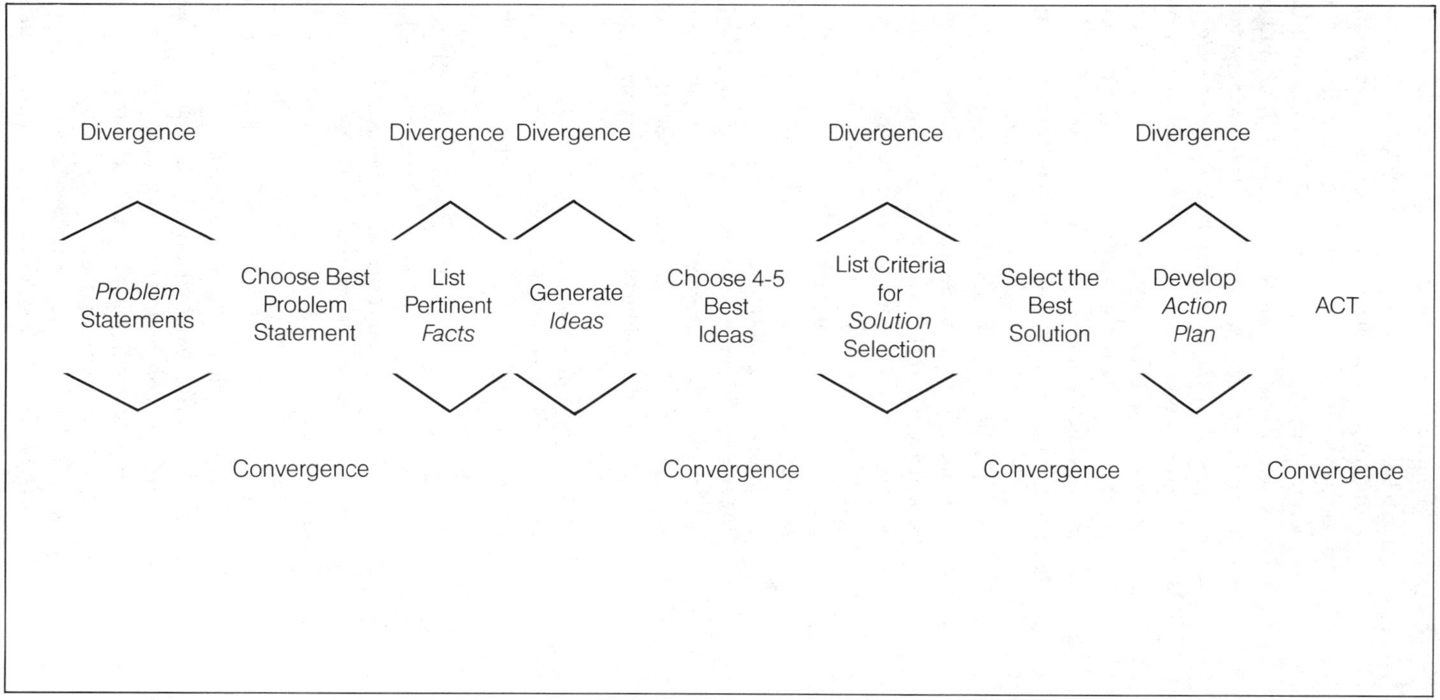

From C. Crealock, "Creative problem solving: A compensatory strategy approach to teaching learning-disabled adolescents," *Teacher Education* 19 (1981).

While CPS has only recently been used for children with learning problems (Crealock 1981), it has been a part of some programs for gifted children. Some gifted children with acting out problems have responded exceptionally well to the CPS technique because it gives them an opportunity to think and have their ideas evaluated and accepted in a positive environment (Brown 1978). Much of the original work in CPS has been done at State University College at Buffalo, where Parnes and his colleagues have been teaching students the techniques for over 25 years. They have been able to demonstrate that college students are able to come up with more and better creative solutions to problems after mastering the CPS technique than students who are not familiar with the technique. Parnes et al. (1977) have also been able to demonstrate maintenance of superior skills over time. In a productive thinking program developed for grade six students, Covington et al. (1972) demonstrated not only that diversity and quality of productive solutions could be increased but also that students enjoyed the experience. Parnes (1978) has attempted to modify CPS for use with mentally handicapped children and this seems to have had some success particularly in the creative arts and movement programs. All of the above programs report positive attitudinal gains for students both in terms of their enjoyment of the actual experience and in terms of improvements in self-concept.

The technique has been extended for use in several types of classroom as an approach to solving both academic and social problems. Teachers who use it with regular classes find that it works best in 30- to 40-minute periods with 10 to 15 children, although it can be used with an entire class (Brown 1978; Crealock 1981). CPS has the flexibility to be used initially with the larger group and subsequently with smaller, more homogeneous groups or with individual students. The latter two approaches seem well-suited to dealing with the few children in many regular classes who have learning difficulties and need to be taught some compensatory skills in dealing with them.

Brainstorming and deferred judgment are two basic strategies underlying the CPS technique. Both are essential for low-achieving children for they stress the importance of forcing oneself, independently or with others, to come up with as many solutions as possible before making any sort of judgment about their usefulness. This process also stresses the concept that anything worth thinking about is worth writing down and talking about. This is important for all children for it trains them to be more flexible and to value their ideas. As more potential solutions appear, ways of combining, changing and building upon them emerge and this, in turn, shows the child that there are different ways of looking at the problem. Learning to defer judgment is perhaps even more difficult than learning to brainstorm since it goes against the tendency to accept or reject statements as soon as they are made. However, deferring judgment is essential to the concept of respecting each other's ideas. In addition, when it is time to evaluate ideas, the criteria for doing so are more likely to be objective and reflect non-personal approaches rather than subjective attitudes. Many low-achieving children suffer censure for their ideas both from their peers and occasionally, from their teachers. This frequently occurs not because the ideas are poor but because the students themselves are often viewed as second class and their ideas, therefore, are viewed similarly. Deferring judgment then, can help low-achieving students risk presenting their ideas and can make the final criteria for accepting those ideas fairer.

One of the major areas of difficulty for many children is their inability to organize their work efficiently. CPS shows them a technique for sorting out the "mess" they are experiencing and finding a way to handle it that makes sense. The very exercise involved in defining their problem can go a long way in helping them understand and cope with it. Another aspect of thinking that they have difficulty with is "going beyond the information given." Many low-achieving students feel most secure with limited amounts of concrete information. CPS, however, stresses the value involved in looking to themselves and stretching their own thinking skills in order to come up with a solution that makes sense to them individually. In time, this can become part of how all problems are handled.

There are five stages in the CPS technique, each of which will be discussed in the following sections.

Stage 1: Choosing the problem.
Stage 2: Determining relevant facts
Stage 3: Generating ideas to solve the problem
Stage 4: Selecting the best solution
Stage 5: Detailing an action plan

Choosing the problem Any problem-solving exercise must start with a problem. When you are working with children you may initially want to set the problem yourself or let them give it to you depending upon how comfortable they are with open-ended, fairly non-structured learning experiences. In fact you may want to combine student initiative with teacher guidance in allowing them to set the basic problem while you refine it for more efficient discussion. For example, there may be a problem concerning the pencil sharpener in the classroom. Rather than let the children ramble all over with a general problem like "How might we improve pencil sharpeners," you might want to make the issue more specific by stating the problem as follows: "How might we make our classroom pencil sharpener easier to use?" They may want to work on something of much greater scope such as "How might I organize my time so that I can do my homework and still have time to do fun things?" Or they might want to work on "How can I do a project on the War of 1812 without getting bogged down by words and writing?" The original problem can come from either the social or academic area. The degree of narrowness or broadness can also vary greatly but teachers who have been successful in using the technique with elementary school-aged children, state that narrower problems are generally easier to work with, especially in the initial stages of training the children.

Determining relevant facts Now that you have decided on the problem, the first thing you have to do is help the children determine the facts that they need to use in order to solve the problem. If they are concerned about doing a group project without getting bogged down with writing, they have to know a) how long it should be; b) if the teacher will allow graphs, pictures, etc. as part of the text; c) if she will allow them to tape the text; d) how well they and the rest of the group read and write; and

e) how much time have they to complete the project. The facts usually involve personal attitudes, knowledge of skills and content, and the specific demands of the problem. If they become bogged down in listing all the facts, you can ask them "who, what, where, when, how" questions. The facts should usually be concrete, but at times opinions can be useful.

Generating ideas to solve the problem This is often the most appealing part of the process to children. However, it is necessary to impress two concepts upon them. The first is that you want as many ideas as possible—quantity breeds quality. The second is the idea of deferred judgment. *Any* idea is accepted at this stage. There is a tendency for many students, and especially exceptional students, to latch on to the first solution and then stop thinking about the problem. Therefore it may be necessary, in order to get a sufficient quantity of ideas, to push the students somewhat. Urging them on to give you a few more ideas, to finish the page or to see how they could combine ideas is frequently necessary in the initial training sessions. Once several ideas have been generated, the children will expand upon them ("piggy back") and give many related ones. This is to be encouraged. Often their ideas are merely repetitive; but occasionally, very slight variations on a theme can be the solutions that yield the best results. Suggesting to students that they think about the problem from different perspectives, e.g., theirs, the teacher's, animals' or younger persons', can encourage unique responses. Using different senses can add variety to their solutions. Ask them to visualize the problem without using words, to hear but not see, to taste or smell, to feel. Much of this may be difficult for the children to do initially, but with time they will approach the problem from varying viewpoints without needing the specific directions.

Selecting the best solution(s) When as many ideas as possible have been generated, the next step involves choosing the best one(s). At this stage, evaluation and selection will be made. The most useful first approach is to get a straw vote on the whole array of ideas by simply asking the group which solutions they want to select for further consideration. This first stage provides for a screening of suggested ideas by weeding out the ones least likely to succeed. The second stage is much more specific and generally involves developing a list of criteria for evaluation. The num-

Figure 10-7

	Criterion Measures					
Ideas	**M**	**N**	**O**	**P**	**Q**	**Total**
a						
b						
c						
d						
e						

ber of discrete criteria will vary with the group. For young children, a few basic ones expressed as questions such as a) "Will it work?" b) "Does it cost much?" c) "Can I do it myself?" will be sufficient. Older children may want to include more advanced criteria. In any case, three to five distinct aspects are usually sufficient to judge the ideas. Once you have established criteria, drawing up a grid and listing the criteria along the horizontal axis and the ideas along the vertical axis creates a matrix format that is easy to comprehend visually and allows for quick tabulation once all the boxes have been filled. (See Figure 10-7.) With most children, evaluating each idea on a three-point scale on which 3 indicates "yes it will work well" and 1 indicates "no, it won't work" will be a sufficient measure. Once students have evaluated each idea using each criterion, they can simply add up the points, record the total, and select the idea which receives the highest score.

Detailing an action plan This final stage in creative problem-solving must be very specific if there is to be action emanating from the process. In the previous stage, the group selected the best solution. Now they must decide, in a deliberate step-by-step fashion, how to go about implementing their decision. If there are some possible negative results to their action plan, they should anticipate them and try to defuse their intensity. If people have to be contacted, equipment obtained, or appointments made, who, how and when should be written down and dates should be set for execution. Tally sheets that list the action plan and indicate when each aspect was completed are very helpful both in ordering the plan and reinforcing the actions as they are checked off. The group may find it useful to brainstorm again at this stage for action ideas. These may resemble ideas presented in the third step but they should be more specific and essentially expand upon the chosen solution. Figure 10-8 demonstrates the use of this process with an academic problem.

The above description of the technique may make it appear too complicated to be useful with large or small groups. In fact, it is possible to go through the entire process in two 30-minute sessions if you put time limits on each stage. However, when you first begin to work with CPS in the classroom you may want either to set a meaningful but simple problem for the children to work through or to concentrate on a couple of stages at a time until they are comfortable with both brainstorming and deferred judgment. To be effective in helping the children think through more of their own problems and improve their self-concept, CPS should have at least one period a week all year devoted to it. After that period of time, students will have used it often enough to have internalized the basic concepts and will be able to use all or parts of the technique with little additional direction.

In summary, we would like to stress the value of CPS in enhancing the self-confidence of learning-disabled children by teaching them to be comfortable with their own problem-solving abilities. Being able to identify themselves as people who are creative and who can solve problems can add greatly to children's feelings of self-worth. There is also some evidence (Parnes, Noller and Biondi 1977) that it can help a child move from a basically fatalistic view of life to a more optimistic view; that is, to move from thinking "what I do or say will have no effect, chance determines what happens to me" to "what I do and say does make a difference."

Figure 10-8

An Example of the CPS Technique Used to Solve an Academic Dilemma

Original Problem: How can I do a project on the War of 1812 without getting bogged down by words and writing?

Problem Finding	Fact Finding	Idea Finding
1. How might I do my share of work on the project without having to write out the text?* 2. How might I look at the effect on family life of the War of 1812? 3. How might I convince my group to act out a short play about the War of 1812? 4. How might I get the teacher to forget about projects? 5. How might I get the teacher to let me tape my ideas and text as well as sound effects, etc.	Project must be 10 pages in length or 1500 words. It should be accompanied by graphs, pictures, tables, etc. It must be handed in four weeks from now. There are four people in the group—the others read and write OK. A tape would be acceptable as part of the total presentation.	*1. Change requirements from 1500 to 150 words. 2. Use pictures instead of written text. 3. Draw a mural that will trace the war over many stages. 4. Have someone do a bugle and drum roll. 5. Make a tape of a bugle and drum roll. 6. Get a film from ETV. *7 Find some slides or a filmstrip in teacher resource centre. 8. Enact a scene between a father and child as he leaves to fight. 9. Interview my brother who is taking local history in high school.

Solution Finding

Criteria					
Possible solutions (from idea finding).	Can I do it?	Will teacher accept it?	Will group think it's fair?	Will we get a good mark?	Total
1	1	1	3	1	6
7	2	3	2	2	9
10	3	3	3	2	11
15	3	3	3	3	12

In this case, recording his share of project on tape seemed to be the best solution. However, number 10 received strong support too, so it was combined with 15 and became a major part of Tom's tape.

Acceptance Finding (Action Plan)

When	What	Done
Oct. 10	Talk to teacher about accepting various modes of presentation.	
Oct. 10	Talk to group about a fair share of work.	
Oct. 12	Set up interview with Mr. X for a specific time.	
Oct. 14	Borrow tape recorder and be sure I can use it.	
Oct. 14	Buy a new tape for interview and one for presentation.	
Oct. 18	Get help from brother and teacher on good interview questions. Write these out in point form.	

Problem Finding	Fact Finding	Idea Finding	Solution Finding	Acceptance Finding
		*10. Interview on tape Mr. X who is our local expert on the subject. 11. Go down to the Canadian History Museum, talk to curator, try to borrow some stuff. 12. Take some pictures of historical plaques that mark battle sites. 13. Do a rubbing of some plaques. 14. Coordinate a presentation rather than just a written essay. *15. Record my share of project on the tape recorder (with some extra sound effects).		Oct. 20 Decide what else I am including in my tape and where interview fits in Oct. 24 Interview Mr. X. Oct. 30. Do final tape.

*The problem and ideas chosen by the student for further consideration.

Reprinted from C. Crealock, "Creative problem solving; A compensatory strategy approach to teaching learning disabled adolescents," *Teacher Education* 19 (1981).

Case Study
Step 1: Background information
This case study will describe a small group of nine students who had been diagnosed as learning-disabled and were receiving remedial help for one hour per day in the resource room.

Steps 2-6: From preliminary hypotheses to IEP goals
Each student had been tested extensively for intelligence, achievement and perceptual difficulties. Much informal assessment had also been done to determine their specific areas of strength and weakness. They were all 13 or 14 years old, they were of low-average to high-average intelligence and had a range of academic difficulties. In addition, they all had some weaknesses in cognitive areas—they could not search effectively for necessary data to solve problems, they had poor cause and effect reasoning skills and they tended to choose the first solution they could find rather than explore the problem for better ones. Thus one of the IEP goals, which they each had was "needs to develop better problem-solving skills."

Step 7: Remediation
Because they needed remediation in the cognitive area, they were taught the CPS technique in eight 40-minute sessions in their resource program. The procedure and purpose of the CPS unit were explained to the students. The same basic worksheet was used throughout the project (see Figure 10-9). Work done in class each session was typed up by a parent volunteer and ready the next session to be modified and/or included in each student's CPS file. This ensured the legibility of the material and resulted in a product in which the student took pride.

Figure 10-9

The CPS Worksheet

Problem Statements	Pertinent Facts	Ideas	Solution	Action Plan

Session 1: After a general introduction to CPS, the class of nine students worked through stages one and two with the teacher. The problem they chose initially was "how to get more gym." However, when they began to gather up facts they realized that they spent approximately three hours per day in some type of physical activity (gym, intramural sports, recess activities, community sports teams) and would not likely be able to successfully argue for more. This result was unexpected but provided an extremely useful way of discussing an appropriate timetable.

Session 2: At the beginning of this session, the students decided to redefine their problem and work on "how they might be more involved in physical activities in out-of-school time." They worked through the idea-generating stage and came up with 26 ideas in a 10-minute period. Next they voted for any idea they liked. Six ideas, each of which had been supported by at least three students, were subjected to more rigorous evaluation. The criteria for judging each were consequences, cost in money, time needed and whether each one really liked the idea. "Getting friends together and doing physical activities spontaneously in the neighbourhood" was judged the best idea. The students then developed an action plan related to that solution.

Session 3: The entire group again worked on the first three stages of a new problem which was "How to ignore kids who were bugging you in class." At the end of the idea stage, they had generated 44 ideas.

Session 4: At this stage the students worked on the solution selection and action plan in groups of three. Each triad short-listed ideas from stage three and then developed criteria for evaluating the selected ideas. In each case, "ignoring them" was chosen as the best solution. Once that solution was selected, each group worked on an action plan to implement it.

Session 5: At this session, the teacher showed the students how they might use the same technique to solve a word problem in mathematics. The entire group worked together to solve the problem illustrated in Figure 10-10. The basic worksheet was reduced to three components since both the problem and action plan aspects are redundant in a mathematics problem. The ideas generated fell into four groups—trial and error, concrete representation, symbolic representation (algebraic solution) and a cognitive, thinking approach.

The students were slow to respond to the math problem and needed considerable prompting to generate the algebraic solution. At the solution selection stage, most chose the symbolic solution (12) and then were made aware of another solution that had been arrived at cognitively (14). The group had just been introduced to algebra in their mathematics class so it provided a useful bridging lesson between old and new methods. It was also valuable for them to appreciate that answers can be arrived at in more than one way and that, occasionally, two answers are possible.

Session 6: The students again worked in triads on a mathematics problem. After they had worked it through in small groups, all nine came together to share ideas and solutions. The concept of there being several ways to solve a mathematics problem was quite foreign to most of them and encouraged them to see mathematics as a subject where thinking counted.

Session 7: In this session each student individually worked on the same problem

Figure 10-10

Using CPS to Solve a Mathematics Problem

Problem: Jumpy is a frog at the bottom of a well 20 metres deep. Each day Jumpy climbs up five metres. Each night he falls back four metres. On which day does the frog get to the top of the well.

Fact Finding	Idea Generating	Solution Selection
well is 20 m deep frog jumps up 5 m each day frog falls back 4 m each night	1. Put a rope down well so frog can jump up. 2. Put steps down so he can walk up. 3. Count his steps. 4. Drop a pail down and scoop him up. 5. Go up 5 and down 4 yourself and see how long it takes. 6. Divide 5 into 100. 7. He jumped one metre a day. 8. Get other frogs to help him. 9. Go see him every day and see how far up he is. 10. Wait and see how long it takes him. 11. Get a measuring tape and see how far up he comes each day. 12. Subtract 5-4=1 x 20 = x. 13. Give him a ladder. 14. Is at 15 m; next jump is 5 m so he is on top on the 16th day.	On a class vote for best solution, number 12 was chosen. "Subtract 5-4=1 x 20 = x" was considered the best solution. Therefore the frog gets to the top of the well on the 20th day. However, solution number 14—which meant the frog got to the top of the well on the 16th day—was also accepted when Kelly explained it to the group.

that was illustrated previously in Figure 10-8. They were asked to go through a complete sequence on their own. After completion, the final product was typed for their own files. Each was able to complete the task and seemed to have internalized the procedure and philosophy of problem solving to some degree, although they were not able to generate more than five or six ideas on their own.

Session 8: The problem presented in session 7 was discussed with the entire group. At a group level, the students generated 16 ideas and chose "put it on tape" as the best solution. For comparison purposes, a grade eight class of 29 students in the same school went through the same problem-solving experience. They generated 68 responses and decided that a combination of "doing your best" and "using visual illustrations and props" would work best. Both the learning-disabled and regular students were approximately the same age and would be going into grade nine the following year. Both groups gave many avoidance and nonsense responses which reflects their age and stage as well as the novelty and flexibility of the technique. However, the regular class group gave many more responses generally and were more likely to mention alternate ways of obtaining information than were the LD students.

Step 8: Evaluation

Students were evaluated on their knowledge of and ability to use the CPS technique. All of the students were able to use it individually but needed to discuss aspects of it with the teacher.

There was good evidence that the students enjoyed this alternate approach to handling academic and social problems, but eight sessions was just enough to introduce them to the technique. They needed frequent opportunities to practice if it was to be internalized and used spontaneously. It is a technique, however, that can be used with regular students and those in special education programs to facilitate their learning programs.

11 ORGANIZATIONAL SKILLS

> The disorganized learner is the student who will drive you wild. His disability is that he literally is unable to organize himself. Further because he appears knowledgeable, his disorganization looks like carelessness, not-caring-ness, and a host of other irresponsible behaviors. (Abbott 1978, p. 39)

Organizational difficulties aggravate whatever other deficits a student may have. This aggravation occurs on two fronts: the first being the student's ability to understand, manipulate and make decisions about the specific material he or she is learning; the second being the effect of this disorganization on those adults who are trying to help him or her learn effectively. Alley and Deshler (1979) have identified the following five specific problems a student may have in organizational skills:

1. Difficulty perceiving organization in materials such as class notes, class assignments and the textbook.
2. Difficulty in appreciating the importance of organizing information.
3. Difficulty realizing that information can be rearranged to produce an organized structure compatible with the learner's own style of understanding, storing and using information.
4. Difficulty dealing with an isolated task in an organized manner.
5. Difficulty dealing with multiple tasks in an organized manner.

Another way of thinking about these areas of difficulty is to view them as difficulties the student has in organizing his or her time, belongings, schoolwork and methods of self-evaluation.

ASSESSMENT

Assessment of organizational difficulties is not as straightforward as assessment of academic difficulties. While aspects of clinical tests (for example, observations of how a student manipulates the blocks in the block design subtest of the WISC-R) may yield some insights into how a student organizes his or her approach to tasks, such insights are usually subjective and not specifically related to the organizational deficit of concern. There is a serious lack of tests that measure skills in this area. Wilgosh (1983) in a survey of tests used by special educators in Alberta, did not list any tests for organizational skills nor did the respondents to her questionnaire spontaneously report using any. The few commercially prepared tests designed to measure organizational skills are listed in Table 11-1. A new test, being developed at the Toronto Board of Education and now in its pilot stage (Wells 1985), includes items that measure awareness of passage organization, ability to logically develop stories from visual sequential stimuli, time management and time awareness skills. Publication of this test will be much appreciated in the field.

Table 11-1 **Commercially Prepared Tests of Organizational Skills**

Test	**Skill and grade level tested**	**Comments**
Survey of Study Habits (Brown and Holtzman 1967)	Grades 7-12	Group administered; untimed; reusable test booklets
Study-Habits Inventory (Wrenn 1941)	Grades 9 and over; allows students to identify their own weaknesses in study skills	Group administered
Assessment of Basic Competencies, Information Processing Subtest (Somwaru 1979)	Kindergarten—grade 12; assesses organizing, observational and relating skills	Also assesses language and mathematics; scores available as percentiles, age and grade equivalents, or standard scores

Available from the Guidance Centre, Faculty of Education, University of Toronto and/or Scholastic Testing Service.

At present, however, the more useful approach to assessment in this area is through teacher-prepared information assessment devices. Teachers who are using informal assessment must themselves be organized and must have planned carefully both how and for what they will assess their students. Slemon et al. (1984) have developed a list of task demands and student potential that contrasts teacher expectations, strategies, activation and grouping with student capabilities, skills, motivation and socialization. In assessing a student's level, they recommend that the teacher carefully match his or her teaching approach with the student's capacity to learn. One example would be matching the teacher's speed of delivering information with the student's speed of processing it. Another example would be matching the teacher's instructions to analyze or organize material with the student's skill in reflecting upon ideas.

Figure 11-1

Twelve-Step Strategy for Assessment for Instructional Planning

1. Decide what to assess.
2. Select or develop a skill hierarchy.
3. Decide where to begin.
4. Select or develop survey instrument.
5. Get ready to test.
6. Administer the survey.
7. Note errors and performance style.
8. Analyze findings and summarize outcomes.
9. Hypothesize reasons for errors and determine areas to probe.
10. Probe.
11. Complete recordkeeping forms and generate teaching objectives.
12. Start teaching: update assessment information.

From N. Zigmond, A Vallecorsa and R. Silverman, *Assessment for Instructional Planning in Special Education* (Englewood Cliffs, N.J. Prentice-Hall, Inc., 1983), 343. Reprinted with permission of the authors and publisher.

Another guideline for teachers in planning an assessment and remediation approach is that recommended by Zigmond, Vallacorsa and Silverman (1983). The unique aspect of their approach to informal assessment is in steps nine and ten. These steps ask the teacher not to be content with the obvious information that comes from achievement testing but to go further and try to make several hypotheses about why the student is doing poorly and then test further to determine which hypothesis comes closest to explaining the behaviour. This approach is especially salient when one is looking at organizational skill deficits. They are rarely as obvious or of as much concern in and of themselves as are the more apparent academic and behavioural problems some students present. However, organizational skill deficits often lie behind many of the academic deficits and are realistic hypotheses for why a student did poorly on a test or project. Examples of hypotheses that reflect organizational deficits and of informal probes to test those hypotheses are presented in Figure 11-2.

Using an hypothesis-probe approach to determining the effect of organizational deficits on student performance is one way to gather informal data. A second approach is to develop a checklist of skills that are required for successful handling of a task and to use that checklist to determine whether a child can in fact do each subtask that is required. Such a checklist is presented in Figure 11-3. Many of the individual skills presented could themselves be broken down into a subskill task

Table 11-2 **Relating Task Demands and Student Potential**

	Task demands	**Student potential**
Background:	Level of language required Type of task	Language background and experience Experience and training with such tasks
Physical:	Concentration and persistence required Degree of manipulation Clarity of figures or sounds	Energy available Coordination Visual or auditory acuity
Developmental:	Proportion and weight of materials Use of physical objects, pictures, or symbols	Size and strength Concrete, pictorial, abstract, thinking
	Strategies vs. skills	
Resources:	Materials used in task	Familiarity with and opportunity to acquire materials
Receptive:	Oral or written instructions and media	Attention to, and comprehension of, words and writing
Expressive:	Oral or written product required	Speaking or writing skill
Integrative:	Speed of information delivery Complexity of figures or sounds Memory or comprehension task Instructions to analyze or organize	Speed of processing Ability to analyze visual or auditory input Skill in memorizing or comprehending Skill in reflecting upon ideas
	Activation vs. motivation	
Reinforcement:	Immediate or delayed Incentives offered	Tolerance of delay Preferences for particular reinforcers
Interests:	Intrinsic interest of activities or materials	Existing needs and interests
	Grouping vs. socialization	
Agents:	Teacher, aide, resource, volunteer, peer	Ease of relating to other people
Groups:	Class, small group, individual	Skill in various groupings

Reprinted, with permission of the authors, from A. Slemon, P. Park, D. Thiessen and M. Sitko, *Assessing Special Needs in Classrooms (London, Ont.: University of Western Ontario, 1984), 112.*

analysis and converted into similar checklists. The advantage of this approach is that the checklists lead both the student and teacher to a clear understanding of the problem and to the required remedial program.

Teacher observation of students as they work also provides valuable information about students' use of time and materials and about their approach to school tasks. Interviews with students themselves also can shed light on the problems for both students and teachers. Asking a student how he or she prepares for a test, taking notes for further analysis and then comparing the student's perceptions of his or her

Figure 11-2
Informal Probes to Test Hypotheses

Observation	Hypothesis	Next Step
Low score group-administered social studies test given during a one-hour time period.	Directions given too quickly.	1. Readminister test at a slower administration pace.
	Testing situation too distracting.	1. Readminister test in a small tutorial room.
	Student doesn't organize time well.	1. Check to see if questions early in test are answered more fully than questions presented towards the end of the test. 2. Readminister test with specific time allotment for each question. 3. Readminister test with no time limits.
	Student does not systematically answer each part of the question.	1. Check to see if student answers only first part of the question and leaves subsequent parts. 2. Readminister test but number each part of each question separately.

From N. Zigmond, A. Vallecorsa and R. Silverman, *Assessment for Instructional Planning in Special Education* (Englewood Cliffs, N.J. Prentice-Hall, Inc., 1983), 343. Reprinted with permission of the authors and publisher.

own behaviour with what you have observed that student actually doing, or with what you have seen in other more successful students, can also lead to specific remedial programming. These approaches to teacher-based assessment will be further illustrated as we look at remediation of organizational deficits in students.

REMEDIATION

In remediating problems in organizational skills, there are at least four agents who play a significant role. These are the parents, the teachers, the peers and the students themselves. Because modelling has such a strong effect on student behaviour (Bandura 1974; Meichenbaum 1982), one aspect of a remedial strategy should be to provide as many models of effective organization as possible.

Parents In a book tracing the experience of three learning-disabled students and their families, Kronick (1981) pointed out that in each case, the students' problems were greatly exacerbated by the lack of organization generally found in the family as a whole. Essential behaviours such as punctuality, routines for meals, bedtimes, special events, identifying appropriate places for things, times for cleaning, etc. were not present in these families. This meant there was no model for the child for these behaviours, nor was there any support mechanism for the child who by nature was disorganized. For example, a disorganized parent could not help a disorganized child accumulate all the gear required for his or her hockey game and get him or her to the arena in time to be ready to play since these behaviours were also not in the parent's repertoire.

Where possible, however, support should be given at home to help a student learn better organizational skills. This can occur on several fronts. The first is his or her bedroom. Clearing it of unnecessary clutter and then finding specific places for the things that are left is a beginning. Helping the child keep his or her bedroom organized on a regular weekly (or even more frequent) basis is often necessary. Finding clothes and keeping them neat and clean may require parental involvement. There must be space to work and the necessary tools and materials must be easily accessible. If students are to work elsewhere in the house, there should be one place established (as much as possible) as his or her work area and, again, necessary materials

Figure 11-3
Checklist of Skills Required for Successful Project Completion

Skill	Student Demonstrates Skill	Student Does Not Demonstrate Skill
Student can narrow topic down to manageable proportions.		
Student can state problem clearly.		
Student can set up reading target.		
Student can use table of contents and index to locate information.		
Student can scan a passage for information.		
Student can discriminate between relevant and irrelevant information.		
Student can evaluate what is read and select or reject as necessary.		
Student can collect information to support or attack a point of view.		
Student can make notes efficiently from a number of sources.		
Student can bring the information collected into an organized format.		
Student can present the information effectively.		

should be allocated to a nearby spot for ready use. This may seem like indulging a child past the point where most children should need help. However, some children must be helped regularly with basic organization of their "things"; otherwise, they have no sense of how materials should be arranged and, as a result, they are too frequently in a mess that detracts from positive effort.

Household routines also help a disorganized student (Maynard 1984). If there is a specific time for meals, bedtime, homework, TV, wash day, etc. for the entire family, it is easier for the disorganized student to attend to these behaviours and be where he should be at the appropriate time. Dates of exams, projects due, special family or recreational events can be posted on the refrigerator or his or her bedroom mirror to provide no-nagging reminders of upcoming events he or she must be ready for. This is very helpful because many disorganized students forget such events only too easily and panic when they realize that time is up.

A final way in which parents can support their disorganized child is to help them to have as much done ahead of time as possible. If the child has a hockey game at 6:00 a.m. the next day, all the gear should be found the evening before and be near the back door ready for picking up and going. If the child has to bring in materials to make a weather indicator, or cookies for a classroom party, or his or her collection of rocks, these things should be organized ahead of time to avoid last minute panic in the morning. If students are not good at telling parents what needs to be prepared for, this is one of the many reasons for teachers and parents to have a regular communication system which includes notes and phone calls.

In time, with dependable support and gradual movement from dependency to independency, these students do become more organized. However, many need more help for a much longer time than the average student if they are to succeed. If they have this support at home, the task is easier for them and for the teacher (Mann et al. 1979).

Teachers Teachers also need to model good organizational skills for students. If the chalkboards, bulletin boards and wall spaces are labelled and organized effectively, the student knows what goes where and where to look for information that he or she needs to complete work or to plan for effective use of time. Chalkboards on which a specific space (far right hand side for example) is reserved for upcoming assignments and special events are easy to point to when you want to remind students of their schedules. Furthermore, it is easy for the student himself or herself to look at the chalkboard when he or she needs to check these things. Everyday activities should be scheduled into a routine that is as regular as possible to help students predict and plan for upcoming subjects. Bathroom routines, material searches, rules for activities that follow completion of seatwork, use of activity centres, etc. should be discussed and then listed, again so the student can be easily reminded of how things are done in the particular class. Writing out the lesson objectives and listing the important points to be covered in a lecture on the blackboard or overhead projector lets the student know what is to be learned and which topics must be covered in his or her notes. Sharing this information has proven to be a most useful organizer for students with note-taking deficits (O'Rourke 1984).

Figure 11-4
Instructional Grouping Grid

Student	Teaching Objectives				Review	
	Long division procedures	Rule for silent *e*	Graphing math worksheet scores	Comprehension of main idea	Spelling errors for last 3 months	Multiplication tables 4, 6, 7, 8
Mike	✓		✓	✓	✓	✓
Molly	✓	✓	✓	✓	✓	✓
Kara	✓			✓	✓	
Blair		✓		✓	✓	
Steve	✓				✓	✓
Jenny		✓		✓	✓	
Mary		✓	✓		✓	
Ken	✓	✓			✓	✓
Mike B.	✓		✓		✓	✓

Adapted from N. Zigmond, A. Vallecorsa and R. Silverman, *Assessment for Instructional Planning in Special Education* (Englewood Cliffs, N.J. Prentice-Hall, Inc., 1983), 335. Adapted with permission of the authors and publisher.

If the teacher is grouping her students to meet individual needs, it is most helpful to have organized what skills need to be taught to whom in order to plan out the remedial unit, to check whether and when students have attained the skill, and, where possible, to form small groups of students who share a deficit. An example of an organization chart for this purpose is given in Figure 11-4. Collecting information from general observation and testing in the early months of the school year and organizing it in this way gives the teacher an easy and fast check of what individual topics need re-teaching and review to which students. And, finally, teachers who are going to mainstream effectively need to keep well organized records on their students. Setting up a file for each student with checklists and observation sheets that are straightforward, simple and quick to use makes this process much easier in the long run, even though it may take extra time early in the year. Teaching a heterogeneous group of students well is professionally and personally demanding. Being well organized helps, however, and provides an excellent model for all the students in the class.

Specific Problem Areas

Students seem to have problems organizing their time, their "things" and their approach to school-related tasks.

Time Time is a problem for students in two major ways: awareness of time and management of time. Kronick (1983) has commented on the many problems some students have with telling time, days, weeks, seasons, holidays, and with the behaviours associated with them. Many have very little sense of where their time goes or how to use it effectively. One of the useful ways to make them more aware of it is through the use of a time-use chart such as that shown in Figure 11-5. Students are asked to fill this out for an entire week and then to add up the total time spent on school work, home routines and recreation. For some students you might want to break down the time periods into shorter blocks; indeed, for some a careful analysis of one hour of class time and/or one hour of homework time could reveal a great deal about how each student uses time. This more detailed observation could be done by the student himself or herself, the teacher or a volunteer. Some students may need to be reminded to fill in their chart by parents and teachers. Once it is complete, both they and their teacher will have a much better picture of how they use their time. The three totals can be graphed initially and then at regular periods to check on the balance of these three activities. As a follow-up activity, students should be encouraged to keep a large monthly calendar with all social and academic events marked. This is a useful check for all students. Calendars on which the teacher has marked the special dates he or she is aware of can be issued at the beginning of each month. The students can then fill it in as more activities are scheduled. In this way the students know at a glance what day it is, what is expected of them and when assignments are due.

Punctuality is a problem that is aggravated by the disorganized student's poor sense of time. Routines help the student to know when he or she should be doing a certain task. Cues and prompts just before a change of activity that remind the student where he or she needs to be and how much time he or she has are helpful. Timers and stop watches are concrete reminders of time passing that students can use both at home and at school. Being paired with a buddy who is punctual provides a good model and a concrete support for the student.

Students can become more aware of time if the length of time available is specified in instructions. The following are several examples of this type of instruction: "This seatwork must be finished in 20 minutes"; "You have until the recess bell to work on your project"; "The assignment must be completed in two days." Using time words regularly also reinforces time awareness in students. This includes saying things such as "There are ten minutes left before art"; "We are going to the library at 10:05"; "Friday the 24th is a professional development day."

Many students (and educators) have difficulty making effective use of time. Once they have become aware of how they presently use their time and how they could better use it, students are ready to plan in terms of specific activities. Writing down all the things they need to do and setting priorities among them is a useful technique. Estimating how long each task will take, how much time they actually have in which to do it and how they can reconcile these two aspects of time also helps students organize their use of time.

Planning sheets such as those illustrated in Figures 11-6 and 11-7 are concrete examples of how students can organize and use their time more effectively. These sheets can be filled out for major projects, such as preparing a speech, that require several weeks of planned work. They can also be filled out as a means of organizing time over a one- to four-week period in which many small and large assignments must be completed. The sheets should be distributed on Monday to those students who need them. The teacher should then ensure that they are completed and added

Figure 11-5
Time Use Chart

Time	Monday	Tuesday	Wednesday	Thursday	Friday	Saturday	Sunday
7-8 a.m.							
8-9 a.m.							
9-10 a.m.							
10-11 a.m.							
11-12 a.m.							
12-1 p.m.							
1-2 p.m.							
2-3 p.m.							
3-4 p.m.							
4-5 p.m.							
5-6 p.m.							
6-7 p.m.							
7-8 p.m.							
8-9 p.m.							
9-10 p.m.							

Total School and Studying:

Totals Meals, Doing Chores:

Total Recreation, TV, etc.:

to during the week as needed. The sheets can be pasted on the students' desks or inside the front covers of either their homework books or some other frequently used books to provide the ready reminder they need to use their time appropriately.

Organizing Things One of the many difficulties some students have in organization is finding the necessary materials for use at a specified time. Figure 11-8 shows a task analysis designed to provide a sense of all the subtasks included in this major skill. This analysis needs to be taught from the bottom up to best ensure that the students can cope with all the subskills. Colour coding workbooks is an easy way to

Figure 11-6
Time Management Checksheet (Single Task)

Task: Preparing and delivering a speech

Completion Date: Wednesday, November 10

Name: Dale S.

Component	Time Needed	Completion Date	Finished
Choose Topic	30 min.	Oct. 26	✓
Research Topic	2 days	Oct. 27, 30	✓
Organize Notes	1 hour	Oct. 31	✓
Write First Draft	1 day	Nov. 1	
Write Second Draft	½ day	Nov. 2	
Make Notes	½ day	Nov. 3	
Rehearse Speech	30 min. for 4 days	Nov. 6-9	
Give Speech	5 min.	Nov. 10	

Figure 11-7
Time Management Checksheet (Multiple Tasks)

Activity	Due Date	Estimated Time Needed	Day(s) Planned to do Task	Finished
Finish mathematics homework	Monday	40-60 min.	Monday	✓
Study for science tests	Friday	3½-hr. periods	Tuesday Wednesday Thursday	 ✓ ✓
Start social studies project	Friday, March 23	3½-hr. periods	Monday Wednesday Friday	 ✓ ✓
Review spelling	Friday	30 min.	Thursday	

Time Period March 12-16 Name: Mary W.

ensure quick access. Also, reinforcing the axiom "a place for everything and everything in its place" can help students realize how important organization of materials, equipment, etc. is.

School Work There are many aspects of school work that require organization. They include reading material effectively, taking notes, study skills, test writing, project planning, daily work completion, appearance of work and self-evaluation.

Reading material effectively Wong (1984) has found that learning-disabled students have considerable difficulty recognizing disorganized paragraphs in a laboratory setting. They concentrate on sentence length, decoding and vocabulary difficulty, and informational load as the major determinants of text difficulty, rather than attending to the logical organization of the text itself. She also found that learning-disabled students neither studied as long nor recalled as much information as normal students. They could, however, be taught how to organize paragraph ideas more effectively and to better understand what constitutes an organized passage, both of which skills facilitate subsequent recall of the passage. In an earlier study (1982), Wong found that learning-disabled students tended to use some form of selective search in choosing retrieval cues for a comprehension passage, but that the form of the search was more primitive and less exhaustive than that used by normal-achieving students in grades five, six and seven.

Alley and Deshler (1979) suggest the need to teach low-achieving students to perceive organization in a set of materials. For textbooks, this would include reading the table of contents and using topic headings and underlining. In the case of notes, perceiving organization includes having an awareness of the amount of information accumulated per topic, clustering that information into general and specific details per topic, and recognizing overlap among topics. In the case of courses, it includes having an awareness of general principles and overlap. Alley and Deshler suggest teaching a student how he or she can manipulate the organization of information from various sources to facilitate understanding, retention and retrieval—a student could organize information, for example, by topic, by attributes, or from easy to difficult.

While there are no programs specifically designed to teach such skills, Case (1982) in his "Reading Critically" materials for grade nine does include a unit on teaching students to read for organization cues.

One of the best-known methods of teaching students how to read material effectively is the SQ3R or PQ4R method (Woolfolk and Nicolich 1984; Alley and Deshler 1979). This approach instructs the student to initially survey or peruse the material, to generate questions about it, and then to read, reflect, recite and review the material. At the survey level, the student is told to glance over the headings in the chapter to identify the few major points which will be developed. This should not take more than a minute and will sensitize the student to the three to six core ideas around which the material is developed. At the next stage, the first heading is changed into a question. This arouses curiosity and subsequently increases comprehension—the student is cognitively involved in the search for answers. Next the student needs to answer the question, reflect on the answer (incorporating information already in

Figure 11-8

Task Analysis for Organization of Materials and Class Readiness

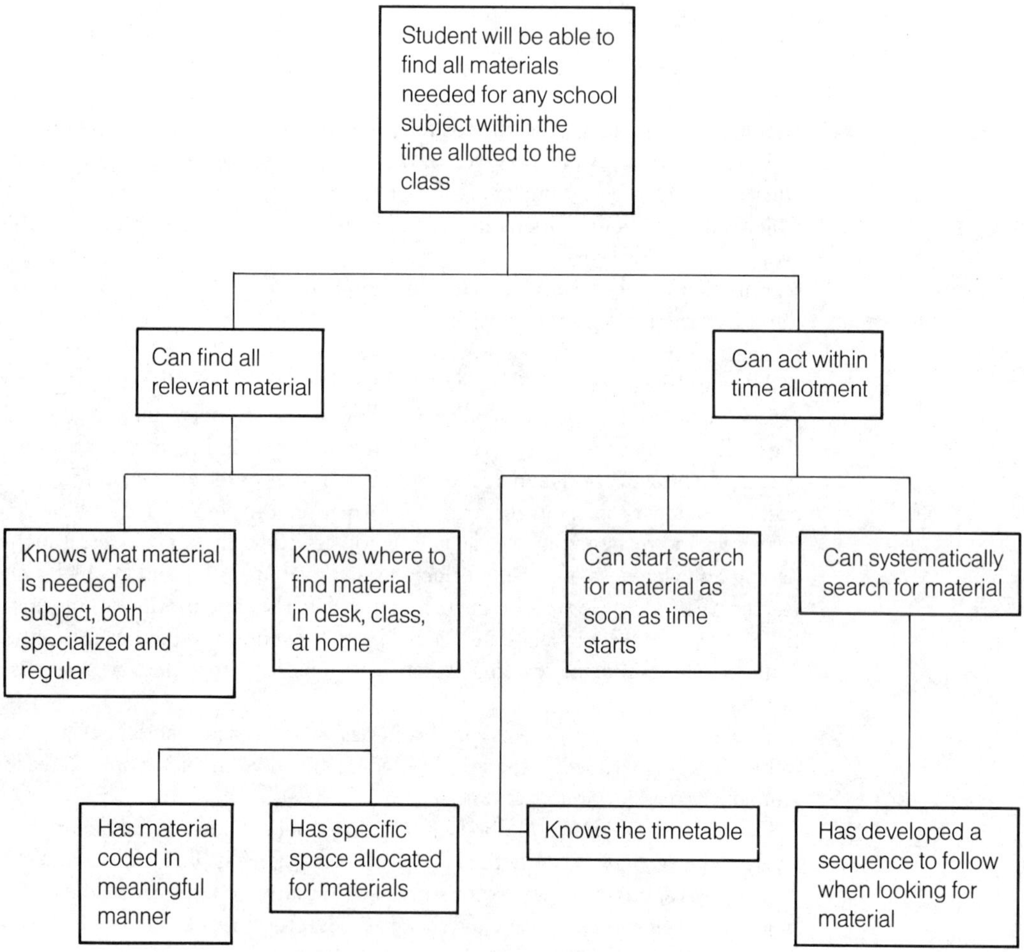

memory with that just discovered in the passage) and recite the answer by paraphrasing it, generating examples, or jotting down cue words. These four stages are repeated for each new heading in the chapter. When the student has gone through the entire chapter, he or she reviews the material by looking over his or her notes to get a sense of the points made and their relationship to each other. Finally, the student checks his or her memory by reciting the major subpoints under each heading. This review procedure can be repeated several times until the student is sure he or she can recall all the main points.

Note taking The student with learning problems is faced with situations regularly where she or he must be able to summarize an oral or written presentation and make quick notes for further reference. Alley and Deshler (1979) have described informal and formal notetaking and suggested specific ideas for improving it. Informal notetaking is usually spontaneous and the information is being taken from an oral source. Informal notetaking can include recording lecture points, assignments, directions and important insights. It requires good listening skills, accuracy and order since the material usually can't be heard again. Formal notetaking, on the other hand, involves deliberate scrutiny of the concept involved and a weighing of material in a logical, organized manner. It is useful to develop a consistent format and to label notes by topic, time and referent so they will be available later for further analysis. The following are several suggestions for improving notetaking.

1. At the beginning of the class, the teacher can give a student a sheet of paper listing the three or four major points to be developed in that class. There should be plenty of room left in which the student may make notes on each topic. Ideally, these notes can be discussed later with the teacher and/or a small group of fellow students.
2. Have each member of the class write a summary of the day's lecture and discuss the summaries later.
3. The teacher can write the major points on the chalkboard to guide the student's notetaking.
4. Have the students read a newspaper article aloud and then answer who, what, why, when and where questions about it.
5. The teacher can give a lecture with carefully prepared verbal clues; for example, "now we will consider" or "the final point is."
6. To emphasize the necessity for order and selection, the teacher should allow time at the end of the class for students to go over their notes.
7. Use the SQ3R method; i.e., survey, question, read, recite and review.
8. When outlining, have students use specific formats that reflect the five "R's" of notetaking—record, reduce, recite, reflect, review.

A mnemonic for notetaking and note makeup is LENS and SMOKE. LENS refers to notetaking from lectures, tapes, etc. and reminds the student to a) *listen*; b) *encode* by selecting relevant points, rejecting irrelevant material and recording in adequate paraphrase; c) *notetake* by organizing and restructuring; and d) *summarize*. Notetaking from printed material—SMOKE—involves a) *skimming* the material, b) selecting the *main* idea, c) developing a system of *organizing* headings and making notes, d) locating *key* points and significant supporting details, and e) *evaluating* notes for the purpose of reorganizing, enumerating and/or rewriting.

Another suggestion for teaching students how to take notes is to assign passages to the entire class and accompany these passages with worksheets with cue headings such as 1) topic, 2) main idea, 3) details, and 4) idea clusters (Lechelt et al. 1980). Another technique is to include key words and phrases in your class presentations; for example, "This is important"; "A key point is..."; "This will be in the test"; "Three principal issues are...." These cues will guide the student in his or her notetaking effort. Checking on the effectiveness of these cues gives the teacher a sense of how easily her or his students will be able to improve their notetaking skills.

Studying Effective approaches to studying are necessary for all students but especially for those with learning problems (Hardwick 1980). A great deal of practice and review of material is necessary if students are to remember essential information when under the pressure of exams. Only what is "overlearned" will be readily retrievable. Bloom and Coulter (1982) suggest nine methods to help students study and remember more effectively.

1. Intensifying Attention. This can be done by underlining main points in texts, attending to the dominant feature of a map, and colour-coding recurrent themes.
2. Mnemonic Devices. Acronyms and other methods described in Chapter 10 should be used frequently.
3. Chunking Information. Chunk according to common attributes to provide manageable units of work.
4. Kinesthetic Rehearsal. Trace over graphic material, either on the material itself or in the air.
5. Visual Rehearsal. Visualize parts of a diagram or a list, close your eyes and picture a sequence, imagine a bizarre association.
6. Auditory Rehearsal. Talk to yourself—out loud or subvocally. Not only does this allow the material to be recorded in a second modality but the throat muscle movements add a third modality to the experience.
7. Number Set. Focus on the exact number of steps, stages, major parts or points to be remembered.
8. Finger Rehearsal. Touch the fingers to trigger memory of each point in the number set.
9. Practice. Do many examples of the type of problem to be tested, formulate questions and answer them, practice and review often for short periods rather than in one massive cram session.

Much of the success in studying comes from effective planning. Spreading it out and doing a little bit at a time makes studying easier (Hewitt n.d.). Setting up a self-reward system is helpful for older students in maintaining good study habits over time. Parents and teachers can suggest appropriate rewards for younger or less mature students who work well with an extra incentive. All students need to be aware that motivation to achieve is important, that there is a limited time to study and that they must make good use of available study time. The time available for studying can be broken down into three periods—before classes, during classes and after classes (Rothwell 1982). Before classes, the student should prepare. This is usually achieved by having homework done and reading any materials that were assigned. Preparation makes the material discussed in class more meaningful, more easily remembered, and it makes the student better able to answer or ask questions in class. During classes, students need to listen and take notes. Even incomplete notes are better than none. One of the frequent concerns about notetaking is that while writing the student is missing the continuing information being delivered by the teacher. While this is a real problem, it seems to be less problematic than not taking notes at all; especially when you consider that material the student has taken notes for is seven times more likely to be remembered than material he or she has only listened to (Woolfolk and Nicolich 1984). After classes, the major task is to consolidate the material in the notes. The student should rewrite them, read them over and think about them. The student should also be encouraged to incorporate them into the notes he or she has made from the text.

A final suggestion for helping students with study skills is to have them complete a questionnaire like the one illustrated in Figure 11-9. This will provide the student and the teacher with valuable information about the positive and negative aspects of his or her present study habits. Older students may be able to fill out the questionnaire themselves but younger students may need an older person to read the items and make sure they answer each one appropriately. Not all the questions would be appropriate for a younger student; however, a revised form that included questions 1, 2, 3, 5, 6 and 8, at a minimum, would yield useful information even from grade three and four students.

Test-taking One of the most important skills in written language involves students' abilities to write tests. Many exceptional students have poor attitudes about their ability to take tests. Also, they are often poorly prepared for tests and quite anxious and disorganized while writing them. Test-taking skills are rarely taught—it is assumed that a student will pick them up over time as he or she writes more and more examinations. Unfortunately, some students often do not pick up these skills and so they must be taught.

At the preparation stage, it is useful to sit down with the student and plan out a daily study routine using a chart much like the worksheets illustrated in Figure 11-7. This will include when, where and how long a particular topic will be studied. In this way a study schedule may be self-monitored. Getting exams from previous years, sharing ideas about exam questions and determining both the content and format of the test will also help the exceptional student at the preparation stage.

Writing objective exams can be made easier if the student is taught to follow certain guidelines. Carman and Adams (1972) have described a method called SCORER that can be applied to any type of exam. Each letter represents a specific step in writing the test.

S — schedule time so all questions are answered.
C — look for clue words.

O – omit difficult questions.
R – read each item carefully.
E – estimate your answers.
R – review your work.

In addition, students should mark questions that they are not sure of and return to those questions at the end of the exam. They should be advised not to change an answer, however, unless they are certain of the second choice. First choices on average are more correct than second choices. This is because first answers often reflect residual graphic features in long-term memory that the student may not be consciously aware of. Students also should be encouraged to guess. Most teachers don't penalize wrong answers and even if they do, the student usually has a better-than-chance likelihood of choosing correctly. Students should be taught to watch for grammatical clues. Often if the correct answer is plural, the item will be written with a plural verb even though several of the choices will be in the singular form.

One of the difficulties of objective questions is the form in which they are written. Those that require conceptual transformation, that is, those that are written in the negative form, use double negatives or ask for multiple answers, require more time to process and can be confusing to a student. So too are questions that ask a student to choose the "best" answer from four or five choices. With items of this type the student should follow these five rules:

1. Analyze—what is the question asking?
2. Eliminate—reject the obviously unsuitable answers.
3. Discriminate—select the best answer from the remaining ones and respect your own intuition.
4. Watch for the syntax structure that makes the item more difficult or gives clues to the answer.
5. Watch for words like "all, always, entirely, never, only and totally."

Short-answer recall questions and completion questions often provide clues to the student in the way they are set up; for example, "The major industries of Moncton are ______________ and ______________." The student should know that the testor expects two industries to be named. At other times, the answer may be indicated by the sentence structure and content, as in the following question: "Mary was home helping her mother ______________ the house."

Essay exams test for content and for the student's ability to organize and present necessary information. The following guidelines can help a student deal quickly and thoroughly with questions and answers (Rothwell 1982).

1. Underline important instructional words in the overall directions given about the exam; for example, "This social studies test is one hour in length;" "All questions must be answered in sentence form."
2. Underline the important instructional words in each question; for example: "Give three reasons why it is important to brush and floss your teeth regularly."

   Many students do not take time to read instructions clearly (if at all) and blindly answer all the questions even when they were only expected to answer a certain percentage of them. They often ignore the cues that could help them

Figure 11-9
Study Skills Questionnaire

1. I study in the same place each day at a desk or table.	yes	no
2. I keep distractions to a minimum (music, desk top clutter).	yes	no
3. I study immediately after I sit down at my desk.	yes	no
4. I know how I learn best (visual, auditory, kinesthetic).	yes	no
5. I want to learn.	yes	no
6. I am aware when I am not concentrating.	yes	no
7. I take short breaks when I get tired.	yes	no
8. I write down all assignments with the date they are due.	yes	no
9. I plan how long an assignment will take to finish.	yes	no
10. I keep a calendar or time sheet of my work schedule.	yes	no
11. I keep dated assignments and tests for study and review.	yes	no
12. I keep an organized binder (subjects separated).	yes	no
13. I spend enough time studying.	yes	no
14. I study each subject one half hour per night or as often as is necessary to get a good grade.	yes	no
15. I do something while studying (write notes, recite).	yes	no
16. I make up illustrations, (diagrams, maps).	yes	no
17. I pay attention to illustrations, graphs, etc. in books.	yes	no
18. I make a list of new vocabulary words and study them.	yes	no
19. I use the SQ3R method of studying.	yes	no
20. I test myself (flashcards, study questions).	yes	no
21. I review often.	yes	no
22. I overlearn the material.	yes	no
23. I actively participate in class (discuss, take notes).	yes	no
24. I use group study techniques.	yes	no
25. I use new information in my daily life (vocabulary, etc.).	yes	no

From W. Rothwell, *Tips for Parents in Helping Students Organize* (London, Ont.: Middlesex Board of Education, 1982).

retrieve information stored in memory or those that suggest the point of view the teacher prefers to have developed. Students should be taught, therefore, to read the questions carefully first.

3. Underline specific cue words and specific content words in a question; for

example: "Describe and compare the hierarchical structure in your school with the hierarchical structure in the provincial government."

4. Budget your time according to the teacher's mark allotment. If a question is worth 10 marks in a 100-mark exam, it should get one-tenth of the total time available to do the test and twice the time a 5-mark question would get.

   Encourage the student to wear a watch and use it. The student should also be sensitive to the fact that he or she should write about twice as much for a ten-mark question as for a five-mark question.
5. When studying before the test, look at the content to be covered, pick out the major topics and prepare answers for questions that could be asked about them. Summarize these answers through key words and develop acronyms for them; for example, if you are studying the use of levers in environmental studies, you should think of about five uses and create a word or sentence that can be used later as a mnemonic to retrieve the information from long-term memory. If you have done this for the major content areas, you can write down each acronym on the back of the test answer sheet or somewhere else as soon as you sit down to the exam. That way, you can then refer to the acronyms when answering specific questions about the material to which they relate. Briefly outlining the points to be addressed in the answer is also useful since it gives you time to reflect and organize both the content and presentation of the answer.

When writing essay-type answers, the SCORER technique described above is also useful. In addition, students generally should be advised to never leave a question unanswered—most teachers will give a student some marks for attempting to answer a question. Students should be encouraged to use a quarter to a third of their time organizing their answers. This is helpful for all students in the sense that having outlined answers to questions, there will be less tendency to repeat ideas or to include non-essential information in the final answer. This is especially helpful for students with handwriting difficulties since they often get tired while writing out long answers and/or write so slowly they don't finish within the the time period.

And finally, students should be encouraged to leave five minutes or so at the end of the exam to proofread their answers. They should check that they have read the directions properly, answered all parts of the question, not confused symbols, answered in the right question slot and that their sentences are complete and logically sequenced.

Doing projects Projects are required in most schools from the early primary grades right through to college. Disorganized students need much specific guidance if they are to complete projects effectively and learn from the experience. There are five major steps involved.

1. Choosing a topic.
2. Collecting and recording information from print and non-print sources. Students should be encouraged to write out the name of the reference book at the top of a card or sheet and to then write all the specific information they found in that book underneath. This should be done for each reference.

3. Organizing the information. If the material is already on cards, this is a relatively straightforward task. The student should decide on the major headings under which he or she is organizing the information and then should sort out the cards accordingly. Both the headings and placement of information should be discussed with the teacher or parent to determine their appropriateness and completeness before the student goes on to write up the work.
4. Determining presentation format. The student must decide whether the project will be reported orally or in written form. Will it be illustrated through photographs, cartoons, magazine pictures, murals, maps, etc.? Who will it be presented to?
5. Evaluating. The student should assess whether he or she did what the teacher wanted done. The student should also decide whether the work turned out as he or she intended.

Finishing assignments Regular assignments are a necessary aspect of teaching and learning. For many students, seatwork and homework are the follow-up that gives them the opportunity to practice and apply the new concept learned in a lesson. The exceptional student often has difficulty finishing this work, however, and needs to be helped both to start assignments on time and to finish them. The following are some suggestions for providing such help. Make the work meaningful and within an appropriate difficulty level for the student. Have the student daily record his or her own seatwork in each subject area. Set up this record so that there is a space in which to check off the work as it is done. If the student does not use time well during the day, make up a daily work record such as that illustrated in Figure 11-10. This record can help focus the student's attention on the importance of each aspect of time use. You should go over it with him or her at the end of each day. Points can be assigned that can later be traded in for a reinforcement that is mutually agreed upon. This technique provides a daily record of how the student is handling assignments. It also points out the areas in which there has been improvement as well as those that still pose problems for the student. Parents might be encouraged to use similar records at home to help their child complete the required work in a reasonable time period.

Pairing the disorganized student with another who does complete seatwork and asking the two to work cooperatively can be useful for some students. Other ways of encouraging students to complete assignments include providing bonus points in subjects for work completed on time; allowing the student to use a calculator for the second half of the questions on a math worksheet; allowing the student to help the teacher or to do other enjoyable activities if he or she finishes the work; and letting the student bank time for preferred activities if he or she finishes ahead of time.

Organizing technical aspects of work Many students have difficulty keeping their notebooks in order and making sure single pages look good. Notebooks are especially important since many students in the elementary grades use them as their primary source of review material for studying. Ideally they should organize notebooks as follows for maximum utility:

1. Have a different coloured notebook for each subject.

2. Start with a table of contents with titles and page numbers.
3. Number each page.
4. Title each note and divide each topic into smaller parts with sub-titles, numerals and letters.
5. Make sure there is an equal space between the notes.
6. Label all maps, graphs, pictures.
7. Make sure each note is legible.
8. Underline, colour or highlight the important parts of the note.

Individual pages should be as uncluttered as possible. Discourage the addition of unnecessary marks or drawings since they can distract the student. If a page contains answers to questions, each answer should be numbered and spaced appropriately. The student's name and date should be clearly evident. Teachers may want to mark for neatness and completeness if these are skills on which the student needs to work. In that case the first piece of work should be marked quickly, before the student goes on, to ensure that it is set up properly. Students may be encouraged to work slowly and neatly, but their attention should be drawn to the time they are taking in order that you can work on time management later. Encourage the student to find the materials he or she needs to do work quickly and to put away all the other things that are around so his or her work space is uncluttered. In this way, the child can concentrate on good form for the work he or she is doing.

Figure 11-10
Daily Work Record

Behaviour	**Points Earned** (5 maximum)	
Starts work on time	5	
Concentrates on work	3	
Finishes work in allotted time	4	
Has homework completed	5	
Works quietly	3	
Total for day	**20**	A good day!
Accumulated Total	**76**	

Self-evaluation The final skill that many disorganized students lack is the ability to evaluate their own efforts. Wong (1982) has found that students with learning problems are much more unlikely than normal students to check, when they have the opportunity, whether they have completed all the requirements of a task. Encouraging self-evaluations in academic areas through the use of checklists described earlier (e.g., in Figure 9-13 and in Figure 11-9) helps in this regard. Figure 11-11 illustrates a checklist more specifically designed to focus the student's attention on organizational issues. This has been used with students as young as eight and nine to help them focus on the essential aspects of presenting a project. Including items that reflect the stage of preparing material would also be helpful for many students. Self-check methods of this type are important in focusing the student's attention on specific organizational aspects of behaviour. They are also important for the opportunity they offer a student to think about how he or she handled a situation.

Figure 11-11:
Self-evaluation of Individual Task

Ratings:
5—super
4—great
3—okay
2—not well done
1—the pits

Directions:
Circle the number that best describes how you feel you organized and presented your project.

Organizing

1. Did I use a good way to organize my information?	5	4	3	2	1
2. How well did I get my information in a reasonable sequence?	5	4	3	2	1
3. How well did I group related information?	5	4	3	2	1

Presenting

1. Did I choose a good way to present my information?	5	4	3	2	1
2. How well was my presentation matched to my audience?	5	4	3	2	1
3. How effectively did I communicate my ideas?	5	4	3	2	1

Comments:

Case Study

Step 1: Background Information

Bob is in grade eight. His teacher has referred him because he is disorganized, is messy, can't find his materials, doesn't keep track of assignments, and is often late. He is below average in science and social studies but reads and does mathematics at grade level. He must be constantly reminded about finishing assignments. Previous testing shows good average intelligence and no serious social, perceptual or emotional difficulties.

Step 2: Preliminary Hypotheses

1. Bob's poor marks in social studies and science are partially the result of his poor work habits.
2. Bob is poorly organized both at home and at school.
3. Bob does not use his time effectively.

Step 3: Individual Testing

Because there were previous testing data available, the resource teacher decided to observe Bob's work habits in class, to carefully analyze his tests, project-planning and final product, and to interview him directly. Through observation, it was found that he started new activities approximately two minutes later than his classmates, that he rummaged through his desk at least twice before finding all the materials he needed, and that he could be distracted easily while doing his work. Examination of his work indicated that he often did not finish it, that some assignments had not been done at all, but that those that were completed showed ability and originality. In the interview, the resource teacher asked the following questions (Doree 1984):

1. How does he feel about the quality of his work? What are the strengths? What needs improvement?
2. How does he think his teacher feels about his work? On what things does she compliment him? What does she suggest needs improvement?
3. How do his parents feel about his work at school? How do they feel about how he handles his home responsibilities?
4. How does he feel when he loses an assignment? When he turns it in late? When it's incomplete? When it's messy?
5. How does the teacher feel when she gets the assignments and they are late, messy, incomplete?
6. How do his parents feel when they see or hear about these assignments?
7. What is his work like in comparison to that of his friends? How is it the same? How is it different?
8. What things might he do to improve the quality of his work and work habits?
9. What things could his teacher do to help him?
10. What could the resource teacher do to help him?
11. What could his friends do to help him?
12. What could his parents do to help him?

Bob was most cooperative in the interview and had many useful reflections to share. He was most aware of how annoyed and frustrated his teacher and parents were about his disorganization; but he had thought that that was how he was and didn't think he could be helped to change. Discussing questions 8 through 12 opened up the possibility that he could improve if he really wanted to and if he would cooperate if a program was set up. He agreed to this, at least in part because he wanted to get into the same high school program as his friends and knew that at present his marks were marginal.

Step 4: Preliminary IEP Goals

1. Bob will improve his work habits.

2. Bob will go from D to a B in social studies and science.
3. Bob will be more punctual.

Step 5: Testing the IEP

Bob's work habits were monitored in his regular class and in the resource room for two one-hour periods per setting to determine his time on task, punctuality and ability to be distracted in each setting. His difficulties were apparent in both classrooms, but they were more serious in the regular class. This was discussed with him and it was recommended that he spend time in the resource setting where he would have an opportunity to learn better study skills and to get specific help organizing his social studies and science assignments. He agreed to this and so was scheduled for three one-hour periods per week.

Step 6: Final IEP Goals

1. Bob will be on time 90 percent of the time.
2. Bob will have materials ready before the start of each class.
3. Bob's work will be neat—no doodles or smudges will be accepted.
4. Bob will organize and maintain his social studies and science notes in an appropriate manner.
5. Bob will improve his study skills and use of time.

Step 7: Remediation

Bob's grade eight teacher set up several positive activities for Bob to be involved in before school started. They included raising the flag, setting up programs for the computer and scheduling floor hockey games at 12:30. This ensured that Bob was at school on time. The teacher also worked with him each week on a calendar of events so that he was aware of what was coming up. His notebooks were colour-coded and his desk was organized.

In the resource room, the teacher helped him become more aware of his use of time through setting up schedules and having him estimate the time required to do everything he needed and wanted to do so. She also showed him how to use the SQ3R method to review his notes and textbook materials and to organize an effective study schedule. All of his projects and tests in science and social studies were monitored in the resource period. Bob was expected to complete specific assignments on a regular basis.

His grade eight teacher, resource teacher and parents agreed on a contract to help him better organize himself. The period of the initial contract was one month. It had both immediate reinforcement (points) and delayed reinforcement (special activities) built into it. Points were given by both teachers each time he was on time, had his materials and books ready as soon as the class started, began seatwork immediately, returned his materials to their proper place when he finished using them, and handed in neat work. The points were exchanged for extra game time, computer time or reading time. His parents reinforced his improvements by using points acquired at school towards a special activity (movie, tape, food) at home.

Step 8: Evaluation

Bob met with his resource teacher three times a week to discuss the program and his progress. He met with the grade eight teacher weekly to discuss issues and with the principal every second Friday for the first month of the program. He made excellent gains in the first month. From November to April he was maintained on the contingency contract and continued in the resource program. In May he was placed in regular class full-time to prepare him for high school, but his grade eight teacher met with him once a week to reinforce the importance of using time well and using his new organizational and study skills. He did very well on his final report in all subjects and was able to generalize his skills to studying in all subjects. He was flagged for grade nine so that the guidance support personnel could continue to periodically reinforce good work habits with him and be alert for any deterioration in his organizational skills.

12 SECONDARY SCHOOL ISSUES

Much of what has been discussed in earlier chapters is applicable to both elementary and secondary school students and teachers. However, elementary and secondary schools are not the same. In secondary schools, the students are more complex, their needs are different, and the expectations of the teachers reflect specific discipline knowledge rather than skill development. Furthermore, the organization and administration of elementary and secondary schools function differently. This is not news to educators; but the many ways in which these differences manifest themselves may not be apparent to those who are trying to meet the needs of individual students.

THE TRANSITION TO SECONDARY SCHOOL

When a student is trying to make the transition from elementary to secondary school, there are several major differences in academic focus which the student, parents and teachers should appreciate (Posno 1983). At the elementary level the emphasis is on mastery of basic skills, while at the secondary level the emphasis is on mastery of subject content. Elementary level teachers tend to be generalists, while secondary teachers tend to be specialists. At the elementary level, the student spends most of his or her day with one teacher who has major responsibility for him or her. At the secondary level, the student may see from four to eight teachers in a two-day period, none of whom see themselves as having the major responsibility for that student's progress and few of whom have much sense of that student as a total person. The elementary school is usually smaller and the student knows and is known by much of the school staff, while at the secondary level, the physical plant is larger and there are many more teachers and administrators—two factors which lead to a sense of anonymity and of being lost in a huge place (Emmons 1984).

Brown (1978) has listed the following 14 secondary school conditions which affect the educational programming of the exceptional student.

1. The administrative problems of allocating staff and materials are more complex at the secondary level and need to be carefully sorted out for students who spend part of their time in a special education resource program and part in the mainstream.
2. Because of the larger number of students and the relatively smaller number of special education support personnel, there are few people available to consider and plan for students with individual needs.
3. Organizing school programs by grade level makes it difficult to individualize programs at the secondary school. However, this is changing dramatically across Canada as completion of credits replaces grades as the requirement for a graduation diploma. In Ontario, even greater flexibility may result from the school boards' exploration of the option to deliver one-half and one-quarter credits in some subject areas.
4. The larger the physical size of the school, the more difficult it is to meet the mobility and scheduling demands of some students.
5. Where school boards use computer scheduling, the exceptional student's timetable usually needs to be hand-scheduled.
6. Any modifications of credit and grade requirements must be negotiated for the student with the individual teacher for each subject.
7. There are usually limited opportunities for alternate educational programs at the secondary level because too many secondary schools are inflexible in dealing with the handicapped.
8. More options ought to be available for both the exceptionally able and the exceptionally disabled.
9. Because some exceptional students are near the end of their academic career, there is a tendency to not bother with a well-planned program to maximize their potential but rather to enrol them in a crash program offering primarily vocational courses or basic life-skill remedial courses.
10. Accommodating instructions, evaluations and expectations for a wide variety of students in a heterogenously organized class is very demanding for the secondary school teacher who may meet with 200 students in any one term. Both the resource teacher and the subject teachers need to work cooperatively to improve the learning experience for the exceptional student, since neither can do it alone.
11. The departmental organization of high schools is foreign to students as they come from the elementary level. It is also foreign to some of the resource teachers who have done remedial work at the elementary schools but only recently followed it through at the secondary level. Departmental organization often results in the resource teacher having no academic or social "home" within the school. This can be alienating for the teacher since it cuts him or her off from needed support and communication channels; and, thus, it can also be alienating for the student who finds that regular teachers and students have little understanding or sympathy for his or her program needs.
12. With more teachers to interact with, the student must learn a variety of ways of interacting with significant others.
13. The competitiveness of many extracurricular activities at the secondary level

makes it hard for many exceptional students to participate since they can't compete at the levels required.

14. There is greater heterogeneity of value systems in large secondary schools. This presents more opportunity for broadening the outlook of both students and teachers and increasing acceptance of minority groups.

APPROACHES TO SPECIAL EDUCATION AT THE SECONDARY LEVEL

Rather than face many of these issues head-on, programs at the secondary level have traditionally handled the heterogenous group that leaves elementary school through curriculum, rather than ability, grouping. This approach has always had a patina of choice and fairness since students could choose whether they wanted to attend a basic level, general level, technical level or advanced level program. In fact, however, students who had failed in elementary school and were not passing at grade level in elementary school were counselled into the basic or vocational program, while students who did well academically were counselled into the advanced programs which opened up the doors of post secondary education to them. Most average students fell in the middle and chose general level courses that usually had a technical or commercial emphasis which was seen as providing a vocational focus. Rosenbaum (1980) argues that this approach is very similar to segregating students on the basis of ability. As a result, this approach has many of the same problems that have been found at the elementary level in the development of self-esteem, in academic expectancy and in achievement. He argues for a more heterogenous approach at the secondary level with the removal of curriculum offerings at different difficulty levels and the inclusion of specific courses that are open to all students. This is starting to happen in some areas in this country, especially at the intermediate level (grades seven to ten), as more schools offer students the option of mixing their programs with courses of varying degrees of complexity in different subject areas. In this way, for example, a strong mathematics student can take the preparatory courses that will allow him or her into the advanced mathematics and science courses at the senior level but can stay with high interest English courses that don't demand such rigorous prerequisites.

Even this type of mixing, however, does not speak to the need for more flexibility within subjects and within course options for the increasing number of exceptional students coming into the secondary system. The cascade model (Crealock and Sitko 1983) shown in Figure 12-1, which has been adapted to individual school systems across the country (Barnes 1984; Posno 1983) illustrates the range of alternatives possible at the high school level. Most school systems offer all six options to their secondary school students, although the majority of student needs are met through the type of programs described towards the top of this model.

Regular Classroom with Minimal Program Alteration This is the most frequent option found in the secondary school. It is useful in those situations in which teachers are aware that some of their students have more than the usual degree of

Figure 12-1

Service Alternatives for Exceptional Students at the Secondary School Level

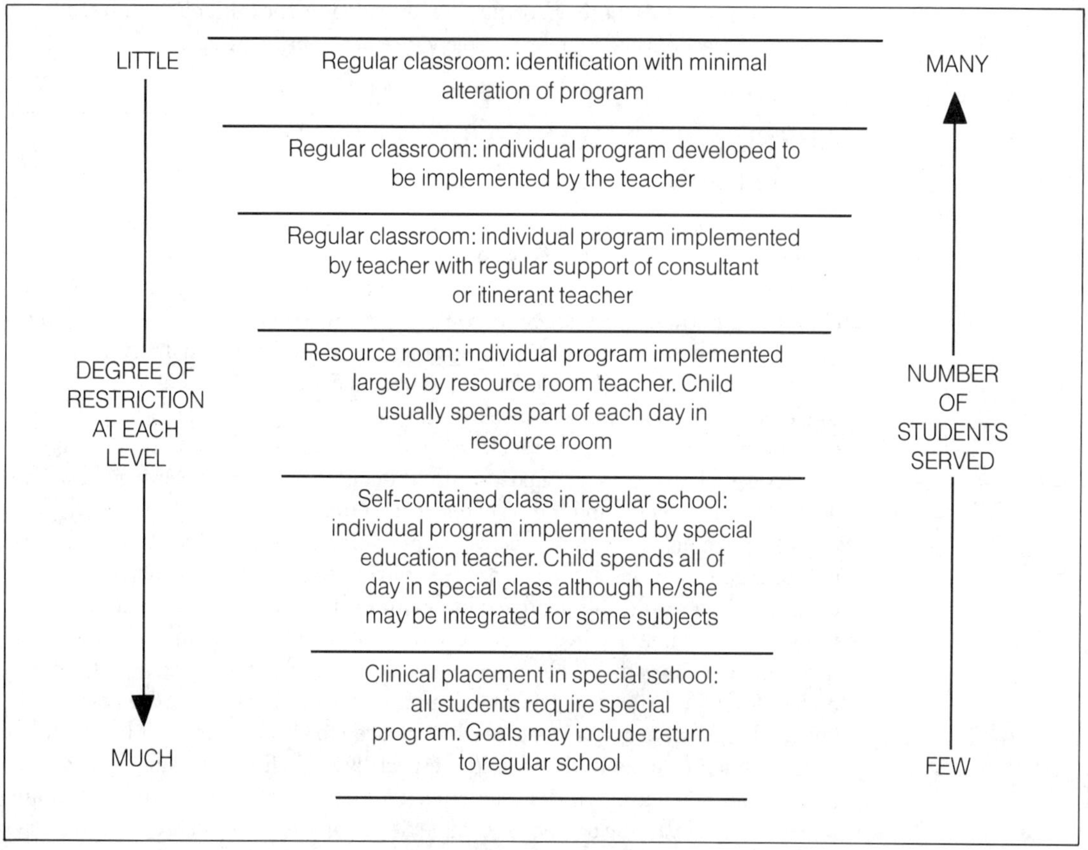

Adapted from C. Crealock and M. Sitko, "Special education in the secondary school: Selected issues," in R. Clark, R. Gidney and G. Milburn (Eds.) *Issues in Secondary Schooling* (London, Ont.: Faculty of Education, University of Western Ontario, 1984), 41.

problems, either because those students were flagged coming from the elementary school or because they have not performed to normal expectancy in the secondary school. The regular class teacher may have discussed this with his or her department head, with the guidance personnel and with the special education resource teacher (if there is one available in the school). Program modifications usually involve the whole class to some extent; for example, establishing cooperative learning opportunities (Johnson and Johnson 1980, 1982; Fader 1976) individualizing projects, or establishing a differential of expectancies and evaluation approaches for individual students. This alternative is the most prevalent in secondary schools at this time.

Individual Program Implemented by Regular Teacher The second alternative involves a special program that may be developed by the resource teacher or consultant but that is implemented by the regular class teacher. An example would be a program in mathematics that has been developed to give a student additional instruction and practice in basic skills. Such a program could include a specific module of a commercially-developed program or materials developed specifically to respond to that individual student's strengths and weaknesses. The program could use either a paper and pencil format or a computer, if the class has access to one. In either case, the program should be set up to allow the student to work independently with only occasional monitoring by the teacher.

Individual Program Involving Regular and Resource Staff Alternative three is similar to the one described above in that the teacher implements a program developed by someone else. However, in this case a special education or guidance person is available to the teacher for regular consultation and to the student for support. In the example described above, the consultant would meet with the student on a regular basis to discuss any difficulties with the program, to modify it as necessary, and perhaps to chart student progress over time. In consultations with the teacher, the consultant might discuss the effects of different teaching techniques on the pupil in question and/or on the class in general and suggest ways in which the classroom program could provide additional support for the exceptional student.

Regular Class and Resource Room Program The fourth alternative listed—a resource room program combined with regular class—is a growing option at the secondary level (Little 1983; Thoms 1981). In this approach the special programming is usually provided and implemented by the resource teacher with varying degrees of carry-over into the regular class. There are several types of resource-room programming: a) remedial instruction, b) regular curriculum support, c) personal support and d) cognitive-strategy instruction.

Remedial instruction usually addresses specific gaps in the student's basic skills in language and mathematics. It may also identify the skills essential to other subjects such as science and history (O'Brien and Pulliam 1981) and show the student how these relate to eventual success in learning by applying them to the content being concurrently taught in their regular classrooms or to future content areas. This application of skills across the curriculum is usually done by the resource teacher. Good programs involve a detailed task analysis (Gagne 1977) which logically sequences terminal skills and required sub-skills and which teaches these skills in a manner that is consistent with the student's preferred learning style.

A second resource-room model involves regular curriculum support. Here the resource teacher helps the student keep up with his or her regular class work by going over homework assignments, discussing difficult concepts in the subject, directing the student to appropriate resources for research, and so on. Basic skills may be taught but only as they relate directly to the subject. This is the most frequent model observed in secondary schools.

Although personal support for the student is part of every resource program, in some programs it is the dominant aspect. This can involve the resource teacher in an advocacy role vis-à-vis the student and his teachers, peers, parents or community. Examples of this would include explaining to the regular teacher that a student needs more time to write an exam and working out the details concerning the time extension; explaining to parents that the student cannot do homework in all his or her spare time but needs recreational activities and then suggesting to the parents a reasonable balance between work and play for the student. Other aspects of the personal support role entail direct involvement with students such as helping them choose courses and choose teachers who will be sensitive to their strengths and weaknesses. In others, the dominant support role is that of "active listener": a safe person to whom the student may express negative emotions or voice personal concerns.

Cognitive-strategy instruction (Alley and Deshler 1979) is especially useful for the mildly handicapped student who often does not have very good academic survival strategies. Many of the skills taught are the organizational skills described in detail in Chapter 11. This approach emphasizes support strategies such as note-taking and study skills that are useful in every subject area rather than only one. This approach is more recent than others but seems to offer a valuable alternative (Wiens 1980).

While most resource-room programs make use of elements of each of the four approaches mentioned above, they usually emphasize one approach more than the others. Most teachers and administrators at the secondary level still see punctuality and good academic attitudes and skills as the most important predictors of school success. They may be pleased with good personal skills, but they are not as likely to view such skills as necessary. In a recent questionnaire completed by 356 high school educators, the following skills were identified by 75 percent or more as being necessary for survival: meets due dates, arrives at school on time, attends class every day, exhibits interest in academic work, accepts consequences of behaviour, behaves appropriately in class, completes assignments, follows written directions, does not talk back to teachers, does not fall asleep in class (Zigmond 1984). Personal strengths such as having friends, offering assistance to others, speaking positively of others, and understanding facial expression were considered necessary for school survival by less than 25 percent of the respondents.

Special Class The fifth alternative, a full-time special class within the school, is relatively rare in most high schools. However, as school boards begin to integrate handicapped students into the mainstream, such classes become more evident. For example, in one London secondary school, a small group of moderately mentally retarded students were placed in a segregated class for their academic courses but were integrated into the main body of the school for social experiences and some physical education classes, members of this class having competed successfully in the olympics for the handicapped.

Special School The final alternative, placement in a special school, is only available to a very small number of students. Some examples include the two residential programs for learning-disabled adolescents supported by the Ontario Ministry of Education and the Step-Up program supported by the Vancouver school district and British Columbia correctional department for low-achieving juvenile delinquents. In most cases, it is hoped that at least some of these students can be re-integrated into the regular system.

While secondary schools are still less flexible in meeting the needs of the exceptional student, they have made great improvements during the past decade. In a recent study, McMurray (1980) asked 165 school principals across Canada about the degree to which their schools offered programs for the learning-disabled. Seventy-one percent of the secondary school principals said they had a program for these students that had been running for at least two years. Such programs were monitored at least monthly in 50 percent of the schools, usually by the guidance or special education teacher. Only six percent felt that a special education teacher was out of place in a secondary school. Sixty-six percent of principals thought the essential role of the special educator was as a consultant/trainer to the regular class teacher, and 70 percent thought special educators should remediate basic difficulties while regular teachers retained responsibility for teaching content subjects. These responses are indicative of the presence of the resource-room models discussed earlier and of the increasing awareness at the secondary level of the responsibility for meeting the needs of the exceptional student.

Attitudes Towards Special Students

One of the issues that must still be addressed involves the attitudes of secondary school teachers towards exceptional students, their experiences with such students and their knowledge about how to teach them effectively. Charleton (1984) asked secondary school teachers who were expected to integrate physically handicapped (PH) students into their program what training they had had in the field, what first-hand experience they had had with PH students, what characteristics of the teacher augured well for successful mainstreaming, and what inservice needs they had. Over 65 percent of the respondents stated that they had no formal or informal experience with PH students themselves or course work designed specifically for that population of students; they were, however, positive about the prospect of teaching them. They felt that a teacher who could most successfully promote integration of these students would be flexible, organized, highly motivated, innovative and creative in programming techniques, competent in varying teaching techniques, knowledgeable of all the content at least to the end of the intermediate year, able to show leadership with volunteers, and willing to adopt a flexible timetable. Their inservice needs list centred around explorations of the responsibilities of teaching staff and realistic expectations for the PH student. Salend and Johns (1983) and Stephens and Brauva (1980) have found that these factors affect the teacher's attitude towards mainstreaming. One could predict, then, that boards which have not prepared teachers for mainstreaming would have less success in adopting this approach to the education of special needs students than would boards who had put effort into inservice and familiarization of staffs with students to be mainstreamed.

Many of the issues that create positive or negative attitudes towards mainstreaming at the secondary level involve concerns about sharing responsibility, teaching in a flexible manner, adjusting expectations and evaluation realistically, behaving appropriately, and sorting through emotional reactions to the handicapped. At the elementary level, these concerns are somewhat less of a problem because the student can be viewed as a young, immature child who needs support. At the secondary level, however, that child is a much more complex person who must be understood and treated like a young adult rather than like a small child. Shaver (1983) has discussed the importance for the teacher of examining his or her attitudes towards the handicapped. He suggests the teacher ask the following six questions both of him- or herself and of the regular students to prepare for the integration of a handicapped student:

1. Do you ever feel awkward in the presence of a handicapped person?
2. Do you act toward them as though you feel they are less able to look after themselves than you are?
3. Would you like to have a handicapped person as a close friend?
4. Can you list some of the problems handicapped people might have physically or socially in getting around the school?
5. Do you feel comfortable offering to help a handicapped person?
6. If you were at a social event the handicapped person was at, would you avoid that person, pay special attention to him or her, or act normally?

Other questions could be added to the list to suit the specific situation. It is essential that the teacher has sorted out his or her own answers to the questions, however, in order for that teacher to be able to help the other students become aware of their attitudes and deal with them effectively. Some of the issues that Shaver suggests need to be shared with students are a) the stress our society places on physical beauty and good ability; b) the fact that few of us have had much first-hand contact with handicapped persons upon which to base our attitudes and behaviours; c) the fact that it is hard to know how a handicap affects someone in specific areas; d) the possibility that being around a handicapped person raises the fear that we too could become handicapped ("don't touch, it's catching"); e) the fact that we avoid handicapped people sometimes because we feel guilty about being able or because we feel that somehow they deserved it; f) the problem that we often don't know how to act with handicapped people and they aren't sure how to act with us; and finally, g) the tendency we have to think in black and white terms—for example, no retarded person could understand a Shakespearian play, no physically handicapped person could play basketball, no socially maladjusted student could be trusted. Discussing answers to the six questions listed above in terms of these issues and any others you feel are relevant, helps students and teachers develop a more positive attitude towards the integration of exceptional students into the mainstream.

Setting Appropriate Goals

Another broad issue for the secondary school is the development of appropriate goals for exceptional students. Barnes (1982) has suggested that the secondary school special education program should emphasize employment, the possibility of specialized training, the wise use of leisure time and the application and use of learned academic skills rather than the continuation of learning academic skills for their own sake. This is confirmed by McMurray (1980) who found that attainment of basic skills, career development, the attainment of good work habits and learning strategies, and job entry skills were the goals favoured by secondary school principals in programs for the learning-disabled. Green (1982) in discussing the recent reorganization of the secondary school system in Ontario stated that he had found consensus on four essential qualities or goals for the secondary school program.

1. The ability to use written and oral language with clarity and understanding.
2. An understanding of and ability to use numbers and commonly used mathematical concepts.
3. The ability to accept some sense of responsibility for one's life and for the general society in which we live.
4. The ability to get along with others and appreciate that we live in a very interactive society.

These general goals are quite consistent with those stated by Barnes and McMurray for special needs students and suggest that it should be quite possible for high school educators to develop appropriate goals for the handicapped within the general guidelines of the provincial and local authorities.

Characteristics of Exceptional Students at the Secondary Level

Exceptional students at the secondary level are usually identified in the elementary grades. In these cases, their strengths and weaknesses are well documented. However, there is always a small group of students who have been able to slide through the identification process and who have problems at high school adapting to the new settings. A large proportion of these students have mild learning disabilities that must be detected if the students are to receive an appropriate program. Some of the characteristics that set these students apart from others include having a chronic history of mild difficulties that can be determined by a careful reading between the lines of school reports; having average intelligence quotients but a variable profile of skills; having poor social communication skills; being easily distracted and having a poor attention span; being impulsive; and showing lower than expected academic achievement (Marsen 1974; Posno 1984; Shumaker et al. 1982). The existence of mild learning difficulties put some students at a high risk of developing further problems, especially if those students are not identified and helped.

Young Offenders One such vulnerable group is the young offenders. As of April 1984, Bill C-61 has defined a young offender as any person between the ages of 12 and 18 who commits a criminal act. There has been considerable work showing that 50 percent or more of young offenders have academic problems (Crealock 1984; Stephenson 1973; Berman 1978) and that the learning problem may have been a predisposing factor in delinquency in some cases (Pihl and McLarnon 1984; Mauser 1974; Schumaker et al. 1982). Others have found that students with learning difficulties are no more likely than other adolescents to get into trouble with the law, but once they do, they are more likely to be convicted because of their poor social skills, inattention and variable memory skills (Zimmerman et al. 1978; Broder 1982). In a monograph written to help the young offender in Canada understand the new legislation, Hardy and Dunn-Worndl (1983) recommend certain behaviours to the adolescent who comes before the courts. They advise the young offender to write down everything he or she can remember prior to and during an arrest, to remember that he or she has rights and can use them, to respond politely to officials, to restrain tendencies to "spill it all out," and to communicate clearly with the court officer about personal wants, ambitions and goals. Most adolescents with learning problems have difficulties in all of these areas—their writing skills are poor, their sequential memory is poor, they tend to be impulsive rather than reflective, and only rarely can they organize their past, present and future hopes in a coherent manner. In addition to this, their predisposition reports often contain information that can work against them such as poor school performance, immaturity and inappropriate behaviour. Court officials should be careful to separate inappropriate court behaviour that is linked to a learning difficulty from that which is generally belligerent and defiant. This is not an easy thing to do in a busy courtroom, but it is more likely to happen if the school report also indicates that the student has been found to have learning problems that are being attended to.

Effects of Adolescence on Learning In addition to the specific problems that exceptional students have in learning situations, there are also the developments of adolescence that complicate attempts to create effective programs (Brown 1978). These developments include increased independence, mobility and unsupervised time; the need to make important choices; social pressures; lessening of parental involvement; growing complexity of personality and the development of secondary problems.

Increased independence Older students are not as likely to simply conform to teacher's or parent's demands. This suggests that any remedial program must be jointly agreed upon if the student is expected to participate in it fully.

Increased mobility On the positive side, this means that the program can include educational opportunities both within and without the schools. On the negative side, it may also mean greater likelihood of absenteeism that can wreak havoc with program continuity. This has become especially evident to us recently as we have tried to formally observe exceptional adolescents in several secondary programs—their behaviour in class is not much different from that of average students but their absentee rates range from 10 to 40 percent, which makes it very difficult to find them when you need them.

Increase in available unsupervised time This occurs both at school and at home. One of the positive aspects of high school programs is that they allow time for the student to study and work during school hours. This becomes a negative aspect, however, if the student is lured out of the building and does not have a good sense of time or strong desire to be in class. This also happens out of school as parents often allow the adolescent more unsupervised time after school and in the early morning. Such periods are not available then to the student for homework and study.

Choices At the secondary school, students are expected to make academic and vocational choices when they are often unprepared or unable to do so.

Social pressures These are much more apparent at the secondary school level. In fact, some adolescents have been heard to say that school is simply the place where they get their social needs met. In any case, remedial classwork may be seen as a negative social experience rather than as a positive academic experience.

Lessening of parental involvement This occurs in terms of actual presence in the school and in terms of direct help with school work, in spite of increasing concern about achievement and post-school planning.

Complexity of the individual The older student has accumulated more life experiences, and so remedial programs have to attend to the whole person rather than concentrate on specific strengths and weaknesses.

Secondary problems The emotional and behavioural problems accompanying learning problems become more pronounced in adolescence. Therefore, an educator at the secondary level may be forced to address these problems before and during academic remediation.

Teaching Requirements

In order to accommodate the structural and administrative aspects of the secondary school, the philosophies and attitudes of the teachers, and the individual characteristics of the exceptional child, both the resource and regular teachers must be very flexible and skillful in a broad range of areas. Zigmond (1978) suggests that the resource teacher must be able to assess and program effectively with students directly; but she also suggests that there is a whole host of indirect services that resource teachers must be able to perform. These include assessing the regular teachers' receptivity to change; communicating effectively with other teachers; analyzing the systems and appreciating the constraints under which she or he must work; analyzing the teaching style of the regular teachers; understanding the content of the curriculum well enough to make suggestions for changes in method of presentation or evaluation; analyzing textbooks used in the curriculum; suggesting, developing and modelling management strategies that could be used to accommodate students with different learning styles and skill levels; assuming the role of student-advocate in interacting effectively with the school staff on behalf of an exceptional student; communicating with parents about the program and helping them deal with their own expectancies for their child; and directing both parents and school staff to community agencies that provide support for special needs students.

Ideally, the resource teacher will have a supportive school staff to work with in both direct and indirect ways. However, many secondary schools do not have a commitment to individualizing educational programming for exceptional students, nor do they always have a well-trained resource teacher available to them. In the latter situation, someone from the guidance staff and/or a few subject teachers become the informal assessment and remedial team in the school. In the next section of this chapter we will look at assessment and remediation alternatives that can be useful to both special education and regular class teachers.

ASSESSMENT

There are very few commercially prepared tests that address the learning problems of secondary school students (Weiderhold and Ysseldyke 1978; Coles 1978; McIntyre et al. 1981). Clinics that serve the adolescent population do test for intelligence, academic, personality, psychoneurological and health-related factors; but the information obtained is of a more global nature than that required by a teacher who needs to develop a specific program. Generally, the most useful information comes from observation of behaviour and informal assessment of academic work. The checklists

Figure 12-2

Checklist for Secondary School Teachers to Use in Assessing Language Encoding Skills

Reading	**Yes**	**Sometimes**	**No**
1. Reading is mechanical.			
2. Guesses words based upon a few letters.			
3. Reads unevenly.			
4. Reads past mistakes without attempting to correct errors regardless of meaning.			
5. Reads very slowly, sounding out words as he goes.			
6. Repeats words, loses place, goes back to find place.			
7. Unable to blend sounds together to get to words.			
8. Moves lips during silent reading.			
9. Does not seem to understand what he or she has read despite being able to read fluently.			
10. Comprehends what others read to him better than what he reads to himself.			
11. Does not read willingly.			
Listening			
1. Does not seem to listen to instructions.			
2. Does not attend to what is happening in class.			
3. Seems to misunderstand language.			
4. Cannot repeat instructions.			

illustrated in Figures 12-2, 12-3 and 12-4 indicate the types of behaviours to look for in language decoding areas (listening and reading), language encoding areas (speaking and writing), mathematics and organization.

These checklists can be used in the initial assessment of a student, during the year, and again at the end of the program to determine change and persisting deficit areas. For each of these checklists, behaviours can be added that reflect the specific goals of a course or subject. For example, in an English class that devotes considerable time to pragmatics (that is, helping students become aware of the appropriate use of language in academic and social settings), the teacher might want to assess his or her students' skills in five major areas: speech acts, sentence utilization, presuppositions, information organization, and conversational constraints or rules of conversation (Kretschmer and Kretschmer 1978). Speech acts facilitate ongoing communication. They include: telling (a declarative speech act), questioning (an interrogative speech act), requesting (an ordering speech act), feeling (an expressive speech act), and promising (a committing speech act). Sentence utilization refers to the intended purpose of the sentence produced; for example, in "Who took the ball?" the intention is to get more information, while in "Did Paul take the ball?" the intention is to verify an hypothesis. Presuppositions involve the assumptions the speaker makes about the listener and the situation in terms of knowledge, attitude and appropriateness. Many exceptional students have problems with several of these speaking skills that influence their overall ability to express themselves well and to communicate with others.

We have spoken often of the need to view the secondary school student as a whole person as well as in terms of his or her specific strengths and weaknesses. One very important aspect of this is careful consideration of affective areas such as interests, motivation and learning style (Zigmond et al. 1983). Emmans (1984) has developed both a checklist on learning styles for teachers to complete after they have observed their students and an interview form on interests and motivation that the student can complete either independently or with the teacher. Items on the learning style checklist include preferred size of instructional group (large group, small group, pairs, individual without teacher, or individual with teacher); preferred mode of instruction and response (oral, written, demonstration, combination of the three); and preferred frequency and intensity of teacher contacts (physical proximity, frequency of contact and frequency of feedback). For each item, teachers indicate whether the student works to criterion under that condition. Checklists can be developed that allow the teacher to use the same sheet over several observation periods (i.e., first term, second term, third term). The student interview form asks students to indicate both general and specific interests in eight categories: entertainment (TV shows and movies), music (favourite type, band, rock video), computers (ones used, favourite game), sports (team and individual), food, books (types, writers), magazines and games. Motivation is assessed through the form illustrated in Figure 12-5.

Using these or some other methods to assess the student's learning preferences will lead directly to both instructional and content recommendations. Regular class teachers might want to use this approach with their entire class to determine the optional topics that could be developed in their subject.

Figure 12-3

Checklist for Secondary School Teachers to Use in Assessing Language Decoding Skills

Writing	**Yes**	**Sometimes**	**No**
1. Does not organize ideas into meaningful paragraphs.			
2. Does not punctuate correctly.			
3. Sentences are incomplete.			
4. Reverses letters in a sequence, e.g., girl = gril.			
5. Spells phonetically.			
6. Erases, crosses out, messes up work with scribbling when it does not please him.			
7. Does not write between lines on the paper, indent lines, write neatly.			
8. Written work seriously deteriorates when under pressure of time testing or when work is long or demanding.			
9. Work shows poor placement on a page.			
10. Avoids written work.			
Speaking			
1. Does not articulate clearly and understandably.			
2. Pronounces ending sounds in words incorrectly.			
3. Has a tendency to confuse words he hears: Propane = profane.			
4. Speaks quickly, nervously, thus is hard to follow or understand at times.			
5. Answers questions tangentially and has difficulty getting to the point.			
6. Has difficulty finding the correct words when speaking.			
7. Interrupts himself when speaking, distracts self and changes the subject, is fragmented and disorganized.			

Figure 12-4
Checklist for Secondary School Teacher to Use in Assessing Mathematical and Organizational Skills

Mathematics	**Yes**	**Sometimes**	**No**
1. Does not understand place value of number.			
2. Has difficulty in spatial concepts and measurement.			
3. Has difficulty with time concepts, seems unaware of passage of time, is always late.			
4. Understands borrowing and carrying in math.			
5. Can remember math facts and recall them automatically.			
6. Language problems cause difficulty.			
Organization			
1. Follows through an assignment in an organized fashion.			
2. Follows through but becomes disorganized and fails to complete work.			
3. Appears lethargic or apathetic, yawns, appears bored and without energy.			
4. Oral performance far excels written performance.			

The questions on computers are of special significance given the emphasis both parents and educators are putting on them. For many special needs students, computers can be used as ways of motivating them to do basic remedial work and of helping them develop good cognitive and problem-solving skills. One of the important caveats, however, is the difference in the effect computers have on male and female students. Collis (1984) has developed a 42-item questionnaire for secondary students that looks at their attitudes towards computers. She has found that students respond to computers in a personal, emotional way and that females are less likely to view themselves as computer users or to show much confidence in their own abilities in using computers. This finding holds in spite of the fact that these same females do use the computer in school and at home and that they themselves state that females in general have as much ability as males with regard to computer use. Thus, although females use computers and feel they are capable of using them, they

Figure 12-5
Survey of Individual Student Motivators

Name: ____________________ Date: ____________________

Instructions:

1. Read each of the following phrases carefully.

2. You are to think about how much each of the items listed would improve your work habits.

3. Use the following rating system:
   3 = a lot **2** = somewhat **1** = very little
   (a) stars on my work ________
   (b) tokens ________
   (c) graphing or charting of my progress ________
   (d) success-o-grams sent home ________
   (e) free time to pursue hobbies or other interests ________
   (f) positive comments from my teacher ________
   (g) work that is interesting and makes me think ________
   (h) topics that I have chosen to study myself ________

Adapted from D. Emmons, "The LD Adolescent," unpublished, University of Victoria, 1984.

do not have much individual liking for them or confidence in themselves as computer users.

Another issue that is important in the assessment of students with learning problems is the manner in which regular teachers share information they gather in class with other subject teachers and with the resource staff. The progress report illustrated in Figure 12-6 has been used successfully at a local secondary school three times a year to provide ongoing communication among all the educators involved with each exceptional student. It encourages all the persons involved to look carefully at the individual student and to summarize their information regarding the behaviours listed. When each of the significant persons involved with the student uses the same form, it makes it easier to reach consensus about the student's needs and about program modifications.

The above approaches to assessment should provide a sense of the student's needs. The next section will look at how these needs can be addressed in the regular and resource classrooms in the secondary school.

Figure 12-6
Student Progress Report

Student: ______________________ Grade: ______________________

Subject: ______________________ Teacher: ______________________

Please observe this student closely in your subject area in order to assist the Special Education Committee in appropriate programming. Thank you.

Does this student:

	Yes	Sometimes	No
1. Follow subject instructions	____	____	____
2. Participate in class	____	____	____
3. Attempt assigned homework	____	____	____
4. Conform to classroom regulations	____	____	____
5. Behave appropriately	____	____	____
6. Complete classroom assignments	____	____	____
7. Listen and attend	____	____	____
8. Display effort	____	____	____
9. Relate well with peers	____	____	____
10. Appear organized	____	____	____

Test or assignment results to date: ______________________

Specific difficulties: ______________________

____

Additional comments: ______________________

____

____

Adaptation of the form used at Essex Secondary School.

REMEDIATION

Remedial approaches will be discussed in terms of strategies that are general in nature and relate to the entire curriculum and in terms of those strategies that are subject-related.

General Approaches

Biochemical One of the general remedial approaches used with elementary students and often continued with secondary students is the biochemical approach. This may involve drugs, diet or both. While this approach does improve the attention and availability for learning of some students, there have been cautions about continued use of drug therapy into adolescence and ambivalence about its use as a major remedial strategy (Amen 1980; Gadow 1983; Karale 1983). Bornstein and Blowin (1978) found that adolescents who had been taking Ritalin (methylphenidate) were more prone to alcohol and drug abuse five years later. Kershner and Grekin report no gains by students on megavitamin therapy. They did report gains, however, by adolescents who were put on a nutritious diet that reduced sugar and useless carbohydrate intake. The gains showed in home discipline, attention span, reduced aggressiveness and hostility, and peer relationships. There were also significant gains in WRAT reading scores and the verbal portion of the WISC-R. It is not always easy to convince adolescents of the importance of nutritious food, but it seems especially important to inform exceptional students about nutrition and support them in attempts to change their eating habits.

Cooperative Goal Structure An approach to classroom instruction that increases both interaction and achievement among normal and exceptional pupils is the cooperative goal structure model. Johnson and Johnson (1980, 1983) have conducted many studies in which they placed handicapped and non-handicapped children in cooperative groups to perform a broad range of tasks. Their results have been very successful. Both increased productivity and positive attitude change were shown by students in these cooperative groups as compared to students in more traditional classroom programs. They suggest that association with a broad range of peers is necessary for all students in order to maximize achievement in both social and cognitive areas. They state eight specific ways in which peer relationships influence social and cognitive development.

1. Contributing to the socialization of values, attitudes, competencies and ways of perceiving the world.
2. Being prognostic indicators of and important influences on future psychological health.
3. Teaching the social competencies necessary to reduce social isolation.
4. Influencing the occurrence or non-occurrence of potential problem behaviors in adolescence such as the use of illegal drugs.
5. Providing the context in which children master aggressive impulses.
6. Contributing to the development of sex role identity.
7. Contributing to the emergence of perspective taking abilities.
8. Influencing educational aspirations and achievement. (1980, p. 91)

Tutoring Tutoring is another example of effective interaction among students (Curtis 1982; Wiens 1980; Gonick, Macdonald and Scarth 1982). Csapo (1976) paired six students whose reading levels were three to five years behind grade level with tutors for a ten-week remedial program. At the end of the period, both tutors and tutees had improved in reading, both groups had increased the number of positive verbal interactions with each other, both had shown improvement in neatness and in meeting deadlines, and both had shown a reduction in adjudicated delinquencies. These effects were significant but, unfortunately, were not followed up for maintenance.

Buddy Model Others have used a buddy model to pair up students with a wide range of skills and skill deficits; for example, socially astute students with socially inept ones, good readers with poor readers, responsible students with irresponsible ones. These matches can provide good models and improve the social skills of exceptional students, but they must be carefully planned, involve frequent meetings, and build in reinforcers for both students if they are to be successful.

Contractual Teaching Contractual teaching is often successful with adolescents. It helps to establish responsibility, incorporates the estimations of the student of her or his own abilities and commits her or him to courses of action. Weber (1982) suggests that establishing a contract should be part of a carefully thought through individualized program. He suggests that establishing a contract should allow these seven steps:

1. Determine the skills to be mastered.
2. Limit the objectives.
3. Prepare the materials for practice.
4. State what is to be accomplished in the contract.
5. Schedule the independent study working time.
6. Complete the progress chart.
7. Assess progress and start a new contract.

Weber further suggests the following six guidelines should be adhered to in setting up independent study programs:

1. The program must be perceived by the student to be an individual one assigned especially for him or her.
2. It is essential for each student to adhere to some form of contract, schedule, or list on which specific objectives are detailed. (Figure 12-7 illustrates a contract.) These should be limited and attainable within the foreseeable future. This is not the place for listing long-range goals, even though these may be the teacher's ultimate purpose.
3. A tangible record of accomplishment generates momentum. This can take the form of a log, a chart, a graph, or any system by which the student can record and, therefore, *see* his or her increments of accomplishment on a day-to-day basis.
4. The materials used in an individualized skill program ideally should be prac-

tice materials which a student can complete independently or, at least, with minimal direction. Instruction should be brief, efficient and done on a one-to-one basis.

5. Timing must be down-graded in importance. Low-achieving adolescents often dismiss, reject or avoid an activity because experience has taught them that they will need more time to complete it than they will probably be allowed.
6. An individualized program should take up only part of the time together. There must also be time for whole-class activities and for teacher-directed lessons which contribute not only to general learning but also to the feeling of belonging to a class or group—an attachment which is vital to the morale of all adolescents.

Subject-specific Approaches

The suggestions discussed above apply to a restructuring of the entire program for all or some students. The following suggestions apply to specific subjects or topics.

Mathematics Remediation in mathematics must be sensitive to the realities of matching ideal competency with reasonable expectancies. It must also be sensitive to the reasons for poor skill development, especially those that relate to the attitude issues discussed in Chapter 7. Cawley (1978) suggests six principles for establishing minimal guidelines for instruction in mathematics at the secondary school level.

1. Programs must include instructional procedures and alternatives that reduce the influence of age-in-grade relationships on curriculum and material development.
2. Individualized programs must be based on the principle that successful achievement facilitates learning.
3. Programs must produce alternatives to the fixed-frequency system. If students are to use texts and workbooks (examples of fixed-frequency materials), they need to be given the extra time and practice necessary to master the concepts being taught.
4. Programs must be capable of providing the teacher with qualitative options. Most commercially prepared materials must be supplemented in order to facilitate the student's mastery of the topic.
5. Programs must be capable of providing instant feedback to both the teacher and student so both know where the student stands in terms of understanding the concept being taught.
6. Programs must be efficient in terms of the learner's and teacher's time. Low-achieving high school students have no time to waste in the effort to bring their skills up to an acceptable level.

With these general guidelines in mind, teachers can look to programs available to them and modify them to suit the needs of the individual students. Resource teachers, who usually have access to a wider range of materials than the regular teacher, may provide the supplementary material that will make it possible to individualize a program in terms of time and quality of input. Some additional methods and materials would include the one-minute precision teaching of skills in multiplication, using coloured pencils to distinguish parts of geometry diagrams, conducting oral discus-

Figure 12-7
Contract for an Independent Study Program

Independent Study Program for

Brenda M. ____________________

I agree to improve my performance in the following areas:

1. Be punctual and ready for math class.
2. Review fraction and ratio concepts to mastery.
3. Do any word problems using these concepts correctly.

I will complete the following work during my independent contract period. Each successfully completed exercise will be initialled by my teacher. Punctuality will be checked daily by the teacher.

1. Exercises 1 through 15 (ratio and fractions).
2. Problems listed on pages 20, 26 and 31.

Each completed exercise and correct problem will count for one point.

____
Signature

______________________________ ______________________________
Date Teacher's Signature

Adapted from K. Weber, *The Teacher is the Key: A practical guide for teaching the adolescent with learning disabilities* (Toronto: Ontario Ministry of Education), 44.

sions of geometry solutions before writing out answers, using summary charts for ever-increasing chunks of a math course, using calculators as a check for computation accuracy, taking more time for exams, and having fewer questions per page.

For many adolescents with math problems, basic computational skills remain the weak link in developing higher order skills. Observing student work and analyzing it through the hypothesis-probe technique described by Zigmond et al. (1983) helps the teacher understand the extent of the student's difficulties and some of the reasons for them. Table 12-1 shows this process in use.

Another suggestion that can be implemented for all students is presenting new concepts first in a concrete or familiar form where possible so students see how a new formula relates to situations they have experienced. Using materials that are very meaningful to most students (e.g., money) to explain decimals, or using lending and borrowing to explain positive and negative integers can also help students make sense of math. Encouraging students to read the mathematics text, controlling for

Table 12-1 **Understanding Error Patterns in Mathematics**

Error pattern	Hypothesis	Next step
Many errors, careless and/or random	Unmotivated	Re-survey with appropriate motivators
Many errors, careless and/or random	Survey too difficult	Re-survey down the math hierarchy
Few errors	Survey too easy	Re-survey up the math hierarchy
Wrong operation	Cannot interpret process words or signs	Probe meaning of computation signs or words
Computation errors	Does not know basic facts	Probe basic computation facts
Faulty algorithms	Does not know procedures	Probe down computation hierarchy to check prerequisite skills
No understanding of concept; particular random responses, item left blank	Never taught	Probe down to test prerequisite skills

Reprinted, with permission of the publisher, from N. Zigmond, A. Vallecorsa and R. Silverman, *Assessment for Instructional Planning in Special Education* (Englewood Cliffs, N.J.: Prentice-Hall, Inc., 1983), 224.

the complexity of the language in the text for poor readers, and relating problems on the board to problems in the text, all help the students to use the materials more effectively (Vance and Baumann 1981).

Reading Sitko and Gillespie (1978) point out that many general problems exceptional students have with language affect their reading skills. These problems include a) limitations in their acquisition and comprehension of syntax, which is most evident in question sentences and sentences with demonstratives and/or possessives, and in relationships with direct and indirect objects; b) quantitative delays in the acquisition of morphology especially in regard to third person singular, possessives and adjectival inflections; c) deficits in comprehending logical relationships between two or more elements; d) deficits in handling analysis, synthesis and generalization demands of language; and e) deficits in longer sentence recall and understanding and in using complex sentence structure. Each of these deficits continues into adolescence and aggravates problems in all areas because of the high reading demand at the secondary level.

One of the most practical approaches to helping students with reading problems is to look very carefully at their vocational goals and secondary school graduation requirements and then counsel them to avoid courses with a heavy reading load where possible. Once they are in a course, make reading lists available at the beginning of the course so they can plan their reading time over the year to take advantage of slack times. Read orally in class while the students follow along in their text and encourage students to listen to tapes of novels or other text passages. Parents, friends and teachers can be asked to tape record materials so there is a good supply of materials available in this form. Be aware of the reading demands of your subject. In a recent analysis of three grade nine science texts using the Fry Readability Index, we found the readability range extended from grade seven to first year college. Texts at the upper range might be appropriate for higher level students, but they are certain to cause difficulties for most average and slower students in a grade nine science class. If students buy their own text, encourage them to use highlighter pens to emphasize main points, key words, etc. They will probably need help initially from you and/or other students to do this effectively, but it does provide a good summary of material later. Use a reading record so both teachers and students can see what has been read and what still needs to be read (Vance and Bauman 1981). And, finally, encourage reading for pleasure—reading anything is better than not reading at all!

Writing It is expected that formal writing skills (handwriting, grammar, spelling, composition) have been mastered at the elementary level; however, many students with learning problems perform at their worst in any subject that requires them to write. Alley and Deshler (1979) state that students with learning problems write at a slower pace, are poorly organized, have a limited word pool to draw from, show a high frequency of mechanical errors and show limited ability to monitor their own writing performance. Their attitudes, skills and creative input each need to be assessed and remediated. Interviewing students directly about their feelings towards writing and their opinions about how writing can be best taught to them encourages the students to think about their own approach. It also gives the teacher valuable information around which to plan a program. Using the grid structure described in Chapter 8 for increasing creativity in writing can lead to improvements in students' written work. General enthusiasm, providing lots of opportunities for writing and providing positive feedback on efforts are also very effective. Many of the specific techniques discussed in the chapters on writing skills and organization are applicable to writing remediation for the secondary student. Some students may find that it is useful to dictate first drafts of a report to a parent or other willing person because this allows them to see their ideas in print (properly spelled and using good grammar) in order to reorganize the material and prepare the final copy.

Handwriting may also be a lingering problem for some students. Hagin (1983) points out that teachers tend to give higher grades to papers with clear handwriting and that even with the advent of word processors with correction programs, most students will rely on their own handwriting skills for the majority of their writing lives. Because superiority of cursive over manuscript has not been proven, she suggests that students be allowed to use whichever form they function best in, but that they consider the compromise she suggests since it looks like writing and thus meets some of the self-concept concerns of students who identify writing with maturity. She has suggested a handwriting technique that combines manuscript letters with vertical downstrokes with connections between the letters. The following is an example of this approach:

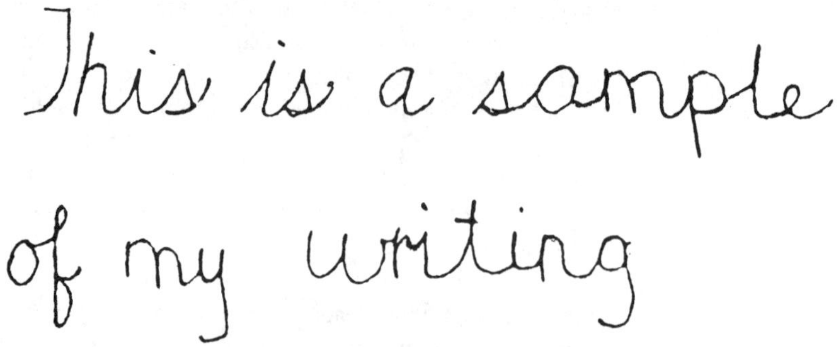

Cognitive Skills Many of the suggestions described in Chapter 10 can be applied directly to the adolescent. Martin (1984) has used Feuerstein's instrumental enrichment program with hearing impaired adolescents and found that they demonstrated improvement in systematic approaches to problems, analysis of problem situations, vocabulary size, analysis of source of error in problem-solving situations, completeness and planning in problem-solving situations, peer cooperation in problem-solving, abstract thinking, precision, and development of multiple strategies to solve problems.

Other approaches to improving the cognitive strategies of secondary school students include providing opportunities for input from various modalities (auditory—Barnes 1982b; visual—Vance and Bauman 1981); encouraging students to talk through conceptual material until it is clear to them and to the listener; helping students focus on the problem at hand; meeting privately with students to discuss their ideas; and using the Socratic method of teaching wherever possible.

Social Skills One of the important requirements for personal survival at the secondary level involves basic social competence. Many exceptional students do not show an awareness of the perspective of "the other" in their attempts to communicate. Kronick (1983) has listed 19 skills in this area that the teacher should assess and provide practice in. The student who has these skills

1. identifies and names another person's feelings;
2. expresses his or her own feelings in an appropriate manner;
3. possesses alternate behaviours for expressing emotion;
4. brings others around to seeing his or her point of view;
5. empathizes with the point of view of others;
6. can imagine and role-play social events;
7. notices non-verbal expressions of mood and emotion;

8. respects physical space of others, uses own body appropriately;
9. uses clues about others (clothes, grooming, posture, etc.) to understand them and to behave appropriately;
10. understands the social meaning of house, office and store layouts;
11. understands single and multiple relationships among people—grandparent, boss, doctor, niece, etc.;
12. understands average capabilities at each stage of life;
13. can tell time, days, weeks, seasons, holidays, their sequence and the behaviours associated with them;
14. shares in turn-taking in conversations;
15. expresses gratitude to those who have put themselves out for her or him;
16. knows when to stop talking and talks appropriately (on topic, moderate pitch, suitable vocabulary);
17. recognizes the impact of his or her behaviour on others;
18. can plan an experience with a friend;
19. can function independently when appropriate.

Remediating these skills is primarily a matter of recognizing the deficit and providing opportunities to practice the skill. A program that focuses on positive interactions among peers is the *Positive, Encouraging, Enjoyable Relationships Program* (PEER) developed by Peterson (1972). Campbell (1984) describes its use with socially maladjusted adolescents in a resource program. It has the advantage of focusing attention on positive behaviours, ignoring negative behaviours and transferring to interactions with students in regular classes. There are three steps involved.

Step 1: Asking the students three questions

1. How many people in this room would be willing to help someone else to be the best person he or she can be? (Usually all want to.)
2. How many people in this room can think of some habits you can change to help yourself be a better person? (Usually all will raise their hands if the person asking the questions raises his.)
3. Would you like to know of a way we, as a group, can help one person at a time be more the person he or she would like to be? (Usual response is "yes.") If a yes response is not forthcoming, abandon the program temporarily and set it up for success the next day or as soon as possible. The purpose of the three questions is to: a) lead the students into a cooperative effort; b) set a desire for helping in the minds of the students; c) elicit the desire of individuals to become better; and d) get a class commitment to helping and not hurting members.

Step 2: Describing how the class can help one person at a time.

1. Choose a "focus person." This person *must* be a volunteer. At the beginning of the program, it may be advisable to pick a better student.
2. The volunteer makes out a list of the good things that he or she is doing, or good things he or she wants to do.
3. Explain to the class why the focus is only on positive things. Point out that a)

negative responses tend to hurt or make a person feel bad; b) attention to negative behaviour tends to encourage the behaviour; and c) the class aim is to encourage or reinforce good behaviour which will make the person feel good.

4. Explain how class focus time will work. The focus person will have a list of good things he or she does posted on the notice board. During class focus time, the focus person will go up to this list and the class will tell him or her good things they have noticed him or her doing. The focus person will place a mark behind each good thing that is stated.

   Class focus time will be of two minutes duration and will be held five times on Monday, four times on Tuesday, three times on Wednesday, two times on Thursday, and once on Friday. Students are reminded that the goal is not points, but changed behaviour.
5. Self Focus. After one student has been on class focus for a week, his list will be transferred to his desk. During class focus time, the person on self focus will score himself as he perceives and remembers his own behaviour.
6. Box Focus. All former focus persons put their goals (one each) on slips of paper with their names on them, and put the slips into "Box Focus" (a box at the front of the room). Each individual pulls a goal and reacts to it in terms of whether the behaviour was demonstrated or not.

Step 3: Putting the program in operation.

1. Pick a focus person who has volunteered. At the start of the program, it is often best to pick a good student because a) it is often easier for the class to notice good behaviours in a good student; b) it gives the class the chance to develop skills of looking for good things in other people; and c) many misbehaving students tend to do better when they have had the chance to observe someone else as a focus person. A second person can be picked to be the focus person for the following week and to be in reserve in case the focus person is absent.
2. A focus person can help to keep the game honest by stating he doesn't feel he should get a point if he hasn't done the behaviour. (He could then be reinforced for being honest.)
3. It is very important that the focus person receive reinforcement if he or she is doing well. The class may need some help in looking for positive things. The teacher may need to reinforce students who are reinforcing the focus person.
4. Ideally, the focus person will be able to state his or her own goals. Class discussions on good behaviour may help students think of goals for themselves. Sometimes a list of goals posted in the classroom will help. The teacher may help the student write his or her goals so that they are specific, recognizable and attainable.
5. Once a person has been on class focus and self focus, he or she then is eligible to put a goal in Box Focus. Goals are drawn out of the box randomly during the week. The goal is then read and the class tells the person positive things they have seen him or her do to work towards that goal.

This program can be used in the resource room or in the regular class if the teacher feels that many of the students would benefit from the experience. The program as described above is more likely to be successful in grades seven through nine. However, the concept could be applied in higher grades in situations in which the teacher would like to change a negative dynamic in the class or to capitalize on a positive support system already in place. Sociology courses, some English courses (especially those discussing novels with adolescent protagonists) or guidance seminars are likely courses for implementation.

Alley and Deshler (1979) have suggested the following seven guidelines for helping adolescents attain social skills through personal interviews between students and teachers.

1. Provide an authentic climate of trust and warmth.
2. Be aware of value differences between yourself and the students.
3. State clearly your limits of confidentiality. If you feel you would need to share information that could be harmful to the student or others, tell the student this before you begin an interview. Most adolescents will still tell you everything, but this clears the air and does not compromise the general spirit of confidentiality.
4. Refer students whose problems are beyond your competence to other professionals.
5. Actively listen to the student.
6. Take notes for later reference and keep cumulative records on the student.
7. Give students feedback using plain language with facts and observations; be specific; try not to alienate student; avoid words like "right," "wrong," or "you should"; don't give your opinion unless asked; and end each session with a positive remark.

Other ways to help students realistically appreciate their own strengths and weaknesses involve pointing out specific indicators of them (especially strengths), discussing information with the student before seeing his or her parents, letting him or her know others have similar problems (and that some are even worse off), and focussing on specific behaviours that can be modified rather than on personality. A technique that can help a teacher start exploring and helping a student with limited social and emotional strengths is to have a student write out his or her first name vertically down a page and then put down a personal characteristic for each letter. For example:

*P*utting off work
*A*bility OK
*U*nmotivated
*L*azy

This would let you know quickly how aware the student is of his or her own characteristics and would be a good place to begin discussing alternatives. Other ways of doing this would be to ask the student to only list good things about himself or herself. You could also work on the list together (this might be necessary for many exceptional students who have not spent time thinking about themselves in a conceptual way).

Evaluation One of the most contentious issues at the secondary level is how to evaluate an exceptional student fairly. Vance and Bauman (1981) suggest that before deciding on a flexible evaluation approach, teachers should ask themselves the following questions:

1. Does the method of evaluation match the method of teaching? For example, it would not be appropriate to ask multiple choice questions in a course on drafting skills that primarily involve demonstration and performance.
2. What skills, beyond the content taught in the class, are required for success in the subject area?
3. Are time and place limits essential to evaluation?
4. Are technical writing skills necessary to demonstrate competence in the subject (i.e., how important is spelling, good sentence structure, etc.).

Once these questions have been explored, evaluation options can be determined. The most frequently used ones include allowing a student to write his or her tests in the resource room with extra time available; allowing someone else to read the question to the student or having the test on tape to help a slow reader; allowing the student to answer in point form or on tape; conducting an oral test individually or with small groups of students; allowing the student to hand in a project typed by someone else and/or using pictures and graphs to illustrate a point rather than verbal discourse. More specific evaluation techniques can be determined with the student and teachers involved.

The suggestions given above are not exhaustive by any means, but hopefully they will provide the practicing teacher with a place to start as he or she attempts to modify the secondary program to accommodate the exceptional adolescent. The following case study will illustrate some of these techniques.

Case Study

Step 1: Background Information

Craig is a grade nine student enrolled in the general four-year high school program. He was referred to the school team because of low achievement in English, social studies and mathematics on his first term report card. His report card stated that he had missed seven days of school and was late eight times. He does not demonstrate good work habits, wastes a lot of time in class, and hands in poor written work.

Step 2: Preliminary Hypotheses

1. Craig's academic work is poor because of poor work habits and poor attendance.
2. Craig has difficulty with written work.
3. Craig has average ability.

Step 3: Individual Testing

In order to determine whether the hypotheses developed in Step 2 to explain his poor performance were accurate, Craig was given the WISC-R to determine his intellectual ability and the WRAT to get a quick estimate of his present academic levels. In addition, he was asked to write a story about anything that came to mind. That story was subsequently analyzed for technical accuracy and overall goodness. Intelligence testing was in the normal range with strengths in general comprehension and social problem-solving and weaknesses in recalling information facts, arithmetic and comprehending social situations. His WRAT scores were at the early grade ten level in reading, mid-grade six level in spelling, and early grade seven level in mathematics. Craig's story went as follows:

> It was the championship game. The score was 1 to 1 and time ran out. The game went into overtime. The frist gole would win it. It went back and forth. The frist overtime period was over so we had to go into the second overtime period. Finally the other team scored. I was very disappointed.

His story showed adequate overall organization, flowed logically, and had an appropriate beginning and ending. The sentence structure was immature, however, relying primarily on simple and compound sentences. There were three spelling mistakes, two reflecting reversals and one a good phonic equivalent. There was good use of nouns and pronouns reflecting a reader-based approach. The handwriting and spacing of the work on the page was poor and the length was less than would be expected given the time available.

Step 4: Preliminary IEP Goals

1. Craig will improve his attendance and punctuality.
2. Craig will write something each day.
3. Craig will improve his writing skills to grade level.
4. Craig will achieve at grade level in all subject areas.

Step 5: Assessment

Craig was observed in several teaching situations using the Emmon's checklist described earlier in the chapter. He was also observed in interactions with other students. His preferred motivational supports were also explored using the checklist illustrated in Figure 12-5.

Step 6: Final IEP Goal Statements

At the end of the second term, Craig will be able to:

1. spell at a grade nine mastery level as measured by special language diagnostic tests;
2. write compositions of appropriate length;
3. attain a 95 percent or better general attendance record (without any subject less than 90 percent full attendance) for the period of time designated by his IEP;
4. develop a more acceptable range of social skills as indicated by his willingness to participate in group activities and by his level of interaction with peers;
5. improve his general academic performance by learning a variety of meta-strategies for organizational and work/study skills; and
6. improve his ability to concentrate on assigned tasks by adopting behavioural modification techniques to increase on-task behaviours.

Step 7: Remediation

Craig was to be scheduled for resource one 70-minute period every two days. He would continue in English and social studies but drop math the second term.

A behavioural contract was set up for Craig with the following specific objectives:

1. For each week of perfect attendance, Craig would receive points that could be exchanged for free time activities. The terms were negotiated in a written contract that also indicated points earned for less than perfect attendance. The objective was to increase attendance to 90 percent overall (with no subject area under 80 percent) by March 15, and to the full criteria stated in Goal 2, above, by May 15).
2. According to the terms of a contract negotiated with his parents, Craig would earn points toward the purchase of new records on the basis of a graduated plan for improved school attendance similar to Objective 1. The criteria for success would be made identical to the performance criteria in Objective 1.
3. Craig would be expected to keep a notebook in each subject area. The teacher would mark the notebook weekly and make specific recommendations for improvement.
4. Craig would put into practice studying strategies suggested by his teachers.
5. Craig would study 30 minutes per week in each subject area in the library with the assistance of a classmate with good study habits.
6. Craig would earn points towards listening to music in a special area according to a contract negotiated between Craig and his teacher for improvement in on-task behaviour. The criteria for success would be a 50 percent increase in overall on-task behaviour. This would be calculated on the basis of the total in-class time allotted for work activities by May 15.
7. Craig would cooperate on team projects with one other student.
8. Craig would cooperate and interact in group projects involving small numbers of students.
9. Craig would be given positions of minor responsibility in small group work.
10. Craig would express his own opinions in small group discussion.
11. Craig would respect fellow classmates' right to express opinions and would listen to them.
12. Craig would be introduced to the grid approach to creative writing and would write a story every day.

Step 8: Evaluation

1. The Resource and English teachers were responsible for administering a minimum of two monthly informal diagnostic tests to ascertain Craig's progress in the writing area. Other teachers (i.e., history, science) were responsible for developing some informal measures of Craig's progress in their subjects.
2. There was a monthly review of Craig's progress in a meeting of all his teachers using these progress reports.
3. Formal standardized diagnostic tests of achievement skills were administered in the first week of May and were used in review sessions to evaluate the IEP program in mid-May.

By May, Craig was attending 96 percent of his classes and had shown marked improvement in study skills and peer involvement. His spelling skills were now at the grade eight level and he wrote compositions of approximately 200 words in length. It was recommended that he stay in the resource program in grade ten to consolidate the gains made and to obtain remedial help in mathematics, but that he pick up a third subject to catch up with his peer group.

13 A CONSULTING ROLE FOR THE RESOURCE TEACHER

Rey A. Carr and R. Vance Peavy
UNIVERSITY OF VICTORIA

While resource teachers will be expected to spend the majority of their time in direct instruction of students, they will also be expected to provide assistance to other adults in assessing, teaching, counselling and planning for the education of special students. In order for this adult interaction to enhance the education of students, it is important for resource teachers to learn the methods and skills associated with consultation—a planned and systematic form of helping others.

In this chapter we present a collaborative, interpersonal model of consultation which we believe is especially useful to the resource teacher. The model, which we will describe in more detail later in this chapter, is a systematic and comprehensive way for resource teachers to work cooperatively with other adults to achieve mutually important educational goals. It consists of seven distinct steps (see Figure 13-1) which enable two or more people to work through a concern or challenge to an effective solution(s) while at the same time enhancing their working relationship.

Figure 13-1
Collaborative Consultation Model

1. Establish a consultation relationship
2. Identify and clarify problem/challenge
3. Set goals
4. Develop ideas and strategies
5. Develop a plan
6. Create an action plan
7. Empower the consultation relationship

In order to get a preliminary idea of what we mean by collaborative consultation, let's listen to a conversation between Michelle and Sabina, a regular classroom teacher and a resource teacher who work at the same school. We have placed numbers in parentheses following some of the statements in the dialogue in order to refer back to these statements later in this chapter.

Michelle: "Sabina, you mentioned on Monday that you were having trouble with one of your students and wanted to talk to me about it. Maybe we can do that now." (1)

Sabina: "Yes, you see I've got a student in one of my classes who seems to be really isolated from the other kids. He is absent a lot. I'm not sure whether his being absent is related to the fact that he often gets teased by other kids or maybe because of other things—I just don't know. I would like to see him cut down on absences and I'd like to see him interact with other kids more."

Michelle: "OK, so your concern is whether or not this student is missing school a lot more than he should mainly because of his social relations with other kids, or whether something else is going on with him, is that it?" (2)

Sabina: "Yes! Actually I'm afraid that he is going to quit school. I think that one of these days he's just going to turn around and say he's not coming back—or just disappear."

Michelle: "Um hm."

Sabina: "I can just feel it coming."

Michelle: "And you want to avoid that?"

Sabina: "Yes, definitely!"

Michelle: "Okay, we can consult on this problem now. In trying to come up with some ideas to solve it we may get further if we go through some specific problem-solving steps. First, we will have to make sure that both of us see just what the problem is, at least as best we can. Then, maybe we can pin down some goals you'd like to achieve. After that, we can look at some different ideas, different things that might help to solve the problem. Once we do that, maybe we can build a plan together that incorporates a solution for this student. (3) How does that sound?" (4)

Sabina: "Sounds great to me—I need some different ideas."

Michelle: "Fill me in a little on what the student has actually been doing in the classroom—what happens specifically that makes you think he is isolated." (5)

Sabina: "He always goes to the back of the room, etc. etc."

In this brief conversation, Michelle (as consultant) has already initiated a consultation process. Michelle has stated her willingness to consult (1), she has confirmed that a consultation is underway and by listening carefully, paraphrasing, and using questions judiciously, she has helped to clarify the issue (2). Further, she has indicated a set of problem-solving steps which she and Sabina (as consultee) can take to move in a goal-directed way through the consultation (3).

Before presenting the collaborative, interpersonal model of consultation, we will now discuss, in a more general way, the nature of consultation and briefly outline other types of consultation.

THE NATURE OF CONSULTATION

There is no single definition for consultation. The needs of persons in different contexts vary, and the consultative resources for helping individuals solve problems and meet their needs also are diverse. Primitive societies contained three prototypical consultative roles: the technological adviser, the healer and the sage (Znaniecki 1965). Technological advisers in primitive societies taught others new techniques and how to solve problems. Healers evolved in all known primitive societies. Their mission was to attend to the disabled and the sick using both practical and supernatural means. Modern day consultation is derived, in part, from both of these prototypes, at least in the sense that consultation is a problem-solving, goal-directed process and also in the sense that consultants often find themselves in an ameliorative role. The sage, on the other hand, was often sought out for his moral and political opinions. This service is more closely related to advocacy than to consultation. These two roles are best kept separate since one of the hallmark features of consultation is that the consultant's values and advice are *not* imposed upon the consultee.

Currently, the word consultation has various meanings depending upon the context in which it is used. For example, one friend may "consult" with another about what movie to attend. A mother may "consult" with her sister about her child's behaviour. Neighbours "consult" with one another about how to vote on a municipal by-law. Everyday usage usually implies an exchange of opinions and advice, and anybody qualifies as a consultant. All that is required is that one have some familiarity with the issue being discussed.

More formally, consultation may also be defined as advice given by experts in a special field. This usage implies that the consultant has been trained and possesses specialized knowledge, either about the consultation process *per se* or about the issue under examination. In this sense, a person may consult a doctor about stomach pain or a teacher may consult a school psychologist about a particularly intractable pupil. However, even this approach is not as straightforward as it might seem. The process by which the advice-giving consultation takes place—that is, the manner in which the consultant interacts with the consultee—may be closely related to whether the person pays attention to or acts on the consultant's advice. Therefore, the effective consultant role must include attention to the process as well as to the content of the consultation. Indeed, some advocates for consultation have gone so far as to say that the process is much more important than the content if the consultee is to achieve his or her own goals and aspirations (Carr and Grove 1982). Because of this emphasis on process in the helping role, the consultant's role is in some ways similar to the roles of teacher, supervisor and counsellor. For our purposes, however, there are some important differences in role which we wish to point out.

Teachers are clearly responsible for instructing and evaluating the progress of pupils. Teachers choose the topics to be studied and they select what seems to be the best method of teaching for the students whom they teach. Teaching is an impositional process in the sense that the teacher clearly guides the process and evaluates the outcomes of the classroom and learning activities. Pupils cannot normally choose to reject either the teacher's method of instructing or what is taught; at least, there are unwanted consequences for doing so.

In contrast with pupils, consultees are free to accept or reject either the process of consultation or the outcome. Consultees choose the matter on which they wish consultation and decide how the results of consultation are to be used, if at all. Consultation is a requested process, as distinct from one which is imposed.

Supervisors usually have the authority to insist, if necessary, that supervisees comply with advice. Supervisors tend to decide on which issues are important and should be discussed. Like teachers, supervisors are often in an evaluative stance with supervisees. Many supervisors do employ a cooperative style of working with supervisees. Nonetheless, supervision, like teaching, is by function an imposition process. This contrasts markedly with consultation which is, as we have just said, a requested process.

Counsellors diagnose and treat clients' personal problems. In contrast, consultants work with peers and focus upon work-related problems. Moreover, in counselling a client tends to focus upon his or her own concerns and troubles. In consultation, a consultee requests discussion on a problem or issue external to self. For example, a teacher may approach another teacher and say: "I want to talk to you about Danny's difficulty with reading and what I should do." This is a request for consultation, not counselling. Consultation stresses problem-identification, problem-solving and coping, while counselling stresses the personal dilemmas and feelings of the help-seeker.

The distinction between counselling and consultation is difficult to make in many situations for several reasons. First, an individual who is facing some work issue, such as what to do about a failing pupil, will have personal feelings as well as an "objective" situation to deal with. Second, the skills which make for effective counselling are also required for good consultation. For example, the consultant and the counsellor must both be active and good listeners, must know how to question effectively, and must grasp the method of problem solving within an interpersonal context. We will have more to say about this later.

The Role of the Resource Teacher

The resource teacher often must provide both a direct or indirect service to students in the school system. When the resource person is working with pupils on a one-to-one, small group or instruction (testing, etc.) basis, it is called a direct service role. However, when the resource teacher is working with any or all of the other persons concerned with a student, it is called an indirect service role. The resource teacher works with these people primarily to improve the learning of the student. For example, when the resource teacher serves as a member of a school-based team or attends meetings of other adults concerned with service to a pupil, he or she is doing so with the hope of helping the student (indirectly, by helping the other adults). Therefore, to ensure the likelihood of benefit to the pupil from these indirect service activities, it is crucial that the resource teacher contribute to identifying and using skills which will enable the interaction to help the student.

A resource teacher may perform teaching roles as in demonstrating for other teachers how to present material to visually impaired pupils. At times, a resource teacher may also perform a counselling role in helping a classroom teacher explore feelings toward a child whose behaviour generates frustration and inadequacy. However, it is the consultation role that the resource teacher is more frequently called upon to play and which we will now discuss in detail.

A MODEL OF THE CONSULTATION PROCESS

While several different approaches to consultation have been described in contemporary literature (Gallesich 1982), we have adopted the model originally developed by Stum (1981). Our choice has been based in part on the success we have had in using this model in our own professional practice and on the success that our students have reported in using this approach. We believe the model achieves three goals which are important in educational organizations: first, persons are helped to identify and mobilize their own and other resources; second, they learn to use an effective problem-solving process in thinking about concerns and challenges; and third, concrete, goal-directed, collaborative solutions are provided. It has been said that this model is illustrated by the parable that if you give a person a fish, they have food for a day; but if you teach a person to fish, they have food for a lifetime.

The resource room teacher acting as a consultant will work with many different people. These will include professional helpers from outside the school setting such as physicians and social workers. At the same time, the resource room teacher will interact frequently with school administrators, teachers, counsellors, psychologists and parents. These persons will often have different concerns about the same pupil—a situation which requires the special educators to be responsible to a variety of different styles and levels of expertise. Because this model has a strong focus on process, it enables the resource teacher to have a valued facilitative role for bringing different experts together to work towards a common goal. The interpersonal model of consultation allows the resource teacher to collaborate effectively and appropriately involve the variety of persons concerned about a pupil.

How does this model work? The model is comprised of seven distinct and sequential steps, each acting as a building block for the step that follows it. We have broken each step into substeps in order to detail the critical ingredients which must be included in order to successfully move on.

Step 1: Establishing a Consultation Relationship

The beginning of this chapter presented a brief dialogue between Sabina and Michelle which illustrated the first step of this seven-step model: establishing a consultation relationship. The first step included four components (and, again, the numbers in parentheses refer to that corresponding number in the dialogue): contact (1), a brief restatement of the consultee's concern (2), a description of the steps involved in the consultation (3), and a check with the consultee as to whether or not to continue (4).

This first step enables the consultant to develop interpersonal rapport. At the same time, the consultee learns the steps ahead so he or she is clear about the procedure involved. All too often persons in helping positions keep their helping procedures secret. This can create an additional problem for the person they are helping: alienation from the helping process. The first step of this model sets the tone for the educative and collaborative process as it enables the consultee to feel listened to and to choose to be actively involved. Sometimes this step can be minimized or eliminated if the two people using it have already become familiar with this process.

To recap this step, the consultant

1. requests a brief statement of the other person's concern;
2. restates or paraphrases the concern to the satisfaction of the other person;
3. briefly describes the steps (orally, although a chart outlining the steps should be visible); and
4. checks with the other person to see if it is appropriate to move on to the next step.

Step 2: Identifying and Clarifying the Problem

The dialogue ended with Sabina beginning the second step of identifying and clarifying the problem by asking Michelle for some background or history about the situation (5). The consultant may ask a variety of open-ended questions while at the same time encouraging the consultee to talk in depth. With the assistance of a consultant who listens in a nonjudgmental manner and probes in a gentle and sensitive way, the consultee can often redefine the initial concern into one which may have even greater meaning or import for him or her. The consultant's role at this point is to summarize and/or identify themes in the consultee's situation. In addition, the consultant asks the consultee to talk about the feelings or emotions associated with this situation. As is the case with other helping professionals, the consultant assists the consultee to identify the depth and variety of these feelings. Being able to discuss these feelings contributes to the motivation, commitment or involvement of the consultees to successfully meet the challenge or resolve the problem. For example, experienced teachers working with difficult and frustrating pupils can often feel discouraged, angry or alone. Although these feelings may be obscured, they may influence the consultee's willingness or ability to mobilize his or her resources to help a student. In this interpersonal consultation model, feelings play an important role in developing an effective strategy for managing the problem. The consultant can also help the consultee narrow the scope of the problem, divide it into workable parts or challenge incongruities in the consultee's presentation.

At the end of this step, the consultant checks with the consultee to see if it is appropriate to move on to the next step. In summary, the consultant

1. requests an elaborated statement of the background/history of the concern;
2. demonstrates understanding by summarizing, paraphrasing and asking appropriate open-ended questions;
3. inquires into feelings associated with the concern and actively listens to ensure understanding of the meaning the concern has for the consultee; and
4. checks with the consultee to determine whether it is appropriate to move to the next step, which is goal setting.

Step 3: Setting Goals

The step of goal setting is often ignored by other models and its inclusion here is one of the reasons we believe this model is both unique and effective. Up to this point, the consultee and consultant have been talking about the situation as it is. This step requires a shift to discussing the situation as the consultee would like it to be. Goal setting is essential in that it sets the stage for developing a plan of action. Goal setting enables the consultee to construct a holistic view of the problem under discussion. We see this step as focusing on the vision, hope and goals of the consultee, thus enabling the use of imagination and expectation in a concrete fashion.

The consultant asks the consultee to describe outcome goals; that is, the consultee is asked how he or she would like the situation to turn out or how he or she would like things to be. Because the identification of these goals will influence the success of later steps, it is important that the consultant focus here on 1) understanding the difference between process goals and outcome goals and 2) ensuring that the consultee is stating his or her own goals and not just making statements about what he or she would like the student to achieve.

In our dialogue between Sabina and Michelle, Michelle makes statements which could at first be seen as outcome goals: "I would like to see him cut down on absences and I'd like to see him interact with other kids more." These two "goals" are actually process goals; that is, they are behaviours which Sabina would like to see her pupil engage in for some, at this point unstated, reason. In other words, she believes that if her pupil demonstrates these two behaviours (interaction and decreased absences) then something important will happen for the student. It is the "something important" which is the outcome goal, and the consultant's role here is to help the consultee identify what that "something" is. This focus on outcome goals generally has at least one desirable result: it enables the consultee to identify a goal which often ties together many separate, or apparently distinct, process goals. For example, both of Sabina's process goals could be united in the outcome goal of improving social skills.

At this stage, the second thing on which the consultant should focus is ensuring that the goal is stated in such a way that it is clear the consultee is related to or has a role in achieving the goal. Let's look at an example. A resource teacher/consultant described a goal which a consultee stated to her in response to the question, "How would you like things to be?" The teacher, Jan, said "Fenton should improve his grades in all subjects." The consultant asked, "What do you see as your role in achieving this goal?" Jan replied, "I would like to create the opportunities for Fenton to improve his academic work, or at least, I'd like to know I had an influence on his academic skills." Thus, the value of this second focus is in targeting a realistic and desirable goal without the consultant having to make judgments about whether the goals are realistic or achievable.

Having stated the goals, the consultant summarizes the discussion to this point, enabling the consultee to clarify or consider the goals. The importance of this step cannot be overstated because often people are not used to thinking about problems or challenges in terms of goals. Therefore, it may take some time to identify and clarify personally relevant or meaningful goals. However, consultees often report that this stage gives them a different perspective on the concern which they initially stated. They often feel more hopeful, encouraged, enthusiastic and determined—almost as if a burden had been lifted from their shoulders.

To complete this step the consultant solicits from the consultee some possible ways that progress toward the goal(s) could be measured. A comprehensive listing is not necessary, but a few suggestions and ideas, contributed by both consultant and consultee, can again ensure the concreteness and clarity of the goals. After these are written down, the consultant checks with the consultee to determine readiness to move on to the next step.

To recap, the consultant

1. solicits goal statements from the consultee by asking questions such as, "How would you like things to be?" (and writes down the goals);
2. summarizes and helps clarify;
3. solicits a brief idea about how to measure progress; and
4. checks with the consultee to determine readiness to move on.

Step 4: Developing Ideas and Strategies

This fourth step begins with the consultant asking the consultee about previous solutions or experiences with implementing the stated goals in similar situations. Once the consultee has described what was tried in the past, the consultant can ask about the merits of previously tried solutions and what, if anything, was learned from them. The consultant and consultee then go on to brainstorm for new ideas (procedures, solutions, alternatives, etc.). This is a specific technique by means of which both contribute ideas in a non-evaluative fashion. The emphasis is on generating many ideas quickly without assessing them. If the consultee has difficulty generating ideas, it is often because the goal is still unclear or has not been stated in a sufficiently concrete fashion. Restating the consultee's goal(s) can often help in generating appropriate suggestions. If the consultee states an idea and then "puts it down," the consultant should remind the consultee that there will be time for evaluation later. The emphasis here is on generating ideas. After soliciting ideas from the consultee, the consultant can add to or elaborate on the ideas suggested by the consultee. At times both the consultee and consultant may contribute ideas equally in a random or alternating order; but the consultant should be aware that, in some cases, the success the consultee will have in putting ideas into practice will depend to some degree on the involvement of the consultee in coming up with the ideas in the first place.

If the consultee has listed some new ideas, it is likely that she or he will have to do some things differently in order to put those ideas into practice. The consultant checks this with the consultee and then determines with her or him whether it is appropriate to move on to step five.

In summary, then, the consultant

1. solicits previous ideas;
2. facilitates brainstorming;
3. asks about new behaviours, and
4. checks whether the consultee is ready to move on.

Step 5: Developing a Plan

This step begins with the consultant requesting feedback from the consultee about the ideas and strategies developed in the previous step. Ideas are selected, rejected, revised or modified, and, when the final selections are made, the consultant asks the consultee what would be necessary to put the ideas into action. This is particularly important because the ideas may call for new or different action on the part of the consultee. To ensure that the potential plan fits the previously stated goals, the consultant restates the goals and asks the consultee to comment on whether the ideas and strategies developed will promote the goals identified. If the answer is yes, then it is appropriate to move on to the next step. If the answer is no, then the consultant asks what would have to happen to make this congruence occur.

Step 6: Creating an Action Plan

In this step the consultee and the consultant work out what will be necessary to put the plan into action. For example, the plan may need a certain sequence of steps, or the plan may require certain events to occur before elements of the plan can be initiated. To help the consultee become specific and to increase the reality of the plan, the consultant asks the consultee to develop a timeline for the plan's steps, assigning specific dates to each of the activities. In addition, both the consultant and consultee identify evaluation criteria. When all the necessary details in the plan have been identified, then it is appropriate to move to the final step in the consultation.

To recap this step, the consultant

1. asks the consultee to break the plan into steps according to what will have to be done first, second, etc. (and this is written down);
2. encourages the consultee to place dates next to steps, if appropriate;
3. brainstorms with the consultee to arrive at potential progress indicators; and
4. checks whether the consultee is ready to move on.

Step 7: Empowering the Consultation Relationship

In the final step the consultant demonstrates the collaborative nature of the interaction by suggesting an opportunity for follow-up and sets a time for a further meeting. The consultant then reviews the process which he or she has gone through with the consultee and focuses on 1) identifying the strengths of the consultee; 2) expressing the consultant's beliefs in the consultee's ability to put the plan into action (and supporting this with specific feedback to the consultee); and 3) soliciting the consultee's perception of the work done together.

THE MODEL IN ACTION

To illustrate the steps described in the previous section we have selected two cases from our experience which demonstrate two different kinds of problems often encountered by special education teachers acting as consultants. The specific goals, solutions and plans may be different for another person with the same concern; therefore, *these illustrations are meant to be a demonstration of the process rather than a prescription for the specific problems described.*

Case Study I
Establishing a Consulting Relationship

Gerry, a new grade six teacher, approached Ted, the resource room teacher, to discuss Emily, a student who had been integrated into her class. As part of Emily's IEP, Gerry was to meet monthly with Emily's parents, and the date for the first conference was fast approaching. Gerry told Ted she was unclear about what type of information she should focus on in the conference and wanted Ted's advice as to what she could tell the parents.

Further Identification and Clarification of the Problem

When Ted asked Gerry for more details, he was aware of the tremor in her voice and her apprehensive demeanor. As she discussed her work with Emily, she elaborated on the details of the IEP, gave brief assessments of Emily's progress, and identified a number of ideas she had for future interactions with Emily. Ted summarized and asked her how she felt about the situation. Gerry said she felt confident and satisfied in her work with Emily, but she felt apprehensive and worried about the upcoming conference. Ted shared his observation with her that she appeared very concerned just talking about it and asked her to elaborate on her feelings. Gerry revealed that parent conferences were "white knuckle" time and that she didn't like to give "negative" information to parents. She had also heard through other teachers that Emily's parents were particularly demanding and difficult. Having clearly stated these three elements as a whole, she also volunteered that she thought that she was beginning to resent both Emily's placement in her class and the IEP for "forcing" her to do something she didn't want to do. She also wondered about, but didn't have any evidence for, the effect this might be having on her ability to be effective with Emily's educational progress.

Comment: This example shows how the initial problem was redefined into a more basic problem, and the consultant paid attention to both the content and feelings of the consultee.

Establishing Goals

Gerry stated she would like to decrease her anxiety about conferences, to feel more confident in conference skills and to increase the likelihood that she would be in a cooperative relationship with Emily's parents. She would like to believe that the parents looked forward to conferences with her and that the parent conference was an integral part of her student's academic success. She was able to summarize these ideas into one overall goal: to be able to use the parent conference as a resource for increasing the effectiveness of her educational planning. Gerry commented to Ted that she was already feeling somewhat relieved and realized that her anxiety level was one way of measuring progress toward her goal.

Comment: Through the consultant's encouragement, Gerry identified several process goals and one outcome goal which included many of the objectives she wanted to achieve. By focusing on what she wanted rather than on her negative feelings, she felt both a release and a rekindling of her energy. In addition to working on the specific goal of helping with Emily, the consultation also helped prepare Gerry for other parent conference interactions.

Developing Ideas and Strategies
Gerry described her previous strategies to achieve her goals as not very effective: she overtalked and overprepared for each conference, scheduled them in a back-to-back fashion with little time in between, and only allowed 15 minutes per conference. She jokingly described these as strategies which prevented her achieving her goal. She and Ted developed about ten ideas and strategies including participating in a parent conference skills workshops series sponsored by the provincial Teachers' Federation, focusing more on setting a mutual agenda with the parents, emphasizing her listening skills, soliciting feedback from the parents, and creating an informal environment for the meetings.

Comment: The consultant also suggested that Gerry could sit in on an up-coming parent conference the consultant was going to have and give the consultant some feedback on what she observed. By giving the consultee an opportunity to talk about past approaches, it prevented the consultant from advocating a solution which had already been tried and found wanting by the consultee.

Developing a Plan
All the ideas generated were considered valuable by Gerry. Several were combined, resulting both in ideas she could use immediately in the up-coming conference and in ideas which would help her achieve her goal over time. She restated her commitment to implementation by recognizing she would have to work on being more relaxed (applying the deep breathing she learned in her yoga classes).

Comment: Gerry's enthusiasm for developing her plan was not unusual and demonstrates her ability to assemble her resources, including applying something from another area of her life in a new way.

Creating an Action Plan
To enroll in the parent conference workshop, Gerry would need to locate the information about it. She also decided to call parents on the phone prior to the meeting to get ideas from them about what they would like to discuss. She organized her other ideas into a set of steps, assigning dates to each. She identified several indicators of progress, including her feelings before and after each conference, the willingness she felt to call and be called by parents on an informal basis, and her willingness to develop detailed and specific educational plans with students in her classes.

Comment: In listening to this tape it is clear that Gerry's voice is more confident sounding. She is speaking more rapidly, and, except for the occasional prompts and non-verbal encouragements from Ted, one would guess she was thinking-out-loud by herself.

Empowering the Relationship

Ted and Gerry set a two-week follow-up date, and Ted expressed confidence in the plan. He told her how he saw a change in her from hesitant and apprehensive to confident and enthusiastic. He also told her that he was sure the plan would work because the goals were clear and achievable and because Gerry had covered all the bases, including her own motivation and commitment. Ted reviewed the steps in the process, highlighting the twists and turns in the problem-identification and goal-setting stages. Gerry spontaneously contributed that she felt much better, that she knew this would work, and that she wasn't going to be discouraged if it didn't work completely at first.

Comment: This step reaffirmed the mobilization of the consultee's strengths and resources. It also emphasized that this is a problem-solving process which can be used again. Because this particular case was taped by a consultant-in-training (with the permission of the consultee), we can also comment on the follow-up which occurred on two additional dates. Gerry stated that her meeting with Emily's parents was very successful. While she would still describe them as demanding and overly concerned about Emily, she didn't feel defensive, there was humour expressed by the parents, they brainstormed some ways to work together, and she was able to encourage them to talk about the trepidation and concerns they had about coming to talk to teachers. In her second follow-up meeting she had already been to the second session of the conferencing workshop and felt her skills had significantly improved.

Case Study II

Establishing a Consulting Relationship

Ardis and Henri were both resource teachers in the same school. Henri had agreed to be the liaison person with the district Health Planning Committee which was being initiated to coordinate the many internal and external specialists working with the same children. Henri had much less experience working with other professionals in this capacity and sought consultation with Ardis to discuss his concerns for the up-coming meeting.

Further Identification and Clarification of the Problem

Henri told Ardis that he was glad to be the school representative on this committee, but that he felt nervous about being able to respond to potential inquiries of other committee members because he lacked depth of experience. He described his motivation for volunteering for the assignment. Ardis asked him, "What worries you the most about this situation?" In response, he detailed his major concern: he felt he would have difficulty describing the steps used by the resource teachers to develop IEPs because he had little experience actually working them through to completion.

Comment: In this example the consultant, by listening and asking open-ended questions, was able to help a colleague become specific about both the feelings about and content of the concern. In addition, the consultant refrained from jumping in immediately with advice or solutions. Thus, she was able to 1) more fully understand the concern from Henri's perspective, and 2) assist in mobilizing Henri's own resources to manage this situation.

Establishing Goals

In response to Ardis asking Henri how he would like things to be, Henri said that he wanted to do a good job at the committee meetings and that he would like to be more knowledgeable about the IEP process. He also said he was feeling somewhat awkward and embarrassed saying he was "ignorant" to a colleague like Ardis. Specifically, he wanted to be proud of his representation and he wanted the support of Ardis.

Comment: In this example, additional feelings were identified in relationship to focusing on the positive goals. This both improved the relationship and enabled Henri to feel some relief and understanding, thus freeing him to be more resourceful.

Developing ideas and strategies

When asked what he had already been doing to prepare himself for the committee meeting, Henri said he had re-read some student files and was refreshing himself by reading notes from courses he had taken. When asked whether he had been in a similar situation previously, he said that he had a role like this on a council at another school and that he found that "doing his homework" really had helped. Ardis showed Henri a card which she uses to summarize IEP steps.

1. Get appropriate background information, both historical and current.
2. Make educated guesses as to why student is having problem.
3. Conduct an assessment.
4. Revise educated guesses about "why" based on assessment.
5. Conduct an instructional assessment.
6. Develop an IEP.
7. Evaluate the IEP for appropriateness and effectiveness.

Henri built on this idea by suggesting he use some anonymous student examples from their present caseload to show how each step works. He also suggested entering the information on the school's computer so that it could be organized in a variety of ways to allow for more accurate up-dating.

Comment: Both Ardis and Henri participated actively in brainstorming solutions, and Ardis was able to integrate her experience at the appropriate time within the problem-solving process. Henri was now motivated to learn what Ardis could teach him, and Ardis' role as an educator was welcomed by Henri. (For a more detailed explanation of IEPs and the decision-making model, see Chapter 2.)

Developing a Plan

At this point, Henri was quite enthusiastic and was feeling very confident about achieving his goal. He organized the solutions into a plan. One of his solutions was designed to enable him to feel comfortable admitting that he did not know the answer to a particular question by saying that he would do his "homework" and get back to the inquirer.

Comment: This step was organized by Henri who showed great initiative in linking ideas together, identifying what he would have to do to implement his plans, affirming that the ideas would promote his goals and spontaneously beginning the next step.

Creating an Action Plan

Henri had decided to send a brief survey to other committee members asking them to let him know in advance what kinds of questions or concerns they might want him to be prepared to discuss at the meeting. He would need an additional session with Ardis in which they could brainstorm about the items to include in the survey. Henri set up specific dates in a timeline ending with the committee meeting. In response to asking how he would measure the progress of his plan, Henri said he would use responsiveness to his survey and solicited/unsolicited feedback from the meeting about his clarity and helpfulness.

Comment: The consultant used minimal prompts at this point and Henri was being quite resourceful. During this step, Henri mentioned to Ardis that he had wondered whether these committee meetings were really going to be worthwhile; that is, was the time he would spend going to really benefit the education of students. Ardis responded by saying that was also a concern of hers and maybe they could use this process and work on that as the next challenge.

Empowering the Relationship

Henri and Ardis agreed to get together just prior to the meeting so that Henri could go over his analysis of the results of the survey and get her reaction. Ardis told Henri that she was confident he would be successful because of the thoroughness and enthusiasm he showed in going through the steps and his willingness to be open about the amount of experience he had. Henri affirmed his own change in feelings and increased confidence and said he believed he was more knowledgeable about what to do.

Comment: This seven-step model took 46 minutes for Henri and Ardis to complete. The relationship was characterized by "non-judgmental receptivity." By this we mean that Ardis as consultant did not make judgments or evaluations about Henri or his "lack of knowledge" and was receptive to Henri's way of describing his concern. At the same time, Ardis' acceptance, listening and open-ended questions increased Henri's willingness to integrate Ardis' ideas into his own thinking.

These two cases have been presented to illustrate the seven steps of the consultation model. Each step is a necessary prerequisite to its following step. We hope these cases, while abbreviated, emphasize consultation as a collaborative process. At the same time, we recognize there are certain specific skills which increase the likelihood of the successful use of this model. Because teaching these skills has become an integral part of our training, we thought we would highlight them in our final section.

SKILLS NEEDED BY THE COLLABORATIVE CONSULTANT

The ability of the resource teacher to consult effectively using a collaborative approach requires a set of interpersonal skills. Each skill is learnable and many resource teachers will, as a result of prior training, already possesses some or all of the requisite skills. The most important skills are:

1. The ability to listen actively.
2. The ability to identify and react positively to feelings.
3. The ability to clarify by asking questions appropriately and sparingly.
4. The ability to guide the problem-solving process.
5. The ability to provide encouragement and positive reinforcement.
6. The ability to give constructive feedback.

Taken as a set, these six skills constitute interpersonal effectiveness and complement the instructional skills which resource teachers have already acquired through preparation as teachers and through teaching experience. Sometimes, the resource teacher may also need to be able to give specialized information or expert advice to the consultee, depending upon the focus of the consultation. However, in using collaborative consultation, the emphasis is less upon the expert knowledge of the consultant than it is upon the consultant's ability to use interpersonal skills to guide the problem-solving process in a cooperative way with consultees.

14 DEVELOPMENTS IN COMPUTER TECHNOLOGY: IMPLICATIONS FOR THE SPECIAL STUDENT

Merrill C. Sitko
THE UNIVERSITY OF WESTERN ONTARIO

Technological advancements and, particularly, the establishment of microcomputers in the schools are having a dramatic impact on classroom teaching as we know it today. Teachers are both excited and threatened by the potential impact of computers in their classrooms. Because of the wide variety of hardware and software applications available, it is extremely difficult for most teachers to determine the potential efficacy of computer-aided or computer-managed instruction (CAI or CMI). Nevertheless, today's teacher must be prepared to teach with the new technologies that are beginning to reshape our culture and schools. All of this puts great pressure on both regular and special classroom teachers to learn about educational and other applications of computers and to put them to effective use in their classes.

With respect to special education, a review of the state of the art of computer applications suggests the computer has a unique *promise* in the assessment and instructional programming of students with special learning needs. Several research studies and literature reviews have already demonstrated the feasibility and potential efficacy of computer-based instruction in assisting teachers to match instructional strategies to pupil needs and abilities and in the development of both positive pupil attitudes towards learning and positive pupil self-esteem (cf. Lindsay and Marini 1983; Mason, Blanchard and Daniel 1983; Budoff and Hutten 1982; Cain 1984; Hofmeister 1982; Taber 1982).

In fact, most research projects and literature reviews exploring the use of microcomputers in special and regular classrooms and in teacher training have reported extremely optimistic findings and applications for the use of computer technology with exceptional children (e.g., Browning and Nave 1983; Hannaford 1983; Bennett 1982; Hasselbring 1982; Thorkilsden 1982; Weir, Russel and Valente 1982; Semmel, Varnhagen and McCann, 1981;Sitko and Semmel 1976). It is clear that with the *appropriate computer educational software* and with *proper teacher training and teacher resource support*, the computer can be a powerful and invaluable *tool* in improving the instructional decision-making of both regular and special classroom teachers for students with special needs.

Modifications of computer hardware can also provide a more accurate assessment of children with physical handicaps or language delays. Such devices as light pens, tablets, touch sensitive screens, joy sticks and simplified keyboards can allow them to enter their responses to assessment tasks with greater ease and accuracy. Computers now emerging equipped with voice synthesizers and video discs will aid those students whose poor reading skills interfere with conventional computer assessment or instruction. These technical applications show how pupil assessment and instruction can more easily be tailor-made to suit the individual needs of the special learner.

It is the purpose of this chapter to provide teachers and other educators with some insight into many of the aspects of computer applications for students with special needs. The five broad goals of the chapter are to:

1. Provide an overview of the research and of the uses and benefits of microcomputers in the classroom.
2. Discuss principles for evaluating and integrating microcomputer software with current classroom curriculum.
3. Outline strategies for effective microcomputer implementation in regular and special classrooms.
4. Discuss innovative uses of computer technology in special education.
5. Provide sources for additional readings and teacher resources for use in the field.

To familiarize the reader with the growing amount of computer software being developed, this first section of the chapter will present exemplary CAI software programs and CAI research with special needs students in the traditional basic content areas of mathematics and language arts instruction. The reader will then be introduced to the more creative, tool and inquiry oriented applications of computer technology for special students in simulations and games, in the creative arts and in student-generated computer programs. Following this, the discussion turns to a consideration of computer applications that can assist classroom teachers and other educators in managing classroom instruction (Computer Managed Instruction, or CMI) for students with learning needs.

COMPUTER-AIDED INSTRUCTION (CAI)

There are several types of computer applications useful in teacher assessment and programming for students with special needs. Computer programs are now available at both elementary and secondary levels which provide a close link between instruction and assessment. CAI refers to instruction of students that is largely controlled by computer programs that determine presentation, sequencing and pacing of instructional content, remediation, drill, practice, testing, monitoring and record-keeping functions. In most of the better CAI educational software or computer programs available, students are assessed to determine initial placement or successful entry level for instruction in such content areas as math, language arts, history or science. The computer analyzes the various instructional tasks or skill levels the student has already mastered and determines which tasks or skill levels will require computer instruction. Throughout the program, the student is given questions or items which monitor achievement or level of mastery for each of the various tasks or subskills. At the end of the lesson unit or program, the student is usually given a post-test to determine level of mastery of specific instructional objectives. Most effective CAI programs depend heavily on accurate computer assessment and can help the teacher present information which has to be task analyzed and broken into small steps or conceptual tasks for individual learners.

CAI is particularly helpful for students with special learning needs who need to review materials with which they have prior familiarity, or who require a great deal of drill to master rudimentary concepts and achieve "automaticity" in such areas as reading decoding and basic computational skills. In fact, "low aptitude" or "deprived" students seem to profit more from CAI than do average or high aptitude students (Chambers and Sprecher 1980; Ragsdale 1982). When CAI and traditional instruction are compared, equal or better achievement using CAI is obtained in less time (cf. Edwards 1975; Forman 1982; Burns and Bozeman 1981;Thorndyke 1982). Other positive effects of CAI include increased speed of learning and more positive attitudes both to computers and to the subject matter studied. Moreover, many students with special learning needs become more active, motivated and interested learners with improved attention span and attendance in courses following CAI instruction. Again the close link between assessment and instruction inherent in CAI software is certainly a critical factor in these findings.

A clear example of this CAI work is the recent research with learning-disabled youth at the Trillium School for Learning Disabled Students, operated by the Ontario Ministry of Education in Milton, Ontario (Lindsay and Marini 1983). In this longitudinal project, the investigators explored the educational use of microcomputers with learning-disabled youth. Computer assisted learning in mathematics with the *Milliken Math Sequences* package for Apple II series computers resulted in a number of students demonstrating accelerated learning rates. In many cases it even demonstrated an improvement of two grade levels during the first year. Given that the typical performance increase per instructional year had averaged three months prior to the computer approach, these advancements represented, in some cases, a six-fold gain. CAI material chosen for the project in the language arts area consisted primarily of the *Milliken Language Arts Sequences*, Edu-Ware's *Spelling Bee* and *Compu-Spell*, and the game-oriented Language Arts disc from the Minnesota Educational Computer Consortium (MECC). To various degrees, each of these computer programs were found to meet the needs of individual students.

This finding is not surprising since the program chosen for the project incorporated many of the recommended principles of effective classroom learning. For instance, the *Milliken Math Sequences* microcomputer program used in the Trillium Computer Project is a commercial program that is designed to enhance any elementary through junior high school math program. Each of the math sequences is divided into several small, well-defined problem levels. Students move up or down these levels according to how well they perform. The higher the problem level, the greater the difficulty. All of the work is completed on the computer monitor screen. The student is immediately informed of correct entries. The program is "success-oriented" so there are generally two or three tries for each problem. When the problem is solved, a correct answer produces one of several positive reinforcements (e.g., "well done!"), and younger students may be given animated reinforcement. When a problem is missed more than once, the wrong answer is flashed on the screen and the correct answer is then displayed step-by-step in the proper sequence. The Milliken program allows the *student* to control the rate at which the solution is displayed and to study the answer as long as desired. When a student achieves the individualized mastery criterion level that has been set, he or she is congratulated and moved up a level. When the session is finished or when the assignment is completed, the student receives a summary of performance which is recorded automatically in the student's file.

Lindsay and Marini (1983) further reported that perhaps the most pervasive gains found in their longitudinal CAI research with the learning disabled youth in the Trillium Project were in the area of student self-concept. Many of their LD students displayed significant gains in important affect domains such as interpersonal interactions, respecting the rights of others, respect of property, and the development of positive self-regard and confidence following their successful experiences with microcomputers and CAI. However, a model of continuing resource support and inservice teacher training was viewed as a critical factor in the successful classroom applications of microcomputers in the residential school used in their project. We will see this point stressed again in other research involving the use of computer technology with special needs students.

CAI and Mathematics Instruction

It is in the area of math instruction in particular that research has demonstrated the computer can have a powerful effect on academic learning (e.g., Burns and Bozeman 1981; Johnson 1983). The analysis of study outcomes has indicated that both the drill-and-practice and tutorial aspects of CAI have significantly improved student achievement as compared with traditional classroom instruction alone. This finding seems particularly true for learning-handicapped pupils and for boys at the intermediate level.

In a recent project, Trifiletti, Frith and Armstrong (1984) studied the efficacy of computer-assisted instruction for teaching mathematics skills to learning-disabled children in grades four to eight. In their project, a computerized instruction group received 40 minutes at Radio Shack computer terminals per day on the *SPARK-80 Computerized Mathematics System*, a microcomputer based instructional and management system. The achievement of learning-disabled students who received CAI was compared with the achievement of learning-disabled students who received the more traditional resource room approach to math instruction. Mid-year assessment revealed significantly greater results for the computerized instruction group in terms of math skills mastered and fluency of problem-solving. Moreover, the computerized instruction sample produced over twice the yearly gains in achievement and in number of math skills learned. The investigators attributed their findings to the fact that the *SPARK-80 Computerized Mathematics System* provided detailed and sensitive information about each learner's growth on a daily basis. The daily speed and accuracy data allowed instructors to select appropriate instruction in the form of tutorial, drill, game or word problems. The investigators further suggested from their data that microcomputers are more efficient for instructional purposes than the more traditional resource or small group approach. However, they stressed the need for more empirical studies using greater numbers of students and teachers in a variety of settings.

An earlier math project with microcomputers and hyperactive children (Kleiman, Humphrey and Lindsay 1981) revealed that hyperactive children, ranging from 6 to 14 years of age, did almost twice as many arithmetic problems on the Commodore PET microcomputers as they did with paper and pencil. This computer project is of particular interest since the computer program used in the study had a number of characteristics suitable for individuals with special learning needs: a) the level of problem difficulty could be individually tailored for each child; b) the display format on the computer was designed to be easily readable; c) the answer format on the computer was designed to be as similar to the paper format as possible (i.e., the student worked from right to left, there was a special key for erasing mistakes, carrying could be marked above the appropriate column of numbers with a special key); d) problem-solving was completely self-paced; and e) motivational features were incorporated such as graphic displays and praise statements following correct answers. One innovative computer application allowed built-in messages that were specifically related to the hyperactive students. Special messages appeared whenever the child answered incorrectly too quickly, took too long to respond, or made too many inappropriate button presses (e.g., "Stop it"). The computer also automatically recorded the child's response, the time taken to request another problem, the number of erasures and the occurrence of any special messages. The authors tentatively concluded that, apparently, hyperactive children are willing to spend significantly more time working on problems on the computer, without any significant loss of accuracy or speed. Informal interviews with the children confirmed their preference for the computer. The authors indicated that their results do encourage optimism as to the potential use of microcomputers as a classroom tool.

In the general math area, many computer programs are now available for a wide range of microcomputers which serve the function of assessing and remediating discrete groupings of mathematical skills. The programs also ascertain through computer error-pattern analyses specific strengths and weaknesses in mathematics achievement. Information on identified weaknesses is then transmitted to other math computer programs for specific drill and practice or tutorial remediation. However, the research on the use of CAI and math clearly indicates that the teacher is the key factor in determining whether the computer math programs will actually be implemented successfully in the classroom for all learners.

In essence, comprehensive reviews of the CAI research in education (e.g., Burns and Bozeman 1981) and such research as the above provide justification for the addition of drill-and-practice and/or tutorial CAI to elementary and secondary math curricula, especially with learning-handicapped students, on a trial basis. However, teachers must plan its introduction carefully so that they can measure the effects on their own students in their own classroom situations in order to maximize the benefits. In my opinion, it is extremely important that teachers see themselves as researchers and evaluators in their own right. That is to say that teachers should systematically learn about educational applications of computers in math and other academic areas in order put them to effective use in their classes. Deciding on what constitutes effective use is not easy. While some courses, workshops or texts are offered to train teachers, many teachers have to rely on their own initiative. Many magazines and journals and even this particular chapter are bulging with testimonials and personal experience that pay glowing tribute to the marvellous contributions computers have made to classrooms and especially to students with special learning needs, but objective measurements obtained by empirical research are not as readily available to classroom teachers. Hence, teachers themselves need to systematically introduce and evaluate the effectiveness of CAI courseware with their own students and be careful to select objective, quantifiable, as well as qualitative, measures of their students' progress. In this way, they will maximize the probability of putting math and other CAI courseware to the best possible use in their classrooms.

Another important issue in the use of computers with learning-handicapped students is the predominant trend to use the computer for drill and practice or for tutorial purposes to improve lower level academic skills such as arithmetic computation. There has been relatively little effort in exploiting computer technology in order to develop higher level thinking processes in learning-disabled students, processes such as inductive and deductive reasoning, concept-relating and problem-solving in mathematics and other areas. O'Brien (1983) terms this type of software, of which the goal is the development of higher-level thinking skills, "software of the second-and-half kind," and emphasizes that it offers the greatest opportunity for education. In his view, software of the second-and-half kind engages students in activities that could be done in the classroom without a computer but, for a variety of reasons, rarely are. He describes various software examples, including, for instance, the program *Teasers by Tobbs* (Sunburst Communications 1983), which was named "Best Software of the Year 1983" by *Learning Magazine* and noted as "exemplary math/science courseware" in the *Classroom Computer Learning* magazine. The program *Teasers by Tobbs* helps students think logically to solve addition and multiplication problems. In two challenging programs, a grid and a character called Tobbs appear on the screen. The grid presents a succession of math problems, each with a missing number. Students use logic to decide which number can't be, might be or must be the number that solves the problem on the grid. The program also helps students invent and practice several problem-solving skills including working backward to solve problems, working within the constraints of a problem, developing chains of thought and following them to their conclusion and deciding how to approach the problem. O'Brien mentions that thought-provoking activities of this kind may be done without a computer but they rarely are, perhaps because they are too novel, too complex or too removed from the "back to basics" movement. In my opinion, O'Brien raises a critical issue that has direct relevance in the education of learning-handicapped students. This important issue and additional examples of this kind of software are explored further in other sections of this chapter.

CAI and Language Arts Instruction

Reading Instruction CAI programs in the reading area are currently of three main types, each of which differs in the complexity and in the amount of interaction required by the student with the computer system. Mason, Blanchard and Daniel (1983) describe the three types as: a) simple linear drill and practice programs which are designed to supplement or reinforce instruction initiated by a teacher for which practice is suggested; b) tutorial programs which are self-contained courses of instruction with capability for machine decisions and program branching and which allow content to be task analyzed and broken into small sequential steps and/or conceptual levels to match entry levels of individual learners; and c) dialogue or inquiry programs which enable students to converse with the computer in their natural language and allow the student to control the pace and level of information sequence. Mason et al. (1983) describe many of the current computer school projects and school applications to reading instruction with students of different ages and developmental reading levels. The computer reading projects reviewed cover broad skill areas and applications such as word attack and decoding skills, including word and letter recognition, syllabication, roots, affixes, inflectional endings, synonyms, homonyms, word classifications and compound words. They deal with recognition of sentences and utilization of key words; labelling and categorization of concepts; seeing the relationship between sentence and sentence order; ordering and sequencing ideas and sentences; finding and understanding supporting details; drawing sound conclusions; and following directions. They provide vocabulary development drill and practice. Finally, they also diagnose reading needs, readability skills, cloze paragraph comprehension skills, reading readiness skills, and interpretive comprehension and word-study skills. Most of the computer reading projects reviewed by Mason et al. report excellent results in terms of student achievement and attitudes towards reading, especially when CAI is used to supplement rather than replace traditional instruction. Moreover, the positive effects of CAI seem to be more pronounced with special needs students and disadvantaged students. In addition, attitude measures have indicated that, in general, teachers, students of varying ages and including handicapped students, and parents have reacted favourably to CAI use over a broad range of reading instruction.

Based on findings such as the above, more major reading book publishers are increasingly marketing extensive CAI reading instructional and management programs, both as supplements to and replacements of existing basal reading programs and other reading materials. In fact, many publishers of instructional learning materials now have special divisions for developing, publishing and marketing educational computer programs in reading and other content areas. Teachers can access these programs through many of the reference sources listed in the appendix to this chapter. An excellent source of recommended uses of computers to improve reading instruction and reading achievement and to diagnose reading problems and prescribe remedial materials is the previously cited text by Mason, Blanchard and Daniel (1984). These authors also stress, however, that it is the classroom teacher, not the computer software, that is the primary determiner of the success of most computer programs. For instance, computer programs which were never intended for reading instruction (such as word processing programs, problem-solving programs, games, data-base software) and computer programming activities can, with the aid of a talented and creative teacher, be adapted to fill a variety of educational reading and other learning needs. Some of these more creative and highly recommended computer programs are described below.

As with math instruction, most educational computer programs in the language arts area tend to be of the drill and practice or tutorial type, which are primarily aimed at teaching or remediating discrete components or subskills. In the reading area, these programs follow a "bottom-up" approach, breaking the reading process into component parts or subskills and emphasizing word identification and words or word fragments presented in isolation. Miller (1984) describes several serious flaws in these types of computer reading programs. First, most of these word identification CAI programs do not teach, but rather, test. In addition, students must often already be good readers to follow the instructions accompanying the exercise. A more serious problem is that students can often demonstrate mastery without actually using the skill being taught or tested. In the area of vocabulary and comprehension, CAI reading programs often resemble exercises found in basal workbooks (Wheeler 1983) and provide little instruction. Moreover, most of the focus in these CAI programs tends to be on the "electronic" test of comprehension. The programs mainly capitalize on the computer's ability to turn pages, control rate of presentation, record performance and provide instant feedback. Instead, Miller (1984) and many other reading experts interested in computer applications in readings (e.g., Mason 1984, Wheeler 1983; Collins and Smith 1982) strongly advocate the use of CAI programs based on a whole language perspective to reading and which tend to involve children in active hypothesis testing, comprehension monitoring and writing, as well as reading.

This whole-language perspective is the same approach to reading advocated by many researchers and educators interested in the education of exceptional students with reading and language difficulties (cf. Blanton, Sitko and Gillespie 1976; Gillespie-Silver 1979, Terry, Cohen and Englert 1978). These authors conclude that emphasis in teaching reading to special needs students should be focused on "conceptual" and "psycholinguistic" aspects of reading more than "perceptual" aspects, and on relations more than on units or single words. Moreover, reading programs designed for exceptional students should help them apply hypotheses/test strategies that involve the integration of word analysis skills and comprehension skills in their reading effort. Hence, computer reading programs designed according to this perspective would treat instruction in interactive comprehension monitoring and in active hypotheses testing as critical features. They would also actively involve students in reading and writing. In addition, reading software of this type would give the student control over the machine but would also combine the creative talents of the classroom teacher with the capabilities of the computer. Current examples of exemplary interactive reading software designed from the whole-language perspective and appropriate for learning-handicapped students are given below.

An interesting reading program of this type is *The Puzzler* (Sunburst 1984), which was also an award winner at the 1984 AMTEC (Association for Media and Technologies in Education in Canada) festival. *Puzzler* presents five intriguing mystery stories that actively engage students in the problem-solving process of reading for meaning. As the stories unfold, students use the reading strategies of predicting, confirming and integrating to solve each story puzzler. More than one answer is possible for each story. The program is based on the "top-down" reading methods of Goodman and Burke (1980). According to Miller (1984), *The Puzzler* breaks many of the rules of traditional CAI programs in that it avoids elaborate graphics, offers no feedback, does not keep performance records and considers the involvement of teachers as important as the disc-based lessons. For instance, *The Puzzler* begins with a whole group introduction to the process of predicting/confirming. This is done using overhead transparencies. Students then "puzzle" through a series of computer-based exercises requiring them to apply the strategies. Small group follow-up work is then used to share ideas about the congruency of student's predictions at a given point in the stories. Miller (1984), who is one of the program authors, indicates that in the near future a new feature of the program will involve students in writing original strategy lesson stories for inclusion on the computer disc. An accompanying teacher's guide with the *Puzzler* program currently suggests discussion ideas and writing lessons for a complete language arts package.

M-ss-ng L-nks (Sunburst 1984) is a series of computer reading programs congruent with a whole-language emphasis in reading software. It provides a series of cloze task puzzles which can serve to improve students' reading, writing, spelling and grammar skills, while giving them an appreciation of syntax, vocabulary and the mechanics of writing. In each program, reading passages are presented on the screen with some of the letters and words missing. Students reconstruct the original passage by filling in the missing letters one by one. Nine different formats are available, including such possibilities as only the first letter of each word printed, only the last letter of each word printed, only consonants printed, only the number of spaces for each word and the punctuation printed, and only the name of the passage printed. Students can allow themselves one to five guesses on a letter before the program provides the missing letter. Students can also move beyond a troubling letter or word to try a word further along in the passage in order to gain enough content to hypothesize or guess the difficult word later. A score is computed at the end of the game and shown on the screen. The *M-ss-ng L-nks* program offers a variety of passages from children's classics, subjects of interest to children, and topics from literature. Specific teaching suggestions and teaching objectives are provided for reading, writing and spelling skills. In addition, the program has a teaching option which helps the teacher customize the learning activity for individual students or groups. A separate editor program is available for teachers, parents or students who may want to create their own texts and puzzles. In a recent review of *M-ss-ng L-nks*, Kleiman and Keyes (1985) particularly remarked on how engaging the activity is and on how much knowledge about the structure of the English language and of spelling is brought to bear while completing the passages. Chomsky (1984), who is one of the co-developers of the program, suggests that the program mobilizes many separate language and reading abilities and capitalizes on students' "implicit linguistic knowledge and on linguistic patterns" (p. 62).

Another useful language arts program which has been highly recommended by several reading experts is *Suspect Sentences* (Ginn 1984). In this computer game program, students try to hide their own original sentence in an existing literature passage and then challenge their peers to find it. Through follow-up discussion, students discover the rudiments of such writing features as style, flow and form. To play the game successfully, students must read critically and model their own writing after a particular author's style (Ohanian 1984). The program gives the "forger" and the "detective" student practice in creativity in reading and writing. It also gives them practice in analysis of style, in composition with different models, and in developing coherence in writing (Chomsky 1984).

An interesting package of reading software which has the potential for facilitating reading comprehension in special needs students is *Microzine* (Scholastic 1984). *Microzine* is actually a bi-monthly computer-based periodical aimed at grades four through eight. It has interesting materials suitable for special needs students at the high school level as well. Each issue consists of a double-sided disc with four programs, a student handbook and a teacher's guide which contains a scope and sequence chart outlining the curriculum areas covered. The computer magazine includes easy-to-use programs such as *Twistaplot*, an ever-changing story/game; *Ask Me*, an interview with a famous young personality. As well, it includes a collection of software tools for creating graphics, colour and sound, or for organizing information (Sickert 1984). The programs can be used by individual students, small groups or an entire class in a lab, resource or classroom situation. One of the first *Twistaplot* stories is "Haunted House" which is actually a branched adventure game whose plot is determined by choices a student makes at various points in the story. Thirteen different endings are possible. Reviewers of *Microzine* remark that the programs hold students' interest by allowing them to control plot development or by replaying stories a number of times. In this way, the reading programs can be used to teach word recognition, new vocabulary, reading comprehension and decision-making (Furst 1984; Sickert 1984). Moreover, Miller (1984) feels the *Microzine* stories provide an interesting way to encourage students to do repeated readings for meaning and, as a result, develop automaticity in decoding. The stories allow students to see how authors make choices in selecting the next sequence in a story. The *Ask Me* program is well-designed and highly interactive and can be effectively used to encourage written and oral expression or to have students do their own interviews based on the language patterns presented in the program (Furst 1984). Hence, teachers can use computer reading programs like *Microzine* and those above to help special needs students improve their language skills, creative thinking and even "pragmatic" or communicative social skills. These few exemplary computer reading programs are emerging as innovative models and feasible alternatives in the language arts field. Teachers should be familiar with them and able to select among them. According to Chomsky (1984), these kinds of programs and many others being developed give students an enhanced sense of "the intrinsic power, fascination and variety of the written world" (p. 63) and will earn microcomputers their place in the language arts curriculum. Hopefully, with creative use, they may assist teachers in effectively matching the linguistic needs and interests of individual students or groups. At the very least, they are dramatic pedagogical improvements over much of the current reading "tutor" style software that solely drills students on discrete language arts skills.

Spelling Instruction Previous research has also documented the successful use of microcomputer-based CAI for diagnosing and remediating spelling problems in learning-handicapped children (e.g., Hasselbring 1984; Hasselbring and Crossland 1981, 1982, in press). In one research program, Hasselbring and his associates have shown that students exhibiting mild learning handicaps can successfully operate microcomputer-based diagnostic programs under test conditions. Their studies indicate that results obtained from computer-based diagnostics are as reliable as those obtained from the more traditional paper-and-pencil spelling methods. Their current work (Hasselbring 1984) is in exploring the use of a microcomputer to provide inexpensive imitation-plus-modelling feedback to improve the spelling performance of learning-disabled students. In this work, an interactive BASIC program entitled *The Computerized Spelling Program* (CRSP) is used with a Radio Shack TRS-80 Model I microcomputer to emulate the imitation-plus-modelling feedback that normally would be provided by a teacher imitating a student's errors first and then providing a correct model. The computer program consists of three parts: "Word List Input," "Word Presentation and Feedback," and "A Performance Summary Printout." According to Hasselbring (1984), their observations suggest that microcomputer-based feedback is as effective as teacher feedback. He believes the use of their type of computer program could improve the spelling performance of large numbers of students, while at the same time freeing the classroom teacher from much of the time-consuming tedium associated with spelling instruction.

In reporting on another project involving learning-disabled students (grades one to six), McDermott and Watkins (1983) suggest that combinations of computerized and conventional remedial spelling instruction may produce higher achievement than either instructional method alone. However, success will probably vary as a function of the learning disorders and differential styles of learning. On the other hand, in their project, which explored the use of microcomputers with learning-disabled adolescents, Lindsay and Marini (1983) reported that a drill/tutorial spelling program for early elementary grades was even found to be useful for those adolescents whose reading skills were still at a primary level and who were very threatened by traditional spelling instruction. They further found that for some learning-disabled adolescents, the experience with the CAI spelling program served as an important initial step leading to more challenging experimentation with words.

Two of the newer applications of microcomputers, sound and colourful graphic effects, have greatly increased the effectiveness of spelling drill and practice software with learning-handicapped students. An example of these applications to spelling courseware is the program *Wizard of Words* (Computer Advanced Ideas 1984). *Wizard of Words* is a double disc package with a collection of spelling games and a dictionary word list of 38 000 words across eight levels. The spelling games use a combination of colourful graphics and sounds and the program contains a built-in authoring system which allows for the creation of personal word lists by the student, teacher, parent or aid. All of the spelling games have a form of potential reinforcement through music and/or graphic activity. In a review of the program, Hummel (1984) notes that the range of included word lists, coupled with a built-in authoring system, makes this package appropriate for all students. He found the program particularly intriguing for junior-high remedial students, even though the program is described as designed for younger students. Hummel commented that the program was very useful as a supplement to other classroom activities since current spelling words could be entered into the personal word lists for two of the games. Programs like *Wizard of Words* are combining some of the best features of microcomputers (e.g., good sound, colour, sprite graphics, animated arcade game format, learner accommodation, immediate feedback, etc.) to turn spelling drudgery into spelling fun.

Unfortunately, most of the spelling computer programs on the educational market still do not capitalize on the unique potential teaching and motivational qualities of the computers. These programs tend to be of the drill and practice variety (or drill and kill) that serve mainly as "electronic workbooks" to drill isolated spelling skills. The better programs, which are more suitable for meeting the spelling needs of learning-handicapped students, capitalize on the unique motivational and creative qualities of the computer. Most of the more innovative programs fall in the category of learning games, creative discovery programs, tutorials and simulations. For instance, one learning game that has become quite popular with learning resource teachers is *Cave of the Word Wizard* (Timeworks 1984). This program combines the challenges and rewards of popular arcade-style adventure games with instructional content in spelling. Teachers report that it motivates the learning-handicapped students to work harder and perservere longer due to its combination of good graphics, sound effects, voice synthesis, fast action and excitement. In this program, students travel through a mysterious cave and are required to find four magic crystals which are needed to reopen the cave entrance. At various times they are challenged by a funny old wizard who requires students to spell a word in order to replenish the energy in their flashlight. During their search they are confronted with spiders, rocks, snakes and other dangerous obstacles that make their quest for freedom even more challenging. This program contains 500 spoken words in 10 spelling levels and makes full use of the sound capabilities of the computer. Programs like the *Cave of the Word Wizard* can be quite useful for learning-handicapped students or for enrichment, but only if teachers are careful to integrate it into their own academic program, or to evaluate and give feedback on progress to their students through appropriate pre- and post-test evaluation activities. However, in this author's opinion, a more creative application of the computer with special needs students—one which has the real potential of combining oral and written language and reading with spelling in an integrated language arts program—lies in the "tool" uses of the computer in the classroom. It is this extremely important classroom application to which we now turn our attention.

Writing Instruction An excellent application of computer technology for language arts instructional programming with special needs students is in the creative use of the microcomputer as a tool in such applications as word processing and expository writing (cf. Geoffrion 1983; Kerchner and Kistinger 1984; Preston 1984; Piper 1983-84). In the previously described exploratory microcomputer project with LD students, Lindsay and Marini (1983) demonstrated that LD students, ranging in age from 13 to 19 years, were able to increase dramatically the frequency and quality of their compositions in creative writing classes over the school year through the application of the word processing capabilities of microcomputers. Students not only added to the length and quality of their compositions in terms of organization and content, but also voluntarily engaged in the writing process more frequently. Based on their observations, Lindsay and Marini reported that many of these students possess poor handwriting and spelling skills. Traditionally, their performance on paper and pen composition revealed "numerous cross-outs, inserts, smudges, and other characteristics which tend to reinforce the inferior image these children tend to have of themselves" (p. 10). Hence, a writing sytem which permits effortless editing and eliminates evidence of previous errors holds a great attraction for learning-disabled students whose severe written language problems are frequently not remediated.

Other research reports that students who use the computer for writing are able to spend more time doing creative writing because they are spending less time in the mechanical aspects of recopying old drafts (Green 1984; Geffrion 1983; Ragsdale 1982). While practice itself is generally found to be the best approach to improving grammatical, organizational and creative writing skills, problems with the physical component in traditional pen and pencil writing often interfere with any attempts to encourage sustained effort and, thus, thwart any sense of enjoyment and skill development in this area for students with language disabilities. However, reducing some of this frustration can go a long way towards increasing both the quality and quantity of writing output.

Some support for this conclusion is provided in a longitudinal writing project developed for mildly handicapped students in the Cupertino School District of California (Preston 1984). In this project, the investigators combined creative teacher prompts and classroom brainstorming with computer word processing technology to yield written compositions that improved in overall literacy quality and in technical appropriateness. Moreover, the investigators reported significant gains for elementary age students in both holistic writing scores and written expression cluster scores on the *Woodcock-Johnson Psychoeducational Battery*.

Another recent project also found support for using word processors to significantly improve the writing skills of elementary age LD students. Kerchner and Kistinger (1984) combined the process or conference approach to writing with the word processing capabilities of the microcomputer to improve the writing skills of fourth, fifth and sixth-grade LD students. Compared with a control group who used no word processing, the experimental LD computer group who used the *Bank Street Writer* (Broderbund Software 1982) to compose, edit and refine their work, made larger writing gains over a seven-month period of instruction. Significant gains favouring the computer experimental groups were revealed on several subtests of the *Test of Written Language* (TOWL) (Hammill and Larsen 1978). The experimental treatment for the computer group consisted of a nine-step process including the following steps: prewriting conference, composing at the keyboard, printing a draft, editing conference, editing at the keyboard, printing final copy, illustrating the composition, and providing an audience. The investigators concluded that their results suggest that improved written language skills on the computer do transfer to pencil-

and-paper tasks. Moreover, they argue for a holistic language arts program which utilizes the word processor with LD students and treats reading and language development and written expression simultaneously, in one integrated program.

In our current work with microcomputers and word processors at The University of Western Ontario, we, like others, find that perhaps the most significant gains with learning-disabled children, adolescents and school dropouts are in the areas of student internal motivation and self-concept. Our teachers report significant student gains over a short period of time in academic motivation, on-task behaviour, risk-taking, confidence and positive self-esteem, following their writing experiences with word processors. Many of these learning-handicapped students regretfully view themselves as being incapable of successful or efficient learning. Our experience indicates that as their writing skills develop, these students begin to recognize that they are succeeding in an area which has traditionally resulted in frustration, failure and avoidance behaviour, and they begin to approach the writing task with a new sense of confidence and a willingness to try. As one of our teachers remarked, "Perhaps the most amazing result of the two months with computers has been in attitude changes in most of the kids. Kids are learning to see with their eyes the results of the sounds that they make, on paper or, in this case, on the screen—and that's turned on." As an example of this motivation, one of our 12 year old learning-disabled students in the computer writing project recently wrote this brief letter to me:

> Dear Dr. Sitko: Thank you for the computer for this year and I hope we have the computer next year! It has helped me to write better sentences and it is fun!
> Your friend,
> Tommy B.

Similarly, in a recent article O'Brien (1984) reports that, when given access to a good word processor, her weakest students revealed a remarkable change. "Where there had been negativity, unreached ability and a disturbing lack of purpose, I began to notice signs of competence and growth" (p. 30). O'Brien indicated that her poorer students sensed that the microcomputer could help them redeem their past academic failures.

In a recent interview, writing expert Donald Graves (Green 1984) indicates that one of the keys in helping students develop good writing skills is publishing their work frequently and getting it to other students and parents. He sees the computer as speeding up and facilitating the vital publishing process, especially for problem writers, by helping the student produce his or her own "camera-ready copy." He feels that since there is no penalty for revising on a word processor (i.e., no erasures, cross-outs or recopying), it may make it easier for students to experiment or play with their writing—with guidance from their teachers. Graves thinks that the word processor might be a better way to help students view their work as part of a drafting process instead of as a finished copy. However, he stresses the critical role of teachers throughout the writing process and the importance of teacher-pupil writing conferences that help students clarify, evaluate and revise their own work through teacher questioning and feedback.

In our current word processing project with learning-disabled children, adolescents and school dropouts (Sitko et al., in progress) we have developed a ten-step training component for our students and teachers in the following sequence:

Prewriting Aspects

1. Development of story grids with groups of students and their teachers in their own classroom (creative brainstorming component).
2. Listing of elements for each student's story.
3. Discussion of story outlining with a peer.
4. Development of an outline.

Writing Aspects

5. First draft of story on computer with instruction to separate each sentence from the ones preceding and following it.
6. First editing of story for organization, comprehension and completion, with teacher consultation.
7. Second editing of story for paragraph structure.
8. Third editing of each sentence for spelling, grammar and punctuation.
9. Final editing for story quality.
10. Printout.

We are also using the *Bank Street Word Processing Program* by Broderbund Software Co. during the writing and editing phases with Commodore C-64 microcomputers and printers. Some of the features of the *Bank Street Writer* include automatic word wrap at the end of the line; the capacity to add, move, insert and erase words or blocks of text; the capacity to perform universal search and to replace target words; automatic centering; automatic identification; disc sorting with private password file capability; page headers and automatic page numbering; highlighting of text; and user-friendly on-screen menus at all times. Student writing is scored holistically according to major strengths and weaknesses in terms of the technical aspects of spelling, grammar and punctuation. Students are also provided with simultaneous training in computer keyboarding and typing skills during the writing aspects of the project. Measures of general self-concept and situational self-concept in writing ability are also obtained, as well as measures of pupil attitudes towards writing and microcomputer technology. We expect at the completion of our current project to develop a teacher writing manual for future inservice and research application for students with special needs.

An interesting finding in the work with special needs students and word processing is the improvement generally found in their spelling and grammatical performance, as well as in the quality of their creative writing and compositions. The improvement in spelling is particularly noted by those teachers who combine the use of a word processor with spelling checkers (or their own checking) and who concurrently integrate keyboarding or typing skills into the total language arts programs. There are many educators who argue strongly that simply "looking" at the correct spelling can facilitate correct spelling. For instance, Fitzsimmons and Loomer (cited in Garvey 1984) provide evidence that suggests that any method allowing the student to see the correct spelling will teach correct spelling. Their research demonstrates that school districts who have taught spelling as primarily a visual exercise have seen a dramatic improvement in their students' spelling. As a result, it appears that a computer spelling checker used in conjunction with a word processor may provide an excellent aid for teaching students with special needs to spell well. This method would allow the students to individualize the visual tracking process, and it would pick out and highlight for students those words they mispell the most in the actual context of discourse writing. Similarly, the repetition of correct spelling inherent in computer keyboarding programs may also facilitate spelling performance by providing both visual reinforcement and practice in correct spelling. It remains to be seen if future research will shed more light on this issue. At any rate, the teachers in our own research have been especially pleased with the positive results of using word processors with spelling checkers and keyboarding practice with their special students.

In terms of writing instruction, there are many computer writing assistance programs now available which are congruent or partially congruent with the holistic writing process approach to composition advocated by current experts in the language arts field. Examples of some of the better designed programs, from both a pedagogical and technical viewpoint, are described below.

At the primary grade level, an interesting computer writing assistance program is *Story Machine* (Spinnaker 1983). This program is actually a learning game that helps teach young children or beginning and remedial readers to write sentences, paragraphs and simple stories and to develop familiarity with the keyboard. Using a vocabulary of about 40 words, including nouns, verbs, prepositions and other parts of speech, children can write sentences, combine them into stories, and see their stories animated with graphics and sound. For instance, if the child or teacher types in "The girl kisses the boy," the computer will present a graphic of a girl doing precisely that. The list of story options allows the child to 1) watch the computer make up a story and act it out on the screen, 2) take turns with the computer in writing a story one word at a time, 3) make up his or her own story, and see it acted out on the screen, and 4) replay and watch a story already saved during a previous session using *Story Machine*. The program emphasizes the rules of grammar and encourages children to write clear, correct sentences. In my opinion, a program such as *Story Machine* may be particularly useful in developing a positive attitude towards creative writing in learning-disabled students whose writing efforts are often extremely frustrating and failure-oriented. In the hands of a creative teacher, this computer program and others like it can combine initial writing literacy with fun and excitement.

A similar computer program, but at a higher academic level, is the recently introduced program *Bank Street Story Book* (Mindscape 1984). This program also allows students to write, edit, illustrate and animate their own original stories. Another useful writing assistance program for special needs students is the creative writing program *Story Tree* (Scholastic 1984). With *Story Tree*, students make choices about how a fictitious story unfolds or branches. They create a story page-by-page, link stories together or write stories that branch by chance. The program allows the student to control the complexity of the story and to gradually work up to complete stories as his or her organizational skills improve. In a review of *Story Tree*, Carter

(1984) stresses that, among its many pluses, the program creates an environment in which students want to read and reread each other's writing and/or choices to experience all of the branches. As a result, students often give each other feedback and suggestions. They can also edit and revise their stories both individually or in groups. Carter emphasizes that *Story Tree* goes beyond just story writing in that it can also be used as an information retrieval system by the teacher to build, for example, a "tree" of books that have been read and reviewed by the class. Teachers can further challenge their students to construct, using *Story Tree*, word games, quizzes or tutorials for their fellow classmates. The program is technically strong, with excellent documentation, and also includes many classroom ideas for the creative teacher. (A program with similar goals of story development, writing and creative thinking is called *Story Maker* (Bolt, Beranek and Newman 1983).

At a more comprehensive level, an excellent set of microcomputer-based language arts programs is *Quill* (Heath and Co. 1984). The goal of the *Quill* computer system is to motivate students to do more writing, revising and editing through exciting, adaptable writing activities that supplement an existing language arts curriculum (Helmstetter and Balunis 1984). *Quill* uses a writing process approach to composition; that is, one based on the processes of planning, drafting and revising. To achieve this, the *Quill* system offers three programs entitled: *The Planner, The Library* and *The Mailbag*. The *Planner* program helps with the prewriting steps and encourages both organized note-taking and general organization. Writing is handled by a word processor, while a text editor facilitates both higher level revisions and editing (Miller 1984). *The Planner* can be used by the classroom teacher in encouraging writing in other content areas such as science and social studies, as well as language arts. *The Library* is actually an information storage or database system with which students can share information; while *The Mailbag* is an electronic message system that encourages students to communicate with others via the computer. Student publishing is aided by a utility program called *Publisher*. According to Helmstetter and Balunis (1984), *Quill* was designed to actually engage students in written language. It does not provide computer drill and practice but a *process* of learning to write, which includes composing, editing and revising. They describe the program as a delightful, creative approach to teaching writing skills which is fun at the same time as it is educational. *Quill* also includes a teacher's guide which provides 20 detailed lesson plans to introduce *Quill* into the classroom.

Miller (1984) describes *Quill* as perhaps the most elaborate computer-aided learning writing program that comes the closest to providing assistance in all five of the writing process steps advocated by Graves (1983) and others; that is, prewriting, writing, revision, editing and publication. Miller sees one of the major advantages of writing assistance programs like *Quill* as the fact that students are producing connected discourse, often for a self-selected purpose, and are not simply manipulating text or isolated words written by someone else. He describes *Quill* as a superior program, combining several language arts activities such as writing, note-taking, discussion, reading, research and publication, among others. However, I can only underscore his conclusion that perhaps the key variable in its success, as with any of the previously described computer language arts programs, is how teachers view and use the program. Computer programs, especially when used with special students, must combine the talents of teachers with the capabilities of computers and must be seen as part of the larger instructional delivery system in which the computer is just one agent in presenting instructional strategies to match individualized learning needs.

COMPUTER MANAGED INSTRUCTION

Computer Managed Instruction (CMI) programs are available to teachers to enable them to plan and monitor individual student academic performance based on assessment of student performance in various content areas (Samuels and Minick 1980; Snodgrass 1980; Hasselbring and Hamlett 1984; Brown 1982; Hayden, Vance and Irvin 1982a,b; Rieth et al. 1984). The term CMI usually refers to diagnostic assessment, monitoring and placement of a student within a particular instructional program, as well as to continual testing as a method of evaluating student progress. The teacher controls both the content and the rate of learning. In CMI, the actual instruction may take place on the computer itself as in CAI or it may take place away from the computer while the computer is used to collect and analyze information and to score tests, interpret results and manage student records, as well as provide feedback to teachers and learners. CMI programs range from those that quickly average test scores to those that ask the teacher to state behavioural goals in the content area (e.g., math, reading). The programs may ask the teacher to develop or choose criteria for multiple-choice tests of content tasks and/or skills, and possibly to use the computers to manage student performance data and carry out graphing. Some computer programs actually print out instructional recommendations for individual students or help the teacher make better decisions about their instruction.

CMI programs usually provide the teacher with print-out records of student achievement across various areas of instruction. Decisions for future instruction are based on the inferences formed from this information. Hence, CMI programs aid tremendously in the scrutiny of students' work. Individual responses to individual items can easily be tallied by a computer providing an analysis of areas in which a particular student or groups of students show mastery or the need for remediation. An analysis that would take a teacher a great deal of time can be done with ease by a computer, thus facilitating instructional decision-making. For instance, programs in the reading area have been designed to quickly analyze samples of instructional material for readability level, thus enabling teachers of special needs students to determine whether or not the materials are at the learner's instructional or independent reading levels. Some of the computer programs recommend changes in wording to make reading passages more readable. Many school systems are now using CMI-based reading programs which are designed to diagnose, prescribe, monitor, and evaluate pupil progress in reading and language arts as well as to provide actual CAI reading instruction (cf. Mason, Blanchard and Daniel 1983). Computer assessment of reading performance is provided in such diverse areas as: finding main ideas; making inferences; noting and recalling details; recognizing sequences; relating concepts; following directions; finding context clues and information; decoding; word analysis; recognizing synonyms, antonyms and homonyms; labelling and categorizing concepts; drawing sound conclusions and cloze task analysis.

CMI and IEP's

Many CMI programs enable teachers to plan and manage exceptional students' Individual Educational Plans (IEP's) or Individual Program Plans (IPP's) as required by current special education mandatory legislation in Canada and the United States. Teachers are able to record student progress towards long-term instructional objectives typically found on student IEP's by storing student performance data collected daily or weekly on sequenced short-range instructional objectives. These short-range objectives are usually chosen from a computer library of objectives and criterion or mastery measurement items. The teacher can test or measure when he or she feels that the student has achieved the short-term objective by giving criterion-referenced or mastery tests to which the student responds by marking on machine-scorable answer sheets or on the computer itself. Many CMI programs now allow teachers to obtain daily or weekly student performance histories in paper or graphic form on any specified sequence of student daily activities. The programs, then, inform both the teacher and student of whether or not a student has met the long- or short-term objectives and which objectives should be taught next. Some of the programs even provide recommended individual prescriptions and instructional activities or lessons to meet specific student objectives. Hence, current CMI programs not only allow teachers to assess, individualize and manage classroom instruction, but also make efficient educational decisions based on student data.

One exemplary CMI program is the *A.M.E.C.U.S.* computer program (Donohue 1983) that won an award at a recent search of software contest, conducted by the Council for Exceptional Children (CEC). The program tracks legal due process data, student demographics, student schedules and program descriptions, health summaries, student IEP goals and objectives (with methods and materials), student performance records (lesson plans), student performance scales and graphs, a master curriculum guide (with goals, objectives and criteria), as well as class lists, costs and other aspects of programs for handicapped students.

Using an excellent computerized instructional management system (CIMS) developed for special needs students at the Centre for Innovation in Teaching the Handicapped (CITH) at Indiana University (cf. Rieth et al. 1984), teachers and administrators are able to plan and monitor student academic performance on both long-term and short-term objectives in the following manner. Either daily or weekly, the teacher enters any or all of the following in the microcomputer: the date; the names and number of students; target long-range and short-range objectives (LRO and SRO) from the student's IEP; the instructional method that was employed; and student performance data, including the number of problems assigned, the number correct, the amount of time required to complete the assignment, and the instructional materials used. Once the performance data are entered, they are instantaneously compared with the mastery criterion specified in the instructional objective, and the teacher is informed of whether or not the student met the objective. In this manner, the teacher receives objective information on whether the instructional teaching method or strategy utilized is successful. Once a student meets a short-range objective, the next objective in the student's need hierarchy is automatically activated. The second feature of the CIMS system is a report generation program which enables teachers to instantaneously generate daily, weekly, monthly or yearly progress reports. These reports include graphic information and can be printed for storage in the student's cumulative folder or portfolio or for sharing with students or parents at parent conferences. The positive aspects of CMI systems such as the Indiana CIMS program are their ability to enable a special or regular classroom teacher to develop individualized programs for *all* students in a classroom and to evaluate their effectiveness in matching teaching strategies to student learning needs.

Another popular use of CMI programs by classroom teachers is in grade-marking and record-keeping. For instance the Apple-compatible *EZ Grade-11* computer program by the Computer Generation Inc. (1984) can handle up to 10 content subjects for a class of up to 35 students. The program provides separate class rosters for each subject or one roster for all subjects and up to 50 assignment records per student per subject. The program allows the teacher to make entries, deletions or corrections and to provide average grades, Z-scores, means and standard deviation statistics, and to print dated reports as well. Marking and record-keeping programs such as the above are becoming commonplace in the schools and provide excellent tool and time-saving functions for classroom teachers. In fact, many CMI marking or data-keeping programs are now being integrated into the better interactive computer-based learning materials. Hence computers can now aid the teacher tremendously in the scrutiny of special students' work. Individual responses to individual items can be tallied by a good CMI marking program, thereby providing a spreadsheet analysis of areas in which a particular student or group of students show mastery or the need for remediation. An analysis that would take a teacher a great deal of time can be done easily by a computer, thus facilitating instructional decision-making. Moreover, CMI programs are now becoming available to regular and special education teachers at every level of instruction and exist for most of the major brands of microcomputers or minicomputers sold in education, including Apple, Atari, Commodore, Digital, IBM, Radio Shack and Texas Instruments.

The reader is especially referred to the article by Gorth and Nassif (1984) for a discussion of the critical issues in the design and evaluation of CMI programs. These authors stress that every CMI system has several major functions (e.g., test-scoring, data-recording and analysis, prescriptions, record-keeping, reports, etc.) but each differs in the strategies used to accomplish these functions and in the degree of operator instruction necessary to use the system. They conclude that in buying or designing CMI systems, educators should take note that a simple CMI system is usually best understood and most widely used.

COMPUTER SIMULATIONS AND GAMES

Computer simulations and games provide good opportunities for teachers to assess a wide variety of student abilities. When combined with the graphics, colour, motion, sound and voice options of the computer, they can provide valuable insight into the learning-handicapped students' motivation, interests and perceptual coordination, as well as their hypothesis-testing, problem-solving, and analytical abilities. Simulation programs can imitate many real or imagined physical situations which can be of either a concrete or abstract nature. Such programs can simulate some activity that might be difficult, costly or time-consuming to actually do in the classroom. Many computer games have been developed to convey academic content or to allow the expression of creative abilities in music and art. At the very least, even recreational games or simulation programs provide an opportunity to develop student hand-eye coordination and strategic decision-making skills and to assess student performance under optimal conditions of interest, attention and motivation. Through replication, teachers can assess how students learn from experience and feedback under both structured and varied role conditions. In many computer simulation and game programs the presentation of content by the computer can be adjusted to match the skill level of difficulty requested by the student, thereby providing a more accurate assessment of cognitive strategy development and creative problem-solving on the part of the student. Some specific examples are in order to illustrate these applications.

A popular simulation game in the public domain (i.e., non-commercial) in our school district (London, Ontario) is *Lemonade Stand*. In this program the student is the proprietor of a neighbourhood lemonade stand for the summer, and his or her goal is to make the business venture profitable. To succeed, students must balance such variables as the weather conditions, the price of cups, sugar and lemon concentrate, and the sale price per glass. By predicting the outcome of their decisions along with random occurrences in the weather (e.g., humidity), they can learn basic concepts of economics as well as problem-solving strategies. According to Friel (1983), *Lemonade*, like all simulations, provides a model of a real-world situation. By studying the simulation, students can make inferences about the simulated model and gather insight into its real-world counterpart. Moreover, students learn various problem-solving skills including observation, data-gathering, hypothesis formation and experimentation. I would probably add the critical skills of evaluation and decision-making to this list. Friel stresses that these skills are transferable to many real-life situations and can be reinforced by studying the simulation from several different angles. The key in any simulation, however, is still the ability of the teacher to bridge the gap between the artificial situation and the real-world event through questioning, feedback and follow-up teaching strategies.

The game is especially popular, as are most computer simulation problem-solving games, with learning-disabled students who view the program as a safe playful game—one with little personal risk or failure involved. The game encourages the student or a group of students to try ideas or strategies without the fear of making a mistake or of receiving negative feedback. (In this case, the computer only provides informational feedback.) In fact, they soon discover that inappropriate solutions to the task can be easily revised through trial and error, and alternative strategies retested for their effectiveness in maximizing "profits" in the game. It is interesting to observe how quickly the LD students develop an intuitive sense of the relationships or principles involved in the simulation.

The commercial program *The Factory* (Sunburst Communications 1983) is another good example of an excellent computer simulation program that can be used to develop spatial problem-solving in special needs students. The program utilizes several strategies in problem-solving, including inductive thinking, working backwards, analyzing a process, determining a sequence and applying creativity. Students are asked to set up a simulated factory (graphically) which contains various machines that can punch, stripe and rotate sheets of "material" to create different "products." The first level of the program allows students to experiment with the three machines. In the second level, students try to design and produce products using combinations of the three machines and then challenge other students to produce them. In the third level, students are asked to look at a final product and then work backwards and determine what process took place to construct the product. The program helps to teach students that there may be more than one solution to a problem. *The Factory* also helps special students develop visual discrimination and spatial perception. The program was awarded "Best Software of the Year, 1983" by *Learning Magazine* and was also an award winner in the 1984 Software Search by the Council for Exceptional Children.

Other excellent problem-solving programs by Sunburst Inc. include *The Pond* and *The Incredible Laboratory*. In fact, Sunburst's software packages in particular contain well-designed programs that address curriculum objectives and provide enjoyable, worthwhile activities for students (Kleiman and Keyes 1985).

In *The Incredible Laboratory*, the students task is to act as a scientist to determine which chemical controls each of six components of a fanciful monster: head, eyes, body, arms, legs and feet. Students select chemicals in a laboratory to form a colourful monster with such features as a blue head, furry legs, sparking feet or scaly body and wings, and then watch as the monster is slowly distilled in a large beaker. At various levels in the program, students repeatedly create monsters, thereby learning to determine the effect of each chemical and exploring how various combinations of chemicals interact. The game is designed to give students experience in "successive scanning" or problem-solving, and in systematic note-taking and organized list-making. The writer was particularly impressed with the way his own children developed cooperative problem-solving strategies in playing the game. The authors of the program also provide excellent ideas for classroom lesson use.

In *The Pond*, a small green frog, lost in a pond of lily pads, helps students recognize and articulate patterns, generalize from raw data and think logically. The student chooses from six ponds, or levels of difficulty, in which lily pads are displayed in increasingly complex patterns. The task is to determine the pattern or strategy that will get a frog across the pond. Emphasis is on the efficiency of the pattern. Students can first practice each level and then participate in a game that will reinforce their skills. *The Pond* should help special needs students to gain experience in understanding the importance of sequence and to learn to use the experimentation process to formulate a hypothesis.

Rocky's Boots (The Learning Company 1982) is another example of challenging computer problem-solving games that require logical reasoning and encourage creativity, analysis and problem-solving, as well as specific skills for success (Hamlett 1984). The object of the 40 different games in the program is to build electronic machines to solve particular problems. Featuring colour graphics, music and sound effects, the animated logic machines operate a kicking boot, and players score or lose points for "booting" particular objects. The program teaches simple concepts such as electricity and connecting wires as well as knowledge of electronic components such as AND-gates, OR-gates, and NOT-gates. Hence, special students are able to learn logic and especially the basics of binary computer logic and circuitry while building the machines. The program is designed to be an "erector set" allowing students seven and above to learn the basics of building computers using simulated computer logic circuits.

An interesting curriculum which uses a discovery-oriented simulation and creative gaming approach to introduce some of the rudiments of problem-solving to children between the ages of seven and twelve, is the computer curriculum *Creative Play: Problem Solving Activities with the Computer*. The course, which was developed at the Lawrence Hall of Science in Berkeley, California, is divided into various conceptual units. Each unit contains a number of educational computer games or simulations and some related off-line activities (Slesnick 1983). The problem-solving skills emphasized in the activities include data collection and evaluation, logical deductive reasoning, creative or divergent thinking, spatial visualization, and pattern identification in language and mathematics. In an excellent review of the curriculum, Slesnick (1983) concludes that the *Creative Play* simulation game model is a new exciting methodology for teaching both content and heuristic strategies in a stimulating and recreational learning environment.

Many of the educational computer simulations and games are in the format of adventure games. These are highly interactive games in which the student assumes the role of a character in an adventure story. Adventure fantasy games are finding a place in education and particularly in special education because success in these games depends on such skills as reading comprehension, creative problem-solving, analytic reasoning, sequencing, risk-taking, perseverance, logical reasoning, prediction, strategic planning, evaluation, critical decision-making, and use of language to communicate intent. Skills such as spelling, reading comprehension and cooperative group decision-making often improves as a result of playing such games. According to Hamlett (1984), a carefully chosen adventure game under the guidance of a thoughtful teacher can be an exercise in "dynamic problem-solving" as well as a source of enjoyment. For instance, in *Snooper Troops* (Spinnaker Software Corporation 1982) the student becomes a detective who must solve a mystery by searching a town to find out who has committed a crime and why. In the course of solving the mystery, students must take notes, draw maps, classify and organize information, use vocabulary and reasoning skills. The program helps the special needs student to think logically and to organize information in a motivating and entertaining fashion. It is interesting that *Snooper Troops* became the first educational game to make a software industry bestseller list.

The adventure fantasy games are especially suitable for small-group activity and for developing cooperation, or teamwork among students. The classroom teacher can add a further instructional component to the task through the application of such hierarchial questioning and feedback skills as concept relating, discrimination, problem-solving, expanding or refocusing, and prompting (Sitko and Slemon 1983).

One of the author's favourite educational adventure games is *In Search of the Most Amazing Thing* (Spinnaker Software Corporation 1983). In this game, students play the role of a B-Liner pilot who goes in search of the "most amazing thing" in the land of "Darksome Mire." From the beginning, decision-making, organization and note-taking are essential parts of the program. To succeed, the pilot (i.e., the student) must deal effectively with the many different cultures he or she will contact. Hence map-making and an understanding of trading and exchange rates are key points in the program. The game has been labelled a non-sexist adventure game and demands a good deal of predicting, remembering, trial and error, shrewd negotiating, creativity and comprehension.

In essence, computer simulations and games combine the best of traditional games and class instruction. They offer considerable logic and problem-solving possibilities and can generate a good deal of oral discussion and group or individual decision-making among special needs students. Our experience indicates that these students are extremely motivated in playing simulation games and engage in considerable risk-taking and perseverance, even in the face of apparent frustration. Moreover, the intense participation of these students appears quite resistant to extinction over time. However, the key is still the ability of the classroom teacher to integrate and transfer the computer simulated experiences to the actual curriculum being taught in the classroom. Teachers must constantly remind the students of the relevance of the simulated experience to other classroom activities. For instance, to enhance the reading and language goals associated with the use of simulations, teachers should have library and other materials available for supplementary reading, writing or discussion activities.

According to Willis and Kuchinskas (1983), most computer instructional simulations are in the mathematics, science or social studies areas. They may be valuable for the reading/language arts teacher or resource special education teacher as well, since all such games require the use of a variety of reading/thinking/problem-solving skills regardless of the subject area. These authors see many advantages to computer simulations which have relevance for the teacher of special needs students. In their opinion, simulations are fun, inexpensive, safe, realistic, provide for transfer, have low development costs, create less threat and anxiety for students, and provide maximum use of learning time. In our experience, we would add that use of well-constructed and pedagogically sound computer simulation games creates less off-task behaviour, more reflective thinking and cognitive awareness, and more opportunities to learn for special need students in the regular or resource classroom. The games provide an opportunity for these students to learn and perhaps experience vicariously a whole range of phenomena that for one reason or another are not provided in their current curricula.

Despite the many advantages of computer simulations and interactive adventure games, the predominant trend, especially with learning-handicapped students, still is to use the computer mainly for drill and practice or for tutorial applications in Mathematics and Language Arts areas. Yet research in special education consistently reveals the critical need to develop within these students more higher-level critical thinking, concept-relating and creative problem-solving skills in addition to self-instructional organizational strategies and meta-cognitive awareness skills, particularly in abstract learning situations. More appropriate uses of the computer with these special students, therefore, should include more tool and dialogue or inquiry applications of the computer in such computer areas as word processing, structured programming, data processing, information retrieval, creative arts, interactive dialogue, as well as simulations and creative problem-solving games. If students who exhibit a wide range of learning disabilities and often inefficient learning styles are to be prepared to meet the demands of the evolving information age, it is critical to develop a flexible and interactive classroom learning environment or "microworld" which fosters the necessary cognitive, affective and attitudinal skills. Most important is that computer technology provides teachers with a workable means for adapting the creative process to suit the individual ability of all students (Gallini 1983). Through its attributes of interactivity, immediate feedback, variability in presentation, neutrality of response and privacy, the computer is a powerful instructional tool that offers hope in fostering a balance of divergent and convergent thinking (Steffin 1983). It is to these creative student-directed and inquiry-oriented applications of the computer that we again direct our attention.

STUDENT-GENERATED COMPUTER PROGRAMS

Student-generated computer programs offer a unique opportunity for teachers of special needs students to assess the conditions under which students learn best. In the course of learning how to program, students develop a wide variety of competencies, including skills in mathematics, reading, spelling, proofreading, typing, logical thinking, planning, organization, evaluation and creative problem-solving, and knowledge of languages. When programming, the student becomes the "tutor" while the computer is the "tutee." The research evidence clearly shows that many students, including special needs students, develop a desire to learn programming after their initial experience with microcomputers (cf. Davis 1983; Lindsay and Marini 1983).

Use of LOGO in the classroom

The structured computer language LOGO in particular allows a disabled learner to build his or her own intellectual structures through estimation, interaction experiences and revision (Papert 1980; Weir 1981; Watt 1982; Weir, Russell and Valente 1982; Russell 1983). The focus of LOGO is on process rather than product. Hence, it can serve as an excellent tool with which learning-handicapped students can explore their own thinking processes and with which teachers can assess students' reflective thinking, hypothesis-testing, creative problem-solving and error analysis (i.e., "debugging"). LOGO's properties were specifically developed to make it useful as a learning language. Its graphics capabilities allow novice programmers to solve interesting problems (not necessarily mathematical) of their own creation from the first session. Because it is interactive (you can see the results of your program right away), user friendly (its messages to the user are very clear), and processes language in hierarchies the way humans do (letters, words and sentences), its use seems natural to the learner, especially the handicapped learner. Its creator, Papert (1980), wishes to enable the child to use the computer as a "mudpie" (1984a), as an environment for natural learning, and as a tool for building his or her own personal models of knowledge.

In the area of mathematics, Papert wants teachers to use the computer to provide an environment in which students are free to explore and discover mathematics on their own terms. LOGO has evolved from a computer language to an "environment," "a philosophy of education," a "culture" (Papert 1984b). Based on the ideas of Piaget, Bruner, Ausubel and other cognitive psychologists who stress the need for all students to incorporate new learning as part of their cognitive structures (Joyce and Weil 1980), LOGO seems to spring from valid theoretical ground. It follows the procedures of "discovery learning," often advocated for learning-handicapped students, in which the student discovers new relationships and links them to what he or she already knows about the world. It follows the trend to individualization and student-centred instruction so critical for students with special learning needs.

Most students and teachers are introduced to LOGO through its "turtle graphics" capabilities, which are particularly suited for teaching concepts of logic and mathematics even to young children. In LOGO the student makes decisions about where to move a small cursor, the turtle, on the computer screen. The turtle cursor can be easily programmed by the student in order to make it move in different directions while leaving a coloured trail on the screen. However, the applications of LOGO go beyond its graphic design, geometric patterns and control capabilities since students can also create animated cartoons (sprites), scenes, video games, educational games, interactive quizzes, poetry, music, art, etc. LOGO offers the observant teacher a unique chance to assess logical thinking and organizational skills since it encourages students to break problems into small steps and then to solve them individually.

Using LOGO as a learning tool, the child takes an active self-guided role in learning. Furthermore, the child begins to program the computer rather than being programmed by it, as is the case with most CAI programs for drill and practice (Bitter and Watson 1983). The graphics mode of LOGO also encourages creativity and artistic skills. An especially attractive feature of LOGO for the student with special needs is that it produces a greater tolerance of errors in its users, since much of the learning is by trial and error. Indeed, errors help the student to see cause and effect and to show how programs can be corrected and improved. LOGO has been found to be particularly effective in assessing and teaching mildly handicapped children and youth (Lindsay and Marini 1983; Turkel and Podel 1984). One reason for the special appeal of LOGO with learning-handicapped students is that it allows the student to discover and explore a new learning world which is less dependent upon facility with language (Russell 1983). In general, it can be said that teaching students to program in multipurpose computer languages such as LOGO, or even a more abstract language like BASIC, can develop essential problem-solving abilities and provide the teacher with extensive opportunities to assess a broad range of cognitive information-processing skills (Sitko 1984).

In one extensive research project, the Brookline Project (Papert, DiSessa, Watt and Weir 1979), researchers investigated the application of LOGO with sixth grade students who were provided with 20 four-hour sessions of hands-on experience with LOGO in a computer laboratory established in their school. The 16 students selected for indepth study represented a broad range of abilities and interests and included several children with learning disabilities. Most of the students (including the student diagnosed as severely learning-disabled) learned various elements of the LOGO computer language. Individual approaches to solving problems with LOGO revealed strengths of the learning-disabled students and suggested possibilities for modifying teaching styles to match these strengths. Overall, the data generated by this study are encouraging and suggestive of many possible advantages of LOGO, particularly in facilitating individualization to take advantage of individual learning styles. In fact, several researchers have pursued such applications with handicapped and other students who have not been succeeding in the traditional classroom environment (e.g., Weir, Russell and Valente 1982, 1984; FireDog 1984; Turkel and Podel 1984). Turkel and Podel (1984) conclude from their investigation of LOGO with mildly handicapped students that computer-assisted learning with LOGO appears to have potential as a valid and exciting means of motivating active problem-solving in special education students. LOGO use has also had a positive effect on student self-esteem, especially when students controlled their own projects. Our own observations with handicapped students using LOGO would certainly support these conclusions. However, as with other computer applications discussed in this chapter, most of the studies using LOGO still point to the critical role of the teacher in the complex interaction of people and environment occasioned by LOGO use. Issues of teaching style (e.g., structural or discovery learning) and teacher support (e.g., leadership, inservice training, materials) have been particularly identified.

CLASSROOM USE OF CREATIVE ARTS HARDWARE AND SOFTWARE

The use of creative arts hardware and software provides an excellent opportunity for the classroom teacher to develop conceptual and creative abilities in special needs students. Current developments in computer applications now allow these students the option of using, developing and designing their own artistic and musical programs to express themselves. Some specific examples especially applicable to special needs students will illustrate this point.

Art Graphics

One particular device which is receiving widespread use by both special and regular class teachers with their special students is the Koala Pad Touch Tablet (Koala Technologies 1983). The Touch Tablet is a white plastic tablet about 15 by 20 centimetres which contains a black, 10 centimetre square, pressure-sensitive drawing surface on its face and two control buttons. Using a software program called *Koala Painter*, students actually draw on the Koala Pad using a stylus or their own finger. Their drawings appear on the computer screen in full-colour high-resolution graphics. *Koala Painter* is a menu-based drawing system which allows a variety of choices on the part of the student, including how to draw, eight different "brushes" to draw with and any of 16 solid colours and patterns. Some of the command options available to the student include: a) freehand drawing, b) making rectangular frames, c) making circles, d) replacing a colour in a design with a different colour, e) creating mirrored images using the current drawing mode, f) making lines that radiate from one point, g) filling specific areas with a selected colour or pattern, h) magnifying designs for detailed work, and i) drawing individual straight lines. There are actually 128 possible pattern combinations. Students can also create designs on two separate viewing screens or copy images from one screen to the other. Another feature allows the student to save designs for future enjoyment. There are also public domain software programs available which allow teachers to "screen-dump" or print their student's artistic products on a peripheral computer graphics printer.

The ability to operate the computer without recourse to the microcomputer keyboard makes the Koala Pad an easy and versatile tool for the student and teacher. Many teachers use the Koala Pad as an introduction to the computer for special students, but more often they use it as a friendly, powerful drawing tool for creating art and graphics that can turn the "most untechnical person" into a computer "lover" (Adam 1984, p. 58). Several teachers have personally informed this author that many of their special students had not been much interested in the microcomputer or were hesitant about using it until they experienced the easy-to-use but multifeatured Koala Pad. Other software is available for the Koala Pad, including a good music composing and playback program called *The Music Construction Set* (Electronic Arts 1984). More recently, Koala Technologies has also produced a powerful computer light pen, The Gibson Light pen, which is appropriate for students at the fifth grade level or above. The pen comes with a library of five programs which allow students not only to draw but also to write and play music, create animation, and design graphs and other layouts. With the light pen, students simply touch the computer screen with the pen to make menu selections and detailed freehand drawings and designs.

There are many other good artistic-graphics drawing programs now available for the special needs student. Besides LOGO which was reviewed previously, another useful LOGO-like program is *Delta Drawing* (Spinnaker 1983). The publishers of *Delta Drawing* describe it as an easy, friendly introduction to computer learning which helps build an understanding of computer concepts and procedural thinking, while encouraging children to exercise their creativity. In *Delta Drawing*, students begin by drawing pictures that produce computer programs, but they soon learn to write programs that draw pictures. They create graphic pictures one step at a time using a minimum of single-key keyboard command strokes. Like the turtle cursor in LOGO, *Delta Drawing* uses an arrow-like "Delta point" which points in the direction it will move and has two major characteristics—position and heading. When students draw with the Delta pointer, they also create a coloured drawing program which consists of a sequence of commands they use to produce the picture. Then, whenever the students wish the computer to redraw their picture on the screen, they simply call the drawing program from memory by executing the program name that refers to it. *Delta Drawing* also allows students to "nest" drawing programs inside another program. Hence, students can gradually construct complex pictures, patterns and designs from simpler programs. They can also save their programs on diskettes or print their drawings on a graphics-type printer for hardcopy storage.

Two more graphics drawing programs recently on the educational and home computer market are *Mr. Pixel's Programming Paint Set* and *Mr. Pixel's Cartoon Kit* (Mindscape 1984). Both of these programs also have good potential for use with special students. The former allows students not only to paint a picture but also to enlarge it, repeat it, move it around, and change its colours. The latter program allows students to create animated cartoons or "sprites" which can move around and even react to each other.

These drawing programs and others like them allow students to develop their imagination, their logical thinking and their programming skills in a situation that is both motivating and non-threatening. The students themselves control the computer, using it as a flexible tool to unleash their own creativity and to explore and learn. Drawing programs are particularly useful for special students who demonstrate language difficulties at either the spoken or written levels. Teachers also remark that student computer drawings often reflect the cultural backgrounds of their students and provide an excellent way for students from a variety of ethnic groups to receive positive recognition for academic work from teachers and students (Ettinger 1983). Drawing programs also help special students discover the elements of design, including line, shape, colour, form and position.

Musical Applications

Besides computer drawing, another exciting medium for stimulating creative exploration in special students is computer music. Teachers are increasingly using microcomputers with music synthesizers for teaching music and letting students compose, analyze, edit, review, and play songs and melodies. In this way, students learn various concepts of sound production, including waveform, chord, octave, timbre, pitch, duration, volume and transformation. The computer offers the special students in particular great control over sound production and enables them to experience musical concepts without traditional notation and terminology. The new generation of relatively inexpensive computerized sound synthesizers with keyboard now allows teachers and students to easily simulate many of the traditional orchestral instruments or to create new instruments that have never been heard before. Moreover, with the addition to the classroom computer of a music sequencer and a music synthesizer, students can even "orchestrate" or combine various simulated instruments into their own musical arrangements.

The available computer-aided music programs greatly facilitate the learning and playing of musical instruments and the creation of sound and music. Some of the programs include "eye-popping" colour graphics that appear across the screen when students replay their own songs and melodies. Many teachers remark that these programs are changing the way their students play, and the way they think and feel about music. They add that even students who don't read music or who reveal poor coordination or dexterity and a poor sense of rhythm are learning to explore creatively and produce excellent sound and music. Many of the available programs allow students to enter notes at any speed and to instruct the computer to resolve the music into the degree of rhythmic accuracy or harmony which pleases them. Hence, the computer, used as a learning tool, can help create a musical "microworld" or musical environment which is highly exciting and lets the special student make musical discoveries on his or her own terms. At the very least, the recent generation of computer software has the potential of making music more fun, more rewarding, more exciting and easier for the special student than it's ever been before.

Most of the extensive music software available for classroom use is geared to specific brands of microcomputer hardware. As a result, teachers need to be cautious in choosing appropriate music software to match their needs and the available computer hardware in their particular classroom or school. A list of selected music software appropriate for special needs students is provided in the teacher resource appendix at the end of this chapter. Specific guidelines for assisting teachers in choosing music and other academic software are presented in the following section.

CURRICULUM IMPLEMENTATION OF COMPUTERS

Probably the most significant aspect of computer technology for classroom teachers is the degree to which the computer can facilitate the match between the curriculum demands of the classroom and the individual learning needs of the students. Teachers are mainly concerned with the efficient use of the computer as a tool which can assist in effectively teaching a particular aspect of the curriculum. The most successful classroom applications of the computer are a direct function of the teacher's ability to determine how best to use the computer's capabilities in the context of existing curriculum demands and diverse learning needs.

Introducing computer applications into the classroom for special needs students requires the same careful considerations that teachers would give to any new instructional technique or curriculum change: the objective(s) and justification for the technique must be clear and teachers must have sufficient knowledge and expertise concerning the technique in order to implement it effectively in their curriculum. A difficult and sometimes confusing step for teachers in this process will be finding and obtaining valid computer software that can be effectively and fully integrated into the existing curriculum. Hopefully, the preceding sections of this chapter have provided some basic knowledge of the types of computer applications and software that may be appropriate for the special needs student and of the kinds of benefits that might be derived from computer use. Throughout the chapter, the author has repeatedly stressed the critical role of the classroom teacher in the decision-making process concerning the appropriate use of computer software. The concluding sections of the chapter will attempt to assist the teacher further in this role by presenting recommended principles for evaluating and integrating microcomputer software with current classroom curriculum.

Implementation of Computer Software in the Classroom

Integrating computer software with classroom curricula requires teachers to plan well and to be flexible decision-makers who can modify their plans according to the demands of the total teaching-learning situation. In essence, implementing computer software for the special needs student requires the same adherence to effective teaching and learning principles and procedures that is necessary for bringing about any change through programming.

Five major beliefs about teaching, program design and the *context* in which teaching takes place have shaped the approach to computer implementation with special needs students that we are presenting (Sitko and Slemon 1983). First, at its best, teaching is seen as an active, ongoing process in which teachers are continually assessing how students are responding to and making decisions among various alternatives. Second, teaching involves a total environment, an ecological system, and decision-making will be most effective if teachers are sensitive to various components of the total environmental context of the classroom. Third, assessment or evaluation is an ongoing process, and it should be considered an integral part of all teaching acts. This belief implies that many events in classroom programming situations furnish data for assessment and decision-making. Analyses of tests, essays and other completed products provide important assessment information; but observations on who students interact with, on how they interact, and on how they respond to different topics, as well as on their questions, requests and their responses to various media are equally important. In this context, the computer is an excellent tool in helping the teacher to assess the conditions under which the student learns best. Fourth, students with special learning needs learn best in situations where the tasks that they are required to master are challenging, but not overwhelming, and where they feel that accomplishing those tasks will assist them in achieving personal goals. Fifth, learning difficulties are assumed to be produced by multiple causes—if teachers focus too narrowly on students, they overlook crucial situational factors. Expectations, teaching strategies and various other factors from the classroom, school and home environment all interact with student capabilities, expectations, and emotional reactions in creating success or failure in school tasks.

Asking, "What computer alternatives or software might be tried in the classroom?" introduces procedures for effecting classroom change through new instruction or programming. In an environmental approach to programming, it is therefore important to ask whether changes should be made in the expectations, demands and the resources of teachers and classrooms, or in the expectations, skills, knowledge and motivation of students. The existing curriculum has been developed to assist students. If special needs students are not progressing well in school, the problem may at least partially lie in the task and in classroom environmental demands. If it is decided that new change procedures should focus on students, two general strategies may be used in selecting computer software for programming purposes: 1) Choose software that teaches to students' strengths, or 2) choose software that remediates or rehabilitates their weaknesses. In other words, selecting effective change procedures for special needs students requires that teachers be familiar with, and select among, a variety of feasible alternatives, including the possible use of computer software, in order to try to match tasks with the special needs of individual students and groups.

Curriculum Task Analysis Phase An important initial procedure in implementing computer technology into the classroom involves a needs assessment by the classroom teacher to determine those areas of the curriculum that might be improved through the use of computer applications (e.g., improvement of creative writing via use of a word processor). Following the procedure the teacher should:

1. establish specific goals and educational objectives involving the computer for a particular student (or group of students);

2. determine the priority of each computer learning objective in the existing curriculum; and
3. plan the teaching sequence for achieving the learning objective.

Figure 14-1

Components of a Task Analysis Procedure for Lesson Planning

1. SPECIFY a main task appropriate for undergoing task analysis.
2. IDENTIFY subtask at the preceding level of difficulty.
3. TREAT each subtask as a main task and identify simpler tasks at preceding levels of difficulty.
4. STOP the analysis when a subtask reaches the pupils' entry level.
5. CONVERT a task analysis into a learning sequence*.
   *Learning sequence: Shows the order in which your learners should proceed through the list of subtasks until they can perform the main task.

   To establish a learning sequence:
   a) Proceed from the easiest to hardest task.
   b) At any given level of difficulty, begin with the subtask that will be most frequently used.

From Semmel et al., *Tips for Teachers Series, Module 15: Task Analysis* (Bloomington: Indiana University, 1974). Reprinted with permission.

These three steps in computer implementation strategy can then be facilitated by means of a task analysis planning procedure (see Figure 14-1). The task analysis will assist the teacher in answering such important questions as: What do I do first in carrying out an effective computer curriculum (e.g., a computer writing program)? What do I do next? What sequence of steps is the student or group to follow? What must my special students know to perform each of the computer tasks or topics in the curriculum? What prerequisites skills or entry skills are required before starting a learning task or topic involving the computer? What benefits will the special needs students obtain from carrying out particular computer tasks or topics? How can a computer task or topic be presented or introduced so students will comprehend its meaningfulness or relevance in their daily school curriculum? How do I know if I have successfully followed a proper teaching sequence in implementing my educational plan for integrating the computer into my existing curriculum?

Task analyses are normally performed by describing the ultimate behaviour or learning outcomes that the teacher wishes a student or group to acquire; by asking what preliminary skills, concepts, attitudes and knowledge the students must already have to achieve the ultimate behaviour with a minimum of instruction; and then by asking, in turn, what sub-preliminary skills and knowledge are necessary for learning the preliminary behaviours with a minimum of instruction. This process should continue until the teacher reaches skills, attitudes and knowledge that a particular student or group already possesses. Thiagarajan, Semmel and Semmel

(1974) review the steps that would be required in a complete task analysis of computer courseware. Figure 14-1 summarizes these steps.

Hence, at the end of the task analysis phase of computer implementation, the teacher should have established a series of justifiable instructional tasks or topics to be taught by a computer program or curriculum; the sequence in which they should be taught; and how, and under what conditions, the appropriate computerized courseware or software should be used. In many cases, completed task analyses containing learning outcomes already exist in computer software that can easily be adapted for the teacher's curriculum objectives. For instance, scope and sequence charts which list instructional objectives for computer courseware are often available in program documentation that can be purchased with computer courseware. The key to their usefulness is in adjusting them to fit a particular class and curriculum.

Learner Analysis Phase Learner analysis is another important step in deciding upon and implementing specific computer courseware that matches student learning needs. The learner analysis phase helps the teacher answer such questions as: What do I know about a particular special student's or group's strengths and weaknesses? What do I know about my special students' preferred cognitive styles of learning (e.g., reflective, impulsive, abstract, representational, concrete, etc. and their preferred sensory modalities (e.g., visual, auditory, tactile, cross-modal)? What teaching strategies seem to work best? Am I being sufficiently flexible in my choice of educational treatments? Under what conditions does this student appear to learn best?

In deciding upon what learning tasks or topics to assign a particular special student or group via computer courseware, it is helpful to consider what is known about students' strengths and weaknesses. If some form of preliminary curriculum task analysis has already been carried out to determine what possible curriculum needs may be met by the use of computer courseware, the next step is to use a systematic learning analysis procedure to assess the match between curriculum and computer task demands and student capabilities, learning skills and motivation. In performing a learner analysis, a teacher will want to evaluate how well the curriculum objectives, teaching strategies and sequences that have been identified in the task analysis phase actually match the entry skills, learning styles and attitudes of specific students. Figure 14-2 presents some of the components which might be considered by the teacher in performing a learner analysis for special needs students in a regular or special classroom setting.

Evaluation of Computer Software Phase In implementing computer courseware, teachers must also be prepared to select, adapt and modify the software and their own teaching methods and formats as they obtain information about how particular students or groups react to them. This computer software evaluation phase is crucial to the successful implementation of computer courseware with special needs students. Unfortunately, much available commercial and public domain software is acknowledged to be of poor quality and often requires considerable adaptation and modification if it is to be used effectively with special needs students. The potential

Figure 14-2

Learner Analysis Components

1. What specific content subjects does the student approach or prefer?
2. What specific content subjects does the student avoid?
3. What specific tasks does the student approach?
   a) oral responding
   b) reading comprehension
   c) listening comprehension
   d) writing activities
   e) problem solving
   f) recall
4. What specific task(s) does the student avoid?
5. In which type(s) of grouping situation does the child appear to work best?
   a) individual-monitored by teacher or aid
   b) individual-independent
   c) small group (instruction), groups of two, three, four or more
   d) small group (game)
   e) large group
   f) other
6. With whom does the student seem to work best?
   a) classroom teacher
   b) aid or volunteer
   c) resource teacher
   d) other adult (specify)
   e) peer (cross or same age)
   f) male or female
7. Are you aware of any students in your class or other classes that may interact successfully with the student?
   a) in a tutoring situation (specify)
   b) in a game situation (specify)
   c) in a creative arts situation
   d) adult-monitored activity (specify)
8. How would you rate the child's attention and comprehension of material presented?

	Adequate	**Inadequate**	**Dependent (be specific)**	**Don't Know**
a) auditory				
b) visual				
c) kinesthetic				
d) tactile				
e) cross-modal				

9. How would you rate the child's attention and comprehension of material presented?

	Adequate	Inadequate
a) through concrete demonstration		
b) through representation material		
c) through use of abstract symbols		

10. How would you rate the child's ability to process information presented?

	Visual		Verbal	
	Adequate	Inadequate	Adequate	Inadequate
a) rapidly (without pauses)				
b) slowly (without pauses)				
c) rapidly (with pauses)				
d) slowly (with pauses)				

11. How many directions do you think the child can process at one time?

12. How would you rate the child's oral, gestural and graphic ability to respond?

	Adequate	Inadequate
a) oral		
b) gestural		
c) graphical		

13. How would you rate the child's reaction to the following reinforcement conditions:

	Unfavourable	Very Favourable	Favourable
a) oral			
b) graphic			
c) immediate			
d) delayed			
e) social			
f) tangible			

14. Does the student depend mainly on sheer rote memory and low-level associative and/or concrete cues in recalling information?

YES NO NOT SURE NEED FURTHER ASSESSMENT

15. Does the student utilize good organizational and grouping or chunking skills to process written or verbally presented information?

YES NO NOT SURE NEED FURTHER ASSESSMENT

16. Can the child describe specific mnemonic, or meta memory, or cognitive strategies when organizing and recalling verbal or visual information?

    YES NO NOT SURE NEED FURTHER ASSESSMENT

17. Rate the student's ability to answer the following questions or requests (see Sitko, Slemon and Bachor 1981).

	Excellent	Good	Average	Inadequate (below average)	Poor
Discrimination					
Verbal					
Written					
Recall-Sequencing					
Verbal					
Written					
Relating Concepts					
Verbal					
Written					
Problem Solving					
Verbal					
Written					

18. a) Would you consider the student to be an impulsive thinker or a reflective thinker?

    Circle I or R

    b) How would you rank the student's skill in reflecting upon ideas?

    EXCELLENT GOOD AVERAGE BELOW AVERAGE POOR

19. Does the student appear to perform better in deductive (from abstract to specific—or whole to part), or inductive (from specific to abstract—part to whole) teaching lessons?

    Circle D or I

20. Does the student appear to be distractable on visual, auditory or motor learning tasks?

	Definitely	Maybe	Not Really	Not Sure
Visual				
Auditory				
Motor				

21. Does the student appear to be more externally or internally motivated in academic learning situations?

    Circle E or I

22. List some of your student's likes or interests as you know them. Be specific, if possible.

23. Does the student reveal figure-ground confusion dealing with worksheets containing many problems or tasks on a page?

    YES NO NOT SURE

24. Does the student require written work to be highlighted or made salient through colour and other cues?

    YES NO NOT SURE

25. How would you rank the student's self-image and self-confidence in undertaking assignments?

    VERY STRONG STRONG ADEQUATE INSECURE VERY POOR

of the computer to assist in the successful learning of these students depends on the pedagogical effectiveness of the computer software. Some computer software has failed to build upon the significant advances that have been made in instructional theory, educational psychology and curriculum design. Hence, teachers and other educators need to critically analyze available software that is either specifically designed for use with special needs students or that could be adapted for use with these students.

After teachers have established tentative instructional needs for students that may involve the use of computer courseware, they should then select computer software methods and classroom management systems which hopefully will match the identified needs of their special education students. Teachers can only consider hardware and software that they can realistically hope to obtain. Other teachers in the system whose curricula and students are similar may be good sources of ideas and information in this regard. If a computer coordinator is employed by the board, valuable practical assistance can also become available. In addition, educational and computer magazines and journals (mentioned in the appendix) contain software advertisements, reviews and teacher reports on their classroom ideas and experiences with computers which may be of valuable assistance. In the final analysis though, classroom teachers themselves must make the ultimate decisions on whether to and how to implement computer courseware with their students.

A try-out should be undertaken to determine whether the computer software and the instructional approach they have selected actually does work with a particular student or group. Evaluating the fit between computer software and teaching approaches incorporating computers is an ongoing process, but in the critical stages of computer implementation, teachers should systematically evaluate the appropriateness of chosen software by trying out the software in simulated or actual teaching lessons. When working through a particular computer program in a simulated situation, the teacher should try to envisage the context in which the eventual instruction will be taking place from the point of view of selected students in the classroom. For example, how will Susan who is impulsive and insecure as well as being a poor reader, react to the program? Mike is an excellent reader and highly motivated to achieve. How will his reactions to the computer program be different? Actual trial teaching lessons with particular students will allow teachers to monitor more systematically the students' daily progress and their interactions with computer courseware. These lessons are the core of the computer implementation process.

A number of educators have developed specific evaluation forms to assist teachers in evaluating and implementing computer software. Figure 14-3 contains one such evaluation form. The form stresses the importance of the interaction between task demands and student capabilities. Figure 14-3 has been purposely developed to assist teachers and other educators in this important task.

Figure 14-3 displays a suggested rating form which presents various characteristics and criteria for evaluating computer software. In using the form, the teacher or other educator analyzes a particular package of computer software across three dimensions: technical qualities, curricular considerations, and pedagogical and psychological considerations. On the basis of these dimensions, the evaluator ranks selected characteristics of each software package for their degree of appropriateness for use with a specific student or group of students. The evaluator gives each characteristic an Appropriateness Rank of (2) if the characteristic is very appropriate, (1) if it is fairly appropriate and (0) if it is inappropriate for a target student or group of students. If a particular characteristic receives a (1) or (0) for its Rank of Appropriateness, it is then ranked a second time for its possible adaptability or modification for use with the target student(s). The evaluator then gives the particular characteristic a Modification Rank of (2) if the teacher or the student(s) can easily modify the computer program to ensure its appropriateness and suitability. For example, if there are directions available for simple modifications of the program in the documentation that is supplied with the computer software package, then the evaluator would give a Modification Rank of (2) to the characteristic being rated, since presumably the program could then be modified by the teacher on site without much delay. On the other hand, the evaluator would give the characteristic a Modification Rank of (1) if the program can only be modified by other support personnel such as an experienced computer programmer, computer consultant, or someone else employed by the school system. This type of modification would necessitate some delay in making the program suitable and available for immediate use by the target student (or group). A modification rank of (0) is given to the characteristic being rated if the program cannot be modified on site or by someone in the school system. The evaluation form also includes a column in which the teacher or other evaluator can check a N.A. if the trait being rated is not applicable for the intended target student or group. After rating all the characteristics across the three dimensions, the evaluator then uses a simple formula to total the ranks and receive a Composite Software Suitability Score (CSSS). The teacher then decides if the total score of the software package is high enough to justify its use with the target student or group. Figure 14-3 presents only one model or type of software evaluation scheme appropriate for use with classroom teachers. Other sources of software evaluation are presented in the appendix to this chapter.

Figure 14-3
Criteria for Evaluating Computer Software

Program Characteristic	**Presence of Trait**	**Degree of Appropri-ateness for Target Student(s)**	**Modification Potential for Target Students**
DIMENSION I Compatibility with Existing Curriculum or Learner Objectives	If present, NA if not applicable	2=Very 1=Fairly 0=Inappro-priate	2=By Teacher or Student 1=By Support Personnel
1. Specific educational goals, objectives and/or learner out-comes are delineated in the accompanying documentation, teacher's guide or in the program itself. (cognitive, affective, or psychomotor domains)			
2. Documentation contains a thorough and accurate description of the content, learning sequence and rationale, structure of the program. Program does not contain any serious errors in the content or fact.			
3. Program makes its goals or objectives clear and evident to the learner. Program content is compatible with instructional goals and objectives.			
4. The suggested target audience, learner group, grade or ability level is specified. (Note, how-ever, that suggested age, grade or ability level may be totally irrelevant to your own objectives for the target learner(s).			

Program Characteristic	Presence of Trait	Degree of Appropri-ateness for Target Student(s)	Modification Potential for Target Students
5. Documentation discussed or one can easily infer the prior knowledge (instructional prerequisites) or entry skills the learner must possess in order to use the program successfully.			
6. Support materials or suggestions are available in my school or system from other teachers who have experience with this software.			
7. Learner's entry level of performance and placement into the computer program is assessed by the program itself.			
8. Program is compatible with, and related well to the learner's current stage of cognitive development, academic achievement and/or interests. Hence, difficulty level should be appropriate for my target student or group.			
9. Knowledge, skills, or attitudes developed in the program can be generalized to other curriculum areas or meaningful real-life situations. Program objectives can easily be tied to other instructional programs or activities for the target learner(s). Hence, the program allows the learner or teacher to easily apply or use newly acquired skills or knowledge.			
10. Documentation contains suggested classroom activities or instructional activities to			

Program Characteristic	Presence of Trait	Degree of Appropri-ateness for Target Student(s)	Modification Potential for Target Students
introduce and integrate the program into a compatible or existing curriculum.			
11. Program documentation is clear, well organized and easy to understand and follow.			
12. Documentation contains suggested classroom activities or instructional strategies for teacher follow-up and supplementary activities.			
13. Documentation contains recommended resource and reference information for additional reading or research.			
14. Documentation contains data for evidence including dates and settings in which the program has undergone field testing and summative evaluation involving specific types of learners. Program has been highly recommended by other teachers or educators who have used it to meet similar goals or learner objectives.			
15. Content is free from racial, ethnic, socio-economic and sexual stereotypes.			
16. Program is compatible for both group and individual use.			
17. Program objectives, content, activities and evaluation are all related and focus on the same knowledge or skills (i.e., program modules are instructionally related to one another).			

Program Characteristic	Presence of Trait	Degree of Appropriateness for Target Student(s)	Modification Potential for Target Students
18. Student mastery of program objectives is adequately tracked or measured by the program. Student progress data is efficiently displayed.			
19. Difficulty level and sequence of presentation is under teacher and/or learner control. (Learner senses that he or she is in control of the learning process).			
20. Program allows learner to branch to easier or less difficult task levels following no more than a few incorrect or inappropriate responses (i.e., failure experiences are kept to a minimum and sense of mastery or success is maximized). Students are consistently on a success base.			
21. Program allows learner to branch to more difficult task levels following appropriate or current responses (learner is adequately challenged so that neither boredom nor satiation set in).			
22. Estimated time for target learners to achieve program objectives is appropriate. Program addresses educational objectives that can easily be met by the target learners.			
23. Program is cost-effective in terms of its estimated usefulness and longevity with different learners or student groups. Program can be exchanged with other teachers or used with similar learners.			

Program Characteristic	Presence of Trait	Degree of Appropri-ateness for Target Student(s)	Modification Potential for Target Students
24. Program meets its intended purpose or goals for the target learner(s). Program objectives or content can be learned better using the computer than by using other instructional methods or programs available in the classroom. (In other words, the computer program provides uniqueness or improvement over other available teaching methods or materials, but is still compatible to my own instructional goals for the target learner(s).			
25. Program stimulates student reflection and high order or higher-level thinking (e.g., problem-solving, divergent thinking, concept-relating, hypothesis-testing, deductive or inductive logic and planning, etc.) Program goes beyond simple recall or discrimination learning.			
26. Program develops or instills a sense of cooperation among students. Program is relatively free from an over-emphasis on student competition.			
27. My subjective reaction to this program in terms of its compatibility to my own values, philosophy and models of teaching is very favourable. I would probably recommend the program to other teachers or parents.			
Sub Total			

DIMENSION II Pedagogical and Instructional Considerations	**Presence of Trait**	**Degree of Appropri-ateness for Target Student(s)**	**Modification Potential for Target Students**
1. Program learning is broken into tasks or a learning sequence that the target student(s) can successfully master. Learning sequence is clear and concepts and skills are organized in a logical progression from the simple to the complex in steps which are manageable for the learner and easy to discriminate.			
2. Program content and format appear interesting and capable of intrinsically motivating students to maintain interest and achieve program goals and objectives (i.e., the program should be fun for the target learner and sustain interest across time and use.			
3. New or abstract concepts, information and other material are introduced or adequately explained in terms of interesting, familiar or meaningful content containing many concrete, perceptual and associative cues and meaningful referents.			
4. Method of presentation seems appropriate or matches the preferred learning style of the target student(s).			
5. Program minimizes information overload and contains relatively little irrelevant, ambiguous, distracting or redundant information, inappropriate to program goals.			
6. Program is versatile. The learner can review, recall or			

Program Characteristic	Presence of Trait	Degree of Appropri-ateness for Target Student(s)	Modification Potential for Target Students
obtain additional instructions, explanations or other content, or obtain help when desired. The learner can also skip unwanted or already completed material and exit from the program when desired.			
7. The learner can load, enter and operate the program easily and independently (or without much external help).			
8. Target learner(s) should not be frustrated by the program content and format. The level of difficulty, pace or rate of presentation, and instructional sequence of information can be controlled by the learner (or teacher).			
9. Learner instructions are included in the documentation or program itself. Program directions do not contain any significant distracting or redundant information. Directions are relatively free from ambiguity or confusion. Directions are clear and concise. Program ensures that the learner understands instructions through use of examples, repetitions, quizzes or practice.			
10. Program directions can be modified to suit target learners' reading level. Approximate program reading level is specified in the documentation.			
11. Program reduces figure ground confusion by placing only a few problems or limited			

Program Characteristic	Presence of Trait	Degree of Appropri-ateness for Target Student(s)	Modification Potential for Target Students
text information on the computer screen at one time. Program has a minimum amount of extraneous stimuli or information on the screen. Written text is presented or organized on the screen in easy-to-read blocks or chunks of information. Graphics, colour, and other cues are used to prevent the screen from being overly crowded with text.			
12. Various hints, prompts, or focussing cues (e.g., under-lining, boxing of important information, colour, sound, music, graphics, animation, voice, flashing lights, pauses, etc.) are used to highlight or heighten information.			
13. Program provides hints, prompts or cues to promote reflective as opposed to impulsive thinking, (e.g., "Think carefully before you answer." "Stop, Think, and Answer"). Where high-level or reflective thinking is the target, the program allows for flexibility in student responses, (i.e., the program assumes a problem or high-level question or request can have more than one appropriate response.)			
14. Program attempts to personalize the learning task through use of personalized names, reinforcers, etc.			

Program Characteristic	**Presence of Trait**	**Degree of Appropri-ateness for Target Student(s)**	**Modification Potential for Target Students**
15. Program provides adequate time for target learner(s) to complete instructional tasks or allows response time to be controlled by the learner (or teacher). Average or estimated time required to complete the program is realistic and practical.			
16. Amount of keyboarding required to execute the program is appropriate for the target learner(s).			
17. Program allows some leeway for incorrect spelling of typed responses (except in spelling programs). However, spelling in the program instructions or content itself in accurate.			
18. Program emphasizes informative positive feedback or reinforcement and avoids or de-emphasizes punitive or demanding feedback. Feedback or evaluation focuses on achievement of mastery or appropriateness of responses, rather than errors.			
19. Feedback on inappropriate or incorrect student responses prompts, probes and guides the student to obtain more appropriate or correct answers. The program does not overly stress guessing to obtain answers. Negative feedback explains why a student response is inappropriate or incorrect.			

Program Characteristic	Presence of Trait	Degree of Appropriateness for Target Student(s)	Modification Potential for Target Students
20. Feedback and reinforcement show spontaneity, variety and credibility for the target learner(s). Reinforcement is immediate or appropriate to the program task.			
21. Rate and type of feedback or reinforcement can be controlled or modified by the learner (or teacher).			
22. The use of sound is appropriate for the target student. In fact, use of voice or sound effectively aids in providing directions, verbal praise and verbal response to keyboard input. Moreover, the sound can be turned off when it is inappropriate for the target student without seriously reducing the efficacy of the program.			
23. Program provides a table, graph or permanent record of student performance over time which is easy for both the teacher and student to interpret.			
24. Student performance or output on the screen can be "screen dumped" or stored on computer disc or tape. Hence, teachers or students can easily obtain hardcopy print-outs of student performance for future retrieval and analysis.			
25. Screen print size is large enough for the target learner to decode. Program			

Program Characteristic	**Presence of Trait**	**Degree of Appropri-ateness for Target Student(s)**	**Modification Potential for Target Students**
will also allow the size of print to be modified to suit student need (i.e., lower vs. upper case text).			
26. Program content or text follows or models established linguistic rules for appropriate use of punctuation, capitaliza-tion, grammar and writing style.			
27. Program allows or accommo-dates voice-synthesized instruc-tions for poor or beginning readers.			
28. Program allows or accommo-dates touch screen, light-pen, joy stick, mouse or voice input for physically handicapped or other special students.			
29. The sound, graphics, colour, etc. appeal to the target learner and can easily be interpreted by the learner.			
Sub Total			

Dimension III Technical Adequacy	**Presence of Trait**	**Degree of Appropri-ateness for Target Student(s)**	**Modification Potential for Target Students**
1. Documentation contains information concerning the type of computer hardware required to run the program. Program is compatible with the available school or classroom computer system and comes in a medium that can be loaded			

Program Characteristic	Presence of Trait	Degree of Appropri-ateness for Target Student(s)	Modification Potential for Target Students
into the existing classroom computer system (e.g., cartridge, cassette, floppy or disc).			
2. Documentation contains information concerning the type of computer peripheral configuration required to run the program (e.g., disc drive, cassette record, printer, joystick, lightpen, mouse, paddles, 40- or 80-column screen, language card, etc.). Moreover, the program is compatible with the available school or classroom peripheral configuration.			
3. Program allows hardware input devices alternative to the standard keyboard (e.g., adaptive keyboard, light-pens, joysticks, touch tablets, etc.)			
4. Amount of computer memory (i.e., random access memory—RAM) required, to load and run the program is stated and is compatible to the classroom computer system.			
5. Technical use and aspects of the program are well explained in the documentation. Documentation is relatively free of difficult technical jargon.			
6. Documentation provides enough information to accurately load and run the program. Moreover, the program is reliable during the loading and running phases.			

Program Characteristic	Presence of Trait	Degree of Appropriateness for Target Student(s)	Modification Potential for Target Students
7. Program options are offered in a menu to the teacher and/or learner. Choices are clear and quick and may be easily revised.			
8. Student response or record-keeping system is easy to use by the student and/or teacher.			
9. The program contains no obvious or serious programming errors. Moreover, listing of the program and hence, program modification, is possible and allowed.			
10. User directions or instructions can be bypassed when not needed by the learner (or teacher).			
11. The program will continue to run when students make inappropriate keyboard responses. Unexpected or unintentional student responses never "abort," "crash," or "terminate" the program. Moreover, the program minimizes unintentional or unexpected keyboard errors by disabling keys which are unnecessary to the execution of the program.			
12 Software company or publisher provides a worthwhile guarantee for the program. A reasonable warranty period is provided (e.g., at least 90 days) and the program can be returned unconditionally or exchanged if proven unsatisfactory. Moreover, the publisher provides a free back-			

Program Characteristic	Presence of Trait	Degree of Appropriateness for Target Student(s)	Modification Potential for Target Students
up copy of the program, or allows a back-up copy of the program to be made without much difficulty. Where a free back-up is not provided, a replacement program can be purchased at a minimum cost (e.g., less than $10) when the warranty expires. (Note: It is highly recommended that the buyer require the publisher to allow purchase on approval (e.g., 30 day approval) on a school or board purchase order. Moreover, the publisher should describe a method for the buyer to receive program updates and should guarantee to allow the buyer to upgrade the program when a new revision is developed for a reasonable fee).			
13. Documentation includes addresses where the program author(s) or consultant(s) can be contacted should questions arise.			
14. Technical features support group as well as individual use. Group progress data or performance scores are adequately displayed.			
	Sub Total		

After ranking the software program on the three suggested dimensions, the evaluator should obtain a Composite Software Suitability Score (CSSS) in the following manner. First, he or she should tally the Appropriateness and Modification Ranks across the three dimensions (Columns 3 and 4) and obtain a Total Rank Score for Degree of Appropriateness and for Modification Potential for the target student(s). Next, he or she should use the following formula to obtain the CSSS for the computer program in question: CSSS equals *twice* the Total Rank Score for Degree of Appropriateness plus the Total Rank Score for Modification Potential, divided by N (where N = the number of characteristics that were actually ranked or were judged to be applicable to the target student or group of students.

$$\frac{2(\text{TRSDA}) + \text{TRSMP}}{\text{N}} = \text{CSSS}$$

Note: In general, the Degree of Appropriateness Rank is more important than the Modification Potential Rank since it requires more time and work to modify an existing program in order to improve its suitability or appropriateness with target learners. Hence, the Degree of Appropriateness Rank is given more weight and its total is doubled in the final calculation of CSSS.

The CSSS should be helpful in comparing various software programs for special needs students. Teachers should establish the minimum CSSS total which is necessary to warrant the purchase or use of a particular computer program with target students in their own classroom.

As part of the software evaluation process, teachers should try out the entire program under conditions that their students will be experiencing in order to determine if the software suits the students and the purposes for which it is intended. As mentioned previously, in trying out the program, the teachers should have certain students in mind and simulate the kinds of things their students would do and how they would react to the various features of the software package. In addition, a well developed list of characteristics and needs of the target students who will be using the software can be a valuable aid in choosing only high quality software (Appel and Hurley 1984). The learner anlaysis technique suggested previously can be of assistance in this regard. The important consideration, however, is that teachers and other educators become astute and critical evaluators and exercise their professional judgments when evaluating computer courseware and other instructional programs.

In addition to the importance of critical teacher evaluation of software at the classroom or school level, it is equally important that teachers become involved with software evaluation at the board level. Most boards now have a computer committee or team responsible for screening computer hardware and recommending high quality software. It is important that teachers with released time have representation on any computer committee and on any committee responsible for designing board-sponsored computer courseware. Using a software evaluation form (similar to the one illustrated in Figure 14-2 or those referenced in the appendix), board level review committees should carefully preview software before purchase. They should also solicit teaching recommendations from both regular and special classroom teachers and consultants from across the system. In addition, boards should be careful to solicit opinions from teachers regarding the types of technical assistance they would find most helpful in implementing computers in their classrooms. Many boards have even established Microcomputer Resource Centres to provide various forms of technical assistance, support and training for teachers, including evaluating reactions to software and hardware applications as well as computer software libraries for teachers.

It is quite possible that teachers may choose to develop or adapt their own computer courseware for special needs students. One recent survey revealed that 40 percent of teachers polled prefer to create their own software or to use programs developed by other teachers (Appel and Hurley 1984). In fact, several microcomputer authoring systems have been developed specifically to enable teachers and other educators to develop their own software (see appendix). However, the same considerations and criteria for computer implementation and for software discussed previously are equally important in the design and development of teacher-made software.

In addition to evaluating computer software per se, teachers must also develop and use systematic observational skills in observing student cognitive and affective behaviour during interactions with computer software. Teachers must also ask objective questions to determine how well special students have profited from the computer lesson. Depending upon the results of trial computer lessons with specific students, teachers may have to "research" and observe several such lessons before a consistent picture of the worth of a particular computer program or courseware emerges. Frequently, the best and most straightforward way to obtain data about how students adapt to a computer task is to try it out and ask them what was simple and what was difficult. By observing their effort closely and by analyzing their completed products, we can infer where problems occur (Slemon et al. 1984). However, if students do not achieve the expected or desired level of performance or learning outcome with a particular piece of software, teachers may still need to gather more information on possible causes of the problem. For instance, they may analyze various components of the computer lesson such as the mode of presentation and the teaching environment in which it is presented. Every computer lesson becomes part of an assessment or data-gathering process, and no computer program once implemented should be followed blindly or rigidly. Such diagnostic teaching is particularly helpful in early stages of computer implementation, but monitoring computer courseware effectiveness by ongoing assessment throughout the teaching program prevents wasting valuable teacher and student time on unnecessary computer instructions or on computer tasks that are not meeting the needs of the students. Continual assessment and decision-making is essential, because procedures that work very well in one situation may be ineffective in another, or computer approaches that initially work well may eventually lose their effect through student satiation or other variables.

In conclusion, remember that in the final analysis your own personal and professional judgment is the best criterion for assessing the appropriateness and usefulness of any computer courseware for individual students. As with other aspects of computer implementation with individual learners, the teacher is still the key decision-maker.

Computer Implementation and Teacher Effectiveness

The large body of research generated in the last decade on teacher and school effectiveness and program implementation provides some valid recommendations for teachers considering the use of computers with special needs students. In a recent article, Bear (1984) makes a very valid point when he cautions that "until we learn what key factors contribute to successful integration of CAI with microcomputers at the school level, one would be wise to apply what we already know about school and teacher effectiveness to the usage of microcomputers" (p. 12). A good overview of this research can be gleaned from the following sources (Bear 1984; Berliner 1984; Bickel 1983; Brophy 1982; Burnett Strother 1983; Leithwood and Montgomery 1982; Rosenshine and Berliner 1978; Stallings 1981).

Among the many important dimensions and descriptors associated with effective schooling and teaching, the following have direct relevance for computer implementation with all students, including those with special learning needs.

1. Goal-focused activities directed towards clear, attainable and relevant objectives at the district, school and classroom levels. Involvement of all staff in the planning and implementation of school policy.
2. Inservice staff training focused on effective teaching and learning. Teaching staff and principals kept abreast of new developments in pedagogy and educational psychology.
3. District-level and parental support for school improvement and innovations. Strong support to increase staff ratios and reduce class size to manageable proportions, especially in basic or core subjects.
4. High and positive teacher achievement expectations for all students with a constant press for student growth and excellence. High teacher enthusiasm and use of praise, encouragement and appreciation for effort.
5. Appropriate levels of difficulty for learning tasks. Emphasis on direct instruction, student mastery and successful learning experiences. Students consistently experience high rates of success (90 to 100 percent).
6. Effective use of classroom instructional time or academic learning time; increased amount and intensity of engagement in school learning for all ability levels. Time spent on-task dominates the instructional period.
7. Continuous diagnosis, evaluation and informative feedback. Frequent monitoring of individual student progress. Students get lots of practice on criterion-referenced or mastery-oriented material and remedial instruction when needed.
8. Well-structured, organized and sequenced classroom activities; classroom climate that's businesslike; classroom management tasks that are simple and routine. More teacher time spent on cognitive instead of managerial tasks.
9. Frequent opportunities for individualized work; number and variety of opportunities to learn for all students; sensitivity to individual differences, developmental levels and cognitive styles. High teacher involvement with students as opposed to paperwork or neutral monitoring.

10. Teachers clearly engage in cross-fertilization of instructional ideas at both inter- and intra-school levels.
11. Enhancement of student motivation and interest through meaningful learning activities. Teacher attempts to relate or match new information to students' prior knowledge, skills and interests.
12. Involvement and encouragement of students in classroom decision-making, management and organization. Enhancement of student self-control, metacognitive awareness and achievement-monitoring.
13. Solid principal support for new teaching practices related to program improvement. Active principal leadership and arrangement of effective professional development activities. Collection of information by principals about promising new practices from professional publications or other sources. Principal as a key factor in setting school-wide policies.
14. A positive, friendly and supportive classroom learning climate; strong teacher focus on promoting pupil self-esteem, confidence, risk-taking and independent learning for all ability levels.
15. Positive attitude toward mainstreaming and integration of exceptional students with special learning needs. Adaptation and accommodation of instruction to meet individual learning needs.

The nature of computer-aided instruction lends itself to the fulfillment of these principles of effective instruction. For instance, the branching, monitoring, feedback and student control features of good CAI software provide students with many opportunities for lots of practice and engaged time on materials appropriate to their individual learning styles, achievement and interest levels. Teachers and their principals should review the above list very carefully and decide how each of the identified components of effective teaching and schooling can be maximized in their own use of computer applications with special students. Bear (1984) concludes that it is likely that future research will find CAI to be effective in classrooms that are characterized by the same elements of instruction that previous research has shown to be associated with effective teachers.

Academic Learning Time One of the most important factors of effective instruction for teacher consideration in implementing CAI is student opportunity to learn, or academic learning time. Teachers can expect greater achievement gains when they allocate more learning time for academic tasks and when they engage their students in direct academic instruction. Effective teachers provide more opportunity to learn, allocate most of their available time to instruction, and organize and manage their classrooms to ensure that the time is actually spent in this fashion (Brophy 1982). Similarly, teachers can expect more academic gains when students have the greatest opportunity to successfully use the computer. However, as indicated by Bear (1984), even having a high student/computer ratio will not be sufficient unless the computers are being used. Bear states that as academic learning time (or academic engaged time) has been found to be a critical factor in effective instruction, so may academic computing time. He cautions that more computing time will be needed in certain subject areas than in others before gains occur. Many schools have attempted to address this issue by establishing specially designed computer laboratories in self-contained resource rooms and school libraries. Other schools have used a more decentralized model but have clustered or rotated their limited supply of microcomputers or terminals in a few classrooms with teachers who are enthusiastic about computer use, know what to teach with CAI, and have received some computer literacy training. Within individual classrooms, effective teachers have maximized computer usage through creative scheduling and adhering to strict computer usage at specified time intervals (e.g., 20-minute intervals) for each student. These teachers have also established computer rules and regulations with consistent follow-through, and have ensured a clear understanding and acceptance of the teacher's expectations in this regard. They have arranged their classrooms to be most conducive to learning with microcomputers, and they have minimized possible distractions or disruptions. Many teachers have particularly increased computer time for their special students by grouping these students in pairs or in small groups (threes or fours) with their peers when the task warranted such creative grouping (i.e., during language arts activities, adventure games, creative problem-solving and brainstorming situations). In these cases, students can use each other as resources and support.

Curriculum Pacing The amount of CAI time a student spends using the computer should be of less concern to the teacher than what the student is actually doing during CAI (Bear 1984). To learn efficiently, students must be engaged in meaningful, interesting and challenging tasks. However, the key variable according to Brophy (1982), seems to be the match between students' "present achievement levels" and the "difficulty levels" of their assigned tasks (p. 2). Students of successful teachers progress or move through the curriculum in small steps, and they experience consistent success along the way. In fact, students require a very high success rate in order to progress efficiently. This finding is especially true of special needs students. Brophy reports recent research which suggests that teachers who aim for success rates of 90 to 100 percent on student assignments produce more learning than teachers who tolerate higher failure rates. Very high success rates seems to be especially important for seatwork assignments or when students are expected to work independently without frequent monitoring by or assistance from the teacher (as is generally the case with students working with computers).

According to Bear (1984), in order to achieve these high rates of success with computers, teachers, with the possible assistance of excellent management program software (i.e., CMI, must continuously monitor student performance and emphasize individualized instruction and self-pacing.

Matching Computer Instruction to Curriculum Objectives In addition to selecting software with high standards of instructional quality, Bear (1984) and others feel it is also critical that "what is being taught, via the selected software, matches or aligns with the objectives of the local curriculum and with tests used in evaluating the program" (p. 13). Moreover, if this is not done by the teacher, then whatever is learned may go undetected and the effectiveness of using microcomputers will be limited. As we have indicated previously in the task analysis and learner analysis phases of computer implementation, it is essential that teachers stress the "cross-fertilization" or linking of computer material both to other content or instruction in the curriculum and to the existing knowledge and experiences of their students. If this is not done, there is the real danger of CAI being taught in isolation and of students wasting valuable academic learning time on objectives inappropriate to their special learning needs. The critical and active role of the teacher in creatively matching the instruction and applications of the computer to the classroom curriculum and to the learning needs and capabilities of each student cannot be underestimated. In this regard, teachers should be careful to include introductory and follow-up classroom activities and discussions which integrate the computer activities with the ongoing classroom curriculum and which demonstrate to their students the relevance of the computer program. Computers can help to improve and develop the academic knowledge, skills and attitudes of special students best when used as a powerful tool to supplement effective planning and instruction already provided by the classroom teacher.

Inservice Training In order to achieve the goal of effective computer instruction with all their students, teachers must continually keep abreast of current and new developments in the educational technology field. Changes in teachers' beliefs about computers and effective implementation of computers in the classroom will most likely result from school-based inservice education experiences which take into account the special conditions of the school and the unique perspectives, attitudes and experiences of each teacher in the school. Usually the school principal will provide the initial leadership for implementing computers into the school curriculum and for providing support and arranging professional development activities for new teaching practices related to computer applications. However, a variety of people may take formal or informal responsibility for inservice education. Also, inservice experiences involving computer literacy may take place in many settings. Regardless of whoever takes the responsibility for implementing computer practices in the school, four activities can help promote computer practices with special needs students throughout the school: 1) involving teachers throughout the school in planning computer programs and applications for special needs students; 2) creating an atmosphere in which teachers are able to support each other and share information about special students and new computer teaching ideas; 3) providing opportunities for encouraging professional development of teachers in computer literacy through adequate computer materials, budget and release time for planning and inservice activities; and 4) actively searching for, and making use of, existing resources in the school board and community (Slemon et al. 1984).

Inservice teacher education in computer literacy may take place in diverse situations including formally planning computer courses and workshops at the university, teacher federation or school district level; informal contacts such as discussions and sharing resources and ideas with colleagues from within and from outside one's school or district; and reading computer and educational journals, magazines, newsletters and books to keep up with the rapid developments and information change in the computer technology and educational field. Computer inservice activities, therefore, can take a number of forms and settings. However, our personal growth as teachers is fostered when we reflect, by ourselves or together with others, on our beliefs about teaching, on our teaching practices, and on the materials that we use for instruction. Hence, being successful in using computer programs with special needs students will probably require us to adapt to the special circumstances of our individual schools and to create a context of mutual trust within our schools that encourages and supports new ideas and practices. The reader is referred to the appendix for additional readings and resources in computer literacy and computer applications for special needs students.

Recommendations for Computer Implementation

Based on previous experiences of teachers in implementing computer practices for all types of learning, the following suggestions are particularly in order.

1. Go slow in the implementation of computers in your classroom. Don't be too ambitious. Start with a single application (e.g., word processing in language arts) and gradually integrate the computer into your curriculum. Continuously evaluate the academic and affective impact on your special needs and other students. As you gain more experience, training and confidence, then you can expand your program further. Develop a plan of action which establishes the role of the computer in your classroom.
2. Try as much as possible through inservice activities to become computer literate in the sense that you understand the role and applications of computers in education and you know how to use microcomputers and other technologies effectively with special needs students, whether or not you have computer programming skills. Placing a microcomputer in the hands of teachers without adequate inservice training is generally ineffective in meeting the educational needs of your students. Get the training you need to become confident, enthusiastic and effective in your computer use.
3. Talk to fellow teachers who already use computers successfully in their own classrooms. Work with a teacher who is already committed and involved in using computers. Use each other as resources and support.
4. Do your best to acquire only high quality computer software and courseware which matches the objectives of your own curriculum. Evaluate potential software critically using your own professional expertise and common sense. Use evaluation criteria based on good learning and pedagogical principles (refer to Figure 14-3).
5. Systematically and periodically evaluate the efficacy and the results of your computer program with all your students. Just because the "process" looks good does not necessarily mean the "product," in terms of student outcomes, will also be good. Consider yourself a researcher as well as a teacher. Examine, experiment and evaluate your computer program as scientifically as possible. Use the same decision-making techniques you use in your own teaching to assess and modify your computer program. Make sure your program is having a worthwhile impact on your special students' learning and attitudes. Do not continue a computer program without concrete and objective evidence of its effectiveness in meeting the learning needs of your students and

in matching the objectives of your curriculum. Do not fall victim to false or unrealistic expectations encouraged by computer hype in commercial media and advertisements. Judge for yourself and be wary of the "band wagon" effect; on the other hand, be equally wary of any backlash effect. Often meritorious programs are abandoned because they have not been adequately researched or sufficiently tried.

6. Remember, the same principles associated with effective teaching probably generalize to effective use of computers. Review these principles (discussed previously) and systematically incorporate them into your total instructional program.
7. Make sure that students understand what is expected of them and how to operate the computer program. Develop a computer training program for students that builds excitement, motivation, curiosity and interest. Show your students how their computer activities relate to their interests and previous curriculum. Make use of available teacher time to set the stage and ensure that special students are appropriately engaged in individualized computer instruction suited to their learning needs. Recall that computer software should allow students to work at appropriate difficulty levels with high rates of success.
8. Use all available personnel to assist in maximizing appropriate academic learning time or academic computer time for special students. Consider using parent or student volunteers, or aides, paraprofessionals and cross/age and same/age tutors to maximize academic computing time. They can provide encouragement and assistance to students using the computer when the teacher is occupied elsewhere.
9. This point is probably best addressed to school principals and board supervisors and consultants: make sure teachers receive ongoing support, inservice training and supervision to facilitate the successful implementation of computers in their classrooms. Many well designed classroom instructional computer programs fail because of the lack of inservice training, ongoing supervision and support from administrators at the school and board levels. Allocate your computers to teachers you know are enthusiastic and committed to using them for all students and who know what and how to teach with the aid of the computer. Develop an efficient system for computer maintenance and troubleshooting. Develop a critical core of computer expertise in the board and in each school by training resource staff who acquire good computer literacy and knowledge in the use of computers for all types of learners. Reduce teaching loads for these resource personnel so they can effectively carry out their computer resource role.
10. Consider all the potential applications of computers with special need students discussed in this chapter. Do not rely solely on the drill and tutorial applications. Rather, consider the tool and other creative uses of the computer in developing learner control and such higher-level cognitive objectives as creative problem-solving, organizational planning skills, reflective thinking, concept relating, application of rules and principles, inductive and deductive logic, and artistic and musical expression. In other words, combine

your talents as a creative teacher with the creative capabilities of the computer in teaching special needs students to become confident independent learners.

11. Obtain parent support and encouragement for the goals and objectives of your computer program. If parents have computers at home, try to integrate their use of the home computer with your own learning objectives for their children. Also include parents when possible in the school's or board's computer literacy inservice programs.
12. Encourage your female students to get involved with computers by choosing computer activities which may be especially meaningful to them (e.g., creative problem-solving, creative writing, artistic expression, etc.).
13. If you are planning to teach computer programming skills to your special students in structured computer languages such as BASIC, LOGO, COMAL or PASCAL, or in authoring languages such as PILOT, make sure you begin slowly with a very few elementary computer procedures and concepts. Use systematic shaping and task analysis procedures to gradually develop more advanced programming skills.
14. Try to approach your first experiences with computers with enthusiasm and an open mind, and without built-in apprehensions and preconceived negative attitudes about what can or can't be done with computers for your learners. Be exploratory, but be creative in your explorations.
15. Final recommendation: keep learning, keep your sense of humour, have fun and GOOD LUCK!

FUTURE CONSIDERATIONS

The increased use of computer technology has significant prospects for increasing the excellence of education provided to special needs students in both regular and special education. Computer-based instruction seems to have the *potential* to provide the richest, most active and most highly individualized interaction between the student and the curriculum of any of the methods of instruction yet developed. The major question is: Will the effective use of computer technology eventually lessen the handicapping conditions of learning-handicapped students by assisting teachers and other educators in providing individualized and effective instruction for all learners? Much work, teacher training and systematic research still lies ahead before computer technology can be utilized with confidence and maximum effectiveness by classroom teachers. Future research and experimentation must still consider the most effective strategies for integrating computer aided instruction with other instruction activities and curricula for students with special learning needs.

It is apparent that we can expect substantial growth in the classroom application of computer technology in the very near future. However, we must be careful that the use of computer technology in educational interventions does not become the next "band wagon" in regular and special education—adopted with zealous vigour, ineffectively implemented, inadequately researched, and viewed as a panacea for meeting the needs of all students, including students with special learning needs.

There is the real danger that the positive benefits and potential of computer technology can be prematurely dismissed because of a backlash which may occur when our unrealistic hopes do not quickly materialize. Teachers in particular must see their role not only as implementers of the new technology but also as researchers and evaluators who can carefully and systematically assess the impact of the new technology on their students' knowledge, skills and attitudes.

At any rate, it is essential that teachers and other educators keep abreast of new developments in technological development. With the rapid advancements in the computer technology field it is easy to fall behind and miss out on the many promising interventions and creative applications offered by computer technology for all learners. In this chapter we have reviewed only the current uses of computers with special learners. Many other exciting applications are on the horizon or have already been developed and are just now beginning to receive wider use in education. Developments in computer technologies are increasingly creating new possibilities for the delivery of educational services to all learners. In fact, future technologies hold the potential to be "normalization agents" for special needs students both in the school and community (Cain 1984). Properly developed, they can overcome many of the physical, communicative and intellectual barriers that prevent special needs students from realizing their full potential as learners and as worthy individuals in society. The following developments are just a few of the many that should be monitored very carefully for their possible educational use for students with special learning needs.

1. Improved voice synthesizers and voice recognition devices to provide verbal directions, praise and content to poorer readers and language-disordered students, and to allow students to enter their responses verbally into the computer.
2. Special and modified keyboards, keyboard commands and key guards to allow physically handicapped students to access the computer and to communicate efficiently with others.
3. More advanced computer input devices including light pens, laser scanner devices, touch tablets, touch screens or panels, etc., which allow special students to bypass the keyboard to input information and communicate with others.
4. Powerful prosthesis adaptations such as "electronic communication boards, keyguards, scanning hardwares and softwares, and pressure touch, suck, and puff, or sound, microswitches" which permit the most profoundly physically handicapped to use computer technology and to communicate with others in school or at home (Cain 1984). One recently invented keyboard called an Eyetyper (Sentinent Systems Technology, Inc. 1985) uses an infrared television camera-equipped keyboard that can be controlled with eye stares. The keyboard can be connected to computers and allows physically handicapped individuals to use their eyes to input information and to do word processing at roughly the same speed as a non-physically handicapped person.
5. Use of modem telecommunications to assess various educational and information database and various electronic mail and bulletin board services such as Special Net, a communication network in the U.S.A. designed to provide

up-to-the-minute information and instant communications for persons concerned with services and programs for handicapped individuals (see appendix).

6. Use of computer-assisted interactive laser video-disc courseware which allows students to combine TV video and microcomputer output. This type of interaction allows the student to visualize the actual environment the learning is intended for instead of working with abstract concepts (Browning and Nave 1983).
7. Computer printers, keyboards and programs that can translate written materials and typed screen inputs into braille (or vice-versa) for blind and visually handicapped students. Electronic microcomputer scanners with speech synthesizers are already available which give these students access to any print material which is available to the general population (Cain 1984).
8. Use of computer-assisted teacher training for training preservice and inservice teachers to use immediate, continuous and delayed computer feedback during and following classroom teaching in order to improve classroom decision-making (Semmel and Sitko 1976; Sitko and Slemon 1981).
9. Use of computer simulation games and computerized courseware in teacher education preservice and inservice programs which focus on teaching skills for special needs students (Semmel, Varnhagen and McCann 1981).
10. Powerful instructional management and data-base programs which greatly enhance educational planning and individualization and monitoring of pupil learning.
11. Speech training devices that enable hearing handicapped students to see visual images of utterances. Microcomputers interfaced with videodiscs to teach reading and language development for deaf and hearing-impaired students (Appel and Hurley 1984).
12. Improved and easy-to-use word processing software and speech synthesizers that can provide severely language-impaired students with successful writing experiences.
13. New microcomputer authoring systems and authoring languages that allow teachers and other educators to develop their own software to suit their own particular teaching and their students' learning needs (Appel and Hurley 1984).
14. Development of more efficient and cost-effective school-based or local networking systems for connecting a whole-set of classroom computers to a host or teacher's computer, and to one or more floppy or hard disc drives (Green 1984).
15. More sophisticated on-line computer data bases and information resources which teachers can assess via computer modem hook-up. The newer data bases will soon contain more specific descriptions and reliable information for educators on "what software is available, what software has been taken off the market, what each program is designed to do, how much it costs, what hardware it runs on, whether it can be networked, whether it has been evaluated and whether it is recommended for purchase" (Komoski 1984).

A few years ago these technological applications in education would have been viewed as science fiction, mere speculation. They are reality today and expanding and accelerating at an amazing rate. They provide alternative and more flexible modes of education delivery which will be increasingly needed in the new information age. They offer tremendous hope and optimism as well as tremendous challenge for all professionals who are obligated to meet the individual learning needs of our special students.

Conclusion

In summary, the reader has been introduced to some of the current and future applications of computer technology for students with unique or special learning needs. This technology will probably take on an increasingly significant role for all educators who work with special learners in regular or special classroom settings. This chapter has purposely focused on the teacher as the key factor in the implementation of the new technology. It is this writer's belief that the new technology, when used creatively and scientifically by teachers and other educators, will greatly enhance our instructional effectiveness with special needs students and help prepare them for life-long learning and the modern information age. With the aid of the new technology, we can provide a secure, motivating and successful teaching and learning environment for our special students, and we can develop and extend their cognitive and affective skills. One thing is certain—computers are here to stay. The challenge is yours. GO FOR IT!

APPENDIX

Additional Sources of Information about the Use of Computers with Special Needs Students

RECOMMENDED COMPUTER JOURNALS AND MAGAZINES FOR INSTRUCTION OF SPECIAL STUDENTS

(Good Sources for Obtaining Practical Information Concerning Educational Software and Hardware for all Students)

Classroom Computer Learning, Peter Li, Inc. 2451 E. River Rd., Dayton, Ohio; (800) 543-4383; (513) 294-5785 (*Excellent Source*). (Monthly except for May, June, July and August.)

Compute! P.O. Box 914, Farmingdale, NY, 11737; (800) 334-0868. (Monthly)

Compute's Gazette for Commodore Personal Computer Users, P.O. Box 901, Farmingdale, NY, 11737; (800) 334-0868) or (919) 275-9809. (Monthly)

Computers in Education Moorhead Publications, 25 Overlead Blvd., Suite 601, Toronto, Ontario, M4H 1B1.

Educational Technology, 140 Sylvan Avenue, Englewood Cliffs, NJ, 07632.

Electronic Education, Suite 220, 1311 Executive Centre Drive, Tallahassee, Fl., 32301.

Electronic Learning, 902 Sylvan Avenue, Englewood Cliffs, NJ, 07632.

Journal of Special Education, Buttonwood Farms, Inc., Neshaminy Plaza II, Suite 101, Bristol Pike and Street Road, Bensalem, PA, 19020. (Quarterly)

Journal of Visual Impairment and Blindness, 15 W. 16th St., New York, NY, 10011.

Logophile College of Education, MacArthur Hall, Queen's University, Kingston, Ontario, K71 3N6.

Microquests, Martin Bearden Inc., P.O. Box 337, Grapevine, TX, 76051; (214) 986-6214. (Monthly publication containing LOGO related materials and articles.)

National Logo Exchange, Posy Publications, P.O. Box 5341, Charlottesville, VA, 29905.

Canadian Journal for Exceptional Children (formerly *Special Education in Canada*), Publication Services, Faculty of Education, University of Alberta, 4-116 Education North, Edmonton, T6G 2G5. (Quarterly)

Teacher Education and Special Education, The Council for Exceptional Children, 1920 Association Drive, Reston, Va., 22091. (Quarterly)

The Computing Teacher Journal of The International Council for Computers in Education. (Excellent Source) Published nine times annually, August to June. Contact University of Oregon, 1787 Agate St., Eugene, Oregon, 97403-1923; (503) 686-4414.

The Education Digest, P.O. Box 8623, Ann Arbor, Michigan, 48107. (Good source on current educational issues and ideas.)

Turtle News, Young People's Logo Association, 1209 Hillsdale Drive, Richardson, TX, 75081.

SPECIAL ISSUES OF JOURNALS CONTAINING ARTICLES PERTAINING TO SPECIAL NEEDS STUDENTS AND COMPUTERS

"*BYTE*," *LOGO in the Schools* (August 1982).

"Classroom Computer Learning," *LOGO* (April 1983).

"Computers and Special Needs" (Special Issue). 1983. *The Computer Teacher, 10*(6).

"Computers and Teacher Education," (Special Issues). 1981, 1983. *Teacher Education and Special Education, 4*(3) and 6(3).

"Computers" (Special Issue). 1983. *Phi Delta Kappan, 65*(2), (October).

"Computing and the Handicapped" (Special Issue). 1981. *Computer, 14*(7).

"Microcomputers." 1984 *Journal of Visual Impairments & Blindness* 78 (9), (November).

"Microcomputers and the Handicapped" (Special Issue). 1982. *Byte, 7*(9).

"Microcomputers' Place in Special Education" (Special Issue). 1982. *Exceptional Children, 49*(2).

"Technology in Special Education" (Special Issue). 1984. *Teaching Exceptional Children, 16*(4).

"Visions of the Future in Educational Computer (Features)." 1984. *Phi Delta Kappan,* 66(4), 239-258.

ADDITIONAL GOOD SOURCES FOR OBTAINING INFORMATION CONCERNING HARDWARE AND SOFTWARE FOR SPECIAL STUDENTS

Alfred I. DuPont Institute, Wilmington, Delaware (voice-activated software).

Center for Innovation in Teaching the Handicapped, 2805 E. 10th St., Bloomington, Indiana, 47405 (Dr. Lou Polsgrove, Director, (812) 335-5847).

Center for Special Education Technology Information Exchange, The Council for Exceptional Children, 1920 Association Drive, Reston, Virginia, 22091; (703) 620-3660 or (800) 345-TECH. (Contact Susan Elting, Project Director.) This center collects and exchanges information about using technology in the education of handicapped children and youth. An excellent source.

ECOO (Educational Computing Organization of Ontario), The Ontario Institute for Studies in Education, 252 Bloor Street West, Toronto, Ontario, M5S 1V6; (416) 923-6641.

ECOO Special Interests Groups: SIGELEM, Elementary; SIGL, LOGO: SIGi for Francophones; SIG-COL-GSI, College; SIG SPED, Special Education; SIGCS, Computer Studies; SIG R & D, Research and Development.

Minnesota Educational Computing Corporation (MECC), 3490 N. Lexington Ave., St. Paul, MN, 55112; (612) 481-3621.

Special Education Technology Resource Center, 108 Babson Street, Mattapan, Massachusetts, 02126, (617) 298-4779. (Contact Madaline K. Pugliese, Project Coordinator.)

TAM, Technology and Media Division of The Council for Exceptional Children (CEC). (Excellent Source, publishes *TAM Newsletter.*) 1920 Association Drive, Reston, Virginia, 22091. (Members also receive *Journal of Special Education Technology* (JSET).

The International Council for Computers in Education (ICCE), University of Oregon, 1787 Agate St., Eugene, OR, 97403-1923, U.S.A. (Organization published *The Computing Teacher* and includes a variety of special interest groups and organization members throughout North America, Europe, Israel, Australia and Japan.)

Toronto PET Users Group (TPUG), 1912A Avenue Road, Suite #1, Toronto, Ontario, M5A 4A1. (Published magazine, *TPUG*, and members have access to the club's extensive library of computer programs on disc and tape for all Commodore computers.)

TV Course by TVOntario, "The Computer Academy," 12 week course, consists of 12 two-part half-hour shows, "Bits & Bytes" and "The Academy," including supporting materials.

SELECTED PUBLISHERS OF COMPUTER SOFTWARE AND MATERIALS APPROPRIATE FOR SPECIAL STUDENTS (INCLUDING EXEMPLARY SOFTWARE)

Addison-Wesley Publishing Co., Microbooks Div., Reading, MA, 01867; (617) 944-3700; (800) 938-3801. (E.G., "Cave Girl Clair," "Jenny of the Prairie.")

Alfred Publishing, 15335 Morrison Street, Sherman Oaks, CA., 91403; (818) 995-8811. (E.g., "Music Made Easy.")

Apple Computer Inc., 20525 Mariani Ave., Cupertino, CA, 95014; (408) 996-1010; (800) 538-9696. (E.g., "Appleworks," "Apple Writer II," "MacPaint."

Batteries Included, 186 Queen Street West, Toronto, Ontario, M5V 1Z1; (416) 596-1405; or 17875 Sky Park North, Suite P, Irvine, CA, 92714. (E.g., "Spell Pack," "Paper Clip Word Processor for Commodore 64.")

Borg-Warner Educational Systems, 600 W. University Dr., Arlington Heights, IL, 60004.

Broderbund Software, 17 Paul Drive, San Rafael, CA, 94903; (415) 479-1170. (E.g., "The Print Shop" (Excellent), "Bank Street Speller," "Bank Street Written," "Lode Runner," "The Music Shop," "The Print Shop Graphics Library, Disc One.")

Canadian Instructional Software Systems Ltd. (CISS), 7707 Hunterview Dr. N.W., Calgary, Alberta, T2K 4P8; (403) 275-5381. (E.g., "Cana-Quiz.")

CBS Software, One Fawcett Place, Greenwich, CT, 96836; (203) 622-2673. (E.g., "Success With Math," "Interplanetary Pilot," "Many Ways to Say I Love You," "Adventure Master.")

Commodore, 1200 Wilson Drive, West Chester, PA, 19380; (215) 431-9100. (E.g., "Reading Professor", "Sky Travel," "Just Imagine," "Story Maker.")

Computer Advanced Ideas. (E.g., "Wizard of the Words.")

Computer Conversations, 2350 N. Fourth St., Columbus, Ohio, 43202; (614) 263-4324. (E.g., "Enhanced PC Talking Program," for use with visually impaired students.)

Davidson & Associates, 6069 Graveoak Pl. #12, Rancho Palos Verdes, CA., 90274; (213) 373-9473; or (800) 556-6141. (E.g., "Math Blaster," "Word Attack", "Speed Reader II.")

D.C. Heath and Co., 125 Spring St., Lexington, MA, 02173; (617) 862-6650, (800) 224-1149. (E.g., "Quill"—writing skills.) Canadian Address: D.C. Heath Canada Ltd., #1600-100 Adelaide Street West, Toronto, Ontario, M5H 1S9; (416) 362-6983.

DesignWare, 185 Berry Street, San Francisco, CA, 94107; (415) 546-1866. (E.g., "Body Transparent," "Creative Creator.")

Developmental Learning Materials, One DLM Park, Allen, TX, 75002.

Electronic Arts, 2755 Campus Dr., San Mateo, CA, 94403. (415) 571-7171. (E.g., "Mule,"—cooperative game, "Pinball Construction Set," "Adventure Construction Set," "Mail Order Monsters," "Cut and Paste Word Processor," "Music Construction Set.")

Fisher-Price, One Kendall Square, Cambridge, MA, 02139; (617) 494-1200. (E.g., "Song Maker," ages 4-8.)

Follett Quality Courseware, 4506 Northwest Highway, Crystal Lake, IL., 60014. (800) 435-6170, (815) 455-1100. (E.g., "Quizagon.")

Freelance Musicworks, Inc., 18 Haviland Street, Boston, MA, 02115; (617) 536-9494. (E.g., "Musicland.")

Ginn & Co., 191 Spring St., Lexington, MA, 02173; (800) 848-9500.

Grolier Electronic Publishing, 95 Madison Ave., New York, NY,10016; (212) 696-9750. (E.g., "Friendly Filer"—database program.)

Hayden Software Company, Inc., 600 Suffolk Street, Lowell, MA, 01854. (E.g., "Music Works," "Word Challenge.")

Holt, Rinehart and Winston, Director of Software Marketing, 383 Madison Ave., New York, NY, 10017. (E.g., "Voyage of the MIMI.")

Houghton Mifflin Canada Ltd., 150 Steelcase Road West, Markham, Ontario, L3R 1B2; (416) 475-1755. (E.g., "Mathematics Activities Courseware 3-8.")

IBM Corp., P.O. Box 1328, Delray Beach, FL, 33444; (305) 998-2000. Or IBM, DRM, Dept. BE1556, 400 Parson's Pond Drive, Franklin Lakes, NJ, 07417.

Infocom, Cambridge, MA. (E.g., "Zork" (adventure) "Hitchhiker's Guide to the Galaxy," "The Witness" (murder mystery), "Suspect.")

Koala Technologies, 3100 Patrick Henry Dr., Santa Clara, CA, 95052; (408) 986-8866, (800) 562-2327. (E.g., "Koala Painter," "Koala Pad Touch Tablet.")

Krell Software, 1320 Stone Brook Rd., Stoney Brook, NY, (516) 751-5139, (800) 245-7355. (E.g., "The Allegory of Plato's Cave" (adventure), "The Devil and Mr. Webster" (reading).)

Learning Trends, Inc., A Simon & Schuster Company, P.O. Box 2987, Dept. 11, New York, NY 10185; (212) 245-6400. (E.g., "Typing Tutor III," "Chem Lab," "Kermit's Story Maker.") Or contact Simon & Schuster, 1200 Avenue of the Americas, New York, NY, 10020.

Litus Software Inc., 3952 Cheswood Drive, Downsview, Ontario, M3J 2W6; (416) 636-9416.

Maryland Computer Services Inc., 2010 Rock Spring Rd., Forest Hill, Maryland, 21050; (301) 879-3366. (E.g., "The Total Talk PC," for use with visually impaired students.)

Math & Computer Education Project, Lawrence Hall of Science, University of California, Berkeley, CA, 97420. (E.g., "Creative Games," "Creative Play: Problem Solving Activities with a Computer," reviewed by D. Carey in *The Computing Teacher, 11*(4): 38-39.

Microsoft Corp., Box 97200, 10700 Northup Way, Bellevue, WA., 98009; (206) 828-8088, (800) 426-9400. (E.g., "Typing Tutor.")

Midwest Software, Box 214, Farmington, MI, 48024; (313) 477-0897. (E.g., "KINDER-KONCEPTS," preschool to grade 1, contains 30 programs.)

Milliken Publishing Co., P.O. Box 21579, St. Louis, MO, 63132; (314) 991-4220, (800) 643-0008. (E.g., "Cloze Plus," "Sentence Combining," "Milliken Word-Processor," "Milliken Math Science," "Milliken Language Arts Sequence.")

Mindscape Inc., 3444 Dundee Rd., Northbrook, ILL, 60062; (312) 480-7667. (E.g., "Bank Street Music Writer," "Bank Street Story Book," "Mr. Pixel's Programming Paint Set," "Halley Project," simulation adventure.) (In Canada, call (613) 523-7272.)

Minnesota Educational Computing Corporation, 3490 Lexington Ave. N., St. Paul, MN, 55113; (612) 481-3500. (E.g., "Taxman.")

Opportunities For Learning Inc., 20417 Nordhoff St., Dept. L101, Chatsworth, CA, 91311; (818) 341-2535.

Prentice-Hall, P.O. Box 819, Englewood Cliffs, NJ, 07632; (201) 592-2641. (E.g., "Ski Writer II," for developing electronic mail.)

Radio Shack, Dept. 85-A-233, 300 One Tandy Center, Fort Worth, TX, 76102; (800) 433-5682. (E.g., "Super LOGO," "Dr. LOGO.")

Reston Publishing Co., Inc., 11480 Sunset Hills Rd., Reston, VA, 22090; (800) 336-0338. (E.g., "Movie Maker.")

Scarborough Systems Inc., 55 South Broadway, Tarrytown, NY 10591; (800) 556-2283. (E.g., "Build-a-Book Refill Kit," "Build a Book," reading skills, "Master Type's Writing Wizard," "Master Type," typing, "Songwriter," "Phi Beta Filer," database program.)

Scarborough Systems, 555 Broadway, Tarrytown, NY, 10591; (914) 332-4545 (800) 882-8222. (E.g., "Master Type," "Pattern Maker," "Run for the Money," "Phi Beta Filer.")

Scholastic Inc., P.O. Box 7503, Jefferson City, MO, 65102; (314) 636-5271, (800) 325-6149. (E.g., "Survey Taker," "Spelldiver," "Story Maker," "Story Tree," "Microzine," "Bank Street Writer," "Mystery Sentences," "Tales of Discovery," twist-a-plot.)

Scholastic Software, 730 Broadway, New York, NY, 10003. (212) 505-3497. (E.g., "Logic Builders," "Operation:Frog," simulation, "PFS:FILE," data base program.)

Sierra On-Line, Sierra On-Line Buildings, Coarsegold, CA, 93614; (209) 683-6858. (E.g., "King's Quest," adventure, "Homeword," with Homeword Speller, "Story Maker.")

Spinnaker Software, One Kendall Square, Cambridge, MA, 02139; (617) 494-1200. (E.g., "Kindercamp," "Kid's on Keys," "Story Machine," "Facemaker," "In Search of the Most Amazing Thing," "Kidwriter," "Snooper Troops," "Delta Drawing.")

Sterling Swift Publishing Company, 7901 South IH-35, Austin, TX, 78744; (512) 282-6840. (E.g., "Computing in LOGOLAND," "Little LOGO," "Writing with a Micro.")

SubLogic Corporation, 713 Edgebrook Drive, Champaign, IL, 61820; (217) 359-8482. (E.g., "Flight Simulator II," "Night Mission PINBALL," arcade simulation.)

Sunburst Communications, 39 Washington Ave., Pleasantville, NY, 10570; (800) 431-1934, (914) 769-5030. (E.g., "M-ss-ng L-nks: Young People's Literature," "The Puzzler," "Sem Calc," "Teasers by Tobbs," "The Factory," "Incredible Laboratory," "The Pond," "Survival Math.")

The Learning Company, 545 Middlefield Rd., Menlo Park, CA, 94025; (415) 328-5410. (E.g., "Gertrude's Puzzles," "Gertrude's Scripts," "Rocky's Boots," "Bumble Games," "Magic Spells," "Reader Rabbit," "Robot Odyssey," "Magic Spells.")

Timeworks, 444 Lake Cook Road, Deerfield, IL, 60015; (312) 948-9200. (E.g., "Cave of the Word Wizards," "Word Writer & Spelling Checker.")

SOURCES FOR REVIEWS OF EDUCATIONAL SOFTWARE*

*See also previous addresses for educational and computing journals and magazines and software reviews.

Canale, M., G. Barker, M. Belanger, K. MacRury, R.S. McLean, and R. Ragsdale. 1984. *Microcomputer Software for Language Arts in Ontario: Survey and analysis*. Toronto, Ontario: The Ontario Institute for Studies in Education.

"Educational Software Buyer's Guide." 1985. *Compute* (Apple Special: Application's Issue), *1* (1) (Spring/Summer); (800) 334-0868, (919) 275-9809.

Kerchner, L.B. 1984 "Software review: Bank Street Writer," *Learning Disability Quarterly,* 7(2), 201-202.

MENU, The International Software Database Corporation, 1520 South College Avenue, Fort Collins, CO, 80524; (800) THE-MENU, or (303) 482-5000.

MicroSIFT News, 300 S.W. 6th Avenue, Portland, OR, 97204.

Software Buyer's Guide. A supplement to *Classroom Computer Learning* (April/May 1985). An index of all educational software CCL has covered in reviews over the past two years.

Software Reports, Software Reports Dept., Trade Service Publications INC., 10996 Torreyana Rd., San Diego, CA, 92121; (800) 854-1527.

SWIFT's Directory of Educational Software for the IBM PC, 1984-85 Edition.

SWIFT's 1984-85 Educational Software Directory, Apple II edition. Austin, Texas: Sterling Swift Publishing Company, 7901 South 1H-35, Austin, TX, 78744; (512) 282-6840.

The APPLE INDEX, BP Publications, P.O. Box 617;, Southbury, CT, 06488; (203) 264-2143.

The Journal of Courseware Review, APPLE Educational Foundation, 20525 Mariana Avenue, Cupertino, CA, 95014.

COMPUTER SOFTWARE REFERENCES

AMECUS, a modular expanding client utility system. Copyright 1983 by John G. Donohue; (315) 564-6213.

CIMS: CITH Instructional Management System. Available from Dr. Lou Polsgrove, Director, Center for Innovation in Teaching the Handicapped, 2805 East Tenth Street, Room 150, Bloomington, Indiana, 47405; (812) 335-5847.

Miller, Larry. *The Puzzler* (computer program.) Faculty of Education, Queens University, Kingston, Ontario, K7L 3N6.

ADDITIONAL READINGS ON GUIDELINES FOR THE EVALUATION OF MICROCOMPUTER SOFTWARES

Caissy, G. A., "Evaluating educational software: A practitioner's guide," *Phi Delta Kappan, 66*(4), 249-250.

Gable, R.K, 1982. "A model for evaluating special education programs: An application to determining the management and instructional benefits of microcomputers, "*Journal of Special Education Technology, 5*(2), 35-43.

Grimes L. 1981. "Computers are for kids: Designing software programs to avoid problems of learning," *Teaching Exceptional Children, 14*(2), 49-53.

Hannaford, A., and E. Sloane. 1981. "Microcomputers: Powerful learning tools with proper programming," *Teaching Exceptional Children, 14*(2), 54-57.

Hannaford, A., and F.M. Taber. 1982. "Microcomputer software for the handicapped: Development and evaluation," *Exceptional Children, 49*(2), 137-144.

Heck, W., T. Johnson, and R. Kansky. 1981. *Guidelines for evaluating computer-*

ized instructional materials. Reston, Virginia: National Council of Teachers of Mathematics; 1906 Association Drive, Reston, VA 22091; (703) 620-9840) (30 pages).

Kamil, M.L., "Software: Evaluations and decisions," *Computers, Reading and Language Arts, 1*(4), 15-16, 26.

Kleiman, G., M. Humphrey, and T. Van Buskirk. 1981. "Evaluating educational software, *Creative Computing, 7*(10), 72-81.

Kohl, H. 1984. "Who should be evaluating software?" *Classroom Computer Learning, 5*(2), 31-33.

MicroSIFT Evaluator's Guide for microcomputer-based instructional packages. 1982. Eugene, Oregon: ICCE.

Ragsdale, R. G. 1982. *Evaluation of microcomputer courseware*. Toronto, Ontario: The Ontario Institute for Studies in Education Press.

E. Sanders, and M. Sanders. 1983. "Evaluating microcomputer software," *Computers Reading and Language Arts, 1*, 21-25.

RECOMMENDED GENERAL READINGS CONCERNING COMPUTERS AND SPECIAL STUDENTS

Brady, H., and T. Slesnick. 1985. "Girls don't like fluffware either," *Classroom Computer Learning, 5*(8), 22-27.

Budoff, M. 1984. *Microcomputers in special education.* Cambridge, MA: Brookline Books; (617) 868-0360).

Goldenberg, E.P. 1984. "Computers in the special education classroom: What do we need and why we don't have any?" In J. Mulick (ed.) *Perspectives in Mental Retardation*. Ablex Press, 107-127.

Green, J.O. 1984. "New ways for special education kids to communicate," *Classroom Computer Learning, 5*(3), 24-29.

Harper, D., and J.H. Stewart. *RUN: Computer Education*. Monterey, CA: Brooks/Cole Publishing Company.

Hasen, D. 1984. *Microcomputer Resource Book for Special Education*. Reston, VA: Reston Publications.

Harper, D.O., and J.H. Stewart. 1983. *Run: Computer Education*. Monterey, CA: Brooks/Cole.

Katz, E.L. 1983. "Microcomputers: A course of study for gifted students," *Focus on Exceptional Children, 15*(6), 1-16.

Lane, A. 1984. *Reading in microcomputers and learning disabilities*. Guilford, Conn.: Special Learning Corporation.

Metzger, M., D. Ouellette, and J. Thormann. 1983. *Learning Disabled Students and Computers: A teacher's guidebook*. Eugene, Oregon: International Council for Computers in Education.

Pagliaro, L.A. 1983. "The history and development of CAI, 1926-1981: An overview," *The Alberta Journal of Educational Research, 29*(1), 75-84.

Topp, R.H. 1985. "Microcomputers in elementary education, 1984: State of the art," *The Computing Teacher*. February, 43-47.

RECOMMENDED COMPUTER IMPLEMENTATION READINGS

Caldwell, R.M. 1980. "Guidelines for developing basic skills instructional materials for use with microcomputer technology," *Educational Technology, 20*, 7-12.

Cory, S. 1983. "A 4-stage model of development for full implementation of computers for instruction in a school system," *The Computing Teacher, 11*(4), 11-16.

Fullan, M. 1981. "Research on the implementation of educational change." In A.C. Kerckhogg (ed.), *Research in Sociology of Education and Socialization*. Greenwich, CT: JAI Press, 195-219.

Fullan, M. 1982. *The Meaning of Educational Change*. Toronto: Ontario Institute for Studies in Education Press.

Fullan, M., and P. Park. 1981. *Curriculum Implementation: A resource booklet.* Toronto: Ontario Ministry of Education.

Topp, R.H. 1984. "State of the art: Microcomputers in elementary education," *The Computing Teacher, 12*, 43-47.

Torgeson, J.K., and K.A. Young. 1983. "Priorities for the use of microcomputers with learning disabled children," *Journal of Learning Disabilities, 16*(4), 234-237.

RECOMMENDED READINGS IN LANGUAGE ARTS AND COMPUTER USE

Burnett, J.D., and L. Miller. 1984. "Computer-assisted learning and reading: Developing the product or fostering the process? *Computers and Education, 8*(1), 145-150.

Devall, Y. 1984. "Using the computer to strengthen spelling," *The Computer Teacher, 12*(2), 19-22.

Knappe, A. 1984. "Word processing: A mini survey," *Computers in Education,* November, 10-11.

Lenk, F. 1984. "Word processing: A mini survey," *Computers in Education,* April, 14-17.

Mackey, K. 1984. "Electronic MAIL in the writing class," *Classroom Computer Learning, 5*(3), 32-33.

Miller, Larry, 1984. *Reading, Writing and Computers: Going beyond the obvious*. Paper presented at the Computers '84 Software for the Classroom, GEMS Conference, Toronto, Ontario. (Contact Larry Miller, Faculty of Education, Queen's University, Kingston, Ontario, K7L 3N6.)

Moorshead, H. 1984. "Word-Processing: An Overview," *Computers in Education*, April, 6-8.

Preston, L. 1984. "Improving written expression skills in Cupertino." Paper presented at the CEC conference on Computers and Special Education, Reno, Nevada, January, 1984.

ADDITIONAL READINGS CONCERNING THE LOGO PROGRAMMING LANGUAGE

Abeson, H. 1982. "A beginner's guide to LOGO," *Byte, 7*(8), 88-115.

Burnett, J.D. 1983. *Logo: An introduction.* Morris Plains, New Jersey: Creative Computing Press.

Thornburg, D. 1983. *Discovering Apple LOGO.* Reading, Massachusetts: Addison-Wesley.

Watt, D. 1983. *Learning with LOGO.* New York: McGraw-Hill.

RECOMMENDED READINGS CONCERNING COMPUTER MANAGED INSTRUCTION

Friars, E., and N. Gelmann. 1981. *Special Education Management by Information: A Design Handbook.* National Association of State Directors of Special Education, Washington, D.C., 20036.

Ragghianti, S., and R. Miller. 1982. "The microcomputer and special education management," *Exceptional Children, 49*(2), 131-135.

RECOMMENDED READINGS AND SOURCES CONCERNING TELECOMMUNICATIONS (ELECTRONIC NEWS SERVICES) NETWORKING AND DATABASES

CompuServe Handicapped User's Database (contains information and addresses concerning computer use for handicapped persons). Contact CompuServe Information Services, P.O. Box 20212, 5000 Arlington Centre Blvd., Columbus, OH, 43220; (800) 848-8199, (614) 457-0802. An excellent electronic mail service.

Hunter, B. 1985. "Problem solving with databases," *The Computing Teacher, 12*(8), 20-27.

Levitan, A. 1985. "Is anyone out there? Getting ready for telecommunications," *Compute!* (APPLE Special: Applications Issue), *1*(1), 24-28. (800) 334-0868; (919) 275-9809.

Newlin, B. *Answers online: Your guide to informational data bases*. Berkeley, CA: Osborne/McGraw-Hill, (415) 548-2805.

Regional Multi-Task Communications Network for Special Education. Northwest Kansas Educational Service Center, 210 South Range, Suite 126, Colby, Kansas, 67701; (913) 462-8282.

SpecialNet, an excellent bulletin board or networking service for special educators or educators interested in sharing, seeking and exchanging information and software concerning special needs students. Membership is obtained from the National Association of State Directors of Special Education, 2021 K. St., N.W., Suite 315, Washington, D.C., 20006; (202) 296-1800. Current basic subscription fee to *SpecialNet* is $200.00 (U.S. funds). Renewal subscription to *SpecialNet* with prepaid *Computer Connect Times* is $500.00. The association also publishes a newsletter for *SpecialNet* entitled, *SpecialNet News*.

Telecommunication references (electronic news services), e.g., Ed-Line National School Public Relations Association's educational electronic news service. Available from the source, or National School Public Relations Association, 1801 N. Moore St., Arlington, VA, 22209; (703) 528-5840.

Telecommunications and the Handicapped: How Technology is Improving Delivery of Educational Services. Order from OSSC, 1787 Agate St., University of Oregon, Eugene, OR, 97403; (503) 686-5045.

RECOMMENDED READINGS CONCERNING MICROCOMPUTERS AND VIDEODISCS

Friedman, S.G., and A.M. Hofmeister. "Matching technology to content," *Exceptional Children, 51*(2), 130-134.

Levin, W. 1983. "Interactive Video: The State-of-the-Art Teaching Machine," *The Computing Teacher, 11*(2).

READINGS CONCERNING MICROCOMPUTERS AND ART INSTRUCTION

Peelle, H.A. 1984. "Computer metaphors: Approaches to computer literacy for educators," *The Computing Teacher 11*(7), 12-13.

Sharp, P., "Using LOGO to shape artistic ideas," *Classroom Computer Learning, 5*(2), 60-62.

READINGS CONCERNING MICROCOMPUTERS AND CLASSROOM TESTING

Salend, S.J., and S.M. Salend. 1985. "Implications of using microcomputers in classroom teaching." *Journal of Learning Disabilities, 18*(1), 51.

READINGS AND SOURCES CONCERNING THE USE OF COMPUTERS WITH VISUALLY IMPAIRED STUDENTS

American Foundation for the Blind Inc., 15 West 16th Street, New York, NY, 10011; (212) 620-2000. (Also publishes journals of visual impairment and blindness.)

American Printing House for the Blind, 1839 Frankfort Avenue, Louisville, KY, 40206; (502) 895-2408.

Braille-Talk, software program which translates information or text stored in disc files into hand copy printouts in braille form. Program also supports various voice synthesizers. Contact Computer Aids Corp., 4929 S. Lafayette St., Fort Wayne, IN, 46806; (219) 456-2148.

Caroll Center for the Blind, 770 Center Street, Newton, MA, 02158; (617) 969-6200.

Computer Aids Corporation, 4929 South Lafayette Street, Fort Wayne, IN, 46806; (219) 456-2148.

Microcomputers for Blind Persons, Box 328, Peabody College, Vanderbilt University, Nashville, TN, 37203; (615) 322-8765.

National Braille Press, 88 St. Stephen Street, Boston, MA, 02115; (617) 266-6160. Publishes *The Second Beginner's Guide to Personal Computers for the Blind and Visually Handicapped.*

National Library Service for the Blind and Physically Handicapped, Library of Congress, 1291 Taylor St. NW, Washington, DC, 20542; (292) 287-5100. Good source of information about computer software and hardware.

Raised Dot Computing Inc., 408 South Baldwin Street, Madison, Wisconsin, 53703; (608) 257-9595.

Sensory Aids Foundation, 399 Sherman Avenue, Suite 12., Palo Alto, CA, 94306; (415) 329-0430.

Telesensory Systems Inc., 455 North Bernardo Ave., Mountain View, CA, 94043; (415) 960-0920.

The National Braille Association Braille Bookbank, 422 S. Clinton Ave., Rochester, NY, 14620. Source of computer-related books translated into braille.

VTEK, 1625 Olympic Blvd., Santa Monica, CA, 90404; (213) 829-6841, (800) 345-2256.

RECOMMENDED TEACHER EFFECTIVENESS READINGS

Brophy, J.E., and T.L. Good. In press. "Teacher behavior and student achievement," in M.E: Wittrock (ed.), *Third handbook of research on teaching.*

Goodman, L. 1985. "The Effective Schools Movement and Special Education," *Teaching Exceptional Children, 17*(2), 102-105.

RECOMMENDED COMPUTER JOURNALS AND MAGAZINES FOR TEACHERS OF SPECIAL NEEDS STUDENTS

Bulletins on Science & Technology for the Handicapped (quarterly publication), American Association for the Advancement of Science, Project on the Handicapped in Science, 1515 Massachusetts Avenue, N.W., Washington, DC, 20005.

Canadian Journal for Exceptional Children (sent free to all Canadian members of the Council for Exceptional Children), Publication Services, 4-116 Education North, Faculty of Education, University of Alberta, Edmonton, Alberta, T6G 2G5.

Closing the Gap (Excellent source), P.O. Box 68, Henderson, MN, 56044; (612) 248-3294 (six issues per year).

Commodore: The microcomputer magazine, Commodore Business Machines Inc., Magazine Subscription Department, Box 651, Holmes, PA, 19043.

Computer—Disability News (quarterly publication), The National Easter Seal Society, 2023 W. Ogden Avenue, Chicago, Illinois, 60612.

Computers in the Schools, The Haworth Press, 75 Griswold Street, Binghamton, N.Y., 13904.

Computers, Reading and Language Arts, CRLA, Dept. M, P.O. Box 13247, Oakland, CA, 94661-0247; (415) 530-9588 (quarterly).

Educational Computer Magazine, P.O. Box 535, Cupertino, CA, 95015.

Educational Technology (monthly), Educational Technology Publications Inc., 720 Palisade Avenue, Englewood Cliffs, New Jersey, 07632.

Educational Technology and Communication, The Far West Laboratory, 1855 Folsum Street, San Francisco, CA, 94103.

Electronic Learning, (8 monthly issues), Scholastic Inc., P.O. Box 2580, Clinton, Iowa, 52735.

Electronic Learning Exceptional Children, (Published 6 times a year) The Council for Exceptional Children, 1920 Association Drive, Reston, Virginia, 22091-1589; (703) 620-3660. Excellent source. Sent to all members of the Council for Exceptional Children, the largest professional association concerned with the field of special education in the world.

Focus on Exceptional Children (9 times a year), Love Publishing Company, Executive and Editorial Office, 1777 South Bellaire Street, Denver, Colorado, 80222; (303) 757-2579.

Journal of Computer-Based Instruction (quarterly), sent free to members of the Association for the Development of Computer-Based Instructional Systems (ADCIS). International Headquarters, Miller Hall 409, Western Washington University, Bellingham, Washington, 98225.

Journal of Learning Disabilities (10 issues, monthly), The Professional Press, 633-3rd Avenue, New York, N.Y., 10003. Excellent source.

Journal of Special Technology, published by the Association for Special Education Technology, ASET.

Learning Disability Quarterly, published by the Council for Learning Disabilities (CLD), P.O. Box 830688, University of Texas at Dallas, Richardson, TX, 75083.

Remedial and Special Education, (6 times a year) PRO-ED, 5341 Industrial Oaks Boulevard, Austin, TX, 78735.

School Courseware Journal, School for Home CourseWare Inc., 1341 Bulldog Lane, Suite C-J, Fresno, CA, 93710.

Teacher Education and Special Education, sent to all members of the Teacher Education Division (TED) of the Council for Exceptional Children, 1920 Association Drive, Reston, Virginia, 22091.

Teaching Exceptional Children, The Council for Exceptional Children, 1920 Association Drive, Reston, Virginia, 22091; (703) 620-3660. Sent to all members of The Council for Exceptional Children.

The Journal of Special Education (quarterly), Buttonwood Farms, Inc., Neshaminy Plaza II, Suite 101, Bristol Pike and Street Road, Bensalem, PA, 19020.

REFERENCES

ABBOTT, J. 1978. *Classroom Strategies to Aid the Disabled Learner*. Cambridge, Mass.: Educators Publishing Service.

ADAM, I. 1984. "The Koala Pad Touch Tablet: How to bear with it. Part I: Overview." *Commander*. April, 56-58.

ALLEY G. 1977. "Grouping secondary learning-disabled students." *Academic Therapy, 13*, 37-45.

ALLEY, G., AND D. DESHLER. 1979. *Teaching the Learning Disabled Adolescent: Strategies and methods*. Denver, Colorado: Love Publishing Co.

ALLINGTON, R. L. AND M. STRANGE. 1977. "The problems with reading games." *The Reading Teacher, 31*, 272-274.

ALSON, M. L. 1977. "Into the mainstream—he's blind, he's in our class, everybody's learning." *Instructor, 87*(2), 222-224.

AMAN, M. 1980."Psychotropic drugs and learning problems: A selective review." *Journal of Learning Disabilities, 13*, 36-46.

AN ACT TO AGAIN AMEND THE EDUCATION ACT. 1980. *Gazette Officielle du Quebec, Part 2, 112*(7).

ANDERSON, P. AND P. GROFF. 1968. *Resource Materials for Teachers of Spelling*. Minneapolis, Minn.: Burgess Publishing Company.

ANDERSON, P. L. 1982. "A preliminary study of syntax in the written expression of learning disabled children." *Journal of Learning Disabilities, 15*, 359-362.

ANDERSON, T. H. 1980. "Study strategies and adjunct aids." In R. J. Spiro, B. C. Bruce and W. F. Brewer (Eds.) *Theoretical Issues in Reading Comprehension: Perspectives from cognitive psychology, artificial intelligence, linguistics, and education*. Hillsdale, N.J.: Erlbaum.

APPELL, L.S. AND K. M. HURLEY. 1984. "Individualizing instruction with microcomputer software." *Focus on Exceptional Children, 16*(5), 1-10.

APPLEBEE, A.N. 1978. "Environments for learning: ERIC resources on the language experience approach." *Language Arts, 55*, 756-760.

APPLEYARD, F., G. LEITENBERG, AND J. GEORGE. 1984. *Basically Right.* Toronto, Ontario: The Ministry of Education.

ARMITAGE, D. 1978. *Special Education in Canada, 52*(3), 23.

ASBY-DAVIS, C. 1985. "Cloze and comprehension: A qualitative analysis and critique." *Journal of Reading, 28*, 585-589.

ASHFORD, J. 1984. "Handwriting." London, Ontario: University of Western Ontario, unpublished paper.

ASHTON-WARNER, S. 1963. *Teacher.* New York: Simon and Schuster.

ASKEY, P. 1982. "Mainstreaming in an elementary school." Personal interview.

AXLINE, V. M. 1977. *Play Therapy*. Revised Edition. New York: Ballantine Books.

AUKERMAN, R. C. 1971. *Approaches to Beginning Reading*. New York: John Wiley.

BACHOR, D. G. 1979. "Suggestions for modifications in testing low achieving adolescents." *The Journal of Special Education, 13*, 443-451.

BACHOR, D. G. 1983, NOVEMBER. "Children's work as a basis for diagnosis." Paper presented at the Canadian Association for Children and Adults with Learning Disabilities Conference, Vancouver, British Columbia.

BACHOR, D. G. 1983. "Personnel preparation for learning assistance teachers: School based instruction." *Special Education in Canada, 58*(1), 16-20.

BACHOR, D. G. 1984, FEBRUARY. "From strategy deficiency to instruction." Invited paper presented at the Learning Assistance Teachers' Association Annual Conference, Vancouver, B.C.

BAILEY, K. 1977. "Psychomotor and fine-motor coordination problems in the mildly handicapped." London, Canada: University of Western Ontario. Unpublished manuscript.

BAKER, L. AND A. L. BROWN. 1980. "Metacognitive skills in reading." In D. P. Pearson (Ed.) *Handbook on Research in Reading*. New York: Longman.

BANDURA, A. 1974. "A behavior theory and models of man." *American Psychologist, 29*, 859-869.

BANNATYNE, A. D. 1966. *Psycholinguistic Color System*. Urbana: Learning Systems Press.

BARNES, D. 1984. "A model for the delivery of special education services." Lunenberg, N.S.: Lunenberg County District School Board.

BARTEL, N. R., J. E. BOYD, AND S. TOBIA. 1984. "Teaching children with reading problems." In D. D. Hammill, N. R. Bartel and G. O. Bunch (Eds.) *Teaching Children with Learning and Behavior Problems*. Canadian Edition. Toronto: Allyn & Bacon.

BASSIN, A., T. E. BRATTER AND R. L. RACHIN. 1976. *The reality therapy reader*. New York: Harper & Row.

BATSTONE, D. 1981. "Models of spelling acquisition." Unpublished Ph.D. Dissertation. London, Canada: The University of Western Ontario.

BEAR. G. G. 1984. "Microcomputers and school effectiveness." *Educational Technology, 24*(1), 11-15.

BELANGER, J. AND R. G. MARTIN. 1984. "The influence of improved reading skill on writing skill." *The Alberta Journal of Educational Research, 30*, 194-212.

BELL, G. L. 1982. "Poetry therapy—changing concepts and emerging questions." *The Arts in Psychotherapy, 9*, 25-30.

BENHAM, K. 1984. "Morphographic Spelling." Victoria, B.C.: The University of Victoria. Unpublished paper.

BENTLEY, H. 1966. "Words in color." *Elementary English, 43*, 515-517.

BEREITER, C. 1978. "Basics of Comprehension." OISE handout.

BEREITER, C. AND M. SCARDAMALIA. 1982. "From conversation to composition: The role of instruction in a developmental process." In Robert Glaser (Ed.) *Advances in Instructional Psychology, Volume 2*. Hillsdale, N.J.: Laurence Erlbaum Associates, Publishers.

BERLINER, D. C. 1984. "Making the right changes in pre-service teacher education." *Phi Delta Kappan, 66*(2), 94-96.

BERMAN, A. 1978. *Delinquent Youth and Learning Disabilities*. San Raphael, California: Academic Therapy Publications.

BICKEL, W. (ED.) 1983. "Effective schools" [Special Issue]. *Educational Researcher, 12*(4).

BILL C.61. 1982. An act respecting young offenders and to repeal the Juvenile Delinquents Act. Ottawa: The House of Commons of Canada.

BIRCH, J. AND M. WANG. 1984. "Comparison of full-time mainstreaming program and a resource room approach." *Exceptional Children, 51*, 33-40.

BITTER, G. AND N. WATSON. 1983. *Apple LOGO Primer*. Reston, Virginia: Reston Publishing Co. (Also *Commodore 64 LOGO Primer*).

BLACKBURN, E. 1984. "Common ground: Developing relationships between reading and writing." *Language Arts, 61*, 367-375.

BLANTON, L. P., M. SITKO, AND P. GILLESPIE. 1976. "Reading and the mildly retarded: Review of research and implications." In L. Mann and D. Sabatino (Eds.) *The Third Review of Special Education*. New York: Grune and Stratton, 143-162.

BLAU, H. AND E. J. LOVELESS. 1982. "Specific hemispheric-routing-TAK/v to teach spelling to dyslexics: VAK and VAKT challenged." *Journal of Learning Disabilities, 15*, 461-466.

BLIESMER, E. P. 1972. "Informal teacher testing in reading." *The Reading Teacher, 26*, 262-272.

BLOOM, B. AND K. COULTER. 1982. "Study strategies for the learning disabled adolescent in the mainstream." *Special Education in Canada, 56*, 24-25.

BODER, E. 1971. "Developmental dyslexia: Prevailing diagnostic concepts and a new diagnostic approach." In H. Myklebust (Ed.) *Progress in learning disabilities, Vol. 2*. New York: Grune & Stratton.

BODER, E. "Developmental dyslexia: A diagnostic screening procedure based on three characteristic patterns of reading and spelling." In B. Bateman (Ed.) *Learning Disorders, Vol. 4*. Seattle: Special Child Publications, p. 297-342.

BOERSMA, F. J. AND J. W. CHAPMAN. 1977. *Student's Perception of Ability Scale*. Edmonton: Psi-Can Consulting.

BOORMAN, J. 1969. *Creative Dance in the First Three Grades*. New York: David McKay.

BOORMAN, J. 1971. *Creative Dance in Grades Four to Six*. Don Mills: Longman.

BORMUTH, J. R. 1967. "Comparable cloze and multiple-choice comprehension test scores." *Journal of Reading, 10*, 291-299.

BORMUTH, J. R. 1968. "Cloze test reliability: Criterion reference scores." *Journal of Educational Measurement, 5*, 189-196.

BORMUTH, J. R. 1974. "Reading literacy: Its definition and assessment." *Reading Research Quarterly, 9*, 7-66.

BORMUTH, J. R. 1975. "Literacy in the classroom." In W. D. Page (Ed.) *Help for the Reading Teacher: New directions in research*. Urbana: ERIC/RCS and the National Conference on Research in English.

BORNSTEIN, B. AND B. BLOUIN. 1978. "A follow-up study of adolescents who have been on medication therapy for learning disabilities." Paper presented at the annual meeting of the Ontario Psychological Association, Ottawa.

BOS, C. S. AND FILIP. 1982. "Comprehension monitoring skills in learning disabled and average students." *Topics in Learning and Learning Disabilities, 2*(1), 79-85.

BOWERS, J. 1981. "Are we prepared to teach all students." *Special Education in Canada, 55*, 19-20.

BRADY, M. E. 1977. "A comparison of CATTS and audio feedback in developing teacher prompting skills in reading." Paper presented at the annual meeting of the American Educational Research Association, New York, N.Y.

BRAMMER, D. 1981. "Desegregation of the handicapped." Lethbridge, Alta.: Lethbridge School District #51, Pilot Project.

BRAMWELL, J. R. 1981. "Opinions of male public school teachers regarding their ongoing educational needs." Toronto: The Ontario Public School Men Teachers' Federation.

BRITTON, J. 1982. "Spectator role and the beginnings of writing." In M. Nystrand (Ed.) *What Writers Know: The language process, and structure of written discourse*. New York: Academic Press.

BRITTON, J. N. 1970. "Now that you go to school." Reprinted from J. Britton *Language and Learning*. Allen Lane: The Penguin Press.

BRODER, P. ET AL. 1981. "Further observations of the link between learning disabilities and juvenile delinquency." *Journal of Educational Psychology, 73*, 838-850.

BROPHY, J. 1982. "Successful teaching strategies for the inner-city child." *Phi Delta Kappan, 63*(4) 527-529.

BROWN, A. L. 1975. "The development of memory: Knowing, knowing about knowing, and knowing how to know." In H. W. Reese (Ed.) *Advances in child development and behavior. Volume 10*. New York: Academic Press.

BROWN, A. L. 1978. "Knowing when, where, and how to remember: A problem of metacognition." In R. Glaser (Ed.) *Advances in Instructional Psychology*. Hillsdale, N.J.: Lawrence Erlbaum.

BROWN, K. G. 1983. "Mainstreaming exceptional students: An evaluation." *Teacher Education, 22*, 16-25.

BROWN, L. 1984. Evaluating and managing classroom behavior. In Hammill, D. D., N. R. Bartel and G. O. Bunch, *Teaching Children with Learning and Behavior Problems.* Canadian Edition. Toronto: Allyn & Bacon.

BROWN, L. L. AND D. G. HAMMILL. 1978. *The Behavior Rating Profile: An ecological approach to behavioral assessment*. Austin: Pro-Ed.

BROWN, M. 1978. "Creative problem solving with the gifted." Paper presented at the Creative Problem Solving Institute, SUCB, Buffalo, N.Y.

BROWN, N. P. 1982. "CAMEO: Computer-assisted management of educational objectives." *Exceptional Children, 49*(2), 151-153.

BROWN, V. L., D. D. HAMMILL AND J. L. WIEDERHOLT. 1978. *The Test of Reading Comprehension: A method of assessing the understanding of written language*. Austin: Pro-Ed.

BROWNE, D. 1984. "Problem solving: Teaching strategies and remediation." Victoria, B.C.: The University of Victoria, unpublished paper.

BROWNING, P. AND G. NAVE. 1983. "Computer technology for the handicapped: A literature profile." *The Computing Teacher, 10*(6), 56-59.

BRUNER, J., J. GOODNOW AND G. AUSTIN. 1956. *A Study of Thinking*. New York: Wiley.

BRYAN, T. H. 1976. "Peer popularity of learning disabled children." *Journal of Learning Disabilities,* 7, 621-625.

BRYAN, T. H. 1978. "Social relationships and verbal interactions of learning disabled children." *Journal of Learning Disabilities, 11*, 58-66.

BRYAN, T. H. AND J. BRYAN. 1978. "Social interactions of learning disabled children." *Learning Disability Quarterly, 1*(1), 33-38.

BRYAN, T. H. AND J. BRYAN. 1981. "Some personal and social experiences of learning disabled children." In B. Keogh (Ed.) *Advances in Special Education. Volume 3*. Greenwich, Connecticut: JAI Press.

BUDOFF, M. AND L. R. HUTTEN. 1982. "Microcomputers in special education: Promises and pitfalls." *Exceptional Children, 49*, 123-128.

BURNETT STROTHER, D. 1984. "The many roles of the effective principal." *65*(4), 291-294.

BURNS, P. AND W. C. BOZEMAN, 1981. "Computer-assisted instruction and mathematics achievement: Is there a relationship?" *Educational Technology, 21*(10), 32-39.

BUROS, O. K. 1978. *The Eighth Mental Measurements Yearbook*. Highland Park, New Jersey: The Gryphon Press.

BURT, M. A. AND R. D. MYRICK. 1980. "Developmental Play: What's it all about." *Elementary School Guidance and Counseling, 15*(1), 14-21.

BYRNE, B. M. 1984. "The general/academic self-concept nomological network: A review of construct validation research." *Review of Educational Research, 54*, 427-456.

CABLE, P. 1976. *Special Education and Services Procedural Handbook*. Halifax: Special Education Department of the Halifax School Board.

CACLD. 1981. "Definition of learning disabilities." Saint John, New Brunswick: Annual Meeting.

CAIN, E. T. (JR.). "The challenge of technology: Educating the exceptional child for the world of tomorrow." *Teaching Exceptional Children, 16*(4), 238-241.

CALDWELL, E. SCHOOLEY AND L. THOMAS (EDS.). 1976. *I Used to Could Spell Wedn-Wenes-Wensday*. Child Service Demonstration Centre, Hillsdale School, Cushing, OK.

CAMP, B. W. AND J. L. DOLECOURT. 1977 "Reading and spelling in good and poor readers." *Journal of Learning Disabilities, 10*, 300-307.

CAMPBELL, J. 1984. "Modifying adolescent behavior." Victoria, B.C.: The University of Victoria. Unpublished manuscript.

CANFIELD, J. AND H. C. WELLS. 1976. *100 Ways to Enhance Self-concept in the Classroom: A handbook for teachers and parents.* New Jersey: Prentice-Hall.

CARMAN, R. A. AND W. R. ADAMS, JR. 1972. *Study Skills: A student's guide for survival*. New York: John Wiley and Sons.

CARPENTER, D. 1983. "Spelling error profiles of able and disabled readers." *Journal of Learning Disabilities, 16*, 102-104.

CARR, R. AND N. GROVE. 1982. "Reducing teacher stress: Using the challenge model to enhance peer relations." *The B.C. Counsellor, 4*(2), 27-36.

CARR, R. 1983. "Theory and practice of peer counselling." In *Natcon 4*. Ottawa-Hull: Canada Employment and Immigration Commission.

CARR, R. AND G. SAUNDERS. 1979. *The Peer Counselling Starter Kit*. Victoria, B.C.: Faculty of Education.

CARTER, L. 1984. "The sustaining effects study of compensatory and elementary education." *Educational Researcher, 13*, 4-13.

CARTER, R. 1984. "Story Tree (Reviews of Software Worth a Second Look)." *Classroom Computer Learning, 5*(4), 16-17.

CARTWRIGHT, G. P., C. A. CARTWRIGHT AND J. E. YSSELDYKE. 1973. Two decision models: Identification and diagnostic teaching of handicapped children in the regular classroom." *Psychology in the Schools, 10*, 4-11.

CASE, R. 1982. *Reading Critically*. Toronto: OISE Press.

CAWALTE, G. 1981. "Descriptors of effective instruction." Address given at the Faculty of Education, The University of Western Ontario, London, Ontario.

CELDIC, 1970. *One Million Children: A national study of Canadian children with emotional and learning disorders*. Canada: Leonard Crainford for the Commission on Emotional and Learning Disorders in Children.

CHALL, J. S. 1967. *Learning to Read: The great debate*. New York: McGraw Hill.

CHAMBERS, J. A. AND J. W. SPRECHER. 1980. "Computer assisted instruction: Current trends and critical issues." *Communications of the ACM, 23*(6), 332-343.

CHAPMAN, R. B., S. C. LARSEN AND R. M. PARKER. 1979. "Interactions of first-grade teachers with learning disordered children." *Journal of Learning Disabilities, 12*(4), 20-25.

CHARLESWORTH, E. A. 1982. "Music, psychology, and psychotherapy." *The Arts in Psychotherapy, 9*, 191-202.

CHARLETON, L. J. 1984. "Teacher survey, kindergarten to grade 13: Basic issues concerning the integration of physically impaired students." London, Ontario: The University of Western Ontario, unpublished manuscript.

CHEONG, L. M. 1980. "The impact of mainstreaming on the handicapped child." *Special Education in Canada, 54*, 30-32.

CHERNOFF, L. 1980. "A summary of the Woodcock-Johnson Psycho-Educational Battery." *Special Education in Canada, 54*, 16-20.

CHESLER, M. AND R. FOX. 1966. *Role Playing Methods in the Classroom*. Chicago: Science Research Associates.

CHOMSKY, C. 1984. "Finding the best language arts software." *Classroom Computer Learning, 4*(6), 60-63.

CLARK, O. 1978. "Mainstreaming: Special education is a part of, not a part from, regular education: Background and guidelines" (ED 169 717).

COLES, G. S. 1978. "The learning disabilities test battery: Empirical and social issues." *Harvard Educational Review, 48*, 313-340.

COLLINS, A. AND E. E. SMITH. 1982. "Teaching the process of reading comprehension." In K. K. Detterman and R. J. Sternberg (Eds.) *How Much and How Can Intelligence be Increased?* Norwood, NJ: Ablex.

COLLIS, B. 1984. "The development of an instrument to measure attitudes of secondary school males and females towards computers." Victoria, B.C.: University of Victoria, unpublished manuscript.

CONNOLLY, A. J., W. NACHTMAN AND E. M. PRITCHETT. 1977. *Key Math Diagnostic Arithmetic Test*. Canadian Edition. Toronto: Psy Can Ltd.

COPEX. 1976. "L'Education de l'enfance en difficulté d'adaptation et d'apprentissage en Québec. Québec, P.Q.: Gouvernment du Québec, Ministère de l'Education.

COSDEN, C. AND D. REYNOLDS. 1982. "Photography as therapy." *The Arts in Psychotherapy, 9*, 19-23.

COVINGTON, M. V., R. CRUTCHFORD, L. DAVIES AND R. OLTEN. 1972. *The Productive Thinking Program*. Columbus, Ohio: Charles E. Merrill Publishing Co.

CRAIK, F. I. AND R. S. LOCKHART. 1972. "Levels of processing: A framework for memory research." *Journal of Verbal Learning and Verbal Behavior, 11*, 671-684.

CRAMER, R. L. 1978. *Children's Writing and Language Growth*. Columbus: Charles E. Merrill.

CREALOCK, C. 1984. "The JD/LD Link: Causation or Correlation." Report prepared for the Solicitor-General of Canada.

CREALOCK, C. 1983. "Teacher and student behaviors in regular and special education settings." *B.C. Journal of Special Education, 7*(4), 321-330.

CREALOCK, C. 1981. "Creative problem solving: A compensatory strategy approach to teaching learning disabled adolescents." *Teacher Education, 19*, 34-40.

CREALOCK, C. 1980. "Can exceptional children learn through problem solving?" Paper presented at the annual meeting of the Canadian Society for Studies in Education, Montreal, Quebec.

CREALOCK, C. 1978. "Juvenile Delinquency: The Canadian Perspective." *Behavioral Disorders, 3*, 309-313.

CREALOCK, C. AND W. NEDIGER. 1983. "Bill 82 and You." Toronto: Ontario Public School Teachers' Federation.

CREALOCK, C. AND M. SITKO. 1983. "Special education in the secondary school: Selected issues." In Clark, R., R. Gidney, and G. Milburn (Eds.) *Issues in Secondary Schooling*. London, Canada: Faculty of Education, University of Western Ontario.

CREALOCK, C. M., M. C. SITKO, A. HUTCHINSON, C. SITKO, AND L. MARLETT. 1985, April "A comparison of paper and pencil and computer technologies to improve the writing skills of mildly handicapped adolescents." Paper presented at the annual meeting of the American Educational Research Association. Chicago, Ill.

CREEKMORE, N. N. AND A. MADIN. 1981. "Uses of sociodrama as a therapeutic technique with behavior disordered children." *Behavioral Disorders, 7*, 28-33.

CROWHURST, M. 1980. "Grammar in the elementary school." *Highway One*. Published by the Canadian Council of Teachers of English.

CROWL, M. 1980. "The basic process of art therapy as demonstrated by efforts to allay a child's fear of surgery." *American Journal of Art Therapy, 19*, 49-51.

CRUICKSHANK, W. 1981. *Mainstreaming: Education for all children*. Montreal: Quebec Association for Children with Learning Disabilities.

CRUICKSHANK, W. M. (ED.) 1980. *Psychology of Exceptional Children and Youth*, Fourth Edition. Englewood Cliffs, N.J.: Prentice-Hall Inc.

CSAPO, M. 1976. "If you don't know it, teach it." *Clearing House, 49*, 365-367.

CSAPO, M. 1981. *Children with Behavior and Social Disorders: A Canadian Focus*. Vancouver: Centre for Human Development and Research.

CTC. 1977, 1978, 1981. *Canadian Achievement Test*. Toronto: Canadian Test Centre, McGraw-Hill Ryerson.

CUMMINS, J. 1984. Paper presented at the Annual Meeting of the Canadian Society for Studies in Education, Guelph, Ontario.

CUNNINGHAM, P. M. 1975-76. "Investigating a synthesized theory of mediated word identification. *Reading Research Quarterly, 11* 127-132.

CURRIE, A. B. 1981. "Instruction and Handwriting in Ontario Schools." Toronto: Ministry of Education, Ontario, 1981.

CURTIS, C. 1982. "Teaching disabled students in the regular social studies classroom." *The History and Social Science Teacher, 18*, 9-16.

DARDIG, J. 1981. "Helping teachers integrate handicapped students into the regular classroom." *Educational Horizons, 59*(3), 124-130.

DAVIS, N. C. 1983. "Yes, they can! Computer literacy for special education students." *The Computing Teacher, 10*(6), 64-67.

DAVIS, N. C. 1983. "Computer literacy for the special student: A personal experience." *Teaching Exceptional Children, 16*(4), 263-265.

DONOHUE, J. G. 1984, April. *AMECUS: A modular expanding client utility system.* Computer program presented at the meeting of The Council for Exceptional Children, Washington, D.C.

DOREE, R. 1984. "IE goals." Victoria, B.C.: The University of Victoria. Unpublished paper.

DAROU, W. 1982. "Canadian Native people and psychological testing." *The School Guidance Worker, 37*(4). Toronto: Guidance Centre, Faculty of Education, University of Toronto.

DE BEAUGRANDE, R. 1982. "Psychology and composition: Past, present, and future." In M. Nystrand (Ed.) *What Writers Know: The language, process, and structure of written discourse.* New York: Academic Press.

DE CHARMS, R. 1968. *Personal Causation.* New York: Academic Press.

DE CHARMS, R. 1972. "Personal causation training in the schools." *Journal of Applied Social Psychology, 2*(2), 95-113.

DE CHARMS, R. 1975. *Enhancing Motivation: A change project in the classroom.* New York: Irvington Press.

DENO, E. N. 1972. "Special education as developmental capital." *Exceptional Children, 37*, 229-237.

DEPARTMENT OF INDIAN AND NORTHERN AFFAIRS. 1979. *Educational Issues.* Ottawa: Office of the Queen's Printer.

DERGOUSOFF, M. J. 1984. "The use of roleplaying to facilitate the growth of social skills in primary school aged pupils." Unpublished master's project, University of Victoria.

DESHARNAIS, W. 1984. "Today... Tomorrow." Special Education Discussion Paper. Victoria, B.C.: University of Victoria.

DEUTSCH, M. 1960. "Minority group and class status as related to social and personality factors in scholastic achievement." *Society for Applied Anthropology Monograph, 2*, 1-32.

DEWEY, J. 1899. *The School and Society.* Chicago: University of Chicago Press.

DINKMEYER, D. 1970. *Developing Understanding of Self and Others*. Circle Pines, Minnesota: American Guidance.

DIVESTA, F., K. HAYWARD AND V. ORLANDO. 1979. "Developmental trends in monitoring text for comprehension." *Child Development, 50*, 97-105.

DIXON, R. AND S. ENGLEMANN. 1976. *Morphographic Spelling Program*. Toronto: Science Research Associates (Canada) Limited.

DIXON, R., S. ENGLEMANN, AND M. MEIER. 1980. *Spelling Mastery: A direct instruction. Level A*. Chicago: Science Research Associates.

DIXON, R. AND S. ENGLEMANN. 1981. *Spelling Mastery: A direct instruction series. Level B*. Chicago: Science Research Associates.

DOCHERTY, D. 1975. *Education: Through the dance experience*. Bellingham: Educational Designs & Consultants.

DOCHERTY, D. 1982. "Organizing and developing movement ideas." *Journal of Physical Education, Recreation, and Dance, 35(3)*, 51-54.

DOLCH, E. W. 1945. *A Manual for Remedial Reading*. Champaign: The Garrard Press.

DOLCH, E. W. 1950. *Dolch Word List*. Morriston, N.J.: General Learning Press.

DOREE, B. AND P. KENNEDY. 1984. "Handwriting." Victoria, B.C.: University of Victoria, unpublished paper.

DOREN, M. 1973. *Doren Diagnostic Reading Test of Word Recognition Skills*. Circle Pines: American Guidance Service.

DOUGLAS, V. I. 1972. "Stop, Look and Listen. The problem of sustained attention and impulse control in hyperactive and normal children." *Canadian Journal of Behavioral Science, 4*, 259-282.

DOWNING, J. 1965. *The Initial Teaching Alphabet*. Chicago: Scott Foresman.

DOWNING, J. 1979. "'i.t.a.' in special education." *Special Education in Canada, 53*(4), 25-27.

DOWNING, J. 1982. "Cognitive clarity and reading disabilities." In J. P. Das, R. F. Mulcahy and A. E. Wall (Eds.) *Theory and Research in Learning Disabilities*. New York: Plenum.

DOWNING, J. AND C. K. LEONG. 1982. *Psychology of Reading*. New York: Macmillan.

DREIKURS, R. 1964. *Children the Challenge*. Des Moines, Iowa: Meredith Press.

DREIKURS, R. 1972. *Coping with Children's Misbehavior*. New York: Hawthorn.

DREIKURS, R., B. GRUNWALD AND F. PEPPER. 1971. *Maintaining Sanity in the Classroom*. New York: Harper and Row.

DREYER, S. 1981. *The Bookfinder: A guide to children's literature about the needs and problems of youth, ages, 2-15.* Circle Pines, Minnesota: American Guidance Service.

DURRELL, D. D. AND J. H. CATTERSON. 1980. *Durrell Analysis of Reading Difficulty.* Third Edition. New York: The Psychological Corporation, Harcourt Brace Jovanovich.

DURKIN, D. 1975. "The importance of goals for reading instruction." *The Reading Teacher, 23*, 380-383.

DURKIN, D. 1978. *Teaching Them to Read.* Third Edition. Boston: Allyn & Bacon.

DURKIN C. S. AND N. D. REPUCCI. 1973. "Learned helplessness and reinforcement responsibility in children." *Journal of Personality and Social Psychology, 25*, 109-116.

EARL, L. M., J. WOOD, AND R. G. STENNETT. 1981. "Morphographic spelling: A pilot study of its effectiveness with grade six students." *Special Education in Canada, 55*, 23-24.

EARLY, G. H., D. A. NELSON, D. J. KLEBER, M. TREEGOOB, E. HUFFMAN AND C. CASS. "Cursive handwriting, reading and spelling achievement." *Academic Therapy, 12*(1), 67-74.

EDGINGTON, R. 1973. "A hard roe to hoe." *Academic Therapy, 9*, 231-233.

EDUCATION AMENDMENT ACT. 1980. Toronto: J. C. Thatcher, Queen's Printer for Ontario.

EDUCATION ORDINANCE OF THE NORTHWEST TERRITORIES. 1984. Yellowknife: Department of Justice and Public Services, Government of the Northwest Territories.

EDWARDS, J., S. NORTON, S. TAYLOR, M. WEISS AND R. VAN DUSSELDORP. 1975. "How effective is CAI? A review of the research." *Educational Leadership, 33*(11), 147-153.

EKWALL, E. E. AND J. L. SHANKER. 1983. *Diagnosis and Remediation of the Disabled Reader.* Second Edition. Boston: Allyn and Bacon.

EMMONS, D. 1984. "The LD Adolescent." Victoria, B.C.: The University of Victoria, unpublished manuscript.

ENGELMANN, S. AND E.C. BRUNER. 1969, 1974. *Distar reading I: An instructional system.* Second Edition. Chicago: Science Research Associates.

ENGELMANN, S. AND S. STEARNS. 1972. *Distar Reading III: Reading to Learn.* Chicago: Science Research Associates.

ERIKSON, J. M. 1979. "The arts and healing." *American Journal of Art Therapy, 18*, 75-80.

ERREY, B. 1984. "Let the children scribble." *Whole Language Newsletter, 2*, 4-5.

ESTES, T. H., J. JOHNSTONE-ESTES, H. C. RICHARDS AND D. ROETTGER. 1981. *Estes Attitude Scales: Measures of Attitudes Toward School Subjects*. Austin: Pro-Ed.

ETTINGER, L. F. 1983. "Talk about teaching computer art graphics." *The Computer Teacher, 11*(3), 16-18.

EVANS, P., P. BROWN AND M. MARCH. 1977. *Criteria for the Evaluation of Student Writing Grades 7 and 8, A Handbook*. Toronto: The Ontario Institute for Studies in Education.

FADER, D. 1976. *The New Hooked on Books: Program and proof*. New York: Berkeley Medallion.

FARR. R. C., G. A. PRESCOTT, I. H. BALOW AND T. P. HOGAN. 1978. *Metropolitan Achievement Tests: Reading instructional tests*. New York: Harcourt, Brace, and Jovanovich.

FEITLER, F. AND V. DUBASIK. 1978. "Attitudes of regular classroom teachers towards EMR students." Paper presented at the 1978 Convention of the American Educational Research Association, Toronto, Canada.

FERNALD, G. 1943. *Remedial Techniques in Basic School Subjects.* New York: McGraw-Hill.

FEUERSTEIN, R. 1979. *Instrumental Enrichment*. Baltimore: University Park Press.

FIREDOG, P. 1984, June. *LOGO effects in Public School Classrooms*. Paper presented at the LOGO 84 Conference, Cambridge, Mass.

FLOWER, L. 1979. "Writer-based prose: A cognitive basis for problems in writing." *College English, 41*, 19-37.

FLOWER, L. AND J. R. HAYES. 1980. "The dynamics of composing. Making plans and juggling constraints." In L. W. Gregg and E. R. Steinberg (Eds.) *Cognitive Processes in Writing*. Hillsdale, N.J.: Laurence Erlbaum Associates.

FORMAN, D. 1982. "Search of the literature." *The Computing Teacher, 9*(5), 37-41.

FORMAN, D. 1981. *Instructional Use of Microcomputers: A report on B.C.'s pilot project*. Victoria, B.C.: Ministry of Education.

FORTE, I. AND M. A. PANGLE. 1976. *Spelling Magic*. Nashville, Tenn.: Incentive Publications.

FOSTER, K. AND J. K. TORGESEN. 1983. "The effects of directed study on the spelling performance of two subgroups of learning disabled students." *Learning Disability Quarterly, 6*, 252-257.

FOULDS, R. A. 1982. "Applications of microcomputers in the education of the physically disabled child." *Exceptional Children, 49*(2), 155-162.

FRANCE, M. H. AND C. MCDOWELL. 1982. "Seniors helping seniors: A model of peer counselling for the aged." *Canada's Mental Health, 30*(3), 17-20.

FRANCE, M. H. AND E. GALLAGHER. 1984. "Peer counselling: An outreach program in a home for the aged." *Canadian Counsellor, 18*(4), 174-179.

FREEDMAN, G. AND E. REYNOLDS. 1980. "Enriching lessons with semantic webbing." *The Reading Teacher, 33*, 677-684.

FRIEL, S. 1983. "Lemonade's the name, simulation's the game." *Classroom Computer Learning, 4* (Jan/Feb), 34-40.

FROEHLICH, M. A. R. 1984. "A comparison of the effect of music therapy and medical play therapy on the verbalization behavior of pediatric patients." *Journal of Music Therapy, XXI*(1), 2-15.

FULMER, S. AND R. FULMER. 1983. "The Slingerland tests: Reliability and validity. *Journal of Learning Disabilities, 16*, 591-595.

FURST, M. 1984. "MICROZINE, A computer magazine." *Journal of Learning Disabilities, 17*(10), 622-623.

GAGNE, R. 1977. *The Conditions of Learning*. Third Edition. New York: Holt, Rinehart and Winston.

GAJRAJ, I. 1982. "Remediation of handwriting difficulties." *Special Education in Canada, 56*, 16-18.

GAJRAJ, I. 1983. "Effective spelling instruction in special education." *Special Education in Canada, 57*, 25-27.

GALLESICH, J. 1982. *The Professional and Practice of Consultation*. San Francisco: Jossey-Bass, Inc.

GALLINI, J. K. 1983. "What computer-assisted instruction can offer toward the encouragement of creative thinking." *Educational Technology, 23* (April), 7-11.

GARREY, I. 1984. "Spelling checkers: Can they actually teach spelling?" *Classroom Computer Learning, 5*(4), 62-65.

GATES, A. I. 1922. "Psychology of reading and spelling with special reference to disability." *Contributions to Education*, No. 129. New York: Teacher's College.

GATES, A. I. 1937. "The necessary mental age for beginning reading." *The Elementary School Journal, 27*, 497-508.

GATES, A. I., A. S. MCKILLOP AND E. C. HOROWITZ. 1981. *Gates-McKillop-Horowitz Reading Diagnostic Tests*. New York: Teacher's College, Columbia University.

GATTEGNO, C. 1968. *Teaching Reading with Words in Color*. New York: Educational Solutions.

GAZETTE OFFICIELLE DU QUEBEC. 1980. Quebec: Ministry of Education.

GENESEE, F. 1981. "A comparison of early and late second language learning." *Canadian Journal of Behavioral Science, 13*, 115-128.

GEOFFRION, L. D. 1983. "The feasibility of word processing for students with writing handicaps." *Journal of Educational Technology Systems, 11*, 239-250.

GEOFFRION, L. D. AND O. P. GEOFFRION. 1983. *Computers and Reading Instruction*. Reading, Mass.: Addison-Wesley.

GERBER, M. 1983. "Learning disabilities and cognitive strategies: A case for training or constraining problem solving." *Journal of Learning Disabilities, 16*, 255-260.

GERBER, M. M. 1982. April. "Effects of self-monitoring on spelling performance of learning-disabled and normally-achieving students." Paper presented at the American Educational Research Association Conference, New York, N.Y.

GIBLON, S. 1981. "New teaching method helps more than slow learners." *Globe and Mail*. 26 Feb. 1981.

GILLESPIE-SILVER, P. 1979. *Teaching Reading to Children with Special Needs*. Toronto, Ont.: Charles E. Merrill.

GILLIE, S. 1984. *Case Study Submitted to Learning Assistance Center*. Unpublished Report, University of Victoria, Victoria, British Columbia.

GILLINGHAM, A. AND H. STILLMAN. 1968. *Remedial Training for Children with Specific Disability in Reading, Spelling and Penmanship*. Cambridge, Mass.: Educators Publishing Service.

GILMORE, J. V. AND E. C. GILMORE. 1968. *Gilmore Oral Reading Test*. New York: Harcourt Brace Jovanovich.

GLASSER, W. 1965. *Reality Therapy: A new approach to psychiatry*. New York: Harper & Row.

GLASSER, W. 1969. *Schools without Failure*. New York: Harper & Row.

GLEASON, G. 1981. "Microcomputers in education: The state of the art." *Educational Technology, 21*(3), 7-18.

GODOW, K. 1983. "Effects of stimulant drugs on academic performance in hyperactive and learning disabled children." *Annual Review of Learning Disabilities, 1*.

GOGUEN, L. J. 1980. "Right to education for exceptional children in Canada: A growing national concern." In M. Csapo and L. J. Goguen, *Special Education across Canada*. Vancouver: Centre for Human Development and Research.

GOLLICK, M. 1973. *Deal Me In*. New York: Jeffrey Morton Publishers, Inc.

GONICK, F., R. MACDONALD, AND R. SCARTH. 1982. Tutor-companion program. *Special Education in Canada, 56*, 13-14.

GOODMAN, K. S. 1975. "Do you have to be smart to read? Do you have to read to be smart?" *The Reading Teacher, 28*, 625-632.

GOODMAN, K. S. 1975a. "Reading strategy lessons: Expanding reading effectiveness." In W. Page (Ed.) *Help for the reading teacher: New directions in research*. Urbana: National Council of Teachers of English and Educational Resources Information Centre.

GOODMAN, Y. M. AND C. BURKE. 1980. *Reading Miscue Inventory.* New York: MacMillan.

GORTH, W. P. AND P. M. NASSIF. 1984. "A comparison of microcomputer-based, computer-managed instruction (CMI) software-programs." *Educational Technology*, January, 28-32.

GOTTLIEB, J. 1974. "Attitudes of Norwegian and American children toward mildly retarded children in special classes." *Journal of Special Education, 8*, 313-319.

GOTTLIEB, J. 1980. "Improving attitudes towards retarded children by using group discussion." *Exceptional Children, 47*, 106-113.

GRAVES, D. 1978. *Balance the Basics: Let them write*. New York: Ford Foundation, April, 30-33.

GRAVES, D. H. 1982. "Break the welfare cycle: Let writers choose their topics." *Eforums*, A newsletter of the English Composition Board, University of Michigan, *3*(2), 75.

GRAVES, D. H. 1983. *Writing: Teachers and children at work*. Exeter, New Hampshire: Heinemann.

GRAY, W. S. 1922. *Remedial Cases in Reading: Their diagnosis and treatment*. Supplementary Educational Monographs, No. 28. Chicago: The University of Chicago Press.

GRAY, W. S. AND H. M. ROBINSON. 1967. *Gray Oral Reading Tests*. Indianapolis: Bobbs-Merrill.

GRAY, W. AND D. ROGERS. 1982. "Mentor-directed enrichment projects for gifted elementary school pupils: Rationale, guidelines, and benefits." *Special Education in Canada, 56*, 24-32.

GREAT BRITAIN DEPARTMENT OF EDUCATION AND SCIENCE. 1975. *A Language for Life: Report of the Committee of Inquiry appointed by the Secretary of State*. Chaired by A. Bullock. London: Her Majesty's Stationery Office.

GREEN, D. 1982. "Reflections on the secondary education review project." *Teacher Education, 21*, 9-15.

GREEN J. O. 1984. "Artificial intelligence and the future classroom." *Classroom Computer Learning, 4*(6), 26-32.

GREEN, J. O. 1984. "Computers, kids & writing: An interview with Donald Graves." *Classroom Computer Learning, 4*(8), 21-28.

GREEN, J. O. 1984. "B. F. Skinner's technology of teaching." *Classroom Computer Learning, 4*(7), 23-29.

GREEN, J. O. 1984. "Straight talk about local networks." *Classroom Computer Learning, 5*(2), 72-77.

GREENWOOD, C. R., H. M. WALKER AND H. HOPS. 1977. "Issues in social interaction/withdrawal assessment." *Exceptional Children, 43*, 490-498.

GRODNER, S., D. BRAFF, D. JANOWSKY AND P. CLOPTON. 1982. "Efficacy of art/movement therapy in elevating mood." *The Arts in Psychotherapy, 9*, 217-225.

GROUNLUND, N. E. 1976. *Measurement and Evaluation in Teaching*. New York: MacMillan.

GRONLUND, N. E. 1985. *Stating Objectives for Classroom Instruction*. Third Edition. New York: MacMillan.

GUIDANCE CENTRE. 1985. Faculty of Education, University of Toronto.

GUMAER, J. 1973. "Peer facilitating groups." *Elementary School Guidance and Counselling, 8*(1), 4-11.

GUNDLACH, R. A. 1982. "Children as writers: The beginnings of learning to write." In M. Nystrand (Ed.) *What Writers Know: The language, process, and structure of written discourse*. New York: Academic Press.

HAGGARD, M. R. 1981. "Woodcock Reading Mastery Tests." In L. M. Schell (Ed.) *Diagnostic and Criterion-referenced Reading Tests: Review and evaluation*. Newark: International Reading Association.

HAGIN, R. A. 1983. "Write right or left: A practical approach to handwriting." *Journal of Learning Disabilities, 16*, 266-271.

HALIFAX BOARD OF EDUCATION. 1976. Special Education and Services Handbook.

HALPERT, N. 1984. "Artificial intelligence and the education of the learning disabled." *Journal of Learning Disabilities, 17* (2), 118-20.

HAMLETT, C.L. 1984. "Microcomputer activities for gifted elementary children: Alternatives to programming." *Teaching Exceptional Children, 16*(4), 253-257.

HAMMILL, D. D., N. R. BARTEL AND G. O. BUNCH. 1984. *Teaching Children with Learning and Behavior Problems*. Canadian Edition. Toronto: Allyn and Bacon, Inc.

HAMMILL, D. D. AND S. C. LARSON. 1978. *TOWL (Test of Written Language)*. Austin, TX: Pro-Ed.

HAMMILL, D. AND J. LEIGH. 1983. *Basic School Skill Inventory*. Austin, Texas: Pro-Ed.

HAMMILL, D. AND G. McNUTT. 1981. *The Correlates of Reading: The consensus of thirty years of correlational research*. Austin, Texas: Pro-Ed.

HAMMILL, D. AND M. POPLIN. 1982. "Problems in written composition." In D. D. Hammill and N. R. Bartel (Eds.) *Teaching Children with Learning and Behavior Problems*. Third Edition. Boston: Allyn and Bacon.

HANNAFORD, A. E. 1983. "Microcomputers in special education: some new opportunities, some old problems." *The Computing Teacher, 10*(6), 11-17.

HANNAFORD, A. AND E. SLOANE. 1981. "Microcomputers: Powerful learning tools with proper programming." *Teaching Exceptional Children, 14*, 54-57.

HANNAN, E. AND G. HAMILTON. 1984. "Writing: What to look for, what to do." *Language Arts, 61*, 364-366.

HANSFORD, B. C. AND J. A. HATTIE. 1982. "The relationship between self and achievement/performance measures." *Review of Educational Research, 52*, 123-142.

HARDWICK, N. 1980. *Children with Learning Disabilities*. Toronto: Ministry of Education.

HARDY, L. AND B. DUNN-WORNDL. 1983. *In Trouble with the Law: A guide to The Young Offenders Act*. Toronto: Justice for Children.

HARDY, M. 1981, FEBRUARY. "School-based special services networks." Paper presented at Special Education Symposium: Special Students, Special Laws, and Special Programs. Richmond, British Columbia.

HARRINGTON, A. 1975. "Teaching parents to help at home." In C. Smith (Ed.). *Parents and Reading: Perspectives in reading*. Number 14. Newark, Delaware: International Reading Association.

HARRIS, A. J. AND M. D. JACOBSON. 1980. "A comparison of the Fry, Spache, and Harris-Jacobson readability formulas for primary grades." *The Reading Teacher, 33*, 920-924.

HARRIS, A. J. AND E. R. SIPAY. 1980. *How to Increase Reading Ability: A guide to developmental and remedial methods*. Seventh Edition. New York: Longman.

HARRISON, S. 1981. "Open Letter from a left handed teacher: Some sinistral ideas on the teaching of handwriting." *Teaching Exceptional Children, 13*, 116-120.

HASSELBRING, T. S. 1984. "Using a microcomputer for imitating student errors to improve spelling performance." *Computers, Reading and Language Arts, 1*(4), 12-14.

HASSELBRING, T. S. 1982. "Remediating spelling problems of learning-handicapped students through the use of microcomputers." *Educational Technology, 22*(4), 31-32.

HASSELBRING, T. S. AND C. L. CROSSLAND. 1982. "Application of microcomputer technology to spelling assessment of learning disabled students." *Learning Disabilities Quarterly, 5*(1), 80-82.

HASSELBRING, T. S. AND C. L. CROSSLAND. 1981. "Using microcomputers for diagnosing spelling problems in learning-handicapped children." *Educational Technology, 21*, 37-39.

HASSELBRING, T. S. AND C. L. HAMLETT. 1984. "Planning and managing instruction: Computer-based decision-making." *Teaching Exceptional Children, 16*(4), 248-252.

HAYDEN, D., B. VANCE AND M. S. IRVIN. 1982a. "A special education management system (SEMS)." *Journal of Learning Disabilities, 15*(6), 374-375.

HAYDEN, D., B. VANCE AND M. S. IRVIN. 1982b. "Establishing a special education management system (SEMS)." *Journal of Learning Disabilities, 15*(7), 428-429.

HEALD-TAYLOR, G. 1984. "Scribble in first grade writing." *The Reading Teacher*, 4-8.

HECKELMAN, R. G. 1969. "A neurological-impress method of remedial-reading instruction." *Academic Therapy, 4*, 277-282.

HELMSTETTER, F. AND E. BALUNIS. 1984. "Quill (Software Review)." *Computers, Reading and Language Arts, 1* (4), 43-44.

HENNINGS, D. G. 1982. "A writing approach to reading comprehension—Schema theory in action." *Language Arts, 59*, 8-17.

HEWETT, F. M. AND F. D. TAYLOR. 1980. *The Emotionally Disturbed Child in the Classroom: The orchestration of success.* Boston: Allyn and Bacon.

HEWITT, G. (N.D.) "The way to higher marks." London, Ontario: Montcalm Guidance Department.

HEYMAN, E. 1977. "Cursive writing begins with chalk." *Teaching Exceptional Children, 9*(4), 106-109.

HIERONYMOUS, A. N., E. F. LINDQUIST, H. D. HOOVER AND E. M. KING. 1978, 1979, 1980, 1981; SCANNEL, D. P., O. M. HAUGH, A. H. SCHILD, G. ULMER AND E. KING. 1978, 1981. *Canadian Tests of Basic Skills*. Toronto: Nelson Canada.

HILL, C. H. AND A. S. MARTINIS. 1973. "Individualizing a multisensory spelling program." *Academic Therapy, 9*(1), 77-82.

HILLERICH, R. L. 1974. "Word lists—Getting it all together." *The Reading Teacher, 27*, 353-360.

HOBBS, N. 1966. "Helping disturbed children: Psychological and ecological strategies." *American Psychologist, 21*, 1105-1115.

HOBBS, N. "The re-education of emotionally disturbed children." In E. M. Bower & W. C. Hollister (Eds.) *Behavioral Science Frontiers in Education*. New York: John Wiley.

HOBBS, N. 1982. *The Troubled and Troubling Child*. San Francisco: Jossey-Bass.

HOBEN, M. 1980. "Toward integration in the mainstream." *Exceptional Children, 47*(2), 100-105.

HOFMEISTER, A. 1982. "Microcomputers in perspective." *Exceptional Children, 49*, 115-120.

HOFMEISTER, A. M. AND R. J. THORKILDSON. 1981. "Video-disc technology and the preparation of special education teachers." *Teacher Education and Special Education, 4*(3), 34-39.

HOMENIUK, F. 1984. Spelling worksheets. St. Leo's School, Brantford, Ontario.

HOPS, H., H. M. WALKER AND S. L. BEICKEL. 1974. *CLASS Program for Acting-out Children: Contingencies for learning academic and social skills*. Eugene, Oregon: Center on Human Development.

HUMMEL, J. W. 1984. "Courseware Review—"Wizard of Words." *Journal of Learning Disabilities, 17*(6), 370-371.

HUMPHRIES, T., M. KINSBOURNE AND J. M. SWANSON. 1978. "Stimulant effects on cooperation and social interaction between hyperactive children and their mothers." *Journal of Child Psychology and Psychiatry, 19*, 13-22.

HUNDERT, J. 1981. "Some considerations of planning the integration of handicapped children in the mainstream." *Special Education in Canada, 55*, 5-13.

INGHAM, A. G. 1978. *The Blended Sound-sight Method of Learning*. Third Edition. Yorkton, Saskatchewan: A. G. Ingham.

ISEN, A. M., B. MEANS, R. PATRICK AND G. NOWICKI. 1981. "Some factors influencing decision-making strategy and risk taking." In M.S. Clark & S. T. Fiske (Eds.) *Affect and Cognition: The seventeenth annual Carnegie symposium on cognition*. Hillsdale, N.J.: Lawrence Erlbaum.

JABEN, T. 1983. "The effects of creativity training on learning disabled students' creative written expression." *Journal of Learning Disabilities, 16*(5), 264-265.

JACKSON, P. A. 1984. Personal communication.

JASKOSKI, H. 1984. "Notes on a competency-based curriculum in poetry therapy." *The Arts in Psychotherapy, 11*, 77-88.

JASTAK, J. F. AND S. JASTAK. 1978. *The Wide Range Achievement Test*. Wilmington, Delaware: Jastak Associates.

JENKINS, J. R. AND K. LARSON. 1978. *Evaluating Error Correction Procedures for Oral Reading*. (Technical Report No. 55), Champaign, Ill.: Center for the Study of Reading, University of Illinois.

JEROSKI, S. 1984. *The 1984 B.C. Reading Assessment General Report*. Victoria, B.C.: Learning Assessment Branch, Ministry of Education, Province of British Columbia.

JOBES, N. K. AND L. W. HAWTHORNE. 1977. "Informal assessment in the classroom." *Focus on Exceptional Children, 9*(2), 1-13.

JOHNS, J. L. 1981. *Basic Reading Inventory*. Second Edition. Dubuque, Iowa: Kendall/Hunt.

JOHNS, J. L. 1981. *Advanced Reading Inventory*. Dubuque, Iowa: Wm. C. Brown.

JOHNS, J. L. 1984. "Equivalence of forms 1 and 3, level E, Gates-MacGinitie reading tests." *Journal of Reading, 28*, 48-51.

JOHNS, J. L. (ND). *Ingredients for a Sound Reading Program*. Unpublished manuscript. Northern Illinois University, DeKalb, Illinois.

JOHNSON, D. D. AND A. J. MOE. 1980. "Current approaches, part two." In P. Lamb and R. Arnold (Eds.) *Teaching reading: Foundations and strategies*. Second Edition. Belmont, California: Wadsworth.

JOHNSON, D. W. AND R. JOHNSON. 1980. "Integrating handicapped students into the mainstream. *Exceptional Children, 46*, 89-98.

JOHNSON, D. W. AND R. JOHNSON, 1982. "The integration of the handicapped into the regular classroom: Effects of cooperative and individualistic instruction." *Contemporary Educational Psychology,* 6, 344-353.

JOHNSON, D. W. AND R. JOHNSON. 1982. "Effects of cooperative, competitive and individualistic learning experiences on cross-ethnic interactions and friendships." *Journal of Social Psychology, 118*, 47-58.

JOHNSON, D. W., R. T. JOHNSON AND G. MARUYAMA. 1983. "Interdependence and interpersonal attraction among heterogeneous and homogeneous individuals. A theoretical formulation and a meta-analysis of the research." *Review of Educational Research, 53*(1), 5-54.

JOHNSON, J. JULY 1983. "Mathematics education and computers: Cause for concern or change?" *The Computer: Extension of the Human Mind Annual Conference*. Eugene, Oregon.

JOHNSON, N. AND L. RAY. 1983. "Adolescent suicide." *Personnel and Guidance Journal, LXII*(3), 131-135.

JOHNSON, R. AND D. W. JOHNSON. 1983. "Effects of cooperative, competitive and individualistic learning experiences on cross-handicap relationships and social development." *Exceptional Children, 49*, 323-329.

JOHNSON, T.D. 1977. "Language experience: We can't all write what we can say." *The Reading Teacher, 31*, 297-299.

JOHNSON, T. D. AND D. LOUIS. 1985. *Developing Literacy through Use of Literature*. North Ryde, N.S.W.: Methuen Australia.

JONES, V. F. 1900. *Adolescents with Behavior Problems: Strategies for teaching, counseling, and parent involvement*. Boston: Allyn and Bacon.

JOYCE, B. AND M. WEIL. 1972. *Models of Teaching*. Englewood Cliffs: Prentice-Hall.

JOYCE, B. AND M. WEIL. 1980. *Models of Teaching* Second Edition. Englewood Cliffs, NJ: Prentice-Hall.

KAGAN, J. 1965. "Information processing in the child." In P. H. Mussen, J. J. Conger and J. Kagan (Eds.) *Reading in Child Development and Personality*. New York: Harper and Row.

KARLSEN, B., R. MADDEN AND E. F. GARDNER. 1976, 1978. *Stanford Diagnostic Reading Test*. New York: Harcourt, Brace, and Jovanovich.

KASDIN, A. 1975. *Behavior Modification in Applied Settings*. Homewood, IL: Dorsey.

KAUFMAN, M., J. AGARD AND M. SEMMEL. (EDS.) 1978. *Mainstreaming: Learners and their environments*. Baltimore: University Park Press.

KAVALES, K. AND S. FORNESS. 1983. "Hyperactivity and diet treatment: A metaanalysis of the Feingold hypothesis." *Annual Review of Learning Disabilities, 1*, 132-138.

KEGAN, R. 1982. *The Evolving Self: Problem and process in human development*. Cambridge, Massachusetts: Harvard University Press.

KENNEDY, P.M. 1984. "Native Indians and Standardized Tests: Are they fair?" Unpublished paper, University of Victoria.

KEOGH, B. K. 1977. "Research on cognitive styles." In R. Kneedler and S. Tarver (Eds.) *Changing Perspectives in Special Education*. Columbus, Ohio: Charles E. Merrill.

KERCHNER, L. B. AND KISTINGER. 1984. "Language processing/word processing: Written expression, computers and learning disabled students." *Learning Disability Quarterly,* 7(4), 329-335.

KERSHNER, J. AND R. GREKIN. N.D. "Nutrition and learning disabilities: A review of treatment approaches and a pilot study of a high-protein and high-vitamin, low carbohydrate, sugar free diet and learning disabled children." Toronto: OISE.

KEVORKIAN, J. 1977. "Stick Reading for Spelling Patterns." *The Reading Teacher*, 154-159.

KIRK, S. A. AND J. J. GALLAGHER. 1979. *Educating Exceptional Children*. Boston: Houghton Mifflin.

KIRSCHENBAUM, D. S. AND R. C. FLANERY. 1983. "Behavioral contracting: Outcomes and elements." In M. Hersen, R. M. Eisler & P. M. Miller (Eds.) *Progress in Behavior Modification*. Volume 15. New York: Academic Press.

KLEIMAN, G. M. AND S. KEYES, 1985. "Sunburst Educational Software." *Compute,* 7(1), 108-112.

KLEIMAN, G., M. HUMPHREY AND P. H. LINDSAY. 1981. "Microcomputers and hyperactive children." *Creative Computing,* 7(3), 93-94.

KNEEDLER, R. D. 1984. *Special Education for To-day*. Englewood Cliffs, N.J.: Prentice-Hall, Inc.

KNOWLES, D. AND N. REEVES. 1981. "Understanding children's concerns about death and dying." *B.C. Journal of Special Education, 5*, 33-40.

KOHL, H. R. 1974. *Math Writing and Games in the Open Classroom*. New York: Vintage Books.

KOMOSKI, P. K. 1984. "Educational computing: The burden of insuring quality." *Phi Delta Kappan, 66*(4), 244.

KOOPMAN, P.R.S. 1983. *Cognitive Disorders and Syntactical Deficiencies in the Inmate Populations of Federal Penitentiaries in Canada*. Report for the Solicitor General of Canada.

KOSC, L. 1974. "Developmental dyscalculia." *Journal of Learning Disabilities, 7*, 164-177.

KRAFT, R. E. 1978. "Can the movement specialist really influence self-concept." *The Physical Educator, 35*(1), 20-21.

KRAMER, E. 1958. *Art in a children's community*. Springfield, Illinois: Charles E. Thomas.

KRAMER, E. 1971. *Art as therapy with children*. New York: Schocken Books.

KRATHWOHL, D. R., B. S. BLOOM AND B. B. MASIA. 1964. *Taxonomy of Educational Objective, Handbook II: Affective domain*. New York: David McKay.

KRETSCHMER, R. R. AND L. W. KRETSCHMER. 1978. *Language Development and Intervention with the Hearing Impaired*. Baltimore: University Park Press.

KRONICK, D. 1979. "The learning disabled adolescent in the secondary school. Paper presented to the London Board of Education, London, Ontario.

KRONICK, D. 1981. *Social Development of Learning Disabled Persons*. San Francisco: Jossey-Bass.

KRONICK, D. 1981. *Five Families*. San Rafael: Academic Press.

KRONICK, D. 1983. "Programming for Social Skills Deficits." *National, 20*, 6.

KROTH, R. 1973. "The behavioral Q-sort as a diagnostic tool." *Academic Therapy, VIII*, 317-330.

KULIK, J. A. 1983. "Synthesis of research on computer based instruction." *Educational Leadership, 41*(1), 19-21.

KULIK, J. A., ET AL. 1983. "Effects of computer-based teaching on secondary students." *Journal of Educational Psychology, 75*(1), 19-26.

LABERGE, D. AND S. J. SAMUELS. 1974. "Toward a theory of automatic information processing in reading." *Cognitive Psychology, 6*, 293-323.

LA PRAY, M. AND R. ROSS. 1969. "The graded word list: Quick guide of reading ability." *Journal of Reading, 12*, 305-307.

LARSON, S. AND D. D. HAMMILL. 1976. *The Test of Written Spelling*. Austin, Tex.: Pro-Ed.

LEBRUN, V. AND P. VAN DE CRAEN. 1975. "Developmental writing disorders and their prevention." *Journal of Special Education, 9*(2), 201-207.

LEIBERT, R. M. AND L. W. MORRIS, 1967. "Cognitive and emotional components of test anxiety: A distinction and some initial data." *Psychological Reports, 20*, 975-978.

LEITHWOOD, K. AND D. J. MONTGOMERY. 1982. "The role of the elementary school principal in program improvement." *Review of Educational Research, 52*, 309-339.

LERNER, J. W. 1976. *Children with Learning Disabilities*. Second Edition. Boston: Houghton Mifflin Co.

LEVY B. AND P. GRAF. 1984. "Reading and remembering: Conceptual and perceptual processing involved in reading rotated passages." *Journal of Verbal Learning and Verbal Behavior, 23*(3), 405-424.

LIEDTKE, W. 1979. "Primary mathematics: Suggestions for diagnosis and remediation." *Special Education in Canada, 54*(1), 13-15.

LIEDTKE, W. 1982. "Learning difficulties: Helping young children with mathematics—subtraction." *Arithmetic Teacher, 30*(4), 21-23.

LINDSAY, P.H. AND L.A. MARINI. 1983. "Microcomputers and exceptional children." *Newsletter* of the Council for Exceptional Children, Ontario Provincial Federation, *19*(8), 1-3.

LINDSAY, P. H. AND L. A. MARINI. 1983. *Microcomputers and Exceptional Children: Releasing the potential*. Unpublished manuscript, Ontario Institute for Studies in Education, Toronto, Canada.

LIPPIN, R. A. 1982. "Poetry and poetry therapy: A conversation with Arthur Lerner." *The Arts in Psychotherapy, 9*, 167-174.

LONDON BOARD OF EDUCATION. 1981. *Printing and Writing Programme*. London, Ontario: Board of Education.

LOTSOF, A. B. 1978. "Behavior related to suicides, runaways, alcoholism, and drug abuse." In D. A. Sabatino and A. J. Mauser (Eds.) *Intervention Strategies or Specialized Secondary Education*. Boston: Allyn & Bacon.

LUM, T. AND L. L. MORTON. 1984. "Direct instruction in spelling increases gain in spelling and reading skills." *Special Education in Canada, 58*, 41-45.

LUSTHAUS, C., E. LUSTHAUS AND H. GIBBS. 1981. "Parental involvement in school decision-making: Parents of children in special and regular education." *Special Education in Canada, 55*, 21-25.

LUTZ, J. 1973. *Expanding Spelling Skills*. Dansville, N.Y.: The Instructor Publications, Inc.

MACGINITIE, W. H. 1978, 1980. *Gates-MacGinitie Reading Tests*. Canadian Edition. Toronto: Nelson.

MACGINITIE, W. H. 1983, OCTOBER. *Vocabulary and Comprehension Discrepancies*. Paper presented at the international Conference on Learning Disabilities, San Francisco.

MACMILLAN, D., R. JONES AND C. MEYERS. 1976. "Mainstreaming the mildly retarded: Some questions, cautions and guidelines." *Mental Retardation, 14*(1), 3-10

MADDEN, N. AND R. SLAVING. 1983."Mainstreaming students with mild handicaps: Academic and social outcomes." *Review of Educational Research, 53*, 519-569.

MALONEY, M. J. AND W. M. KLYKYLO. 1983. "An overview of anorexia nervosa, bulimia, and obesity in children and adolescents." *Journal of the American Academy of Child Psychiatry, 22*, 99-107.

MANGIERI, J. N. AND M. S. KAHN. 1977. "Is the Dolch list of 220 basic sight words irrelevant?" *The Reading Teacher, 30*, 649-651.

MANN, P. H., P. A. SUITER AND R. M. MCCLUNG. 1979. *Handbook in Diagnostic-Prescriptive Teaching.* Second edition. Boston: Allyn and Bacon Inc.

MARKELL, C. 1976. *Exceptional Students in Regular Classes: Interviews with 43 North Dakota Elementary Teachers*. North Dakota: Minot State College. Ed 117 912.

MARTIN, D. 1984. "Cognitive modification for the hearing impaired adolescent: The promise." *Exceptional Children, 51*, 235-242.

MASON, G. E. 1981. "The computer as teacher of the disabled reader." *Journal of Research and Development in Education, 14*, 97-101.

MASON, G. E., J. S. BLANCHARD AND D. B. DANIEL. 1983. *Computer Applications in Reading*. Second Edition. Newark, DE: International Reading Association.

MAUSER, A. J. 1974. "Learning disabilities and delinquent youth." *Academic Therapy, 9*, 343-344.

MAYNARD, R. SEPTEMBER 1984. "New hope for learning disabled kids." *Chatelaine, 38*

MCCANN, S. K. AND E. J. KELEMEN. 1984. "Microcomputers: New directions and methods for the preparation of special education personnel." *Teacher Education and Special Education, 7*(3), 178-184.

MCDERMOTT, P. AND M. WATKINS. 1983. "Computerized vs. conventional remedial instruction for learning-disabled pupils." *The Journal of Special Education, 17*, 81-87.

MCGRATH, P. 1984. "Warnings about classroom computers." *Macleans*, August 13, 1984.

MCINTYRE, B. 1974. *Creative Drama in the Elementary School*. Itasca: F. E. Peacock.

MCINTYRE, C. W., M. E. MURRAY AND S. L. BLACKWELL. 1981. "Visual search in learning disabled and hyperactive boys." *Journal of Learning Disabilities, 14*(3), 156-158.

MCMURRAY, J. G. 1980. "Learning disabled adolescents: Perspectives from Canadian principals." *Special Education in Canada, 55*, 10-14.

MEICHENBAUM, D. 1972. "Cognitive modification of test anxious college students." *Journal of Counselling and Clinical Psychology, 39*, 370-380.

MEICHENBAUM, D. 1976. "Cognitive behavior modification." In J. T. Spence, R. C. Carson and J. W. Thibaut (Eds.) *Behavioral Approaches to Therapy*. Morristown, N.J.: General Learning Press.

MEICHENBAUM, D. 1977. *Cognitive-behavior Modification: An integrative approach*. New York: Plenum.

MEICHENBAUM, D. 1980. "Cognitive behavior modification with exceptional children: A promise yet unfulfilled." *Exceptional Education Quarterly, 1*, 83-88.

MEICHENBAUM, D. 1982. "Stress-innoculation training: A cognitive-behavioral approach." In D. Meichenbaum, and M. Jaremko. (Eds.) *Stress Prevention and Reduction*. New York: Plenum Press.

MEICHENBAUM, D. AND M. GENEST. 1980. "Cognitive behavior modification: An integration of cognitive and behavioral methods." In F. H. Kanfer and A. P. Goldstein (Eds.) *Helping People Change: A textbook of methods*. Second Edition. New York: Pergamon.

MEICHENBAUM, D. AND J. GOODMAN. 1971. "Training impulsive children to talk to themselves: A means of developing self-control." *Journal of Abnormal Psychology, 77*(2), 115-125.

MEISELS, S. 1977. First steps in mainstreaming. *Young Children, 33*(1), 4-13.

MICHEL, D. E. 1971. *Music Therapy: An introduction to therapy and special education*. New York: Day.

MILLAR, G. AND E. ROAD. 1983. "Academic visiting privileges: A successful enrichment mentorship program in practice." *Special Education in Canada, 51*, 12-14.

MILLER, G. 1956. "The magical number seven, plus or minus two: Some limits on our capacity for processing information." *Psychological Review, 63*, 81-97.

MILLER, G., E. GALANTER AND K. H. PRIBRAM. 1960. *Plans and the Structure of Behavior*. New York: Holt, Rinehart and Winston.

MILLER, L. NOVEMBER, 1984. "Reading, writing and computers: Going beyond the obvious." Paper presented at the Computers '84—Software for the Classroom, GEMS Conference, Toronto, Ontario.

MILLER, R. D. 1984. "Organizing an individualized evolving word list spelling program." Victoria, B.C.: The University of Victoria, unpublished manuscript.

MILONE, M. N. AND T. M. WASYLYK. 1981. "Handwriting in special education." *Teaching Exceptional Children, 14*, 58-61.

MINISTRY OF EDUCATION, PROVINCE OF BRITISH COLUMBIA, 1977. *Guide to the Core Curriculum*. Victoria.

MINISTRY OF EDUCATION, PROVINCE OF BRITISH COLUMBIA. 1982. *Language arts for Native Indian Students*. Victoria.

MINISTRY OF EDUCATION, PROVINCE OF BRITISH COLUMBIA. 1983. *Special Education Core Curriculum Supplement: A guide for teachers of the mentally handicapped*. Victoria.

MOE, A. J. AND D. D. JOHNSON. 1980. "Current approaches, part one." In P. Lamb and R. Arnold (Eds.) *Teaching Reading: foundations and strategies*. Second Edition. Belmont, California: Wadsworth.

MOORE, R. W. 1983. "Art therapy with substance absurers: A review of the literature." *The Arts in Psychotherapy, 10*, 251-260.

MORENO, J. 1943. "Sociometry in the classroom." *Sociometry, 6*, 425-428.

MORRIS, L., M. DAVID AND C. HUTCHINGS. 1981. "Cognitive and emotional components of anxiety: Literature review and a revised worry-emotionality scale." *Journal of Educational Psychology, 73*, 541-555.

MORRIS, N. T. AND W. D. CRUMP. 1982. "Syntactic and vocabulary development in the written language of learning disabled and non-learning disabled students at four age levels." *Learning Disability Quarterly, 5*, 163-172.

MORSE, W. C., J. ARDIZZONE, C. MACDONALD AND P. PASICK. 1980. *Affective Education for Special Children and Youth*. Reston, Virginia: ERIC Clearinghouse on Handicapped and Gifted Children.

MORSINK, C. 1981. "Perspectives of an ivory tower realist." Kentucky: College of Education, Kentucky University at Lexington. Ed 202 836.

MOYER, S. B. 1983. "Repeated reading." In G. M. Senf and J. K. Torgensen (Eds.) *Annual Review of Learning Disabilities, 1*, 113-117.

MULCAHY, R., L. WILGOSH, B. CRAWFORD AND B. WATTERS. 1984. "Gifted education for Inuit youngsters: Problems and issues. *Special Education in Canada, 58*, 127.

MULLINS, J. AND F. JOSEPH. 1970. "A script to supplant cursive writing or printing." *Teaching Exceptional Students, 3*(1), 23-32.

MUTTART, K. 1984. "Assessment of effects of instrumental enrichment cognitive training." *Special Education in Canada, 58*, 106-108.

MYKLEBUST, H. R. 1964. "Learning disorders: Psychoneurological disturbances in childhood." *Rehab. Lit. 25*, 354-360.

MYKLEBUST, H. 1965. *Development and Disorders of Written Language: Picture story language test. Volume 1.* New York: Grune & Stratton.

MYKLEBUST, H. 1973. *Development and Disorders of Written Language: Studies of normal and exceptional children. Volume 2.* New York: Grune & Stratton.

NARROL, H., H. SILVERMAN AND M. WAKSMAN. 1982. "Developing cognitive potential in vocational high school students." *Journal of Educational Research, 76*, 107-112.

NATIONAL ASSESSMENT OF EDUCATIONAL PROGRESS. 1980. *Writing Achievement, 1969-79: Results from the third national writng assessment, Volume III—9 year olds.* Report No. 10-W-03. Denver, Colorado: National Assessment of Education Progress.

NATIONAL ASSESMENT OF EDUCATIONAL PROGRESS. 1982. "Building effective schools—Here's how. *NAEP Newsletter, 15*, Fall, 6.

NEIMEN, A. S. 1971. "Handwriting, spelling and creative writing in the primary grades." *Elementary English, 48*(8), 963-969.

NELSON, C. M. AND L. POLSGROVE. 1984. "Behavioral analysis in special education: White rabbit or white elephant?" *Remedial and Special Education, 5*(4), 6-17.

NORTH WEST TERRITORIES, 1984. Consolidation of Education Ordinance ONWT 1976 as amended. Yellowknife, N.W.T.: Territorial Printer.

NULMAN, J. A. AND M. M. GERBER. 1984. "Improving spelling performance by imitating a child's errors." *Journal of Learning Disabilities, 17*, 328-333.

O'BRIEN, P. 1984. "Using microcomputers in the writing class." *The Computing Teacher, 11*(9), 20-21.

O'BRIEN, T. C. 1983. "Five essays on computers in education." *Phi Delta Kappan*, October, 110-112.

O'BRIEN, T. C. 1983. "Software of the second-and-a-half kind." *Classroom Computer Learning, 4*(2), 33-36.

OHANIAN, S. 1984. "How today's reading software can zap kids' desire to read." *Classroom Computer Learning, 5*(4), 26-31.

OLLILA, L. 1981. "A primary reading attitude inventory." Unpublished manuscript, University of Victoria.

ONTARIO MINISTRY OF EDUCATION. 1981. *Special education: Information Handbook.* Toronto, Ontario.

O'BRIEN, L. AND W. PULLIAM. 1981. "Success or failure: Skills can make the difference." *The History and Social Science Teacher, 16*, 139-143.

O'ROURKE, R. 1984. Yellowknife Catholic Board of Education, N.W.T., personal communication.

ORTON, S. T. 1937. *Reading, Writing and Speech Problems in Children*. New York: W. W. Norton.

OSBORN, A. F. 1963. *Applied Imagination*. New York: Charles Scribner's Sons.

OWENS, M. F. 1981. "Mainstreaming in the every child a winner program." *Journal of Physical Education, Recreation, and Dance, 52*(7), 16-18.

OWINGS, R., G. PETERSEN, J. D. BRANSFORD, C. D. MORRIS, AND B. C. STEIN. 1980. "Spontaneous monitoring and regulation of learning: A comparison of successful and less successful fifth graders." *Journal of Educational Psychology, 72*, 250-256.

OWL. 1984. Toronto, Canada.

PAGE, W. D. AND G. S. PINNELL. 1979. *Teaching Reading Comprehension: Theory and practice*. Urbana: National Council for Teachers of English.

PAPERT, S. 1984. "Computer as mudpie." *Classroom Computer Learning, 4*(6), 36-40.

PAPERT, S. JUNE, 1984. Introductory panel at the LOGO '84 Conference, Cambridge, MA.

PAPERT, S. 1980. *Mindstorms: Children, computers and powerful ideas*. New York: Basic Books.

PAPERT, S., A. DISESSA, D. WATT AND S. WEIR. 1979. *Final Report of the Brookline LOGO Project, Part II: Project summary and data analysis* (Artificial Intelligence Memo No. 545). Cambridge, MA: M.I.T. Artificial Intelligence Laboratory. (ERIC Document Reproduct Service No. ED 20780.)

PARNES, B. 1978. Personal communication.

PARNES, S., R. NOLLER AND A. BIONDI. 1977. *Creative Actionbook*. New York: Charles Scribner's Sons.

PEARL, R. 1982. "Learning disabled children's attributions for success and failure: A replication with a labeled learning disabled sample." *Learning Disability Quarterly, 5*, 173-176.

PEARL, R., T. BRYAN AND M. DONAHUE. 1980. "Learning disabled children's attributions for success and failure." *Learning Disability Quarterly, 3*, 3-9.

PEARSON, P. AND D. JOHNSON. 1978. *Teaching Reading Comprehension*. New York: Holt, Rinehart and Winston.

PELANDER, J. AND B. WILLIS. Branching cursive penmanship program. Clover Park School District No. 400.

PETERSON, W. O. 1972. "P.E.E.R. on focus." University of Oregon, unpublished manuscript.

PIHL, R. D. AND L. MCLARNON. 1984. "Learning disabled children as adolescents." *Journal of Learning Disabilities, 11*, 96-100.

PIPER, K. 1983-1984. "Word processing." *The Computing Teacher*, December-January, 82-83.

PIPER, K. L. 1983. "Separating wheat from chaff: Educating word processing programs for language arts instructions." *Computers, Reading and Language Arts, 1*(3), 9-14.

POLSGROVE, L. AND H. J. REITH. 1983. "Procedures for reducing children's inappropriate behavior in special education settings." *Exceptional Educational Quarterly, 3*(4), 20-33.

POPEIL, P. S., J. D. HOLLINGER, G. A. LOSCHI AND D. R. CRAWFORD. 1983. "A sociometric approach to understanding interpersonal relationships in a classroom for emotional disturbed pupils." *The Pointer, 27*(3), 5-12.

POPLIN, M. S., R. GRAY, S. LARSEN, A. BANIKOWSKI AND T. MEHRING. 1980. "A comparison of components of written expression abilities in learning disabled and non-learning disabled students at three grade levels." *Learning Disability Quarterly, 3*(4), 46-53.

POSNO, R. 1981. "Who's afraid of individual educational plans (IEPs)?" *Special Education in Canada, 56*, 15-20.

POSNO, R. 1983. "Eagles fly and lions walk alone: The learning disabled adolescent." Paper presented at the annual CEC Conference, Vancouver, B.C.

PRECISION TEACHING MATERIALS AND ASSOCIATES. N.D. *Language Arts Notebook 1*. Great Falls, Montana.

PRESSLEY, M. 1984. "The generalizability of the key word method in instruction." Paper presented at the annual meeting of the American Educational Research Association, New Orleans, Louisiana.

PRESSLEY, M., J. R. LEVIN, G. V. NAKAMURA, D. J. HOPE, J. G. BISPO AND A. R. TOYE. 1980. "The keyword method of foreign vocabulary learning: An investigation of its generalizability." *Journal of Applied Psychology, 65*, 635-642.

PRESTON, LYNN. JANUARY, 1984. *Improving Written Expression Skills for the Learning Disabled Student: A model for effective use of direct instruction, microcomputers and software programs*. Paper presented at the Technology in Special Education Conference of the Council for Exceptional Children, Reno, NV.

PRIDDLE, F. 1981. *Step-by-step Volunteer Services for Emotionally Disturbed Children*. Ottawa, Ontario: Ottawa Board of Education.

QUEBEC MINISTRY OF EDUCATION. 1983. *Activité's éducatives*. Québec: Gouvernement du Québec.

RAGSDALE, R. 1982. *Computers in the schools: A guide for planning*. Toronto, Ont.: The Ontario Institute for Studies in Education Press, Informal Series/45.

RAGSDALE, R. 1982. *Evaluating of Microcomputer Courseware* . Toronto: The Ontario Institute for Studies in Education Press.

RAPHAEL, T. 1984. "Teaching learners about sources of information for answering comprehension questions." *Journal of Reading, 27*, 303-311.

REDL, F. AND D. WINEMAN. 1952. *Controls from Within: Techniques for the treatment of the aggressive child.* New York: The Free Press.

REDL, F. 1966. *When We Deal with Children*. New York: The Free Press.

REGAN, E., K. LEITHWOOD, P. SULLIVAN AND A. MCCONNELL. 1978. *Stimulating Cognitive Growth*. Toronto: The Ontario Institute for Studies in Education, Informal Series/4.

REICHMAN, M. S. AND W. C. HEALEY. 1983. "Learning disabilities and conductive hearing loss involving otitis media." *Annual Review of Learning Disabilities, 1*, 39-45.

RHODES, W. C. 1967. "The disturbing child: A problem in ecological management." *Exceptional Children, 33*, 449-455.

RHODES, W. C. 1970. "A community participation analysis of emotional disturbance." *Exceptional Children, 36*, 309-314.

RICHARDSON, S. A. AND J. ROYCE. 1968. "Race and physical handicap in children's preference for other children." *Child Development, 2*, 467-480.

RIETH, H.J., G. HAUS, T. HASSELBRING, AND L. POLSGROVE. JANUARY, 1984. "CITH Instructional Management Systems (CIMS)." Paper presented at the Council for Exceptional Children Technology in Special Education Conference, Reno, NV.

RIETH, H. J. AND M. I. SEMMEL. 1979. "The use of microcomputer technology to prepare and to enable teachers to meet the educational needs of handicapped children." *Teacher Education and Special Education, 2*, 56-60.

ROBINETT, R. F., P. W. BELL AND P. M. ROJAS. 1970. *Miami Linguistic Reading Series*. Lexington, Mass.: D. C. Health.

RODGERS, D., H. COVELL AND K. SLADE. 1978. *I Can Spell: Super guide to teaching spelling*. Toronto: D. C. Health Canada Ltd.

ROGERS, S. 1981. "Acquisition and maintenance of spelling skills." London, Ontario: University of Western Ontario, unpublished research paper.

ROLLINS, N. AND E. PIAZZA. 1978. "Diagnosis of anorexia nervosa: A critical reappraisal." *Journal of the American Academy of Child Psychiatry, 17*, 126-137.

ROSE, C. D. 1979. "The social acceptance of severely handicapped children in a regular school and the utilization of peer tutors to improve their social interactions." *B. C. Journal of Special Education, 3*(4), 399-409.

ROSENBAUM, J. 1980. "Social implications of educational grouping." In D. Berliner (Ed.) *Review of Research in Education*. American Educational Research Association.

ROSENSHINE, B. 1982. "How time is spent in elementary classrooms." In C. Denham and A. Lieberman (Eds.) *Time to Learn*. Washington, DC: National Institute of Education.

ROSENSHINE, B. AND D. BERLINER. 1978. "Academic engaged time." *British Journal of Teacher Education, 4*, 3-16.

ROSS, J. AND F. MAYNES. 1982. *Teaching Problem-solving*. Toronto: OISE Press.

ROSS, R. R. AND E. FABIANO. 1983. "The cognitive model of crime and delinquency, prevention and rehabilitation. II. Intervention techniques." Toronto: Ministry of Correctional Services.

ROSS, R. R. AND E. FABIANO. 1983. "The cognitive model of crime and delinquency prevention and rehabilitation. I. Assessment procedures." Toronto: Ministry of Correctional Services.

ROSS, R. R. AND E. FABIANO. 1981. *Time to Think. Cognitition and Crime: Link and remediation*. Ottawa: Department of Criminology.

ROTHWELL, W. 1982. *Tips for Parents in Helping Students Organize*. London, Ontario: Middlesex Board of Education.

RUSSELL, S. J. 1983. "LOGO in special education." *Classroom Computer Learning, 4*(3), 34-39.

SACHS, A. 1984. "Accessing scripts before reading the story." *Learning Disability Quarterly, 7*, 226-228.

SALEND, S. J. AND J. JOHNS. 1982. "A tale of two teachers: Changing teacher commitment to mainstreaming." *Teaching Exceptional Children, 15*, 82-85.

SALSBERG, J. 1982. "Information processing models of memory." London, Ontario: University of Western Ontario, unpublished manuscript.

SALVIA, J. AND J. E. YSSELDYKE. 1981. *Assessment in Special and Remedial Education*. Second Edition. Boston: Houghton Mifflin.

SAMSON, R. 1976. *Thinking Skills*. Stanford, Conn.: Innovative Sciences, Inc.

SAMUDA, R. J. 1978. "Can Testing Serve Minority Needs?" *The Social Guidance Worker, 33*(4). Toronto: Guidance Centre, Faculty of Education, University of Toronto.

SAMUELS, M. AND B. MIMICK. 1980. "A computer managed instructional program for teachers of learning disabled students." In *Proceedings of the 1980 Conference of the Association for the Development of Computer-Based Instructional Systems*.

SAMUELS S. J. 1979. "The method of repeated readings." *The Reading Teacher, 29*, 403-408.

SAMUELS, S. J. 1983. "Diagnosing reading problems." *Topics in Learning and Learning Disabilities, 2*(4), 1-11.

SAPON-SHEVIN, M. 1978. "Another look at mainstreaming: Exceptionality, normality and the nature of difference." *Phi Delta Kappan, 60*(2), 119-121.

SASKATCHEWAN EDUCATION, 1981. *Special Education: A manual of legislation, regulations, policies and guidelines*. The Saskatchewan Department of Education.

SCALDWELL, W. 1984. *The Wesdiand Project*. Ottawa: Ministry of Indian and Northern Affairs.

SCARDAMALIA, M. 1981. "How children cope with the cognitive demands of writing." In C. H. Frederiksen and J. F. Dominic (Eds.) *Writing: The nature, development and teaching of written communication*. Hillsdale, N.J.: Laurence Erlbaum Associates.

SCARDAMALIA, M., C. BEREITER AND B. FILLION. 1981. *Writing for Results: A sourcebook of consequential composing articles*. Toronto: OISE Press.

SCARDAMALIA, M., C. BEREITER AND H. GOELMAN. 1982. "The role of production factors in writing ability." In M. Nystrand (Ed.) *What Writers Know: The language, process, and structure of written discourse*. New York: Academic Press.

SCHAEFER, C. 1979. *The Therapeutic Use of Child's Play*. New York: Jason Aronson.

SCHEIER, M. A. AND R. E. KRAUT. 1979. "Increasing educational achievement via self concept change." *Review of Educational Research, 49*, 131-150.

SCHEIER, M. F. AND C. S. CARVER. 1981. "Cognition, affect and self-regulation." In M. S. Clark and S. T. Fiske (Eds.) *Affect and Cognition: The seventeenth Carnegie symposium on cognition*. Hillsdale, N.J.: Lawrence Erlbaum.

SCHIEVEN-PARR, R. 1978. *Handwriting*. Toronto: Green Tree Publishing Co. Ltd.

SCHONELL, F. J. 1942. *Backwardness with Basic Subjects*. Edinburgh: Oliver and Boyd.

SCHUMAKER, J. B., D. D. DESCHLER, N. J. ALLEY, AND M. M. WARNER. 1983. "Toward the development of an intervention model for learning disabled adolescents: The University of Kansas Institute." *Exceptional Education Quarterly, 4*(1), 45-74.

SCHUMAKER, J. B., J. S. HAZEL, J. A. SHERMAN AND J. SHELDON. 1982. "Social skill performances of learning disabled, non-learning disabled and delinquent adolescents." *Learning Disability Quarterly, 5*, 388-397.

SCHURR, E. L. 1975. *Movement Experiences for Children: A humanistic approach to elementary school physical education*. Englewood Cliffs, N.J.: Prentice-Hall, Inc.

SEMMEL, M. ET AL. 1974. *Tips for Teachers Series; Module 15: Task analysis; Module 16: Lesson planning through task analysis*. Bloomington, Indiana: Center for Innovation in Teaching the Handicapped, Indiana University.

SEMMEL, M., J. GOTTLIEB AND H. ROBINSON. 1979. "Mainstreaming: Perspectives on educating handicapped children in the public schools." In Berliner, D. (Ed.) *Review of Research in Education*. American Educational Research Association.

SEMMEL, M. I., M. C. SITKO, L. HESHUSIUS AND D. COHEN. 1976. *The Effectiveness of a Computer-assisted Teacher Training System (CATTS) in the Development of Reading and Listening Comprehension Strategies of Preservice Special Education Trainees in a Tutorial Classroom Setting*. Final report (Bureau of Education for the Handicapped). Bloomington, IN: Indiana University Center for Innovation in Teaching the Handicapped. Indiana University (ERIC Document Reproduction Service No. 1 ED 162 467) (265 pages).

SEMMEL, M. I., M. C. SITKO, D. S. SEMMEL, T. FRICK AND T. HASSELBRING. 1976. *The Effects of Computer-assisted Teacher-training System (CATTS) Feedback in a Preservice Special Education Teacher Training Program*. Bloomington, IN: Center for Innovation in Teaching the Handicapped. Indiana University, Final Report to the U.S. Office of Education, Bureau of Education for the Handicapped. Grant #OEG-9-242178-4149-032, August, 1976. Volume 1—120 pages (No. 53317; Volume II) (Trainee Manual)—161 pages (No. 5332).

SEMMEL, M. I., S. VARNHAGEN AND S. MCCANN. 1981. "Micro-games: An application of microcomputers for training personnel who work with handicapped children." *Teacher Education and Special Education, 4*(3), 27-33.

SHAVER, J. 1983. "Teaching the handicapped in social studies: Attitudes toward disabilities." *The History and Social Science Teacher, 18*, 129-134.

SHOOK, S. 1982. *The Relationship between the Written Language of Grade One and Two Children and Their Reading Ability, I. Q., and Attitudes*. Unpublished Masters Thesis, University of Victoria.

SICKERT, J. "Microzine (Software Review)." *Computers, Reading and Language Arts, 1*(3), 43-45.

SILVERMAN, R., N. ZIGMOND, J. M. ZIMMERMAN AND A. VALLECORSA. 1981. "Improving written expression in learning disabled students." *Topics in Language Disorders, 1*(2), 91-99.

SIMON, D. 1975. "Spelling—a task analysis." Learning, Research and Development Center, University of Pittsburg: ERIC Number ED 102 606.

SIMPSON, S. AND Y. AVIEZER. 1980. "Variability and instability in perceptual and reading functions of brain-injured children." *Journal of Learning Disabilities, 13*(6), 327-333.

SINICROPE, R., AND L. MICK. 1983. "A comparison of the proportional reasoning abilities of learning disabled and non-learning disabled children and adolescents." *Learning Disability Quarterly,* 6, 313-320.

SITKO, C. *The Effectiveness of LOGO: What does the research say?* Unpublished manuscript, The University of Western Ontario, Faculty of Education, London, Canada.

SITKO, M. C. MAY 1977. "The effectiveness of a computer-assisted teacher training system (CATTS) in the development of hierarchial questioning strategies of preservice special education trainees." Paper presented at the twenty-second annual meeting of the International Reading Association, Miami Beach, Fla.

SITKO, M. C., C. CREALOCK, C. SITKO AND A. HUTCHISON. 1985. *Creative Writing Competency for All. Using Computer Technology to Improve the Writing Skills of Mildly Handicapped Learners*. London, Ontario: Faculty of Education; The University of Western Ontario (Interim Project Report).

SITKO, M. C., A. H. FINK AND P. H. GILLESPIE. 1977. "Utilizing systematic observation for decision-making in school psychology." *School Psychology Monographs, 3*(1), 23-44.

SITKO, M. C. AND F. B. W. HARPER. 1984. A centre of specialization in communicative exceptionalities. Funded grant proposal from the Secretary of State, Government of Canada.

SITKO, M. C. AND P. H. GILLESPIE. 1978. "Language and speech difficulties." In R. Mann, L. Goodman, and J. L. Weiderholt (Eds.) *Teaching the Learning-Disabled Adolescent*. Boston: Houghton-Mifflin.

SITKO, M.C. AND A. G. SLEMON. 1978. *The development and evaluation of a training package for inservice teachers on hierarchical questioning strategies for teaching children with learning disabilities*. London, Canada: Faculty of Education, The University of Western Ontario. Final report to the Ministry of Education.

SITKO, M. AND A. SLEMON. 1982. "Developing teacher's questioning skills: The efficacy of delayed feedback." *Canadian Journal of Education,* 7, 109-121.

SITKO, M. C. AND A. G. SLEMON. NOVEMBER 1983. "Programming: The teacher-pupil match." Paper presented at the International Symposium on "Assessment for the Teaching/ Learning Process," Ontario Institute for Studies in Education, Toronto, Ontario.

SITKO, M., A. SLEMON AND D. BACHOR. 1983. "A questioning and feedback system for developing comprehension skills." *Special Education in Canada, 57*(2), 29-32.

SITKO, M., D. BACHOR AND A. G. SLEMON. 1980. "Teacher questioning techniques: The state of the art." Paper presented at the annual meeting of the Canadian Society for Studies in Education, Montreal, Quebec.

SLEMON, A. G., P. B. PARK, D. THIESSEN AND M. C. SITKO. 1985. *Assessing Special Needs in Classroom*. London, Ontario: The University of Western Ontario.

SLESNICK, T. 1983. "Creative play: An alternative use of the computer in education." *Simulation & Games, 14*(1), 11-20.

SMITH, F. 1978. *Understanding Reading: A psycholinguistic analysis of reading and learning to read*. Second Edition. Toronto: Holt, Rinehart and Winston.

SMITH, R. 1978. "Investigation of the needs of handicapped children to be served in regular classrooms." ERIBC Report No. 78:14, p. 1-22.

SMITH, S. 1980. Paper presented to the Conference on Computers in Education. In *Highlights: The Canadian Psychological Association,* 6, 10E-11E.

SMITH, S. 1984. *Futures in Science Education*. Ottawa: Science Council of Canada.

SNODGRASS, G. 1980. *Computerized EP and Management Information System.* Washington, DC: National Association of State Directors of Special Education.

SOKOLYK, O. 1981. *Mainstreaming in Quebec: Teacher competencies in an integrated classroom*. Montreal: The Quebec Association for Children with Learning Disabilities.

SPACHE, G. D. 1963. *Toward Better Reading*. Champaign: Garrard.

SPACHE, G. D. 1981. *Diagnostic Reading Scales*. Revised Edition. Monterey: CTB/McGraw-Hill.

SPACHE, G. AND E. SPACHE. 1973. *Reading in the Elementary School.* Second Edition. Boston: Allyn and Bacon, Inc.

SPINA, D. J. 1983. "Gifted learning disabled and gifted underachievers: A literature review." Unpublished research paper, The University of Western Ontario.

SPINA, D. J. AND C. CREALOCK. (1985.) "Identification and programming for the gifted and the gifted underachiever: A survey of the current situation in Ontario's schools." *Canadian Journal for Exceptional Students*, 2, 8-13.

SPIVAK, G. AND M. SHURE. 1981. "The cognition of social adjustment: Interpersonal cognitive problem solving thinking." In B. Lahey, and A. E. Kazdin (Eds.) *Advances in Clinical Child Psychology, Volume 5*. New York: Plenum.

STALLINGS, J. A. 1981. "What research has to say to administrators of secondary schools about effective teacher and staff development." In K. Duckworth, E. Kehop, W. DeBevoise and F. Donovan (Eds.) *Creating Conditions for Effective Teaching*. Eugene, OR: Center for Educational Policy and Management, University of Oregon.

STANLEY, J. C. AND K. D. HOPKINS. 1972. *Educational and Psychological Measurement and Evaluation*. Englewood Cliffs: Prentice-Hall, Inc.

STEER, M. 1981. "Update on services for exceptional children in Newfoundland and Labrador." *Special Education in Canada, 55*, 21-23.

STEFFIN, S. A. 1983. "Using the micro as a weapon fighting against convergent thinking." *Childhood Education, 59*(4), March/April, 255-257.

STEIN, N. L. AND T. TRABASSO. 1982. "What's in a story: An approach to comprehension and instruction." In R. Glaser (Ed.) *Advances in Instructional Psychology. Volume 2.* Hillsdale: Lawrence Erlbaum.

STENNETT, R. G. AND L. ISAACS. 1979. "The distar reading program in junior opportunity classes: Phase III—Impact of two full years of instruction." *Special Education in Canada, 53*(2), 7-10.

STEPHENS, T. M. AND B. L. BRAUN. 1980. "Measures of regular classroom teachers' attitudes towards handicapped children." *Exceptional Children, 46*, 292-296.

STEPHENSON, P. S. 1973. "Myths about juvenile delinquency." *Canadian Journal of Criminology and Corrections, 15*, 83-92.

STERN, M. 1952. "Free painting as an auxiliary technique in psychoanalysis." In G. Bychowski (Ed.) *Specialized Techniques in Psychotherapy*. New York: Basic Books.

STEVENS, R. AND B. ROSENSHINE. 1981. "Advances in research on teaching." *Exceptional Education Quarterly, 2*(1), 1-9.

STIPEK, D. J. AND J. R. WEISZ. 1981. "Perceived personal control and academic achievement." *Review of Educational Research, 51*, 101-137.

STOTT, D. H. 1970. *Programmed Reading, Kit 1* and *Kit 2*. Toronto: Gage.

STOTT, D. H. 1971. *The Flying Start Learning-to-learn Kit*. Guelph: Brook Educational Publishing.

STOTT, D. H. 1971. "Behavioral aspects of learning disabilities: Assessment and remediation." *Experimental Publication System*, April, *11*, Ms. No. 400-36.

STOTT, D. H. 1972. *The Parent as Teacher: A guide for parents of children with learning difficulties*. Toronto: New Press.

STOTT, D. H. 1978. *The Hard-to-teach Child: A diagnostic-remedial approach*. Baltimore: University Park Press.

STRATTON, V. N. AND A. ZALANOWSKI. 1984. "The effects of background music on verbal interaction in groups." *The Journal of Music Therapy, XXI*(1), 16-26.

STUM, D. 1982. "DIRECT: A consultation skills training model." *Personal and Guidance Journal, 60*(5), 296-302.

SUMMERS, E. G. 1984. "A paradigm for developing quick word tests for reading assessment in the secondary grades." *B. C. Journal of Special Education, 8*, 11-33.

SUMMERS, G. C. 1984. *Learned Helplessness, Test Anxiety, and Learning Disabilities: An assessment with intermediate-age children*. Unpublished master's thesis, University of Victoria.

SURER, B. AND J. RIZZIO. 1983. *Special Children: An integrative approach*. Illinois: Scott, Foresman & Co.

SUSSMAN, P. S. 1983. "Peer intervention technology." In *Natcon 4*. Ottawa-Hull: Canada Employment and Immigration Commission.

SWABY, B. E. R. 1984. *Teaching and Learning Reading: A pragmatic approach*. Boston: Little Brown.

SWIFT, M., G. SPIVACK, A. DANSET, J. DANSET-LEGER AND F. WINNYKAMEN. 1972. "Classroom behavior and academic success of French and American elementary school children." *International Review of Applied Psychology, 20*, 1-11.

TABA, H. 1966. Teaching strategies and cognitive functioning in elementary school children. Cooperative Research Project, San Francisco State College.

TABER, F. M. 1983. *Microcomputers in Special Education*. Reston, VA: Council for Exceptional Children.

TABER, F. M. 1982. *Microcomputers in Special Education: Selection and decision-making process*. Arlington, VA: Council for Exceptional Children.

TABER, F. M. 1981. "The microcomputer—Its applicability to special education." *Focus on Exceptional Children, 14*(2), 1-13.

TAYLOR, B. M. AND S. J. SAMUELS. 1983. "Children's use of text structure in the recall of expository material." *American Educational Research Journal, 20*, 517-528.

TAYLOR, S. E., H. FRACKENPOHL AND C. E. WHITE. 1979. *A Revised Core Vocabulary*. New York: Educational Developmental Laboratories, McGraw-Hill.

TERRY, P. R., D. A. COHEN, AND C. S. ENGLERT, 1978. *The Hypothesis Test Reading Program*. Bloomington, IN: Indiana University, Center for Innovation in Teaching the Handicapped.

THIAGARAJAN, S., D. S. SEMMEL AND M. I. SEMMEL. 1974. *Instructional Development for Training Teachers of Exceptional Children*. Reston, VA: The Council for Exceptional Children.

THOMAS, A. 1979. "Learned helplessness and expectancy factors: Implications for research in learning disabilities." *Review of Educational Research, 49*, 208-221.

THOMAS, V. 1979. *Teaching Spelling: Canadian word lists and instructional techniques*. Second Edition. Toronto: Gage Publishing Co.

THORKILDSEN, R. 1982, MARCH. "Microcomputer/videodisc authoring system for instructional programming."—Paper presented at the annual meeting of the American Educational Research Association, New York, N.Y.

THORKILDSEN, R. 1982, JULY. "Microcomputer/videodisc system for delivering computer assisted instruction to mentally handicapped students." In *The Computer/ Extension of the human mind*. Proceedings from Annual Summer Conference, College of Education, University of Oregon.

THORNDYKE, G. G. AND R. D. SYMMACH. 1982. "CAI and the microcomputer." *Education Canada, 22*(4), 4-7.

TIERNEY, R. J., J. E. READENCE AND E. K. DISHNER. 1980. *Reading Strategies and Practices: Guide for improving instruction*. Boston: Allyn and Bacon.

TOUZEL, S. 1978. "Secondary L.D. curricula—A proposed framework." *Learning Disability Quarterly, 1*(4), 53-55.

TOWLE, M. 1978. "Assessment and remediation of handwriting deficits for children with learning disabilities." *Journal of Learning Disabilities, 11*(6), 43-50.

TREMBLEY, P. W. 1982. "Vertical word processing: A new approach for teaching written language to the learning disabled adolescent." *Journal of Learning Disabilities, 15*, 587-593.

TRIFILETTI, J. J., G. H. FRITH, AND S. ARMSTRONG. 1984. "Microcomputers versus resource rooms for LD students: A preliminary investigation of the effects on math skills." *Learning Disability Quarterly,* 7(1), 69-76.

TRITES, R. 1977-78. *Proceedings of the Subcommittee on Childhood Experiences as Causes of Criminal Behavior*. Senate of Canada, Third Session, Thirtieth Parliament, Issue No. 8.

TRYON, G. S. 1980. "The measurement and treatment of test anxiety." *Review of Educational Research, 50*, 343-372.

TUINMAN, J. AND J. R. KENDALL. 1980. *British Columbia Reading Assessment, 1980, Summary Report*. Victoria: British Columbia.

TURCIC, R. AND C. RAILTON. 1983. "The integrated community classroom: Mainstreaming in action." *Special Education in Canada, 57*, 8-11.

TURKELL, S. B. AND D. M. PODELL. 1984. "Computer-assisted learning for mildly handicapped students." *Teaching Exceptional Children, 16*(4), 258-262.

TWAY, E. *Reading Ladders for Human Relations.* Sixth Edition. Washington, D.C.: American Council on Education.

ULMAN, E. 1975. *Art Therapy in Theory and Practice*. New York: Schocken Books.

UNESCO, 1973. *Present Situation and Trends in Research in the Area of Special Education*. Paris: Unesco Publication.

VANCE, P. AND R. BAUMANN. 1981. "What do you do with a kid like that?" Toronto: Ontario Educational Television Production.

VANDER ZANDEN, J. W. 1980. *Educational Psychology in Theory and Practice*. New York: Random House.

VARENHORST, B. 1978. *Curriculum Guide for Student Peer Counseling Training*. Palo Alto, California: Palo Alto Unified School District.

VENEZKY, R. L. AND L. WINIFIELD. 1979. *Schools that Succeed Beyond Expectations in Teaching Reading*. Newark: University of Delaware Studies on Education, Technical Report No. 1.

VERNON, M. D. 1978. *Reading and its Difficulties: A psychological study*. Cambridge: Cambridge University Press.

VYGOTSKY, L. S. 1978. *Mind in Society: The development of higher psychological processes*. (Edited by M. Cole, V. John-Steiner, S. Scribner and E. Souberman). Cambridge, Mass.: Harvard University Press.

WADWORTH, H. G. 1971. "A Motivational Approach Toward the Remediation of Learning Disabled Boys." *Exceptional Children*, 33-42.

WALLACE, G. AND S. C. LARSEN. 1978. *Educational Assessment of Learning Problems: Testing for teaching*. Boston: Allyn & Bacon.

WALKER, H. M. 1983. "Applications of response cost in school settins: Outcomes, issues, and recommendations." *Exceptional Education Quarterly, 3*(4), 47-55.

WALKER, H. M., S. MCCONNEL, D. HOLMES, J. TODIS, AND N. GOLDEN. 1983. *The Walker Social Skills Curriculum: The ACCEPTS program*. Austin, Texas: Pro-Ed.

WANG, C. M. 1981. "Mainstreaming exceptional children: Some instructional design considerations." *Elementary School Journal, 81*(4), 195-221.

WATKINSON, A. AND Y. PETERS. 1982. "Teaching about the rights of disabled Canadians." *The History and Social Science Teacher, 18*, 17-21.

WATT, D. 1983. "LOGO: What makes it exciting?" *Popular Computing, 2*(10), 106-112, 148-166.

WATT, D. 1982. "LOGO in the schools." *Byte, 7*(8), 116-134.

WATT, M. 1982. "What is LOGO?" *Creative Computing, 8*(10), October.

WATTS, N. 1981. "A dozen uses for the computer in education." *Educational Technology, 21*(4), 18-22.

WEBER, K. 1982. *The Teacher is the Key: A practical guide for teaching the adoles-*cent with learning difficulties. Toronto: Ministry of Education.

WEES, W. R. 1970. "24 hour schools used in apartments, parks and plazas." *Toronto Telegram*, March 24.

WEIDERHOLT, L. 1978. "Educating the learning disabled adolescent: Some assumptions." *Learning Disability Quarterly, 1*(4), 11-23.

WEIL, M. AND B. JOYCE. 1978. *Information Processing Models of Teaching*. Englewood Cliffs, N.J.: Prentice-Hall, Inc.

WEINER, B. 1981. "The emotional consequences of casual attributions." In M. S. Clark and S. T. Fiske (Eds.) *Affect and Cognition: The seventeenth annual Carnegie symposium on cognition*. Hillsdale, N.J.: Lawrence Erlbaum.

WEINER, E. 1980. "Diagnostic evaluation of writing skills." *Journal of Learning Disabilities, 13*, 48-53.

WEINSTEIN, G. AND M. D. FANTINI. 1970. *Toward Humanistic Education: A curriculum of affect*. New York: Praeger.

WEIR, S. 1981. "LOGO and the exceptional child." *Microcomputing, 5*(9), 76-83.

WEIR, S. AND D. WATT. 1980-1981. "LOGO: A computer environment for learning-disabled students." *The Computer Teacher, 8*(5), 11-17.

WEIR, S. R., S. RUSSEL, AND J. VALENTE. 1982. "LOGO: An approach to educating disabled children." *Byte, 7*(9), 342-360.

WELLS, L. 1981. Paper prepared for Inservice Special Education. Toronto: Toronto Board of Education.

WELLS, L. 1985. "Test of Organisational Ability." Toronto: Toronto Board of Education.

WELLS, N. F. AND T. STEVENS. 1984. "Music as a stimulus for creative fantasy in group psychotherapy with young adolescents." *The Arts in Psychotherapy, 11*, 71-76.

WENER, A. AND R. O'SHAUGHNESSY. 1981. "Mainstreaming in action: A model." Montreal: The Quebec Association for Children with Learning Disabilities.

WHALEY, J. F. 1981. "Story grammars and reading instruction." *The Reading Teacher, 34*, 762-771.

WHEELER, F. 1983. "The Puzzler: An answer to the reading riddle?" *Classroom Computer Learning, 4*(4), 46-47, 50-51.

WIENER, J. AND J. LIDSTONE. 1969. *Creative Movement for Children: A dance program for the classroom*. New York: Van Nostrand.

WIENS, E. 1980. "Secondary special education: A practical tutorial approach for high school students with learning disabilities." *Special Education in Canada, 54*, 9-11.

WIIG, E. AND E. SEMEL. 1975. "Productive language abilities in learning disabled adolescents." *Journal of Learning Disabilities, 8*, 45-53.

WIIG, E. H. AND E. M. SEMEL. 1980. *Language Assessment and Intervention for the Learning Disabled*. Columbus, Ohio: Charles E. Merrill.

WILGOSH, L. 1983. "Teacher views on usefulness of special education assessment instruments." *Special Education in Canada, 54*, 7-9.

WILLIS, J. AND G. KUCHINSKAS. 1983. "Simulations = Fun + Learning." *Computers, Reading and Language Arts, 1*(3), 24-30.

WILSON, A. 1984. "Opening the door—a key to resource models." Ontario Public School Teachers' Federation.

WILSON, W. R. 1982. "Uses of portable microprocessor-based communication systems and personal computers in the instruction of persons with severe physical disabilities." *Journal of Special Education Technology, 5*(4), 40-41.

WINNE, P. H., M. J. WOODLANDS AND B. Y. L WONG. 1982. Comparability of self-concept among learning disabled, normal, and gifted students." *Journal of Learning Disabilities, 15*, 470-475.

WONG, B. Y. L. 1979. "Increasing retention of main ideas through questioning strategies." *Learning Disability Quarterly, 2*(2), 42-48.

WONG, B. Y. L. 1980. "Activating the inactive learner: Use of questions/prompts to enhance comprehension and retention of implied information in learning disabled children." *Learning Disability Quarterly, 3*(1), 29-37.

WONG, B. Y. L. 1982. "Understanding learning disabled students' reading problems: Contributions from cognitive psychology." *Topics in Learning and Learning Disabilities, 1*(4), 43-50.

WONG, B. Y. L. 1982. "Strategic behaviors in selecting retrieval cues in gifted, normal achieving and learning-disabled children." *Journal of Learning Disabilities, 15*, 33-37.

WONG, B. Y. L. AND W. JONES. 1982. "Increasing metacomprehension in learning disabled and normally achieving students through self-questioning." *Learning Disability Quarterly, 5*, 228-240.

WONG, B. Y. L. AND M. WILSON. 1984. "Investigating awareness of and teaching passage organisation in learning disabled children." *Journal of Learning Disabilities, 17*, 477-482.

WOODCOCK, R. E. 1973. *Woodcock Reading Mastery Tests.* Circle Pines: American Guidance Service.

WOOLFOLK, A. E. AND L. M. NICOLICH. 1984. *Educational Psychology for Teachers.* Second Edition. Englewood Cliffs, N.J.: Prentice-Hall, Inc.

WORMELI, JR., C. T. 1983. *B. C. QUIET (British Columbia quick individual educational test for reading, spelling and arithmetic).* Vancouver, British Columbia: Education Clinic, Faculty of Education, University of British Columbia.

WRIGHT, E. N. AND A. WINTER. "Teaching styles and pupils' academic and behavioral changes in learning disabilities self-contained classrooms." Paper presented at the Annual Conference of the Canadian Society for the Study of Education, Guelph, Canada, 1984.

ZEY, K. 1981. "Before you mainstream...." *Special Education in Canada, 55*(4), 11-12.

ZIGMOND, N., E. LEVIN, AND J. BIRCH. 1984. "A follow-up study of 52 learning disabled adolescents." *Journal of Learning Disabilities, 18*(1), 2-7.

ZIGMOND, N., A. VALLECORSA, AND R. SILVERMAN. 1983. *Assessment for Instructional Planning in Special Education.* Toronto: Prentice-Hall Canada Inc.

ZIMMERMAN, J., W. RICH, I. KEILITZ, AND P. BRODER. 1978. *Some Observations on the Link between Learning Disabilities and Juvenile Delinquency.* National Centre for State Courts, Northwestern University.

ZNANIECKI, F. 1965. *The Social Role of the Man of Knowledge.* New York: Octagon Books.

CREDITS

Our thanks to those publishers and authors who granted us permission to use material from their books and articles.

D. Armitage: From (1978) "Neurological Impress Method of Reading Instruction," *Special Education in Canada* 52 (3). Reprinted with permission of Canadian Journal for Exceptional Children.

Canadian Association for Children and Adults with Learning Disabilities: Definition of learning disabilities from CACLD Annual Meeting (1981). Reprinted with permission.

R. de Charms: From (1972) "Personal causation training in the schools," *Journal of Applied Social Psychology* 2 (2). Reprinted with permission.

A. J. Harris and E. R. Sipay: From *How to Increase Reading Ability* by A. J. Harris and E. R. Sipay. Copyright ©1980 by Longman Inc. Reprinted with permission of the publisher.

D. G. Hennings: From (1982) "A writing approach to reading comprehension—Schema theory in action," *Language Arts* 59. Reprinted with permission of National Council of Teachers of English.

N. Hobbs: From (1982) *The Troubled and Troubling Child*. San Francisco: Jossey-Bass, Inc., Publishing Co. Reprinted with permisison.

A. G. Ingham: From (1978) *The Blended Sound-Sight Method of Learning*. Third Edition. Yorkton, Saskatchewan: A. G. Ingham. Reprinted with permission.

D. Johnson and R. Johnson: From (1980) "Integrating children into the mainstream," *Exceptional Children* 46. Reprinted with permission.

D. Meichenbaum and J. Goodman: (1971) "Training impulsive children to talk to themselves: A means of developing self-control," *Journal of Abnormal Psychology* 77 (2). Reprinted with permission.

M. Scardamalia, C. Bereiter, and B. Fillion: Reprinted from *Writing for Results: A Sourcebook of Consequential Activities* by Marlene Scardamalia, Carl Bereiter, and Bryant Fillion, 1981, with permission of the publisher OISE Press, The Ontario Institute for Studies in Education.

M. Semmel et al.: From (1974) *Module 15: Task Analysis*. Tips for Teachers Series. Bloomington, Indiana: Centre for Innovation in Teaching the Handicapped, Indiana University, 47401.

D. H. Stott: From (1978) *The Hard-to-teach Child: A Diagnostic-Remedial Approach*. Reproduced with permission of Pro-Ed, Inc., Austin, Texas.

SUBJECT INDEX

AUTHOR INDEX

Printed and bound in Canada by
Best Gagné Book Manufacturers